T0135291

Intelligent Road Transport Systems

WANG Yunpeng • YAN Xinping •
LU Guangquan • WU Chaozhong
Editors

Intelligent Road Transport Systems

An Introduction to Key Technologies

TSINGHUA UNIVERSITY PRESS

Editors

WANG Yunpeng
School of Transportation Science
and Engineering
Beihang University
Beijing, China

LU Guangquan
School of Transportation Science
and Engineering
Beihang University
Beijing, China

YAN Xinping
Intelligent Transportation Systems Research
Center
Wuhan University of Technology
Wuhan, China

WU Chaozhong
Intelligent Transportation Systems Research
Center
Wuhan University of Technology
Wuhan, China

ISBN 978-981-16-5778-8   ISBN 978-981-16-5776-4   (eBook)
https://doi.org/10.1007/978-981-16-5776-4

Jointly published with Tsinghua University Press.
The print edition is not for sale in China (Mainland). Customers from China (Mainland) please order the print book from: Tsinghua University Press.

This Springer imprint is published by the registered company Springer Nature Singapore Pte Ltd.
The registered company address is: 152 Beach Road, #21-01/04 Gateway East, Singapore 189721, Singapore

Preface

Intelligent transportation system (ITS) is the application of a variety of advanced science and technology (including information technology, computer technology, data communication technology, sensor technology, electronic control technology, automatic control technology, and artificial intelligence technology) to service, management, and control for the transportation system. ITS coordinates vehicles, transportation facilities, and behaviors of road users to improve traffic safety, improve traffic operation efficiency, reduce adverse effects on the environment, and reduce energy consumption.

Nowadays ITS has become an indispensable part of transportation. Transportation majors in domestic colleges and universities have offered intelligent transportation-related courses due to the importance of ITS in transportation. Some domestic universities and scholars have made great exploration and efforts on the course system construction and textbooks of ITS and achieved fruitful results. However, with the rapid advance of computer technology, information technology, wireless communication technology, artificial intelligence, and other high and new technologies, the teaching contents related to ITS must be updated in time.

ITS involves a broadening of applications. The components of ITS are changing with the advance of computer technologies and the change of transportation demands. Intelligent transportation technology provides a common technology for building ITS. It is the product of the deep integration of modern high-tech and transportation industry. It also develops with the development of modern high tech.

This book tries to break through the general discussion mode that regards the system as the core. The knowledge is introduced from four aspects: sense (perception and management of traffic information, Chaps. 2 and 3), transmission (interaction of traffic information, Chap. 4), cognition (prediction of traffic state, Chap. 5), and use (intelligent transportation applications and systems, Chaps. 6–10). This book focuses on displaying common, necessary, and basic knowledge among subsystems in ITS, building a basic intelligent transportation technology knowledge system for the reader.

This book belongs to *Internet of Things in China* series. Professor WANG Yunpeng at Beihang University and Professor YAN Xinping at Wuhan University of Technology are the chief editors. Professor LU Guangquan at Beihang University and Professor WU Chaozhong at Wuhan University of Technology are the associate editors. In addition, more than 10 experts and young scholars devoting to intelligent transportation researches from Beihang University and Wuhan University of Technology also participated in the preparation of this book.

Professor WANG Yunpeng and Professor YAN Xinping are responsible for chapter planning, overall coordinating, and unifying draft of the book. Professor WANG Yunpeng and Professor LU Guangquan are in charge of the organization and the unifying draft of Chaps. 3, 4, 5, 8, and 9. Professor YAN Xinping and Professor WU Chaozhong are in charge of the organization and the unifying draft of Chaps. 1, 2, 6, 7, and 10.

Chapter 1 was prepared by HE Yi of Wuhan University of Technology; Chap. 2 was prepared by CHEN Zhijun of Wuhan University of Technology; Chap. 3 was prepared by MA Xiaolei of Beihang University; Chap. 4 was prepared by WU Xinkai and WANG Pengcheng of Beihang University; Chap. 5 was prepared by MA Xiaolei of Beihang University; Chap. 6 was prepared by ZhANG Hui of Wuhan University of Technology; Chap. 7 was prepared by ZhANG Cunbao of Wuhan University of Technology; Chap. 8 was prepared by YU Guizhen of Beihang University; Chap. 9 was jointly prepared by CHEN Peng, DING Chuan, and LU Guangquan of Beihang University; Chap. 10 was prepared by CHU Duanfeng of Wuhan University of Technology. In addition, the contributors of this book also include: LUAN Sen, TAN Erlong, YU Xiaofei, YAN Haoyang, LI Yujie, WEI Lei, DAI Rongjian, ZHANG Junjie, CAI pinlong, LIU Qian, HAN Xu, ZHOU Bin, WANG Zhangyu, LI Han, LIAO Yaping, and LIU Pengfei of Beihang University; YANG Xinwei, XIONG Shengguang, GAO Lin, CAO Bo, FENG Qi, LI Jipu, FAN Yixiong, LENG Yao, YU Jinqiu, CHEN Qiushi, SUN Yifan, ZHANG Qi, LI Shaopeng, HOU Ninghao, ZhANG Yijun, CHEN Feng, QIN Ruiyang, LI Xuemei, WANG Houyi, and CAO Yongxing of Wuhan University of Technology.

ITS covers all aspects of the transportation system, and its content is very extensive. High and new technologies such as computer technology, information technology, and artificial intelligence are developing rapidly, so it is challenging to arrange intelligent transportation technology in a book systematically. Although the writing team has made great efforts, it is inevitable that some mistakes may exist in the book. We welcome comments from peers and readers.

The publication of this book is strongly supported by the office of the national gold card project coordination leading group. We also sincerely acknowledge all the support from Tsinghua University Press.

Beijing, China	WANG Yunpeng
Wuhan, China	YAN Xinping
Beijing, China	LU Guangquan
Wuhan, China	WU Chaozhong
August 2021	

Contents

Chapter 1
Introduction

HE Yi, YAND Xinwei, XIONG Shengguang, GAO Lin, CAO Bo, FENG Qi, LI Jipu, and FAN Yixiong

With the rapid development of information and intelligent technology, there are more and more applications in the field of intelligent transportation, such as new-generation perception technology, artificial intelligence technology, communication technology, mobile internet services, energy management, vehicle–road collaboration, and intelligent networked vehicle technology. The intelligent transportation system has entered a new era. New technologies, new concepts, and new models are subverting the previous transportation system. New technologies have promoted the comprehensive upgrade of intelligent transportation systems in terms of perception, storage, sharing, interaction, and integrated services. The system and content of the original intelligent transportation system are undergoing major changes, and the connotation of the intelligent transportation system is constantly enriched and improved.

1.1 Introduction to Intelligent Transportation System

1.1.1 Basic Concept of Intelligent Transportation System

The basic elements of the transportation system are people, vehicles, roads, and the environment. Humans are intelligent, but they have shortcomings in perception and exception, such as insufficient viewing distance under poor light conditions, insufficient response ability when people are tired and distracted, etc. If we can enhance people's abilities in these aspects, and at the same time make cars, roads, and the

HE Yi (✉) · YAND Xinwei · XIONG Shengguang · GAO Lin · CAO Bo · FENG Qi ·
LI Jipu · FAN Yixiong
Intelligent Transportation Systems Research Center, Wuhan University of Technology, Wuhan,
China
e-mail: heyi@whut.edu.cn

© Tsinghua University Press 2022
W. Yunpeng et al. (eds.), *Intelligent Road Transport Systems*,
https://doi.org/10.1007/978-981-16-5776-4_1

environment intelligent, then all the elements in the transportation system will be intelligent. All elements of an intelligent transportation system (ITS) should be intelligent. The difference between ITS and traditional transportation systems lies in the enhancement of human perception and execution capabilities, as well as the intelligence of transportation tools and the environment. Regarding the intelligent transportation system, the more recognized definition is the efficient integration and application of advanced information technology, communication technology, sensor technology, control technology, and computer technology to the entire transportation management system, thereby establishing kind of a real-time, accurate, and efficient integrated transportation and management system that works in a wide range and all-round way.

The connotation of ITS is gradually expanded. The following will discuss the connotation of ITS from some characteristics and attributes of ITS.

1. **Advancement**

    Before the concept of ITS was formed, many countries were seeking to transform and arm the transportation system with modern advanced technologies such as telecommunication, computer, and electronic technology and to improve the management and operation of the transportation system with advanced theoretical methods. The ITS subsystem proposed by the United States clearly adds the attributive "advanced" to the name. Advancement is a vague concept. Generally speaking, advancement should refer to the development of products and systems with some technologies that have emerged in recent years.

2. **Synthesis**

    The key technologies involved in ITS include information technology, communication technology, computer technology, electronic technology, traffic engineering, system theory, artificial intelligence, knowledge engineering, etc. It can be said that ITS is the intersection and synthesis of these technologies and the integrated application of these technologies in the transportation system.

3. **Informatization**

    People obtain the status information of the transportation system through various means and provide timely and useful information for the users and managers of the transportation system. Only with information can it be intelligent. Moreover, when the level of traffic informatization reaches a certain level, it will change the behavior of traffic travel, traffic management methods, etc. and then cause changes in traditional traffic theories. Therefore, informatization is the foundation of ITS.

4. **Intelligence**

    The word intelligence is used more and more widely, there are more and more people studying intelligence, and the application of smart technology is also increasing. Terms such as smart robots, smart instruments, and smart buildings appear frequently. The intelligence of products has brought vitality and vitality to many traditional technologies, including intelligent transportation systems. Many subsystems in the intelligent transportation system are different from the traditional transportation system because of the realization of intelligence, and the electronic toll collection system (electronic toll collection, ETC) is a typical example. The traditional road toll system sets up toll stations. Vehicles pass through the toll stations and hand in cash manually. Vehicles have to line up, which is not conducive to statistics. The electronic toll collection system uses electronic settlement, automatic vehicle identification technology, microwave communication technology, etc., which can achieve automatic toll collection without stopping. It not only saves time but also improves accuracy.

It can also provide information such as traffic flow statistics data, reflecting intelligence. Another example is the automated highway system (AHS), which can realize fully automatic driving of vehicles. Once drivers enter the system, they can reach it safely and quickly as long as they enter the destination, which shows high intelligence.

The essence of ITS is to use high and new technology to transform the traditional transportation system, thereby forming a new type of informatization, intelligence, and social transportation system. It enables the transportation infrastructure to maximize effectiveness, improves the quality of services, and makes the society use the limited road traffic facilities and resources most effectively. At the same time, it promotes the development of related communication, computer, network, and other industries, so as to obtain huge social and economic benefits.

At present, the understanding of intelligent transportation systems at home and abroad is not the same. But no matter from which point of view, one thing is common: ITS is a technical and economic system that uses various high-tech, especially electronic information technologies, to improve traffic efficiency, increase traffic safety, and improve environmental protection. Therefore, based on a relatively complete transportation infrastructure, the intelligent transportation system is a real-time, accurate, and efficient transportation system that effectively integrates advanced information technology, communication technology, control technology, sensor technology, and system integration technology and applies these technologies to the ground transportation system, so as to work on a large scale.

1.1.2 History of Intelligent Transportation System

With the development of the economy, the society's demand for transportation continues to grow, and the transportation industry has developed rapidly. Developed countries and regions in the world have vigorously developed road infrastructure and automobile industry since the 1950s, which has promoted the rapid development of road traffic. With the development of road traffic, it also brought serious problems such as frequent traffic accidents, serious environmental pollution, and traffic congestion. The "China Artificial Intelligence Series" white paper pointed out that vehicles consume the most nonrenewable energy among various traffic modes, which results in environmental pollution being dozens of times that of other traffic modes. In traffic accidents, the accidents caused by road traffic are also dozens of times higher than other ways. Traffic congestion is a common phenomenon in road traffic, especially urban road traffic. The increase in transportation infrastructure still cannot keep up with the increase in traffic volume, and road traffic problems have become a traffic problem that plagues countries all over the world. In order to solve a series of problems brought about by the development of road traffic, people engaged in traffic engineering research have thought of ways to improve the intelligence of vehicles and roads early. If the traffic information of the intersection can be detected in time and the control strategy can be displayed dynamically, the traffic capacity of the intersection will be greatly improved. The study found that during the peak period of traffic, the urban road system and highway system will not all have traffic congestion, and a considerable part of the road is still very smooth. If the traffic information of the road network can be told to the drivers in time and prompt them to use these road sections reasonably, the resources of the road network can be fully utilized. If the car can detect surrounding information in real time and can make correct decisions or even drive fully automatically, traffic accidents will be greatly reduced and efficiency will be greatly improved. This idea was proposed in the 1960s and 1970s. But the issues such as how to collect traffic flow information at intersections, what algorithms are used to process this information to obtain a reasonable control strategy, how to collect real-time traffic status data on main roads, how to transmit and process these data, how to transmit information to traffic participants, how to detect the surrounding information in real time, how to make the correct decision after processing this data, and how to execute the decision made have become the subject of traffic engineering researchers at that time.

According to these ideas, people try to make the transportation system intelligent and carry out a lot of work. From the perspective of the development history of the international intelligent transportation system, all countries generally believe that the computerization of traffic management, which started in the 1960s to the 1970s, is the budding of the intelligent transportation system.

What is an intelligent transportation system? To understand an intelligent transportation system, we must first understand intelligence. Intelligence refers to the ability to recognize, analyze, judge, process, and invent and create things. If many systems or products in engineering have some kind of intelligence, it can be called an artificial intelligence system. The artificial intelligence system uses sensors, CPUs, and actuators to simulate human facial features, brains, and limbs. In a broad sense, an intelligent transportation system is also an artificial intelligence system. It uses traffic sensors, a CPU with traffic knowledge, and an executive body that can perform traffic functions to simulate the human five senses, brains, and limbs to achieve the purpose of traffic intelligence. Taking smart traffic lights as an example, the corresponding relationship between human intelligence, artificial intelligence, and smart traffic lights is shown in Fig. 1.1.

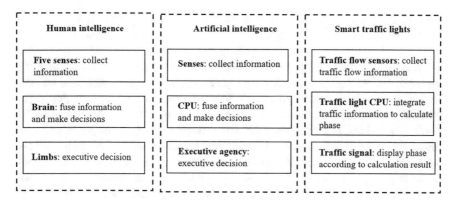

Fig. 1.1 Correspondence between human intelligence, artificial intelligence, and intelligent traffic lights

1.2   Development of Intelligent Transportation System

1.2.1   Development History and Present Situation of Intelligent Transportation System in the United States

1.2.1.1   Development History

The United States is a large country of intelligent transportation, not only researched and used early but also widely used. However, the United States has likewise experienced a period of exploration in the research of intelligent transportation. From the late 1960s to the 1970s, the United States was committed to developing the electronic route guidance system (EGRS), which used two-way communication between roads and vehicles to provide road guidance.

Throughout the development process of research on intelligent transportation technology in the United States, it can be roughly classified into two stages according to its research goals, characteristics, and focus of attention. The first stage is from the 1990s to the end of the twentieth century. The main feature is that the scope of research is comprehensive and wide. The research and projects were broad and scattered. The second stage is the twenty-first century; the United States has made strategic adjustments, from the "full-scale research" in the first stage to the "major special research, focusing on vehicle safety and vehicle–road coordination technology" strategy, and from the perspective of the integrated transportation system, research on intelligent transportation and safety technology is carried out. The research content includes integrated transportation coordination technology, vehicle safety technology, etc. The characteristic is to pay more attention to practical results and promote the industrialization of related technologies.

In order to accelerate the development of ITS, in April 2001, the United States held a high-level national seminar attended by 260 experts and related personnel from the ITS industry. After the meeting, the ITS development plan for the first 10 years of the twenty-first century was formulated, and the future of ITS was outlined. The mission and development goals clarify the actions that must be taken to achieve the development goals of ITS in the future. The plan provides for the establishment of a coordination committee representing the government's relevant public institutions, private enterprises, and academic groups to organize and coordinate the implementation of the 10-year plan, formulate a series of comprehensive development policies, and determine and initiate a series of construction and research projects, including the necessary institutional transformation to promote the application of ITS technology, so that the future ground transportation system will gradually be transformed into an advanced system with efficient and economic management through ITS. This system will be truly endowed with the safe, effective, and economical transportation of personnel and materials. The basic functions can meet the diverse needs of users to a great extent and have good compatibility with the natural environment.

In December 2009, the United States Department of Transportation (USDOT) released the "ITS Strategic Research Plan: 2010–2014" to provide strategic guidance for ITS research projects from 2010 to 2014. The core of the plan is Intelligent Driving (IntelliDrive), which establishes a wireless network between the vehicle, the control center, and the driver and communicates information in a timely manner through monitoring and prediction, alleviating traffic jams, reducing crashes, and reducing exhaust emissions. Realize safety, flexibility, and environmental friendliness. From 2010 to 2014, ITS research projects will receive 100 million US dollars of funding each year and carry out research in multiple fields. IntelliDrive's research content includes the safety of vehicle-to-vehicle (V2V) communication and vehicle-to-infrastructure (V2I) communication, real-time data collection and management, dynamic mobile applications, etc. In addition, the 5-year plan also supports active traffic management, international borders, electronic payments, maritime applications, technology transfer in the field of intelligent transportation, knowledge and skills research and development, and the formulation of related technical standards.

In 2014, the US Department of Transportation and the US Intelligent Transportation System Joint Project Office jointly proposed the "ITS Strategic Plan 2015–2019", which clarified the direction for the development of the United States in the field of intelligent transportation in the next 5 years. The strategic plan established two strategic priorities, that is, the realization of automobile and interconnection technology and the promotion of vehicle automation, and has formulated five strategic themes: (1) through the development of better risk management and driving monitoring systems, create safer vehicles and roads; (2) through exploring management methods and strategies, improve system efficiency, alleviate traffic pressure, and enhance traffic mobility; (3) transportation is closely related to the environment—through optimized management of traffic flow and the use of Internet of vehicles technology to solve actual vehicle and road problems, the purpose of protecting the environment is achieved; (4) for better cater to the needs of future transportation, comprehensively promote technological development, and promote innovation; and (5) by establishing system architecture and standards, apply advanced wireless communication technology to realize communication and inter-action between automobiles and various infrastructures and portable devices and promote information sharing.

1.2.1.2 System Framework

In 1992, IVHS American (formerly known as ITS American) officially recommended a set of ITS system structure development methods to mobilize several state-owned and private institutions to jointly tackle key problems at the US Department of Transportation. In 1993, US Department of Transportation officially launched the ITS architecture development plan. Its purpose is to create a detailed planned national ITS architecture. This architecture will guide rather than direct the configuration of ITS products and services. Features and flexibility provide guarantees for compatibility and coordination across the country. Its progress is divided into two stages: the first stage mainly consists of four companies respec-tively proposing preliminary development plans for the system framework, and the second stage is to work with the above four companies. On the basis of this, two companies were selected to cooperate in the development of the framework of the US federal ITS system.

The main principles and goals of its construction are, considering economy as the basic principle, maximizing the use of existing facilities to provide ITS services; low fees so that most people can enjoy information services while providing a variety of alternative service methods; increasing private enterprises to accelerate the imple-mentation and application of ITS; encouraging cooperation between the state and individuals; strengthening the safety of travelers; and providing local management space.

Its progress is guided by a process-oriented approach, using system analysis and software engineering methods to provide user services, logical frameworks, physical frameworks, and their standards. The fifth edition has been revised so far.

1. **User service**

    US ITS involves multiple user entities such as investors, builders, users, and managers. Through discussions and other methods, the needs of these participants are summarized, and 8 types of service areas and 32 user services are obtained, as indicated in Table 1.1.

2. **Logical framework**

    The ITS logic framework in the United States is guided by a process-oriented development method to refine how to implement various user services and give a hierarchical logic function element table and data flow connections between each element, including 9 logic functions, 57 item sub-functions, etc.

Table 1.1 US national ITS system framework user service hierarchy

Service sectors	Services
Travel and traffic management	Pre-trip information; information of the driver on the way; route induction; joint ride and reservation; traveler service information; traffic control; event management; travel demand management; tail gas emission testing and mitigation; road/rail intersections
Public transport management	Public transportation management; bus information on the way; personalized public transport; public travel safety
Electronic payment	Electronic payment
Commercial vehicle operation	Electronic clearance of commercial vehicles; automatic roadside safety inspection; vehicle safety monitoring; commercial vehicle management; hazardous goods incident response; commercial fleet management
Emergency management	Emergency notification and personal safety; emergency vehicle management
Advanced vehicle safety system	Longitudinal collision avoidance; lateral collision avoidance; intersection collision prevention; expanded visual field; safety preparation; implementation of measures before collision; automatic vehicle control
Information management	Archive data management
Maintenance and construction management	Maintenance and construction operation management

3. **Physical framework**

The ITS physical framework in the United States divides the intelligent transportation system into 4 major categories and 19 subsystems. The subsystems of the physical framework are central system (emergency management subsystem, exhaust management subsystem, freight management subsystem, plan management subsystem, Toll management subsystem, traffic management subsystem, rapid transportation management subsystem, service information provision subsystem, commercial vehicle management subsystem), field equipment system (commercial vehicle inspection subsystem, parking management subsystem, toll management subsystem, field equipment subsystem), remote access system (personal information access subsystem, remote traveler access subsystem), and vehicle-mounted system (commercial vehicle subsystem, emergency vehicle subsystem, construction and maintenance vehicle subsystem, transportation vehicle subsystem). The physical framework of ITS in the United States is given in Figs. 1.2 and 1.3.

The United States has used equipment packages and market packages to design the ITS physical framework. The equipment package is the component module of the subsystem. It follows certain rules and is obtained by grouping and combining parallel logical processes in a given subsystem. It can be used as the basis for predicting the implementation cost. The market package is a new content added in the revision process of the third edition of the US ITS system framework. It consists of one or several equipment packages, which can be used to guide the construction of ITS projects alone and communicate information with other ITS market packages through the framework flow.

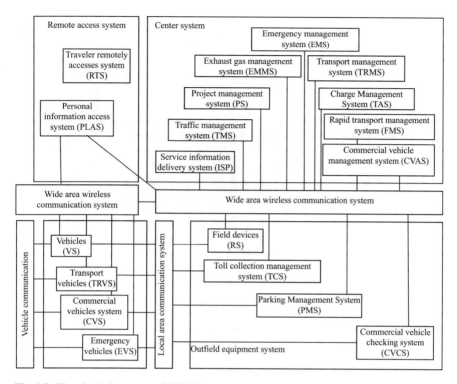

Fig. 1.2 The physical structure of US ITS

The United States is constantly improving the revision of the ITS system framework and at the same time strengthening the application and promotion of the system framework. On the basis of the national ITS system framework, the United States has developed the Turbo Architecture, a support system for the local ITS system framework, which is updated simultaneously with the national ITS system framework to facilitate the development of the local ITS system framework. Moreover, the Federal Highway Administration and the Federal Transportation Administration stipulated in April 2001 that all federally funded ITS projects must be carried out on the basis of the national ITS system framework and standards, and each locality needs to formulate a regional ITS system framework under the guidance of the national system framework. For areas that have already built ITS projects, the regional ITS system framework needs to be completed within 4 years of the promulgation of the regulations. For areas that have not implemented ITS project construction, their regional ITS system framework must be 4 years after the completion of the design of their first ITS project. It will be scheduled for completion within the year to make the systems coordinate with each other and reduce duplication of system construction.

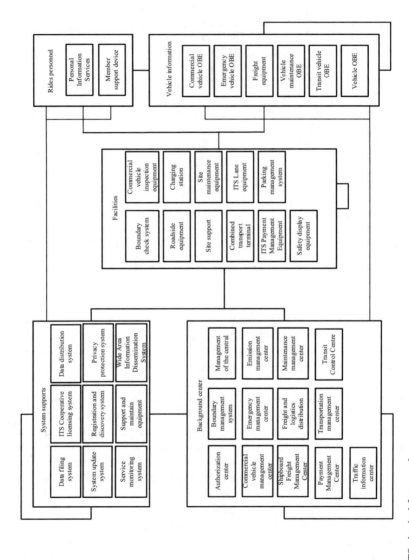

Fig. 1.3 US ITS physical framework

1.2.1.3 Main Technical Features

In order to encourage the development and application of intelligent transportation technology, the US government has promoted some large-scale research projects, which are mainly reflected in the following aspects.

1. *IntelliDrive*. The United States launched the IntelliDrive project in 2009. The project moved from a single 5.9-GHz dedicated short-range wireless communication technology (dedicated short-range communication, DSRC) to consider other approaches, such as establishing open communication through mobile phone broadband wireless communication, WIMAX, satellite communication, etc. The platform provides seamless communication services for vehicles. The service provided by IntelliDrive focuses on the active safety of vehicles, while taking into account solutions for multiple modes of transportation and travel modes to provide drivers with dynamic and continuous services. Among them, vehicle active safety services with high real-time and reliability requirements will be mainly realized through DSRC dedicated communication technology, while travel services with relatively conventional real-time requirements will be realized through 3G, 4G, Wi-Fi, and other public communication technologies. The convenience of connection between the vehicle and the vehicle, the vehicle and the roadside, and the vehicle and the management center is ensured.

2. *Connected vehicles*. Connected vehicle research (CVR) is the core content of the current research on intelligent transportation systems in the United States. In the Internet of vehicles, the vehicle has the information perception function and can perceive ego vehicle and surrounding environment information through a series of mobile information collection technologies such as radio frequency identification (RFID), telematics (Telematics), wireless location technology (wireless location technology, WLT), and exchange. Surrounding environment information, and through dedicated short-range wireless communication technology (dedicated short-range communication, DSRC) to enable information exchange between vehicles and vehicles, vehicles and infrastructure through dedicated short-range wireless communication (DSRC) technology. The US "Intelligent Transportation System Strategic Research Plan (2010–2014)" expands the application range of the early Internet of vehicles from light vehicles to all models and addresses the problems caused by the high-speed movement of vehicles and the influence of buildings and trees around the road during the communication process. In the case of the unstable quality of the wireless channel of the Internet of vehicles, the communication method has been expanded from a single 5.9-GHz DSRC communication technology to multiple forms, such as mobile broadband wireless communication, WiMAX, satellite communication, etc., and an open communication platform has been established. On February 3, 2014, the US Department of Transportation made a statement, deciding to promote the application of V2V technology in light vehicles. The statement emphasized the operational effect of using vehicle-to-vehicle (V2V) communication technology on light vehicles to avoid collisions and improve driving safety, indicating that V2V safety applications can solve most vehicle collision problems. US Secretary

Fig. 1.4 Vision of new generation of Internet of vehicles applications

of Transportation Anthony Foxx regards V2V technology as the third-generation safety technology after seat belts and airbags, and this has an important role in keeping the United States as a leader in the global automotive industry.

3. *Vehicle networking test*. US Department of Transportation has invested a lot of resources in the development of Internet of vehicles technology and established a large number of test bases. In 2016, the US Federal Highway Administration proposed a new generation of vehicle networking application development vision, covering the three aspects of safety, environment, and mobility in the vehicle network environment (see Fig. 1.4). Among them, safety applications include violation driving reminder system, deceleration zone/work zone reminder system, street pedestrian reminder, bad weather driving reminder, left turn assist system, deviating lane reminder, etc.; environmental applications include environmentally friendly driving reminder, environmentally friendly parking management, dynamic environmental protection route navigation and environmental protection signal timing, etc.; and mobility applications include queue length warning, emergency guidance, dynamic public transportation scheduling, and adaptive cruise control. City Proving Ground is an unmanned virtual city led by the University of Michigan in the United States and supported by the Michigan Department of Transportation. It locates in Ann Arbor, Michigan, United States. This is a simulated town built for testing driverless car technology (see Fig. 1.5). As the world's first dedicated test site for intelligent networked vehicles, one of its design features is to use the idea of intensive testing to test smart cars. A variety of road emergencies can occur in a concentrated manner. Therefore, test distance per

Fig. 1.5 Mcity test site

kilometer can be represented as a journey of tens of kilometers or even hundreds of kilometers in the real environment. Another feature of Mcity test site is the flexible design concept. The road has no fixed markings and can change the lane layout at any time. A variety of traffic elements (such as building exterior walls, dummy, etc.) can be moved, and traffic signs can be replaced at any time according to the test requirements. Moreover, a large and flat asphalt pavement area is reserved for the design and layout of scenes not included in the existing site, such as large parking lots. These can greatly facilitate the testers to adjust the test scenarios as needed, thereby greatly reducing the subsequent upgrade costs.

4. *Car-to-car intelligent communication system.* The car-to-car scholarly communication system is also well developed in the United States. This system can realize the communication between cars at any time, and it is convenient to understand the distance between cars, so as to prevent traffic accidents in time. Since the technology was approved in the United States on December 31, 2013, it has been trial-operated on multiple streets in Ann Arbor, Michigan, and other cities. For automobile manufacturers, the research and development cost of the car-to-car communication system is too generous, but for traffic safety considerations, relevant departments believe that the system is very necessary. The National Transportation Department of the United States NHTSA stated that the annual cost of solving traffic congestion is as high as 88 billion US dollars. After utilizing this technology, 80% of traffic accidents can be prevented. The car-to-car communication system was fully utilized in the United States in 2017.

5. *Driverless cars.* Unmanned vehicles are essential research on the safety of vehicles in intelligent transportation in the United States. Unmanned driving of vehicles is realized by introducing mature robots and automatic control, artificial intelligence, visual computing, and other technologies. A typical example is Waymo, a subsidiary of Google, which applies the above-mentioned technology to practice. The self-driving car it develops is based on cameras, radar, and laser

rangefinders to perceive the surrounding environment of the vehicle and uses onboard sensors to measure the distance, relative speed, and speed of the vehicle ahead. Obstacles and additional data information are transferred to the onboard main control computer, the data is processed through computer software, and the processing results are fed back to the main control computer. The automatic driving control software issues action commands to the steering wheel, accelerator, and brake controllers based on the feedback information to control the vehicle's steering, acceleration, deceleration, overtaking, lane change, and other behaviors so that the vehicle can safely and reliably drive on the road. The development of autonomous driving technology is accelerating. As of July 2018, the total road test mileage of Waymo's autonomous driving fleet has reached 8 million miles (about 12.87 million kilometers). It is worth mentioning that it only took Waymo 1 month to increase the total road test mileage from 7 million miles (about 11.27 million kilometers) to 8 million miles (about 12.87 million kilometers). Waymo only took one month.

6. *Intelligent traffic management.* The United States adopts a large number of advanced monitoring and management technologies in traffic management and provides drivers with real-time road conditions and vehicle violations through systems such as road monitoring centers and variable signs on the roadside. At present, the application of ITS in the United States has reached more than 80%, and related products are also more innovative. US ITS is used in vehicle safety systems (51%), electronic toll collection (37%), highway and vehicle management systems (28%), navigation and positioning systems (20%), and commercial vehicle management systems (14%). The development is relatively fast.

1.2.2 Development History and Present Situation of Intelligent Transportation System in Japan

1.2.2.1 Developmental History

The development process of Japan's intelligent transportation technology research has also gone through two stages. In the first phase, from the 1990s to the beginning of this century, although the research fields involve traffic safety assisted driving, navigation systems, electronic toll collection, traffic management optimization, road management efficiency, public transportation support, truck efficiency, pedestrian assistance, and emergency vehicles. However, the focus is on navigation systems, automatic toll collection systems, and advanced vehicle systems, and breakthroughs have been made in these technologies. In particular, navigation systems and automatic toll collection systems have been widely used. The second phase started from the beginning of this century, and the research focus has shifted to the improvement of road traffic safety, the smoothing of traffic and the reduction of environmental load, the improvement of personal convenience and comfort, the development of local vitality, the construction of public platforms, and the promotion of international standardization. Pay greater attention to system integration and humanized transportation services and the promotion and application of technology.

The first stage: In January 1994, Japan established the Vehicle and Road of Traffic Intelligence Society (VERTIS) with the participation of private enterprises and organizations. As the support of the association, a liaison meeting of five ministries and departments was established with the participation of heads of five government agencies, including the National Police Agency, the Ministry of International Trade and Industry, the Ministry of Transport, the Ministry of Posts, and the Ministry of Construction to improve the unified system for ITS research and development in Japan. In June 1995, the Japanese government determined the "basic policy for an advanced information society" at the cabinet meeting. In accordance with this policy, the applicable five provinces and agencies, mainly the National Police Agency, have issued the "Information Implementation Policy in the Road, Transportation, and Vehicle Fields" and formulated the basic national policy for the development of ITS. The document clarified 9 development areas and determined 11 measures to promote it. It is expected that the ITS construction goal will be completed in the early twenty-first century.

The second stage: After the first stage of development, Japan made a summary, combined with Japan's frequent traffic accidents, serious traffic congestion, rapid growth of mobile phones and the Internet, more attention to the environment, the advent of an aging society, and personal privacy protection. A new development strategy has been formulated for the enhanced social environment during the economic downturn and the traffic environment where some ITS systems (car navigation systems, information service systems, and electronic toll collection systems) have been widely used: (1) constructing a safe and reliable "ITS area," (2) promoting the development of autonomous driving of logistics and transportation vehicles, (3) commercialization of navigation systems to make traffic more comfortable, and (4) specific measures such as building an ITS integrated platform.

1.2.2.2 System Framework

Japan began to develop the national ITS system framework in January 1998 and completed it in November 1999. The biggest feature of Japan's ITS system framework is to emphasize the interaction and sharing of ITS information. Entire ITS construction is part of e-Japan. The overall content of the system framework is the same as that of the United States and Europe. It consists of three parts: user service, logical framework, and physical framework. It absorbs the characteristics of the United States and the European Union system framework. The Japanese ITS system framework adopts an object-oriented method to establish the logical and physical framework of the system. The object-oriented method for the development of the Japanese ITS system framework is mainly reflected in the construction of the logical framework. Through the abstraction of ITS, an information model is established to describe the information relationship between the objects involved in ITS (such as inheritance, etc.), and the control model is established to achieve each user service. The development process of Japan's ITS system framework is shown in Fig. 1.6.

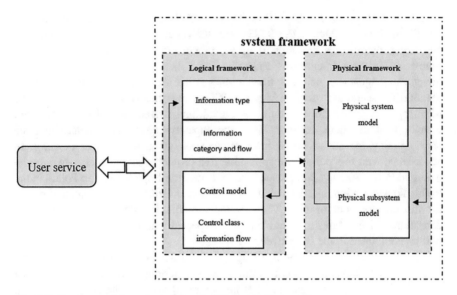

Fig. 1.6 Development flow chart of Japan's ITS system framework

Japan's ITS system framework is also constantly improving. In recent years, they are committed to the application and promotion of this system framework. In 2003, a local ITS system framework development auxiliary support system was launched and applied in Tokyo as a demonstration.

A detailed object model has been established in the Japanese ITS logical framework, including the overall model (for the overall ITS or the content shared by several service areas) and the detailed model (for a single service), which are based on the overall and dynamic perspectives. The two models were analyzed from overall and dynamic perspectives. In addition, the core model and detailed model corresponding to the two models were applied. Among them, the core model (overall) gives the relationship between the object classes involved in the service, and the detailed model (overall) analyzes the object classes in the core model in detail and gives the inheritance and other relations of the attributes of each object class. The detailed model is a deepening of the object classes in the core model; the core model (dynamic) is to establish an information interaction model between objects with dynamic information requirements, and the detailed model (dynamic) is for the deepening of the object classes in the core model (dynamic).

The Japanese ITS physical framework includes high-level subsystems, subsystems, low-level subsystems, a single independent physical model, an overall physical model, and information flow. Among them, the high-level subsystem is separated by location; the bottom-level subsystem is proposed based on the control model in the logical framework. The basic principle is to provide an unbiased bottom-level subsystem for each control module in the control model, and there is also a bottom-level subsystem. The system is commensurate with the situation where multiple control modules are included. Among them, the logical function corresponding to the 172 subservices in the ITS system framework is matched with the realization location through the method selection table, which is to complete the positioning of the bottom subsystem in the high-level subsystem. The system is obtained from the combination of the high-level subsystems and the similar low-level subsystems, which is a classification method and has no practical significance; the physical model is proposed for user services and consists of the basic units of the low-level subsystems. In fact, the high-level subsystems and the system framework flow together form an overall physical model, and the low-level subsystems and the framework flow together form a single autonomous physical model through different combinations. It is like the physical framework of the United States in that it assumes that dividing the physical system is based on people, vehicles, roads, centers, and the environment, and a corresponding physical model is proposed for user services.

1.2.2.3   Main Technical Features

Currently, Japan is stepping up the research and development of the Internet of vehicles and automation technology and aims to create a world-leading intelligent transportation system by 2020. In 2017, the Japanese government first joined automakers to test autonomous vehicles on highways and areas with low traffic and people. In 2020, the government planned to commercialize this service. This year is of great importance to Japan because Tokyo is the site of the Summer Olympics. After this, the country hopes to further demonstrate the strength of its new generation of technologies, such as fewer cars, environmentally friendly transportation transmission systems, and so on. In 2025, the Japanese government and automakers will hope to popularize autonomous driving technology across the country so that vehicles can go on the road without requiring a driver. In addition, the Japanese government also hopes to promote new technologies to promote local employment in 2030 and maintain the overseas competitiveness of the country's automakers. The government stated that traffic fatalities will also be significantly reduced by then, and even reduce this number to zero.

At present, Japan's ITS is mainly used in traffic information provision, electronic toll collection, public transportation, commercial vehicle management, and emergency vehicle priority.

1. *Electronic toll collection system "ETC2.0."* The ETC system is commonly used in countries all over the world, but no country or region has so far used ETC as many vehicles as Japan. One of the reasons why ETC is so popular in Japan is that the specifications of Japanese ETC are unified nationwide. There are various toll road companies in Japan, but once you install an ETC vehicle and insert an ETC card in your car, you can use almost all toll highways in Japan. In addition, the company can guarantee basic security; the company can develop the business of ETC in-vehicle devices and ETC cards, achieve low prices through free competition, and promote the popularization of ETC.

Japan's ETC is composed of an onboard device and an ETC card. In this way, it is impossible to distinguish between the owner of the car (onboard device) and the payer of the ETC fee (ETC card). In other words, using ETC to pay for fees does not necessarily have to be the owner of the car. If you have an ETC card, you can use the ETC card on an onboard device that does not have a vehicle such as a rental car.

Since 2016, Japan has upgraded the previously utilized ETC electronic toll collection system in some areas, and ETC 2.0 has appeared. ETC 2.0 is the world's first I2V collaboration system. It provides driving support services through smart antennas installed next to highways. ETC 2.0 can provide drivers with valuable information services, such as traffic jam avoidance, safe driving assistance, traffic accident rescue, and the original ETC highway toll. In addition, it is also more useful in promoting the diversification of urban parking fees and the management of vehicle receipts in the future. With the support of ETC 2.0, the resources of the road network can be utilized more effectively.

2. *Advanced safety vehicle (ASV)*. Advanced safety vehicles can prevent accidents before they occur and use various sensors on the vehicle and the road to grasp the driving environment information such as the conditions of vehicles around the road and provide it to the driver in real time.

There are five stages in the ASV implementation plan. The first stage is a 5-year plan from 1991 to 1995. The purpose is to test the technical possibilities of passenger cars and to verify the reduction of the impact of accidents; the second stage is from 1996 to 2000; the third phase is from 2001 to 2005; the fourth phase is from 2006 to 2010; the fifth phase is from 2011 to 2015. As of 2016, ASV has been extensively used in Japan and has significantly reduced traffic accidents.

3. *Efficient logistics system*. Japan also attaches great importance to the development of the Internet of things technology. In 2007, the government and 23 well-known private enterprises jointly initiated the Smartway program, which is used to promote the development of road infrastructure, transportation, tourism, and advanced safety cars. The focus of Smartway development is to integrate existing ITS functions such as ETC, online payment, and VICS on the onboard unit OBU so that roads and vehicles can be connected in two directions to become Smartway and Smartcar to reduce traffic accidents and alleviate traffic congestion. In January 2011, based on the research results of Smartway, a new ITS service "ITS Spot Service" was implemented on the Tokyo Bay Shore Routes. As of March 2011, this service has been extended to 1600 points, mainly on highways. At the same time, the Japanese government also actively supports the development of the transportation Internet of things. In 2009, Japan formulated the "i-Japan Strategy 2015." While realizing transportation e-government, it is committed to reducing traffic congestion, improving logistics efficiency, and reducing CO_2 through the Internet of things technology. In 2010, Japan formulated a "new IT strategy" to promote green travel. The short-term plan goal is to utilize vehicle detection technology to ensure the smooth flow of traffic, improve logistics efficiency, and use bus priority systems and bus positioning systems to

enhance the convenience of the public transportation system and increase its utilization rate.

Japan is also making efforts to build a new generation of the traffic management system UTMS. UTMS uses optical beacon sensors to achieve two-way communication between vehicles and the traffic command center. Through information exchange, it can improve driving safety, reduce traffic congestion, reduce traffic pollution, and build a traffic society with the goal of safety, comfort, and low environmental pollution. Japan mainly promotes the work system in UTMS in the following aspects:

(a) *Traffic information service system AMIS.* AMIS utilizes variable information boards and traffic broadcasts to provide traffic information services to onboard equipment through optical beacon sensors to achieve the purpose of dispersing traffic flow and alleviating road traffic congestion. The system was promoted nationwide in Japan in 2012 and was still improved and promoted in subsequent periods.

(b) *Emergency rescue vehicle support system FAST.* In areas where emergency rescue vehicles are frequently dispatched and passed, FAST uses optical beacon sensors to detect emergency rescue vehicles performing tasks and implement signal priority control to them to shorten the time for the vehicles to reach their destinations and prevent the traffic accidents resulted by the high-speed emergency rescue vehicles.

4. *Road traffic information and communication system (vehicle information and communication system, VICS).* The VICS system is widely used in Japan as part of the application of the Japanese scholarly traffic management system. Traffic managers and road managers (road public corporations, etc.) provide traffic information free of charge and then collect information to the VICS center through the Japan Road Traffic Information Center. Then, the VICS center transmits it to the driver and the onboard device through a variety of methods. This system has a high penetration rate in Japan, mainly due to a successful business model. In Japan, the service of the VICS system is free. Users only need to purchase a car navigator with the VICS system to enjoy the free service provided by the VICS system, and there is no need to pay additional fees for daily use. However, the navigator equipped with the VICS system is usually more expensive than the ordinary one. Every time a navigator with VICS function is sold, the VICS center will receive the technical support fee returned by the navigator manufacturer or the car factory. The annual VICS center's account is enough to support the center's operating expenses.

Japan is also putting into practical use the DSSS, an auxiliary safe driving system based on vehicle–road collaboration. DSSS uses roadside detectors to detect risk factors and provide information services, "DSSS-I (Information Service Type)," which has been implemented in Tokyo, Saitama Prefecture, and other places. Judgment type DSSS vehicle-mounted device judges whether it is necessary to provide information services and the timing of the service to the driver based on the information and reminds the driver to pay attention through sound and images. The system has undergone large-scale empirical tests. The system could not only prevent traffic accidents in the section with road test detector, but also improve driver's behavior through learning function. In 2012, empirical experiments were performed on the simple DSSS, which uses radio waves to continuously provide information services to the onboard equipment and is not connected to the traffic command center. The simple DSSS is characterized by not having to be connected to the traffic command center and can reduce costs.

In order to reduce fuel waste and CO_2 emissions caused by the frequent starting of vehicles in traffic jams, Japan has started research and development of unmanned driving and positioning and the establishment of methods for evaluating CO_2 reduction effects since 2008. Autonomous driving and platooning research and development are some of the countermeasures to deal with energy-saving vehicles. ITS technology is used to organize cars to platoon, and the interval between multiple vehicles is shortened as much as possible to make them drive in platoons, as well as the research and development of key technologies needed for the ecological driving on urban roads. In 2012, four trucks were successfully tested, driving in a line at a speed of 80 km/h and 4-m intervals. The project was supported by four truck manufacturers, and a test vehicle of the "car-vehicle communication distance system CACC" was produced.

1.2.3 Development and Status of Intelligent Transportation Systems in Europe

1.2.3.1 Development History

The research on intelligent transportation and safety in Europe started in the same period with the United States and Japan, and the development process also experienced two stages. The first stage is from the 1980s to the early twenty-first century, the research field involves advanced traveler information system (ATIS), advanced vehicle control system (AVCS), commercial vehicle operations system (CVOS), electronic toll collection system, etc. focusing on road and onboard communication equipment and vehicle intelligence and public transportation. Its characteristics are the same as the first stage of the United States; that is, the scope of research is relatively wide and the projects are relatively scattered. The second stage starts from the beginning of the twenty-first century and focuses on the research of security issues. It pays more attention to the research of the system framework and standards, traffic and communication standardization, integrated transportation cooperation, and other technologies and promotes the practicability of intelligent transportation and security technologies.

In November 2016, the European Commission adopted the European cooperative intelligent transportation system (C-ITS) strategy; the goal is to deploy cooperative intelligent transportation systems on the roads of EU countries on a large-scale by 2019 to realize intelligent communication between vehicles and between vehicles and road facilities. Cooperative intelligent transportation system is characterized by multiple communication technologies to enable communication between vehicles and between vehicles and road facilities so that road users and traffic management personnel can share information and effectively coordinate. For example, vehicles can automatically send warning information such as emergency braking and front congestion or can automatically receive the speed limit information sent by a certain section of facilities. At the same time, the European Union has also carried out the comprehensive development of telematics. It proposes establishing a special traffic (mainly road traffic) wireless data communication network in Europe. It is developing an advanced travel information service system (ATIS), advanced vehicle control system (AVCS), advanced commercial vehicle operation system (ACVO), advanced electronic toll collection system, etc. Some European countries are also developing traffic communication information superhighway (TIH) and video information superhighway (VIH). At present, the development of intelligent transportation in Europe mainly includes automatic vehicle positioning system, variable information system, intelligent parking system, travel information highway, video information highway, national traffic control center, urban traffic management and control system, scoot system, electronic toll system, digital traffic law enforcement system, radio frequency identification technology, Internet of things, etc.

1.2.3.2 System Framework

In April 1998, the European Union started the project code named Karen (Keystone architecture required for European networks), which lay the foundations of the EU ITS system framework. In August and October 1999, Europe completed the logical and physical framework of ITS and then supplemented and improved the contents of other parts, forming the overall framework of the European Union. Compared with the American ITS system framework, which is all-encompassing and comprehensive content, the EU ITS system framework selects typical systems for detailed analysis in content and does not aim at "all."The structure analysis method is also used in the development of the ITS framework in the EU. The overall structure of its framework is comparable to that of its framework in the United States, which mainly includes the functional framework, physical framework, communication framework, cost–benefit analysis, etc. Figure 1.7 shows EU physical structure example system structure diagram.

After the introduction of the EU ITS system framework, EU countries (such as Italy, France, etc.) have built a system framework suitable for their own national conditions on this basis to further provide guidance for the EU and ITS construction.

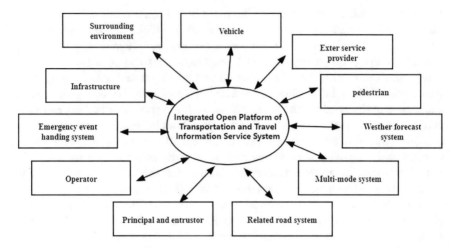

Fig. 1.7 EU physical structure example of system structure diagram

1.2.3.3 Main Technical Features

In order to encourage the development of intelligent transportation, Europe has mainly promoted the following research projects in recent years.

1. *ITS4rCO$_2$ (2015).* ITS4rCO$_2$ (building an intelligent transportation system for CO$_2$ emissions) is a pilot measure implemented by the European Intelligent Transportation Association to reduce CO$_2$ emissions. Based on this background, the European Association of automobile manufacturers has provided a large amount of data for the European Intelligent Transportation Association, which strongly supports a series of measures taken to reduce CO$_2$ emissions. This research mainly focusses on the following two aspects:

   (a) CO$_2$ emission of cars is reduced through intelligent transportation measures.
   (b) CO$_2$ emission of cars is reduced through intelligent transportation-related infrastructure.

2. *EcoDriver (2011–2016).* EcoDriver (driving style supporting energy conservation and emission reduction) project brings together engineers, behavior analysts, and economists, aiming to reduce CO$_2$ emissions and energy consumption by optimizing the driver power composition environment system to feedback and guide drivers to drive green. In the process of driving, drivers will receive various ecological driving suggestions according to their own and vehicle characteristics, including driving habits, energy consumption mode, and various information, and data including vehicle types will also be analyzed and discussed. On this basis, driving suggestions on the most suitable ecological driving mode in any environment will be put forward.

3. *UDRIVE (2012–2016)*. UDRIVE (European Natural Driving Research Project) is about cars and trucks and is the first large European naturalistic driving research project for electric bikes to observe and analyze the driving behavior, using the latest information and communication technologies, with the aim of improving road safety and reducing energy consumption. In 2012, the European Intelligent Transportation Association (ITSA) prepared for large-scale collection of natural driving data in seven member regions of Europe. The objectives of the project include the following: using quantitative methods to analyze the behavior of road users in different regions of Europe in the normal situation and at the time of traffic accidents, and quantitatively evaluating the safety-related driving behavior; quantitative analysis and emission levels; energy consumption-related road user behavior; and researching new methods to improve the safety and sustainability of transportation systems.

4. *VRA project (2013–2016)*. VRA is an active network of European road and vehicle automation experts and stakeholders, which helps to maintain common European positions and issues on vehicle and road automation. VRA is determined to international cooperation and simultaneous development of road and vehicle automation in the EU, the United States, and Japan. VRA discusses the automation settings of roads and vehicles in different fields, including deployment paths and solutions, legal and regulatory issues, road adaptability testing, connectivity, digital infrastructure, human factors, benefit evaluation, decision-making and control algorithms, etc. VRA union meetings for different discussion groups support the ongoing projects and activities (CityMobil2, Autonet2030 iGames, and others), but also by summing up the past and the present in view of the road and vehicle automatic transmission.

5. *CARTER project (2016–2018)*. CARTER (European Road Traffic Autopilot Collaborative Development) is a project that coordinates and supports the development and deployment of automated road transportation under the H2020 plan. Through EU Member States' more clear and consistent policies for industrial partners, CARTER aims to ensure the consistency and coordinated development of its systems and services in Europe. CARTER held a forum on road automated transport systems at European and international levels to support current research and activities in specific areas of automated transport systems. Currently, CARTER has 36 partners working together to achieve the following goals:

   (a) Establish a leader in the development of the European road automated transportation system through the cooperation of the government and enterprises.
   (b) Support international cooperation activities in the field of road automated transportation, especially with the United States and Japan.
   (c) Establish a forum with stakeholders to coordinate and support the development of road automated transportation systems at European and international levels.
   (d) Provide a platform for data, experience, and achievement exchange.
   (e) Support field testing and provide guidance for national and EU development.

6. *INLANE project (2016–2018)*. The purpose of the INLANE (lane-level navigation and automatic driving map generation through the fusion of low-cost navigation satellites and computer vision) project is to develop a new generation of
low-cost, lane-level precision navigation systems based on the fusion of navigation satellites and computer vision, which will enable a new generation of the
system that can realize real-time update of information based on groundsheet
technology. INLANE will develop low-cost EGNOS/EDAs + GNSS
(GPS/GLONASS) + system + positioning module based on computer vision,
make full use of low-cost elements, improve the positioning performance, reduce
the cost, and achieve lane-level positioning. Lane-level vehicle positioning can
make navigation and traffic management more detailed and efficient. This line-
level positioning module can provide an interface for smartphones. INLANE will
also develop new, computer vision-based road models and traffic signal recognition information, which can achieve line-level real-time traffic management based
on groundsheet. INLANE will consider the integration of complex GNSS, a
computer vision signal, IMU, hybrid algorithm to achieve submeter accuracy of
the map, and ultimately achieve the goal that the relative error between the map
and the line position is not more than 5 cm.

   The project started in January 2016 and is currently in the process of identifying requirements. During its first year, the project was exhibited at various high-
profile events, including the Spatial Information Day and the European and
World Conventions.

7. *European cooperative ITS corridor project*. This project is the world's leading
intelligent transportation project jointly deployed by three countries. In
November 2014, the five test vehicles of the project accomplished the 1300 km
driving experiment through the vehicle–road cooperation technology, including
the test sites in Munich, Vienna, and Helmand. The demonstration functions
include road construction reminders, surrounding vehicle information, and traffic
signal prompts at intersections. After the first phase of the project, it is planned to
connect intelligent roads built in various countries with the corridors. France,
Poland, and the Czech Republic will be the first countries.

8. *OPTITRUCK (2016–2019)*. OPTITRUCK is an optimization project for heavy
vehicles, dedicated to reducing the impact of heavy vehicles on the environment.
The impact of road traffic on energy efficiency is a major global policy concern,
which has stimulated large numbers of innovations, which are mainly reflected in
the improvement of basic vehicles, traffic management technology, and public
policy actions. Currently, many approaches to reducing energy consumption and
CO_2 emissions are applicable to light vehicles but are not feasible for heavy
vehicle applications. Reducing fuel consumption and other consumables without
sacrificing emissions is an important challenge for heavy vehicles. However, in
actual driving conditions and actual transport tasks, there are many optimization
possibilities, especially when balancing fuel efficiency and emissions with specific vehicle applications and operating conditions. Through extensive development of big data using advanced optimization and calibration technologies, the
integration of telematics and communication technologies into smart architecture

drive train systems could further reduce the environmental impact of heavy vehicles. OPTITRUCK aims to achieve the best global fuel consumption (at least 20% reduction) by applying the most advanced technologies from powertrain control to intelligent transportation systems, and at the same time achieve Euro VI emission standard for heavy road transportation (40).

9. *NEMO (2016–2019).* NEMO will create a super network including tools, models, and services to build an open, distributed, and widely accepted electric travel system. NEMO's super network is in a distributed environment based on standard interface and open architecture, in which all parts involved in electromigration (charging stations, power grids, system operators and service providers, vehicles and car owners/drivers) can connect and interact seamlessly, thus exchanging data and providing ICT services in a fully integrated and interoperable environment. In terms of data connection, data-based dynamic conversion and service interfaces will be used. In order to meet specific local scenarios and stakeholder needs, existing services and data repositories will be integrated in a seamless and efficient manner.

The purpose of NEMO is to reduce digital (interface) and physical (location) barriers by providing back-end data and services suitable for participants, to improve the availability of electric travel services and achieve better planning and safer grid operation, which is expected to promote the market development of electric vehicle charging facilities, ICT services, and wider B2B interconnection. The project started in October 2016, and the current work focuses on the investigation of use cases and needs of electric travel participants and the determination of specifications and reference architectures.

1.2.4   Development History and Present Situation of Intelligent Transportation System in China

1.2.4.1   Development History

At the end of the 1970s, China began to apply electronic information technology in transportation and management. Over the past 20 years, with the support of the government and the insistence on independent development, China has carried out preliminary theoretical research, product development, and demonstration application in the field of ITS through extensive international exchanges and cooperation and achieved certain results. A few research centers and production enterprises engaged in ITS research and development are growing through the combination of theory and practice. The Ministry of Science and Technology of the People's Republic of China approved the establishment of the national ITS engineering technology research center (ITSC) in 1999 and the construction of the National Railway Intelligent Transportation System Engineering Technology Research Center in 2000. Many universities and research institutions have set up ITS research centers to engage in ITS theoretical research and product development, such as ITS center of Southeast University, ITS research center of Wuhan University of technology, ITS research center of Jilin University, ITS research center of Beijing Jiaotong University, ITS research center of Tongji University, and ITS research center of South China University of technology. Many Chinese enterprises in transportation and IT industries are attracted by the huge market of ITS and involved in the field of ITS product development research and application. In order to coordinate and guide the development of ITS in China, in 2001, the Ministry of Science and Technology, together with the National Planning Commission, the Economic and Trade Commission, the Ministry of Public Security, the Ministry of Railways, and the Ministry of Communications at that time, jointly established the National ITS Coordination Steering Group and Office and established the ITS Expert Advisory Committee, which was responsible for organizing, researching, and formulating the overall strategy, technical policies, and technical standards for the development of ITS in China. It actively supported relevant ministries, local governments, enterprises, and research units, carried out the key technology research and application demonstration projects of ITS according to the characteristics of industries and regions, and promoted the industrialization of ITS research results. In July 2012, the Ministry of Transport of China issued the "Intelligent Transportation Development Strategy for Transportation Industry (2012–2020)," which pointed out the direction for the future development of intelligent transportation in China. In 2020, China will basically realize the intelligent management and service of urban public transportation at and above the prefecture level. In conditional large- and medium-sized cities to promote construction and the application of integrated passenger transport hub collaborative management and service system, the overall efficiency of urban traffic has been greatly improved. In addition, the strategy also puts forward a clear development goal of intelligent transportation: by 2020, China will reach the international advanced level in traffic information collection, traffic data processing, urban traffic signal control, container transportation, port automation, and other aspects; the overall technical level will reach the level of developed countries at that time and realize the independent development and scale application of main intelligent transportation technology and equipment, application software, and control software.

1.2.4.2   System Framework

Although China's intelligent transportation system started late, the rapid development of ITS in the development process also caused many problems, such as repeated construction and system incompatibility. Government and scientific workers have deeply realized the importance of carrying out research on the framework of the intelligent transportation system. In 1999, the national intelligent transportation system coordination leading group and office were responsible for organizing and implementing the formulation of China's intelligent transportation system framework. China officially published "Framework of China Intelligent

Transportation System" in January 2003 and revised "Framework of China Intelligent Transportation System" in 2005 (2nd edition).

1. The goals and steps of ITS framework in China.

The national ITS system framework is the highest-level overall system architecture, which has important guiding significance for ITS planning and construction nationwide. It mainly expounds the function of the system, determines the subsystems and elements that constitute the system, as well as the information flow between them, and highlights the relationship between various parts. The design objectives of the ITS framework in China are as follows.

(a) Define the overall requirements of ITS and describe the main contents of ITS in China. Fully understand, analyze, and summarize the needs of ITS users in different fields and at different levels, and formulate user services that meet the national conditions and actual needs.
(b) It defines the system framework of ITS, analyzes the overall framework structure of ITS in China based on user needs and user services, and puts forward the basic components of the system and the interconnection relationship of each component.
(c) Analyze the technical and economic factors that affect the development of ITS and reduce the impact of technology on the frame structure.

The framework of China's ITS adopts a process-oriented approach, and the specific development steps are as follows.

(a) Determine user service. Based on the investigation of traffic-related government departments, traffic participants, and researchers to clarify the needs, define the user and service subjects; combined with the international experience of ITS system framework and China 's actual situation, determine ITS service areas and give the meaning of China.
(b) Establish a logical framework. Starting from the analysis of users, the main functions that the system should have are determined, and the functions are divided into several levels such as system functions, processes, and sub-processes. On this basis, the logical structure of ITS and the interaction between various functions are analyzed, and the main information of interaction between functions and processes is defined in the form of data flow.
(c) Establish a physical framework. The structure of the actual intelligent transportation system should be analyzed from the perspective of the physical system. The structure of the system is analyzed at the levels of system, subsystem, and module, and the interactive information between ITS physical systems is analyzed. The information is defined in the form of frame flow, and the realization relationship of the system function and the inclusion relationship of frame flow to data flow are clarified.
(d) Define the standardization content. Determine the standards of its related technology, the interface standards of its related equipment, and the interface standards among its subsystems.

2. Service definition of ITS framework in China.

ITS user service defines the main content of ITS system and describes what ITS should do from the perspective of system users. User service is divided into three levels: user service domain, user service, and user subservice definition. ITS user service should not only conform to the reality but also be forward-looking.

The determination of its user service in China is based on the detailed investigation and analysis of China's transportation infrastructure, transportation status, traffic travel and management needs, traffic management related laws and regulations, traffic development planning, and socio-economic, political, cultural, scientific, and technological development background. The first edition of the national system framework includes 8 service areas, 34 services, and 138 subservices.

3. Logical framework of ITS architecture in China.

The logical framework of China's intelligent transportation system architecture includes 10 functional domains. The codes of each functional domain in the name of data flow are as follows.

Traffic Management and Planning, TMP
Electronic Payment Service, EPS
Traveler Information System, TIS
Vehicle Safety and Driving Assistance, VSDA
Emergency and Security, EM
Transportation Operation Management, TOM
Inter Modal Transportation, IMT
Automated Highway System, AHS
Transportation Geographic Information and Positioning System, TGIPS
Evaluate, EVAL

For data flows in the same functional domain, the naming format is functional domain code _ data flow name. For data flows, starting and ending terms belong to different functional domains; the name format is functional domain code of starting point _ functional domain code of destination _ data flow name. For data flows in and out of the terminal, the name formats are as follows:

Data flow from the terminal: f terminal name _ function domain code _ data
    flow name
Data flow into the terminal: t terminal name _ functional domain code _ data flow
    name, where the terminal name is defined in Chinese

A set of data streams that flow out or flow into the same terminal can be merged into a data stream group, whose name format is as follows:

Data flow group outflowed from the terminal: fr terminal name
Data flow group into terminal: to terminal name
Terminal bidirectional data flow group: to/fr terminal name

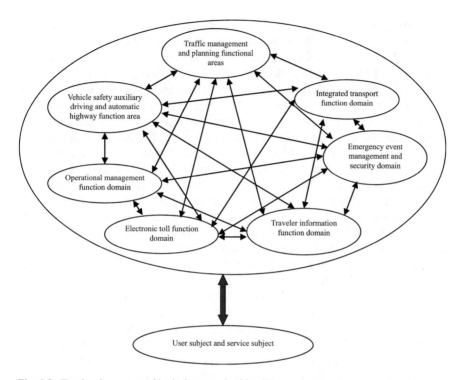

Fig. 1.8 Top-level structure of logic framework of intelligent transportation system in China

The top-level structure of China's ITS logic framework is shown in Fig. 1.8. The main content of the logical framework is to describe the data flow among the system functions. The system function and data flow can refer to the relevant function description table and data flow description table (data dictionary) in the research report to be published.

1.2.4.3 Main Technical Features

In order to promote the development of intelligent transportation, China has mainly promoted the following research projects in recent years and achieved a series of results.

1. National "863" Plan—"Urban Ground Traffic Network Coordination and Environmentally Friendly Traffic Control Technology" (2014–2016).

   The overall goal of the project is to alleviate the prominent problems of traffic environmental pollution and inefficient bus network service faced by urban traffic in China. Through the research and development of key technologies and systems such as intelligent low-emission traffic control and multimode and multilevel bus coordination control, it provides technical support for the low-carbon environmental protection development of urban traffic development and the efficient operation of urban bus network. The project breaks through the key technologies of multilevel traffic and environmental data sharing, dynamic monitoring and prediction of regional traffic emissions based on driving trajectory characteristics, and intelligent control of vehicles and traffic signals for environment and traffic efficiency. Aiming at the coordination problem of urban ground bus network, the key technologies of efficient configuration and passenger flow organization of multimode bus transfer hub under the competition and cooperation relationship of ground bus network are developed. A traffic operation and environmental data sharing system, a regional traffic emission dynamic monitoring and forecasting system, a multiobjective traffic signal control prototype system, a collaborative vehicle speed assistance prototype system, a multimode bus network coordination scheduling system, and a trunk bus intersection-station-signal collaborative control equipment are developed.

2. National Science and Technology Support Project—"Integration and Demonstration Application of Road Traffic Intelligent Networking Control Technology in Medium-Sized Cities" (2014–2017).

The main goal of this project is to focus on the application needs of the development of urban road traffic management in medium-sized cities in China. In view of the traffic operation characteristics of medium-sized cities and the difficulties faced by traffic networking and joint control technology, this paper studies the key technologies such as traffic organization optimization and dynamic and static traffic coordination and control suitable for medium-sized cities, breaks through the bottleneck of the application of intelligent networking and joint control technology such as urban traffic state perception, signal control, and traffic guidance, integrates and constructs the urban traffic intelligent networking and joint control platform, organizes and implements demonstration projects in several representative medium-sized cities, forms relevant technical specifications and application guidelines, standardizes and guides the construction and development mode of intelligent transportation system in medium-sized cities in China, improves the intelligent level of urban traffic control and service, and improves the level of road smoothness and mass travel satisfaction.

3. National Science and Technology Support Project—"Highway Air-Ground Integrated Traffic Behavior Monitoring and Information Law Enforcement Technology and Equipment Research and Development" (2012–2016).

The project aims to comprehensively improve the information-based law enforcement capability of expressway traffic safety in China and focuses on five research topics: the development of air-ground integrated monitoring technology and equipment for traffic behavior, the development of traffic behavior analysis technology and system for key drivers, the development of forensics technology and equipment for typical traffic violations, the development of intervention technology and equipment for typical traffic violations, and the information-based law enforcement technology and system integration of traffic safety. In 2016, the project completed the key technical research and equipment development. Focusing on comprehensively improving the safety supervision level of expressway vehicles, the intelligent sensing system of expressway traffic behavior with air-ground integration and the humanized law enforcement system of "illegal warning, illegal evidence collection, and illegal law enforcement" were constructed. The development of three-dimensional monitoring equipment such as airspace monitoring of expressway traffic behavior and comprehensive perception of vehicle characteristics was carried out. The information law enforcement system of expressway traffic safety was integrated and developed. The functions of three-dimensional monitoring of expressway traffic behavior, visual warning of key illegal behavior, humanized law enforcement of typical illegal behavior, and information management of traffic vehicles were realized.

4. National Science and Technology Support Project—"Research and Development of Key Technologies and System Integration of Highway Traffic Safety Networking" (2014–2017).

Guided by the demand of highway traffic safety information management, this paper focuses on the hot and difficult issues such as the low integration of traffic safety information sharing and business resources in highway network and the weak traffic safety control ability under sudden traffic accidents such as bad weather and large traffic flow on holidays. It focuses on the research and development of road network traffic safety information integration system, multiscale traffic safety risk assessment system, large-scale traffic flow intervention control system, police command and road police linkage joint control system, and traffic safety information socialization service system. It integrates and establishes the three-level highway traffic safety active prevention and control platform of ministries, provinces, and prefectures and constructs the technical support system of all-time and space active prevention and control of highway network traffic safety.

5. National Science and Technology Support Project—"Highway Major Emergencies Disposal and Emergency Rescue Technology and Equipment Research and Development" (2014–2017).

This project aims to comprehensively improve the disposal and emergency rescue ability of major highway emergencies in China. The rapid response and decision support technology and system development of major emergencies, emergency rescue technology and equipment development of major emergencies, field disposal technology and equipment development of major emergencies, infrastructure support technology and equipment development of major emergencies, analysis and evaluation technology, and system development of major emergencies are studied from five research topics of highway emergency decision support, emergency rescue, field disposal technology and equipment development of major emergencies, analysis and evaluation technology, and system development of major emergencies. The mechanism, technology, system, and equipment of emergency disposal and rescue are systematically studied, which breaks through the concept and idea of highway emergency disposal, innovates disposal methods and means comprehensively improving the disposal efficiency, and achieves good results.

6. The National Natural Science Foundation of China (NSFC) project "Cognitive Computing of Visual and Auditory Information" (2008–present).

The project takes the cognitive mechanism and computational model of images, voices, and texts (languages) related to human visual and auditory information in the fields of social, economic, and national security as the research object and aims to improve the computer's understanding of complex perceptual information and the processing efficiency of massive heterogeneous information. Specific performance in the basic theory of visual and auditory information research has made important progress. Significant breakthroughs have been made in the calculation of visual and auditory information and related key technologies of brain–computer interface. Based on the above main research results, the verification platform for driverless vehicles with natural environment perception and intelligent behavior decision-making ability and the brain–computer interface verification platform for brain function and limb rehabilitation are developed. A new key technology of intelligent assistant safe driving based on a comprehensive analysis of the human–vehicle–road state is provided. Since its inception in 2008, 65 nurturing projects, 26 focused support projects, and 4 integrated projects have been funded. In 2015, four research directions of driverless vehicle intelligence (environmental cognition, path planning, behavior decision-making and control, etc.) testing and evaluation and brain–computer interface were funded.

1.3   Development Trend of Intelligent Transportation System

As the basic, leading, and service industry of the national economy, transportation is the basic industry of the country and the infrastructure that supports economic and social development. It is in terms of optimizing the national industrial layout, promoting economic restructuring, reducing development costs, and reducing environmental pollution, and has an extremely important strategic role.

Facing the development needs of "safe transportation," we have a long way to go in research on traffic safety facilities, vehicle safety performance, safety supervision of transportation enterprises, and traffic safety management methods. The focus is to study and promote the use of information technology and advanced management methods. Improve the level of road traffic safety management technology, especially the use of intelligent means to improve the ability to guarantee traffic safety. From the perspective of intelligent transportation technology, the focus should be on the research and development of the following cutting-edge technologies and key core technologies.

Pay attention to the development of smart car technology. Research and develop distributed vehicle body network control technology based on CAN/LIN bus, realize vehicle-mounted intelligent human–computer interaction and vehicle–vehicle communication based on network and cloud identification methods, and breakthrough vehicle artificial intelligence control technology. At present, the United States has completed the development and actual road tests of smart vehicles, and some research institutions in my country have also initially developed prototypes and conducted preliminary tests. The development of smart car technology will comprehensively integrate various new traffic safety technologies and significantly improve the level of road traffic safety.

Improve the technical level of active and passive safety of vehicles. Make breakthroughs in vehicle safety control technology, study the precise identification technology of people, vehicles, and environment-related factors in the integration process, solve the communication and information fusion technology between multiple dynamic stability control subsystems, and improve the active safety performance of vehicles. Provide basic support to reduce the incidence of automobile traffic safety accidents. Breakthrough the monitoring and early warning safety technology, research and develop real-time detection technology for illegal driving and dangerous driving behavior of drivers in the actual driving environment, and realize accurate judgment of drivers' dangerous driving behaviors.

In order to breakthrough the integrated technology of passive safety protection, it is necessary to form pedestrian collision protection design technology under the premise of meeting the requirements of automobile safety performance. The design and development of intelligent occupant restraint systems based on collision prediction technology and research on key technologies such as safe car seats and headrest systems are also important directions.

Research on the active prevention and control technology of road traffic safety in the environment of the Internet of vehicles. With the goal of improving the level of road traffic safety, research on the perception of the safety status of the human–vehicle–road–environment system in the state of the Internet of vehicles, the accident risk assessment based on the coordination of elements, the active prediction, and intelligent judgment of accident risk based on big data, and the safety of the Internet of vehicles Information push and service technologies form a road traffic active prevention and control technology system under the Internet of vehicles environment.

Promote the development and application of road traffic network optimization control technology. Facing both normal and abnormal situations, research the multimode traffic trajectory and operation feature extraction in the human–vehicle–road collaborative environment, self-organization and collaboration-based vehicle operation auxiliary control and fleet dynamic control, and multimode traffic flow at intersections. Target optimization control and other core technologies are used to establish a networked multimode traffic flow active control system that integrates lane use, signal control, information guidance, and individual guidance.

1.4 Summary

With the in-depth research and development and practical application of intelligent transportation systems, many intelligent transportation technologies have been developed and become mature, and they have been widely used in actual transportation systems, such as public mobile communication technology, vehicle–road collaboration technology, electronic toll technology, path guidance, and navigation services, vision extension technology, intersection collision avoidance system, horizontal and vertical collision avoidance system, occupant protection system before collision, safety status detection system, intelligent headlight follow-up technology, safe distance warning, active collision avoidance, adaptive cruise, intelligent parking assist system, etc.

Many intelligent transportation technologies need to be further studied, such as vehicle–road coordination, car networking, automatic highways, parking guidance systems, fleet management systems, transportation demand management systems, bus priority systems, emergency management systems, tourism and traveler information service systems, electronic customs clearance systems for commercial vehicles, onboard safety monitoring systems, automated systems for roadside safety inspections, emergency response systems for dangerous goods, driver vision enhancement systems, onboard route guidance systems, collaborative driving systems, and vehicle intelligence-based anonymous autopilot, highway control autopilot based on intelligent highway infrastructure, etc.

In short, an intelligent transportation system, as a representative of modern transportation technology, is an important support for realizing transportation modernization. Intelligent transportation has achieved remarkable results in improving transportation efficiency, ensuring safety, serving the public, and promoting the sustainable development of the transportation system. In the future, based on national conditions, using new technology and combining it with the construction of smart cities, building a new generation of intelligent transportation systems with Chinese characteristics will be an important direction for the development of intelligent transportation in my country.

Bibliography

1. Yan X, Wu C (2012) Intelligent transportation system-principle, method and application. Wuhan University of Technology Press, Wuhan. (Published in Chinese)
2. Jin M (2012) Status and development tendency of intelligent transportation systems in China. J Transp Inform Saf 30:1–5. (Published in Chinese)
3. Wu Z, He Y (2015) Enhancing the road traffic safety based on ITS technologies. J Transp Inform Saf 1:1–8. (Published in Chinese)
4. Li R (2011) Introduction to Japan's intelligent transportation system and its reference. China ITS J 4:142–144. (Published in Chinese)
5. Xu H, Xia C, Sun L (2013) System and application of ITS intelligent transportation system in Japan. Highway 9:187–191. (Published in Chinese)

6. Wang T (2013) Brief talk on American intelligent transportation. Northern Commun (S2):4–6. (Published in Chinese)
7. Cao JL (2012) This decade—report on the development of science and technology in the modern transportation. Science and Technology Literature Press, Beijing. (Published in Chinese)
8. Yan X (2016) Research on information fusion of transportation system. Science Press, Beijing. (Published in Chinese)
9. Lu H (2002) Intelligent transportation system. People's Communications Press, Beijing. (Published in Chinese)
10. Zhang G (2003) Introduction to intelligent transportation system. People's Communications Press, Beijing. (Published in Chinese)
11. Yang Z (2005) Basic traffic information fusion technology and its application. China Railway Press, Beijing. (Published in Chinese)
12. Chu H, Yang X, Wu Z (2005) Traffic mobile collection technology and its applicability analysis. In: 2005 National Doctoral Academic Forum (Transportation Engineering Discipline) proceedings. (Published in Chinese)
13. Jiang G (2004) Malfunction identifying and modifying of dynamic traffic data. J Traffic Transp Eng 1:121–125. (Published in Chinese)
14. Li C, Yang R (2004) Data-fusion prediction of traffic information based on artificial neural network. Syst Eng 3:80–83. (Published in Chinese)
15. Weng X (2000) Expressway electromechanical system. People's Communications Press, Beijing. (Published in Chinese)
16. Jiang G (2004) Road traffic state discrimination technology and application. People's Communications Press, Beijing. (Published in Chinese)
17. He Y, Yang XW, Wu B, Zhong Y, Yan X (2018) A comparison of statistical survey methods of traffic accident data between China and the United States. J Transp Inform Saf 36:1–9. (Published in Chinese)
18. Tang K, Yao E (2006) Example of ITS development and application in Japan—an introduction on nagoya probe-vehicle-based dynamic Rout guidance system. Urban Transp China 3:74–76. (Published in Chinese)
19. http://www.itsc.com/
20. Liu X, Xu P (2006) Transportation geographic information system. Science Press, Beijing. (Published in Chinese)
21. Wang Y, Yuan K, Li T (2005) Transportation GIS and its application in ITS. China Railway Press, Beijing. (Published in Chinese)
22. Zhang X, Chu X, Mao Z (2005) Research on geometric distortion correction in road marking image collection. J Transp Inform Saf 6:58–61. (Published in Chinese)
23. Chu X, Yan X, Zhang X (2005) Design of a virtual testing system for the visibility of road signs and markings. J Wuhan Univ Technol (Inform Manage Eng) 4:135–138. (Published in Chinese)
24. Chu X, Wang R (2004) Asphalt pavement surface distress imagerecognition based on neural network. J Wuhan Univ Technol (Transp Sci Eng) 3:373–376. (Published in Chinese)
25. Lü Z (2001) Edge detection based on minimum and maximum of luminanceon normal line of image's edge. J Image Graphics 4:16–19. (Published in Chinese)
26. Xiao W, Zhang X, Huang W, Yan X (2005) One new algorithm of automatic classification for pavement distress. J Highway Transp Res Dev 11:79–82. (Published in Chinese)
27. Zhang L, Yan X, Chu X (2006) Methods of the road sign size picture measure. J Wuhan Univ Technol (Inform Manage Eng) 5. (Published in Chinese)
28. Yan X, Chu D (2013) State-of-the art ITS in Chinese Mainland and Taiwan—general introduction to the cross-strait symposium on ITS. J Transp Inform Saf 6:1–5. (Published in Chinese)
29. Peng L, Wu C, Huangzhen (2013) Situation assessment of vehicle collision risk based on variable precision rough set. J Transp Syst Eng Inform Technol 5:120–126. (Published in Chinese)

30. Zhang L, Wu C, Huang Z, Wang B (2012) Design and implemention of information collecting module embeded in a real-vehicle experimental system for driving behavior study. J Wuhan Univ Technol 5:76–81. (Published in Chinese)
31. Yang Z, Mo X, Yu Y (2013) Estimation of travel time under abnormal state. J Jilin Unive (Eng Technol Ed) 6:1459–1464. (Published in Chinese)
32. Chinese Society of Artificial Intelligence (2017) China artificial intelligence series white paper-intelligent transportation. (Published in Chinese)
33. China Intelligent Transportation Association (2017) China intelligent transportation industry development yearbook (2016). Electronic Industry Press. (Published in Chinese)

Chapter 2
Information Collection Technology in ITS

CHEN Zhijun, LENG Yao, YU Jinqiu, XIONG Shengguang, and CHEN Qiushi

2.1 Characteristics and Classification of Intelligent Transportation Information Technology

The intelligent transportation system (ITS) is a large system that integrates many advanced technologies and contains many subsystems. In these subsystems, a variety of technologies are applied such as sensor technology, information processing technology, and database technology, intelligent control technology, communication technology, network technology, and transportation engineering, etc. Only by combining these technologies, the realization of the various subsystems of ITS can be achieved, leading to the realization of the whole intelligent transport system. From the perspective of the whole system, ITS is the embodiment of the combination of many technologies.

2.1.1 Characteristics of Intelligent Transportation Information

With the continuous development of ITS, enormous and complex intelligent transportation information has been generated. The traffic flow information, pedestrian information, and signal light control strategy are detected and collected by the urban road system during rush hours at intersections. The bus timetable, route information, and operating status are displayed in the traffic information service system. The GPS information, vehicle information, and driver information are gathered by the traffic

CHEN Zhijun (✉) · LENG Yao · YU Jinqiu · XIONG Shengguang · CHEN Qiushi
Intelligent Transportation Systems Research Center, Wuhan University of Technology, Wuhan, China
e-mail: chenzj556@whut.edu.cn

monitoring system. All this information demonstrates the complexity of intelligent traffic information. In order to enhance the efficiency of the ITS and make it better serve the public, appropriate methods for collecting intelligent transportation information should be taken. First of all, the characteristics of intelligent transportation information should be fully understood, which are mainly reflected in the following aspects.

1. Large in quantities

A large amount of structured and unstructured historical data has been generated during the long-term operation of an ITS. The data sources are various, so a huge information database has been constructed after long-term collection and storage. The huge amount of historical data provides data support for the development of ITS and is the foundation of solving intelligent transportation problems. However, the huge amount of data also brings new challenges for data processing.

2. Wide in varieties

There is a wide range of information sources that generate intelligent traffic information. According to the source of information, traffic information can be roughly divided into traffic flow data obtained by fixed detectors, traffic flow data, and localization data obtained by mobile detectors, unstructured video data, and multisource Internet data. Besides, the information can be divided into real-time detection data and historical traffic information data according to timeliness. There are many types of traffic information, which are not only reflected in the daily travel, but also various aspects of traffic control, traffic management, and decision-making.

3. High in timeliness

Intelligent traffic information has strong timeliness. Traffic flow is usually time-varying. Traffic management is time-sensitive, and traffic information services require timely, reliable, and accurate information. When such information loses timeliness, its value shrinks rapidly, which puts forward requirements for the real-time performance of data analysis results. Therefore, it is necessary to increase the analysis speed of historical traffic data, periodic data, random data, behavior habits, meteorological data, etc., while ensuring faster data processing speed to ensure the timeliness of information.

4. High in values

Intelligent transportation information has multidimensional features such as time, space, and history due to its wide range of sources. On the one hand, the information is the foundation of multiple services and contains great value. But on the other hand, the data value density is low, and there are abnormal phenomena such as data missing, data error, and data redundancy.

5. Strong subjectivity

The traffic information generated by different user needs has different values. Thus, traffic data needs to be stored and flowed according to different user needs. On the one hand, the same traffic information holds different values for different travel purposes. For example, extreme storms and blizzard weather have a great impact on users traveling by plane, and the value of such traffic information is very high beyond doubt. As for users who take the subway to work, the impact of weather is almost negligible, and its value is greatly reduced. On the other hand, its subjective value also changes with the attitude of the user for the same traffic information. In reality, some users are more tolerant of bumpy journeys but cannot accept long-term traffic jams. At the same time, some groups hold the opposite opinion. They would rather be stuck in traffic than tolerate poor road conditions. All these show that the value of traffic information is highly subjective.

6. Strong periodicity

The traffic information data presents obvious periodic changes at the intersection or the whole road network in a city. Among the dynamic data in intelligent traffic, traffic flow data is one of the data most concerned by road managers and users. The traffic flow has a certain periodicity at different times of the day and at the same time on different days. It also shows strong time characteristics during working days, nonworking days, major holidays, and abnormal weather. Besides, traffic flow and other information also have seasonality and stability. In an adjacent period, the changing trend and cycle of traffic flow data at the same place are the same under the premise that no special event occurs.

2.1.2 Classification of Intelligent Transportation Information Technology

2.1.2.1 Frequency of Information Change

Real-time and accurate traffic information collection is the key and prerequisite to realize traffic control, traffic management, traffic flow guidance, and other applications. In general, traffic information can be divided into static traffic information and dynamic traffic information according to the frequency of change. Static traffic information refers to traffic information that will not change violently in the short term, such as road network information, traffic infrastructure information, etc. Dynamic traffic information refers to traffic information that changes over time, such as traffic flow information, traffic accident information, environmental status information, etc.

1. **Static traffic information**

Static traffic information mainly refers to some fixed information, which is usually related to road traffic management and will not change much in the short term. Static traffic information mainly includes land information, urban geographic information, urban road network information, vehicle ownership information, and traffic management information. The static traffic information is relatively stable with little or irregular change frequency. Therefore, static traffic information does not need to be collected in real time. It is generally imported once for all and does not need to be modified until the data changes. The main methods of collecting static traffic information are shown below.

(a) **Survey**

The surveys are performed manually or by using measuring instruments. This method can obtain all-around basic urban geographic information and urban road network basic information.

(b) **Import from other systems**

In order to reduce unnecessary repetitive work and reduce the possibility of data inconsistency, information collection can also be gathered by imports from other systems. Static traffic information can be obtained from other departments, such as the planning department, the urban construction department, and the traffic management department. After gathering this basic information through investigation, it is generally stored in the static traffic information database through a one-time manual entry. The data in the static traffic information database need to be updated only when the system changes.

2. **Dynamic traffic information**

The main concern in the information collection technology is the dynamic traffic information, such as traffic volume, average speed, vehicle type, vehicle positioning, travel time, etc. There are many collection techniques for different types of traffic information. The dynamic traffic information collection can be divided into two categories, including nonautomatic collection and automatic collection. Nonautomatic collection requires manual intervention to complete the collection of traffic information, which requires a lot of manpower and material resources. Thus, the nonautomatic collection is not suitable for long-term observation. And the dynamic traffic information obtained manually is difficult to meet the real-time requirements of ITS.

Dynamic traffic information mainly includes traffic flow status feature information (such as flow, speed, density, etc.), traffic emergency information (event information obtained by various means, such as road detector information, manual report information, etc.), the real-time information of vehicles in transit and driver (such as various vehicle positioning information, etc.), environmental status information (such as atmospheric conditions, pollution status information, etc.), and traffic dynamic control management information. Dynamic traffic information is significantly different from static traffic information, which is mainly manifested in its real-time nature. That is to say, dynamic traffic information reflects the changing traffic conditions over time. Therefore, the collection of dynamic traffic information must be timely and accurate. Dynamic traffic information collection technology includes traffic detection technology, floating car technology, vehicle identification technology, vehicle positioning technology, weather collection technology, and road environment information collection technology.

2.1.2.2   Traffic Information Demand

Intelligent transportation information can be divided into government department demand information and social public demand information. Among them, government departments not only collect, process, and release information related to the traveler information system but also are the users and demanders of such information. Different levels of traffic managers have different focuses on information needs, and there are differences in the depth and breadth of information required. As the main body of travel, the public's demand for information during the entire travel process is constantly changing with changes in travel purpose and travel time. According to the needs of travel subjects for traffic information, intelligent traffic information can be divided as follows.

1. **Information for government demand**

   The ITS involves multiple government departments such as transportation planning, transportation construction, and transportation management. Each department collects part of the basic information closely related to itself according to its own business needs. For example, the transportation management department needs to focus on the collection of road infrastructure information, road network historical traffic flow data, road network real-time operation information, and road network emergency information. All traffic information is necessary to improve traffic efficiency and traffic safety. However, the transportation operation department has a high demand for dynamic and real-time traffic information and needs to focus on operating status information and road network emergency information to improve transportation efficiency.

   Otherwise, government departments need to exchange and share information among different departments. For example, the traffic planning department needs to obtain road traffic operation data (such as the flow of major roads, vehicle speeds, etc.) from the traffic management department to formulate a reasonable road traffic planning plan. The traffic management department needs to obtain road construction plans and schedule information from the transportation construction department to better implement road traffic management measures.

2. **Information for public demand**

   The public demand for traffic information often comes in significant differences due to different travel purposes and travel modes. However, in the overall travel process, the required information mainly includes the following aspects.

   (a) **Information before travel**

      Information that travelers need before traveling includes real-time information such as bus schedules and routes, transfer stations, fares, and ride-sharing matching and information on traffic accidents, road construction, congested sections, speeds of individual sections, arrangements for special events, and climate conditions. Travelers can use this information to formulate travel modes, travel routes, travel time, etc., and plan the best travel.

(b) **Information in travel**

The information needed by travelers during trips includes dynamic optimal route information, road network traffic operation status information, traffic incident information, parking lot information, traffic control information, toll station information, weather information, and roadside service information. The above information can be released to the driver through variable information boards, in-vehicle navigation equipment, etc. so that the driver can choose the best travel route.

(c) **Personalized information**

Personalized information refers to the information that meets the individual needs of travelers. The personalized information required by travelers includes the geographic distribution of services and facilities along the way such as gas stations, auto repair shops, catering services, police stations, hospitals, etc.

(d) **Parking guidance information**

Parking guidance information refers to the information provided to travelers with the location, usage, route, and related road traffic conditions of the parking lot to guide the driver to find the parking lot effectively. It mainly includes real-time data related to each parking lot in the area (such as the location of the parking lot and the number of empty cars, etc.). Through the variable information display board in the road and vehicle, this information can provide the driver with real-time, accurate, and comprehensive parking information, guiding them through the optimal path to the most suitable parking lot.

2.1.2.3   Information Collection Method

According to the source of traffic information data, intelligent traffic information collection can be divided into indirect collection and direct collection. Indirect collection refers to the integration of various information in the transportation sections through various collection data nodes of different intelligent transportation subsystems. Direct collection refers to obtaining related traffic information directly through various collection devices in different collection methods. According to the difference of information collection methods, there are mainly the following types of information collection.

1. **Collection by magnetic signal**

The magnetic signal is passively received by the magnetic sensor, and the magnetic detector is the main equipment that uses magnetic signal acquisition. The magnetic detector detects the passage of vehicles by detecting the abnormality of the magnetic field intensity. When ferrous objects pass through the earth's magnetic field, they will cause disturbances in the earth's magnetic field. When the vehicle enters and passes through the detection area of the magnetic detector, the detector detects the abnormality of the geomagnetic field caused by the ferrous material of the vehicle.

2. **Collection by microwave signal**

   The microwave radar detector receives the reflected radar beam by a vehicle that passes through the radar wave coverage area through the radar antenna. The radar completes the vehicle monitoring through the receiver and calculates traffic data such as flow, speed, and body length of the vehicle. The microwave radar detector can be installed in the mid-air in the center of the single-lane road to measure the traffic data of the incoming or outgoing vehicles.

3. **Collection by ultrasonic signal**

   Similar to the collection of microwave signals, most ultrasonic detectors emit pulse waves and provide traffic information such as vehicle count, presence, and road occupancy. The detection area of the ultrasonic detector is determined by the amplitude of the ultrasonic transmitter. By measuring the waveform of the pulsed ultrasonic wave reflected by the road surface or the vehicle surface, the distance from the detector to the road surface or the vehicle surface can be determined.

4. **Collection by infrared signal**

   The infrared detector obtains a lot of traffic information by collecting infrared signals. For example, it can be installed above the road to observe the oncoming or leaving traffic flow and on the roadside for signal control and the measurement of flow, speed, and vehicle type. Besides, the infrared detector can also monitor pedestrians on crosswalks and issue traffic information to drivers. Active infrared detectors can provide information such as vehicle appearance parameters, traffic, road occupancy, vehicle speed, vehicle length, vehicle queue length, and vehicle classification at signalized intersections.

2.2   Fixed Traffic Information Collection Technology

2.2.1   Information Collection Using Geomagnetic Sensor

The geomagnetic sensor is a widely used detector for detecting traffic flow in intelligent transportation. It is made of soft magnetic materials with high magnetic permeability. The geomagnetic information collection is based on the principle of electromagnetic induction and the rule that the magnetic field changes around the detector. The sensor is a toroidal coil buried under the road with a certain working current. When a vehicle with ferrous materials approaches the sensor, the sensor senses the change of the surrounding magnetic field relative to the earth's magnetic field and then analyzes and calculates the microprocessor to determine the vehicle's existence and passing status at the sensor location.

2.2.1.1   Composition

The detector is mainly composed of a sensor, central processor, detection card, input, and output (Fig. 2.1).

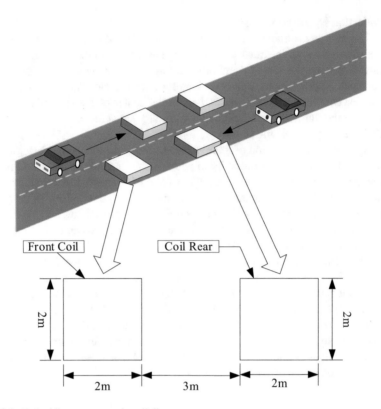

Fig. 2.1 Embedding geomagnetic coil diagram

1. Sensors

The sensor adopts a modular design with a small body and is installed vertically to the ground. A groove about 2 m × 1.5 m with a width of about 10 mm is excavated under the ground 10–18 mm, and the inductance coil is formed by using a wire to wind several times along the groove. Through the underground trench, low-resistance wires can be used to lead the two sections of the coil to the processing box. When a motor vehicle passes through the coil section, the inductance of the coil changes. The vehicle type can be estimated according to the magnitude of the decrease in inductance when the vehicle passes through. The change of the inductance causes the oscillation frequency to change. The inductance decreases when the oscillation frequency increases, and the relative amount of the frequency change is half of the inductance change.

When the vehicle passes through or stops on the toroidal buried coil, the ferrous part of the vehicle cuts the magnetic flux lines, causing a change in the inductance of the coil loop. The detector can detect the existence of the vehicle by detecting the inductance. There are two ways to detect the inductance change. One is to use a phase latch and a phase comparator to detect the change of the phase. The other is to use a coupling circuit composed of a loop geographic coil to detect the oscillation frequency.

2. **Central processor**

The central processor is a module that performs calculations of the collected signals. It is generally a single-board computer with an embedded operating system and has strong digital calculation, storage capabilities, and communication interfaces. By scanning the ports, the time of level change is captured to calculate the corresponding traffic data.

The communication interface of the general detector includes RS232/485, and the more advanced one has an Ethernet interface and a GPRS module. In most domestic applications at present, system integrators generally use modem point--to-point connections to transmit data or transfer data through PLC due to the distance between the monitoring road and the monitoring center.

If online debugging and online programming are required, PLC series single-chip microcomputers can be used.

When any abnormal working conditions caused by the occurrence of unexpected circumstances occur, such as processor crashes, malfunctions, etc., the central processor should be able to restart within a short period, which should not exceed 30 s.

3. **Detection card**

When the detecting vehicle passes the detection domain or is stationary in the detection domain of the induction coil, the inductance passing through the induction coil will decrease. The function of the detection card is to detect this change and accurately output the corresponding level.

The detection card used by the coil-type vehicle detector is generally a European standard card interface. As far as the angle of coil induction is concerned, the detection card should have stable response time and high-precision level jump performance consistent with the actual situation of the vehicle passing by. The one reason is that the detection time is very short when the vehicle passes at high speed. The time between the coils is generally one or two hundred milliseconds, and the response time of a single coil is shorter. The weight of the chassis of various vehicles and the height of the distance from the ground will affect the level response time of the detection card. The response time and start time are the main parameters for calculating the length and speed of the vehicle. This explains why the length and speed measured by some test cards are inaccurate when the vehicle passes at high speed. Thus, certain compensation must be used. Only by adjusting the sensitivity correctly can the accuracy of the detector be guaranteed.

4. **Input and output**

Sensor signal input: the coil is connected to the oscillating circuit, and the single-chip microcomputer determines whether a vehicle passes by measuring the frequency of the oscillating signal.

The signal output is the optical isolation switch output and the working status indication (Fig. 2.2).

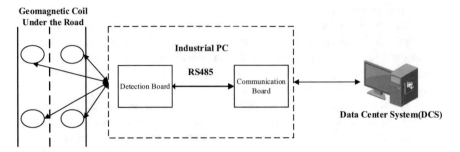

Fig. 2.2 The framework of the geomagnetic coil detection system

2.2.1.2 The Structure of the Information Collection Protocol

Each vehicle detector should have its address to distinguish it from the other. The method of data uploading is generally responsive. The outdoor system (vehicle detector) does not actively send data, and the data is not sent until the indoor system (monitoring center) issues an instruction for data collection. The communication commands between the vehicle detector and the monitoring center are usually divided into two types, including query and setting. Query commands generally query the current working status and failures of the vehicle detector, which can classify different vehicle captain types according to the preset vehicle captain type. The setting command is to adjust the parameters of the detector to participate in the calculation; the parameters include time, vehicle captain classification, etc.

2.2.1.3 Information Collection Algorithm

In the traffic control system, the commonly used traffic parameters include traffic flow, lane occupancy, queue length, speed, average car length, average car spacing and density, etc. Some of these parameters can be directly measured, and the other need to be calculated based on other detection data. The following will discuss how to use geomagnetic coils to detect and calculate various traffic parameters [1] (Figs. 2.3 and 2.4).

1. **Traffic flow**
       Suppose the counting period of the geomagnetic coil detector is T and N_i is the count value of the ith lane detector during the observation period. Then, the traffic flow q of the i lane in this period is

$$q_i = \frac{N_i}{T} \qquad (2.1)$$

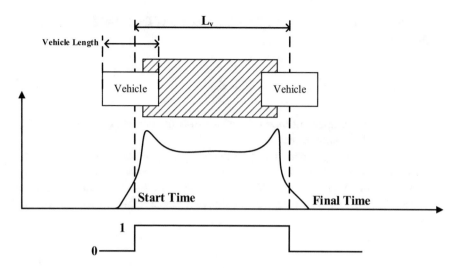

Fig. 2.3 Detection principle

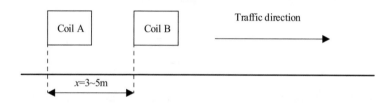

Fig. 2.4 Principle of speed detection by coil

2. Speed

The travel speed is used to judge the smoothness of the road. In order to accurately measure vehicle speed, two loop coils with the same performance are usually buried in the direction of traffic flow. And the distance between the coils on the same side is 3–5 m.

When the vehicle enters coil A, the pulse counts. When the vehicle enters coil B, the pulse count ends. So, the microprocessor gets a reference time pulse, such as a time pulse of P (ms). When the vehicle enters coil A, the pulse count starts and the pulse count ends when the vehicle enters coil B. Therefore, the pulse count required for the vehicle to pass the distance s is n. The speed of the vehicle is

$$v = \frac{s}{pn/1000} \tag{2.2}$$

where the unit of v is m/s. Suppose that in a certain observation period, there are a total of n vehicles passing the observation point and the speed of each vehicle is v_1, v_2, v_3...v_n. Then, the average speed of the traffic flow during this period is

$$\bar{v} = \frac{1}{n} \sum_{i-1}^{n} v_i \qquad (2.3)$$

where v_i is the speed of the ith speed and n is the number of vehicles passing during the observation time.

The space average speed refers to the average speed of all vehicles passing through a section of road in a certain period of time. Suppose the length of the road section is Δ, and there are N vehicles passing through the road section in an observation period. The average travel time of N vehicles through the road section is

$$t_m = \frac{1}{N} \sum_{i=1}^{N} \frac{\Delta}{\bar{v}} \qquad (2.4)$$

Then, the average speed in space within the observation is

$$v_m = \frac{\Delta}{\bar{t}} = \frac{N}{\sum\limits_{i=1}^{N} \frac{1}{v_i}} \qquad (2.5)$$

That is, the space average speed is equal to the harmonic average of the speeds of all passing vehicles.

3. **Vehicle occupancy rate**

Vehicle occupancy rate is defined as the ratio of the total road length occupied by vehicles to the length of the road section in the total road section. But this parameter is difficult to measure and is usually replaced by time occupancy. To calculate the occupancy rate with a geomagnetic coil detector, it is necessary to set the detector into a square wave mode of operation. Set in a certain observation period T, a total of N vehicles passing through the coil. And measuring the square wave broadband of the vehicle j in lane i passing through the loop coil. During this period, the occupancy rate of vehicles in lane Z:

$$\sigma_i = \sum_{j=1}^{N} \frac{t_{ji}}{T} \times 100\% \qquad (2.6)$$

4. **Traffic density**

Traffic flow density refers to the number of vehicles present at a certain instant in a unit length lane and can also be measured by the number of vehicles per unit length of a certain driving direction or a certain section of road. As long as the space average speed of traffic flow and vehicle flow is measured, the traffic density in the observation period T can be measured.

$$\rho_i = \left(\frac{N}{T}\right)/v_s = \left(\frac{N}{T}\right)\frac{\sum\limits_{i=1}^{N}\frac{1}{v_i}}{N} = \frac{1}{T}\sum_{i=1}^{N}\frac{1}{v_i} \tag{2.7}$$

5. **Vehicle type**

The collection of vehicle type information is mainly judged by the vehicle length. First, collect the signal output of the coil and pulse time to calculate the vehicle speed. The vehicle length is calculated based on the obtained vehicle speed. To be more specific, the calculation principle is based on the time average value of the vehicle passing through the front and rear coils. Suppose the time average value of the entire vehicle body passing through the front and rear coils is $\bar{t}$, the vehicle speed is v, and the coil width is l. According to $L + l = \bar{t}v$, the estimated length of the vehicle L is

$$L = \bar{t}v - l = \frac{\bar{t}s}{pn/1000} - l \tag{2.8}$$

Vehicles types can be divided into the following categories: small passenger cars, coaches, small trucks, medium trucks, large trucks, extra-large trucks, and trailers. It is mainly based on vehicle height, vehicle length, and vehicle profile (the mean square deviation of vehicle height of trucks is significantly larger than that of passenger cars). The height of the vehicle's chassis directly affects the change in inductance caused by the vehicle passing through the coil. The change caused by the high chassis is small, and the change caused by the low chassis is large. When the chassis is higher than a certain range, it may not be detected. Under normal circumstances, the height of the chassis of the car is 150–200 mm, and the height of the chassis of small and medium passenger cars is generally about 400 mm. The detection accuracy of vehicles of different models is different. In addition, the increase in the length of the feeder will cause the proportion of the inductance of the feeder to increase. And the ratio of the change relative to the inductance of the coil is relatively small, which has a certain impact on the detection accuracy. According to the chassis height data set DATA, many data can be obtained. They include the maximum chassis height MAXH, the minimum vehicle height MINH, the average chassis height AVGH, the mean square error of the chassis height MSEH, and the difference between the maximum height and the minimum height DIFFH. Carrying out classification statistics and summarization of the above-mentioned large amount of data can realize vehicle model judgment.

Compared with the current loop coil detection equipment, the geomagnetic coil has the following characteristics [2]:

(a) The length of cutting pavement can be reduced by more than 70%. The installation and debugging time can be reduced by 50%. And the installation cost is reduced by 50%.
(b) Long service life. With an average daily vehicle flow of 10,000 to 15,000 vehicles, the service life reaches 5—10 years, which is twice that of the traditional coil.
(c) Modular design, without field production, with high stability and consistency.
(d) Diversity in function. It can detect stationary and moving vehicles, speed, and statistics flow.
(e) High detection accuracy. The detection accuracy of vehicle flow is greater than 99.6%. The detection accuracy of average vehicle speed is 98%, and the accuracy remains when the vehicle speed is lower than 10 km/h.
(f) The temperature adaptability is strong, and the alpine environment at $-40\,^{\circ}\mathrm{C}$ can still work stably.
(g) Strong anti-interference, not affected by ferromagnetic, rain fog environment, can be used for tunnel and bridge environments.
(h) Good compatibility, vehicle control detection box using standard 86CP11 interface, compatible with the traditional vehicle detector.

2.2.2   Information Collection Using Ultrasonic

With the development of urban construction and the improvement of people's living standards, more and more vehicles are running on the road. The vehicles not only bring convenience to people but also bring heavy loads to traffic. In order to meet the needs of modern traffic management, urban managers and decision-makers must achieve traffic information in real time. Therefore, information collection using ultrasonic is proposed.

Traffic flow and average velocity are the two basic information elements of road traffic, which can basically reflect the road traffic conditions. Ultrasonic detectors emit sound waves at frequencies ranging from 25 to 50 kHz beyond the range of human hearing. Most ultrasonic detectors emit pulsed waves, which contain traffic information.

The basic principle of ultrasonic measurement is based on the reflection properties of waves. The ultrasonic generator emits a certain frequency of ultrasonic wave, which produces reflection wave after it encounters obstacles. The ultrasonic receiver receives the reflected wave and converts it into an electrical signal. The wave experiences t from its departure from the generator to its reception by the receiver. According to the formula $R = (t \times v)/2$, we can get the distance change of the vehicle (v is the velocity of ultrasonic wave). Then, the velocity of the vehicle can be calculated.

The information collection using ultrasonic is widely used on expressways. This skill belongs to noncontact active detection technology. As shown in Fig. 2.5, the ultrasonic detector is installed above the road.

Fig. 2.5 Installation
diagram of ultrasonic
detector

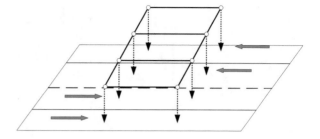

The ultrasonic measurement has the advantages of high resolution and low cost. So, it is often used in industrial measurement. In recent years, it has been gradually promoted to civilian use, showing a strong momentum of development.

Commonly used ultrasonic ranging methods include transit time method, frequency difference method, amplitude method, and so on. Among them, the time-of-flight method is widely used because of its simple principle and convenient implementation [3].

The ultrasonic transit time method is used to measure distance. The controller sends a pulse signal of a certain frequency to excite the ultrasonic transmitter to generate ultrasonic waves. The ultrasonic waves pass through the medium and reach the surface of the object to form reflected waves. The reflected wave propagates through the medium and returns to the receiver, where the acoustic signal is converted into an electrical signal by the receiver. The distance can be calculated according to the time required for the ultrasonic wave from emission to reception and the propagation speed of the ultrasonic wave in the medium:

$$S = \frac{1}{2} \times 331.4\Delta t \sqrt{\frac{T}{273} + 1} \tag{2.9}$$

In formula (2.9), S is the distance between the object and ultrasonic, Δt is the time interval from transmitting ultrasonic wave to receiving echo pulse, and T is the ambient temperature.

As shown in Fig. 2.6, first, a high-level pulse is given to the transmitting end of the ultrasonic module. After receiving the high-level pulse, the module transmits a 40-kHz pulse acoustic signal. When the ultrasonic is transmitted, the receiving pin of the module is at a high level. When the transmitted acoustic wave meets an object and returns to the module, the receiving signal turns to a low level.

The transmission time interval of a high-level pulse is related to the accuracy of vehicle detection. In general, when the set launch time interval is 40–50 ms, the vehicle speed does not exceed 200 km/h. It is impossible to be missed, which meets the actual road situation in China.

Using the principle of distance measurement, the object position coordinates in the space coordinate system can be calculated. The localization function of local

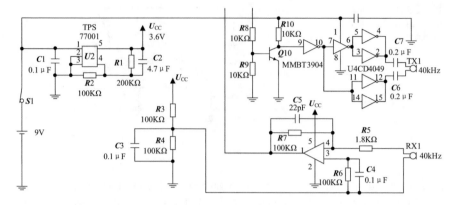

Fig. 2.6 Installation diagram of ultrasonic detector

Fig. 2.7 The schematic
diagram of ultrasonic
positioning principle

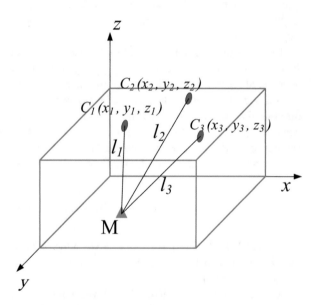

space can be realized. Figure 2.7 is the schematic diagram of the ultrasonic positioning principle.

In the space indicated by the red solid line, if we want to locate the moving object M, we need to establish a rectangular coordinate system as shown in Fig. 2.7. And set up three ultrasonic receiving points above the space, and the coordinates are $C_1(x_1, y_1, z_1)$, $C_2(x_2, y_2, z_2)$, and $C_3(x_3, y_3, z_3)$. If the distances l_1, l_2, and l_3 between the moving object M and the three receiving points can be measured, the coordinates (x, y, z) of the moving object M can be expressed by the following formula:

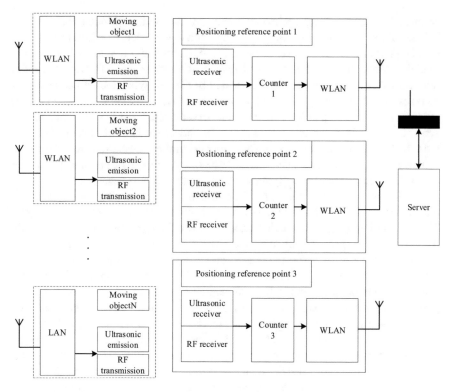

Fig. 2.8 The diagram of ultrasonic positioning system based on WLAN

$$\begin{cases} (x - x_1)^2 + (y - y_1)^2 + (z - z_1)^2 = l_1^2 \\ (x - x_2)^2 + (y - y_2)^2 + (z - z_2)^2 = l_2^2 \\ (x - x_3)^2 + (y - y_3)^2 + (z - z_3)^2 = l_3^2 \end{cases} \quad (2.10)$$

According to Formula (2.10), the positioning and tracking of target M can be realized. x, y, and z are the coordinates of the moving object M. l_1, l_2, and l_3 are the distances from the moving object M to the three reception points, respectively. (x_1, y_1, z_1), (x_2, y_2, z_2), and (x_3, y_3, z_3) are the coordinate values of the three ultrasonic receiving points in the spatial coordinate system, respectively. As the position of the moving object M is constantly changing, the values of l_1, l_2, and l_3 are also changing, and their coordinate values are constantly updated so that the localization tracking of the target is achieved [4].

The ultrasonic positioning system based on WLAN is shown in Fig. 2.8.

The movement object (vehicle) is equipped with a wireless WLAN module communicating with the server, an ultrasonic transmitter, and a radio frequency transmitting device. The positioning reference point is composed of a WLAN module, counter, ultrasonic receiver, and RF receiver. The server sends out the command of transmitting ultrasonic waves to the vehicle through the wireless WLAN. After receiving the command, the vehicle will send the starting signal of counter counting to each reference point s through the radio frequency transmitting module. When the signal is received by the reference point s through the radio frequency receiving module, the counter will be started. When the reference point s receives the ultrasonic signal, it will count the counter stops counting. The counter module sends the value of the counter to the server through the WLAN module. The server converts the value into time and further converts it into the distance between the vehicle and the reference point. The server determines the coordinate value of the vehicle by calculating the distance value from the three reference points. The above digital logic circuit can be easily realized by complex programmable device.

The feature of this program is that the initial information of ultrasonic emission is transmitted to the positioning reference point through the radio frequency signal. The propagation speed of the radio frequency is very high (about 3×10^8 m/s), while the propagation speed of the ultrasonic wave is low (about 341 m/s),, so the propagation time of the radio frequency can be ignored. This solves the synchronization problem caused by the fact that the transmission and reception of ultrasonic waves are not on the same side. Although the propagation delay of the radio frequency can be ignored in the calculation process, a frame of serial data (address code, data and synchronization code of the receiving end, etc.) sent by the sender still needs a certain amount of time. The time is related to the serial clock selected by the system. The decoding end also needs a certain amount of time to decode these data, and the system must consider these times for some compensation when calculating. Since these times are known, the compensation can be done more accurately.

According to the above ultrasonic information collection method, the distance threshold is first set. When the distance detected by the ultrasonic module is less than this value, it is considered that there is no vehicle. When the distance is greater than the threshold, it is considered that there is a vehicle. When counting the number of vehicles, because there is a distance between the front and rear probes, the rear probe should be used to judge. When the rear probe detects the passing of a vehicle, the vehicle counter increases by one. When calculating the vehicle speed, the method used is as follows: when the front probe detects the vehicle, the current system time is saved. When a vehicle is detected later, the current system time is saved. Calculate the time difference. Since the distance between the probes is known, the speed of the vehicle can be calculated based on this.

At present, pulse ultrasonic detectors and constant frequency ultrasonic detectors are widely used.

In the system of pulse ultrasonic detectors, the method of target measuring can be described as follows: at least one pulse of high-frequency energy is transmitted from the transducer to the target, and then the transducer receives the energy transmitted from the target direction to form at least one echo signal, and then the amplitude of the return signal is repeatedly sampled within a certain interval to form a digital database. The database links the amplitude of the echo signal with the time recorded by the system. Finally, at least one target area is found from the database, in which the amplitude of the echo signal is larger than the background noise. At present, there

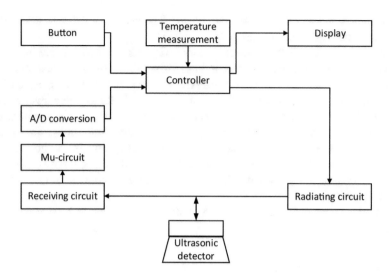

Fig. 2.9 The structure diagram of pulse ultrasonic detector system

are three major problems in the system of pulse ultrasonic detectors: (1) how to design the high-power transmitting circuit, (2) how to design the low-noise programmable receiving amplifier circuit, and (3) how to design the intelligent echo receiving software.

In this system, it is necessary to apply a specific excitation pulse to the circuit and achieve the purpose of measuring distance by processing the received pulse in time series. At present, the excitation pulse mainly includes sinusoidal signal, square wave, or instantaneous high-voltage sharp pulse. For the measurement of a large target, the instantaneous high voltage continuous excitation pulse is mainly used.

In the receiving circuit, the quality of the system analog signal echo mainly depends on the shape of the object's surface and the distance between the detector and target. However, interference is ubiquitous, so the design of an analog circuit must have strong adaptability and the ability to suppress interference signals. The structure diagram of the pulse ultrasonic detector system can be described in Fig. 2.9.

The detection area of the ultrasonic detector is determined by the amplitude of the ultrasonic transmitter. The distance from the detector to the road or vehicle surface can be determined by measuring the waveform of the pulse ultrasonic reflected by the road or vehicle surface. That is to say, the signal measured by the detector is different when there are vehicles or no vehicles on the road. It can be used to determine the presence of vehicles. The received ultrasonic signal is converted into an electrical signal, which is analyzed and processed by the signal processing unit. Because the pulse energy wave reflected by the ultrasonic sensor is known and divided into two beams at a small angle, that is, the distance between the two pulse energy waves is known, the vehicle speed can be determined by measuring the time of the vehicle passing through the two beams. The best installation position of the pulse ultrasonic detector is overhead the road [5].

The system of constant frequency ultrasonic detector always measures the vehicle velocity by Doppler effect; its cost is higher than the system of pulse ultrasonic detector. The constant frequency ultrasonic detector is installed on an elevated platform, facing the oncoming traffic at an angle of 45°. It has two converters: one is used to transmit the ultrasonic wave, and the other is used to receive the ultrasonic wave, which detects the frequency change of the received ultrasonic wave to determine whether the vehicle passes. The electronic system in the constant frequency ultrasonic detector can generate the internal pulse signal whose pulse width is proportional to the measured vehicle speed, and the vehicle speed can be determined by a series of calculations.

The advantages of using ultrasound to collect information can be summarized as follows:

1. Safety
2. Good real-time performance
3. High accuracy
4. Strong function scalability
5. Modular design
6. Low cost

2.2.3   Information Collection Using Video

Information collection using video is a new road information acquisition method developed gradually on the basis of the traditional TV monitoring system. It is mainly composed of surveillance cameras, microprocessors, and computer processing technologies. It involves computer vision, video image processing, signal processing, pattern recognition, pattern fusion, and other fields of knowledge. It analyzes traffic data through closed-circuit television and digital technology.

A vehicle detector based on video mostly uses CCD (charge-coupled device) camera. CCD camera takes charge as the signal, which is different from most detectors taking current or voltage as the signal. The basic functions of CCD are charge storage and charge transfer. Therefore, the working process of CCD is mainly the generation, storage, transmission, and detection of signal charge. CCD has two basic types: one is that the charge packet is stored at the interface between semiconductor and insulator and transmits along the interface, which is called surface CCD (SCCD). The other is that the charge packet is stored in a body at a certain depth from the semiconductor surface and transmitted in a certain direction in the semiconductor. Such devices are called body CCD (BCCD). Through the electron beam or CCD self-scanning system, the CCD camera decomposes the image distributed according to the spatial position into the time signal corresponding to the pixel. The monitor uses the same way of the electron beam from left to right, while scanning from top to bottom. The image is displayed on the screen to publish relevant traffic information.

The basic principle of a video detector is that in a very short time interval, two images are continuously captured by a semiconductor charge-coupled device camera. And this image itself is a digital image; it is easy to compare the whole or part of the two images. If the difference exceeds a certain threshold, it indicates that there are moving vehicles.

After the development of digital, high-definition, and networking of video detectors, as well as the development of infrared complementary light, star-level imaging, and other technologies, the basic problems of video image acquisition have been well solved.

An information collection system that uses video images for intelligent transportation has been basically established. Video information collection technology has been developed rapidly, and it has played a great value in video information collection. The efficiency of intelligent transportation information collection has also been continuously improved.

In recent years, there are three forces promoting the video image information acquisition technology into a new round of development and evolution.

1. Deep learning technology makes video intelligent analysis technology continuously practical, and big data analysis system for vehicle characteristics, face recognition, especially static face recognition technology begins to enter the practical stage.
2. Driven by the development of deep learning technology, video information acquisition is rapidly entering the development stage of video image structure. Security card device, electronic police equipment, face card device, smart camera, and back-end video batch structure system are constantly producing a large number of video clips, images, and their structural description information and other massive video image information, reaching the record level of 10 billion or even 100 billion. Big data technology has been widely used in video information acquisition.
3. The combination of video intelligent analysis technology with deep learning and big data is promoting the comprehensive deepening of the application of video information acquisition technology. The demand for intelligent transportation differentiation is emerging. Multisource information fusion analysis combined with other IoT systems has become a consensus.

   With the continuous integration of big data, video analysis, and processing technology based on deep learning and video information acquisition technology, a series of new problems have emerged:

   (a) Due to the lack of unified standards and specifications, the data format of each manufacturer after video structured processing is not uniform and cannot share applications with each other.
   (b) Since there is no unified framework for the sharing and application of video image information networking after structured processing, vehicle big data, face recognition applications, video image applications, and so on are

relatively independent. There are multiple intelligent application platforms in the same system, and the effective sharing of data cannot be between platforms.

In the continuous integration process of big data, intelligent video analysis, and processing technology based on deep learning and video image information acquisition, many new concepts and terms have emerged in the industry, which go far beyond the concept category of traditional video surveillance. Moreover, these concepts and terms are not unified; sometimes, the same concept will have different terms, often causing confusion in practical work. Therefore, the Science and Technology Information Bureau of the Ministry of Public Security of China officially approved the release of GA/T1400 and GA/T1399, two series of six standards, which were unified and standardized, and the following core concept terms were proposed [6].

2.2.3.1   Video Image Information

Video clips, images, files related to video clips and images, and related description information are collectively referred to as video image information.

2.2.3.2   Video Image Information Object

Video image information described by object-oriented method includes video clips, images, files related to video clips and images, and objects such as personnel, vehicles, objects, scenes and video image labels, etc. Video clips, images, files related to video clips, and images are called basic objects of video image information. People, vehicles, objects, scenes, and video image labels contained in basic objects of video image information are called semantic attribute objects of video image information.

According to the different collection sources of video image information objects, they can be divided into automatic collection of video image information objects and manual collection of video image information objects.

The video image information object without manual intervention and triggered by trigger events in the acquisition process is called the automatic acquisition of video image information object, including video clips, images, and information objects presented in the form of files except for video clips and images. In the absence of manual intervention, the continuous and automatic acquisition of video image information object content is compatible with manual acquisition content, and the object feature attributes that can be automatically collected are relatively less. Because most of the collected content has nothing to do with the case event, so generally only save a certain time, beyond the save time, the system will automatically delete or loop over. The automatically collected video image information object relationship diagram is shown in Fig. 2.10.

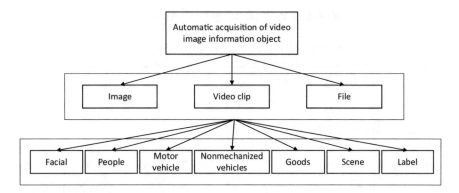

Fig. 2.10 Object relation graph of automatic video image collection

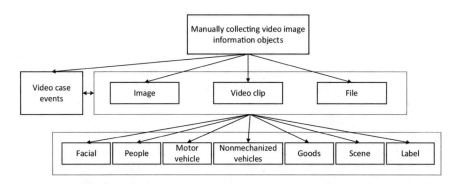

Fig. 2.11 Video image information object relation graph manually collected

The video image information objects that need to be manually screened in the acquisition process are called the artificially collected video image information objects, including video case events, video clips, images, and information objects presented in the form of files except for video clips and images. The video image information objects directly collected by humans or automatically collected video image information objects after manual screening and supplementing belong to the video image information objects collected by humans. Automatically collected video image information objects can be transformed into manually collected video image information objects after manual screening and supplementing, and their storage location and time will change accordingly. The manually collected objects are generally associated with special events. Only the authorized users can delete the manually collected video image information objects that exceed the specified storage time. A manually collected video image information object relationship diagram is shown in Fig. 2.11.

2.2.3.3   Feature Attributes of Video Image Information Objects

The specific content information of the video image information object can be described in the form of key-value pair, and a video image information object can contain multiple feature attributes.

2.2.3.4   Video Image Labels

According to certain rules, semantic attributes are collected such as object type, quantity, motion behavior, and its related space-time information of people, vehicles, objects, and other video image information from the basic objects of video image information such as video clips and images.

2.2.3.5   Triggering Event

The factors that cause online video image information acquisition devices to automatically collect video image information, such as the sensing coil, radar or video trigger events, alarm events and intelligent video analysis events, etc., generally have the properties of time and place and will be accompanied by relevant grabbing and license plate recognition. The relationship between the above concepts is shown in Fig. 2.12.

Vehicle detection and vehicle identification technology based on video image collection is a noncontact passive detection technology. Through the analysis of continuous video images, this method can track the vehicle behavior process and capture the speeding vehicle by analyzing and controlling the photograph. This

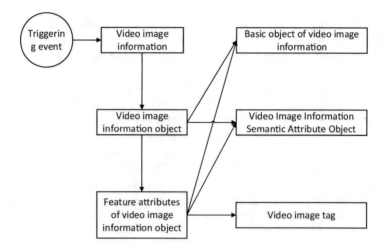

Fig. 2.12 The relationship diagram of the basic concepts

detection method is sensitive to the change of light at the intersection, so the superiority of the image algorithm is the fundamental factor affecting the detection effect.

Traditional vehicle detectors such as magnetic induction coil have many short-comings and limitations. In view of this situation, people continue to propose new alternatives, such as suspended sensors using radar, ultrasonic, infrared, microwave, audio, and video image technology. In recent years, with the continuous develop-ment of computer and image processing technology, vehicle detection based on computer vision has been widely used.

In 1978, the Jet Propulsion Laboratory in Pasadena, California, first proposed the method of vehicle detection using computer vision and pointed out that it is a feasible alternative to traditional detection methods. A few years later, researchers at the University of Minnesota developed the first video-based vehicle detection system that can be put into practical use. The system used the most advanced microprocessor at the time. Tests show that the system has good performance in different scenarios and environments, indicating the feasibility of using video sensors to detect vehicles in real time. Video-based vehicle detection has also been widely studied in Europe and Japan in the next decade, video-based vehicle detec-tion technology has made considerable progress.

In 1991, the US University of California Technology evaluated the detection technology of using video method on the highway. In the evaluation report, different video-based vehicle detection technologies were classified in detail. Three years later, Hughes Aircraft Company of the United States evaluated several existing detection technologies, including video detection technology. The evaluation results pointed out that the vehicle detection system based on video image processing had the potential to be put into practical use. In 1994, MnDOT (Minnesota Transport Department) carried out a more detailed and rigorous evaluation for FHWA (Federal Highway Administration). The results show that the accuracy and reliability of the video detector can reach a satisfactory level. With the development of video vehicle detection technology, people are not satisfied with only detecting vehicles. FHWA further uses this technology to extract traffic parameters, such as traffic flow, vehicle steering information at crossroads, etc. Compared with other vehicle detection methods, the method based on video image technology has the advantages of intuitive, wide monitoring range, more types of traffic parameters, and low cost, so it can be widely used in the traffic monitoring system of intersection and highway trunk. In the traffic monitoring system, the sensors for video detection (i.e., camera) are placed above the road to obtain road and passing vehicle information. The installation height is generally 5–6 m to ensure a good viewpoint for the whole traffic scene, and the obtained video image sequence can provide sufficient infor-mation for vehicle detection and tracking.

Vehicle detection and tracking system usually includes three modules: region of interesting (ROI) extraction, vehicle detection, and vehicle tracking, as shown in Fig. 2.13. First, the video sequence image of real-time traffic scene is captured by the camera, and then the ROI of the sequence image is extracted, and the extracted ROI is sent to the vehicle detection module to determine whether a ROI area is a vehicle according to certain image processing methods and criteria. After the vehicle is

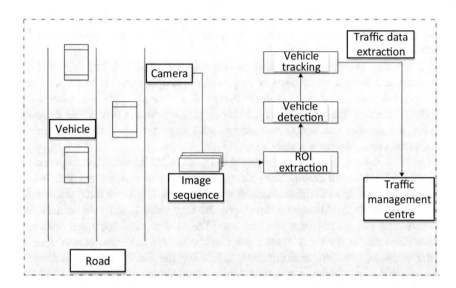

Fig. 2.13 Principal diagram of video detection for vehicles

detected, the vehicle can be tracked in the tracking module. Traffic flow parameters can be analyzed and extracted from the results of detection and tracking, such as vehicle speed, traffic density, and steering information. This kind of real-time road traffic information and all kinds of service information are aggregated to the traffic management center and centralized processed, then the results will be transmitted to each user of the traffic system so that the public can use the traffic facilities efficiently and improve the road load capacity and driving efficiency and save energy. The real-time detection and tracking of vehicle objects in traffic scenes is the most important and basic step in the video-based traffic monitoring system, and it is the core of the video detection method. The correctness of the detection and tracking directly affects the correctness of the decision-making of the intelligent transportation system. For the three key modules of video image detection, many video image processing and analysis methods are proposed, including frame difference method, edge detection method, unsupervised video segmentation method, etc. for the ROI module; for vehicle detection module, diaphragm method, detection line method, model method, etc.; and model method and region method for vehicle tracking module [7].

The camera is installed above the road section (supported by structure) or on both sides, as well as high road intersections or high buildings, and then the captured scene images of the traffic situation are transmitted to the on-site treatment unit. The on-site treatment unit is responsible for real-time processing of these images, and real-time calculation of road traffic flow, vehicle speed, vehicle type, road occupancy, and other road traffic parameters. The calculation results and scene images are transmitted to the traffic control center for generalization. The traffic control center stores the collected road traffic data or takes corresponding road traffic

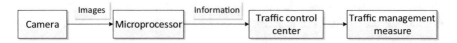

Fig. 2.14 Structural diagram of video information collection

management measures according to the collected information. The structural diagram is shown in Fig. 2.14.

The video-based vehicle detection system has more advantages and better development prospects than other vehicle detectors. The main advantages are as follows.

(a) The installation of this equipment is convenient and does not affect the operation of road traffic. The camera is installed above the road section, which is convenient and will not destroy the pavement structure, affecting road traffic. However, the installation needs structures to support. In addition, the equipment is installed in a place far from the collision of the vehicle, which greatly ensures the safety of the machine itself and improves the convenience of daily maintenance compared with the coil detector.

(b) A wealth of road traffic information can be obtained. The video detection method can not only obtain road traffic information such as vehicle speed, vehicle type, road traffic volume, and road occupancy, etc. but also detect other information such as vehicle trajectory lines and large-scale road traffic scene images that cannot be obtained by other detectors.

(c) Large coverage. A video detector can simultaneously detect several lanes on the road cross-section but also detect traffic parameters on adjacent road cross-sections. Moreover, the observation of the road length is also large, which can greatly reduce the installation number of cameras.

(d) Easy to use for road traffic management department. The detection method based on video image processing can provide the road traffic management department with visual images. At the same time, the device has the function of road traffic monitoring and management. For example, running a red light at road intersection, pressure line, parking space, and speed detection can all be monitored.

2.2.4   Information Collection Using Microwave Radar

Doppler proposed a Doppler-type radar based on the Doppler effect in 1842, while the world's first radar, the "native chain" radar, was born during World War II nearly a century later. Radar supports all-weather, all-day work, which has strong detection capability and can effectively detect the location, speed, size, and other information of enemy aircraft and enemy ships. It can be seen that radar plays a huge role in warfare. Thus, governments have invested heavily in the research and development of radar, which has resulted in a range of advanced radar technologies. The United States developed the first anti-aircraft fire control radar SCR-268 in 1938. In the same year, the world's first warfare-type naval radar XAF was also developed in the

United States. British used airborne radar for the first time in 1939. It can be said that the war not only gave birth to radar technology but also gave great impetus to its development.

After World War II, radar technology, a key military project, was even more booming. Continuous wave radar, pulsed Doppler radar, phase control radar, and moving target display radar were born one after another. With the demand and driven of the market, radar technology gradually moved toward civilian. A new wave of research was triggered from the 1950s.

Microwave radar sensing technology, which is a special means of distance measurement, can solve the measurement problem of near-range targets. Thus, microwave radar is also called near-range radar. It has been extended from military to civilian fields and has been gradually applied in level measurement, target identification, velocity measurement, vibration monitoring, displacement monitoring, etc. Compared with other sensing technologies, radar sensing technology has unique advantages such as noncontact and resistance to rain, fog, and dust. Thus, radar sensing technology gradually becomes an important sensing means in the field of near-range target displacement measurement.

Nowadays, microwave radars have been widely used in many nonmilitary fields such as industry, meteorology, medical treatment, and transportation. The main principle of so many microwave radars in different fields is still the application of fluctuation effect and Doppler principle to identify the target's distance, speed, angle, shape, size, and other information.

Microwave radars are divided into several bands according to their frequency, and the operating frequency of a microwave detector is usually 24 GHz or 10 GHz. The performance of each band varies greatly; thus, their application areas are also very different. The civil radar are mainly divided into X-band (8–12.5 GHz) and K-band (12.5–40 GHz). The wavelength of K-band is close to the resonance wavelength of water vapor, which is easily absorbed by it. Thus, the radars work in K-band should not be used on foggy days. The radars that work in X-band are commonly used for severe weather because X-band is not affected by water vapor. In addition, the bandwidth of radar sensors operating in X-band is large, which makes radars easy to achieve miniaturized designs. Therefore, X-band is often chosen as the operating band for microwave radar sensors to collect traffic information [8].

A typical X-band microwave radar sensor consists of a front-end circuit and an RF front-end, as shown in Fig. 2.15. The RF front-end is composed of a transceiver antenna, microwave oscillator, circulator, and mixer. The front-end circuit includes the transmitting front-end circuit and the preamplifier.

As shown in Fig. 2.15, a microwave oscillator generates a microwave signal of frequency f_s, which is looped through end 1 of the circulator to end 2 and emitted by the transmitting antenna (with partial leakage to end 3 of the circulator). When there is a relative velocity v between the target vehicle and the sensor, the reflected wave will be frequency shifted according to the Doppler principle. Assuming the frequency shift is f_d, the reflected wave received by the receiver antenna has a frequency of $f_s + f_d$, and the reflected wave is looped through the circulator to the mixer. Finally, with the help of the mixer, the transmitting wave and the reflected wave output an

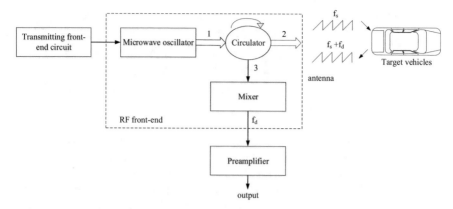

Fig. 2.15 The structure schematic of microwave radar sensor

electrical signal with a frequency f_d, which is equal to the Doppler frequency shift. According to the Doppler principle, it is easy to obtain the following equation.

$$f_d = \frac{2v}{\lambda_s} \tag{2.11}$$

where v is the relative speed of the target vehicle and λ_s is the wavelength of the microwave signal. Therefore, the relative speed of the target vehicle can be obtained by Eq. (2.11) as long as the frequency of the mixing output signal is detected.

The microwave oscillator is the wave source of the radar sensor and is the core part of the whole sensor. Microwave oscillators can be divided into two categories according to the selection of semiconductor active devices, including electric vacuum oscillators and solid oscillators. Most of the early radar devices use such vacuum devices as klystron, traveling wave tube, magnetron, forward wave tube, etc. Although they can meet the system requirements for source power, they are gradually replaced by solid-state oscillators because of their large structure, high supply voltage, high power consumption, and expensive cost.

A mixer is an essential key component in radar sensors. Its basic function is to transform the frequency of the signal. In radar sensors, the mixer makes a difference between the frequency of the transmitted signal and the frequency of the received signal.

Radar detection is the earliest technology used to detect vehicle speed, and the detection equipment is called a radarscope speedometer (or Doppler radar). The radarscope speedometer transmits a certain frequency of radio waves to the moving object (such as vehicles) and then detects the difference between the reflected and emitted frequencies of the object to calculate the speed of the moving object. Through this principle, the radarscope speedometer realizes the detection of a vehicle. At the same time, the vehicles are counted to achieve the purpose of statistical traffic flow.

Fig. 2.16 Principle diagram of Doppler effect used by radar monitor

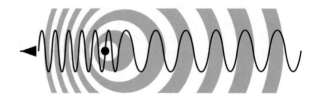

The principle of microwave radar sensors' velocity and distance measurement mainly involves the Doppler effect of wave and the energy theory of wave. The radar detection of vehicles is based on the Doppler principle. During the movement of the source point, the frequency away from the source point becomes lower, and the frequency close to one end of the source point becomes higher, as shown in Fig. 2.16.

Radar equipment is usually composed of transmitting antenna and transmitting receiver. The transmitting antenna mounted on the door frame or on the roadside pillar emits a microwave beam to the pavement detection area. When a vehicle passes, the reflected beam returns to the antenna at different frequencies. The transmitting receiver of the detector measures the frequency shift caused by the vehicle movement, which can generate a vehicle inductive output signal to determine the passing or the existence of this vehicle [9].

Radar detector has the characteristics of multidetection area, which can detect traffic volume, vehicle speed, occupancy, and other traffic flow information. At present, it has great advantages in traffic detection. Compared with video detection, its disadvantage is that it cannot provide visual surveillance capability and record the visual characteristics of traffic vehicles or traffic conditions.

In the motion mode, when a moving object in the radar beam exceeds its own carrier, the radar will simultaneously sense the Doppler signal of its own carrier velocity and the relative velocity between the carrier and the moving object. Through the broadband amplification of the circuit, the lock-in filter amplification, separation, and software identification, the appropriate software gate is selected to count the pass number of the two Doppler signals, and then the carrier velocity and the relative velocity can be measured.

Microwave radar irradiates the target by transmitting electromagnetic wave and receives its echo so as to obtain the information of distance, distance change rate (radial velocity), azimuth, and height from the target to the electromagnetic wave transmitting point. Microwave radar is not only an indispensable military electronic equipment but also widely used in social and economic development, such as meteorological forecast, resource detection, environmental monitoring, etc., and scientific research, such as celestial research, atmospheric physics, ionospheric structure research, etc. In the intelligent transportation system, microwave radar is used to detect vehicle targets and collect traffic information, including traffic flow, vehicle speed, models, and so on.

The microwave radar detector can be installed in the half-space of a single-lane road to measure the traffic data of approaching or leaving traffic flow. It can also be installed on the roadside of a multilane road to measure the traffic parameters of vehicles on multiple lanes.

The installation methods of a microwave radar detector can be divided into forward installation and lateral installation. According to the principle of microwave operation, the velocity detected by lateral installation is the average velocity of each vehicle within a certain distance. The forward installation uses the Doppler effect to detect the real-time velocity of each vehicle, and other measurement parameters are the same as the measurement parameters of lateral installation. One device of forward installation can only detect one lane of information. The detection of multilane vehicle information requires multiple detection devices. Forward installation needs to install a suspended door frame, which is in the middle of the road construction and needs to interrupt traffic. Lateral installation is recommended considering installation cost and convenience. Factors that need to be considered in lateral installation include the number of lanes to be detected and the position of the column, as well as the impact of the middle isolation zone and road width.

The radar with forward installation and wide beam can collect traffic parameters in one traffic flow direction on multiple lanes, and the radar with forward installation and narrow beam can collect traffic parameters in one traffic flow direction on single lane. The detection area of radar with roadside installation and multiple detection areas is perpendicular to the direction of traffic flow. This radar sensor can provide traffic parameters of multilane traffic flow, but its accuracy is lower than that of the radar detector with forward installation. The radar detector installed by the roadside and in the single detection area is generally used to detect the appearance of vehicles in single-lane or multilane vehicles at signalized intersections. There are two types of radar installed by the roadside, including continuous wave Doppler radar and frequency modulated continuous wave radar. The radar detector can obtain traffic parameters through waveform. The continuous-wave Doppler radar can be used for vehicle speed measurement on urban main roads and highways [10].

The principle of microwave radar detecting road traffic information is described as follows. When the vehicle passes through the radar wave coverage area, the radar beam is reflected back from the vehicle to the radar antenna and then enters the receiver. The vehicle monitoring is completed through the receiver, and the traffic data such as flow, speed, and body length are also calculated through the receiver.

The ranging principle is described as follows. The frequency of the transmitted wave is mixed with the signal frequency of the current antenna when it comes back to the antenna after the reflection of the front object. The farther the distance between the transmitting antenna and the reflector is, the higher the frequency of the echo signal is. After mixing, the output frequency range is 1–100 kHz. The received mixing signal is sent to the DSP signal processing unit to calculate and judge the target distance. According to the time and waveform change of the vehicle passing through the radar irradiation area and the numerical value recorded by the internal timer of the signal processor, the vehicle flow, vehicle speed, lane occupancy, and vehicle type are calculated, and the data are transmitted to the client terminal through the existing communication network in real time.

In the actual detection, the microwave radar detector can be installed on the street lamppost near the lane to be detected, and the installation height range is 4—8 m. The microwave radar transmits continuous frequency modulation wave down to each lane, and the wave is reflected back from each lane and then mixed with the emission signal to produce intermediate frequency signal with a certain frequency difference. Due to the different distance between each lane and microwave radar, the center frequency of intermediate frequency signal is also different. The radar judges whether there is a vehicle passing according to the spectrum difference of intermediate frequency signals generated by different lanes at adjacent times and then analyzes the information of vehicle flow, vehicle speed, and vehicle type.

Microwave radar transmits a series of continuous frequency modulation waves through the antenna and receives the reflected signal of the target. The frequency of the transmitted wave varies with time according to the modulation voltage. Generally, the general modulation signal is a triangular wave signal, and the frequency of the transmitting signal and the receiving signal varies according to certain rules. The echo frequency and the transmitting frequency reflected from the target are the same. The difference between the transmitting frequency and the echo frequency can characterize the distance between the target and the radar. The calculation formula of target distance R is as follows.

$$R = \frac{\Delta t \times c}{2} = \frac{c}{2} \times \frac{T}{2\Delta F} \times f \tag{2.12}$$

where R is the distance between the target and the radar, c is the speed of light, T is the period of triangular wave modulation signal, ΔF is VCO modulation bandwidth, and f is the intermediate frequency signal frequency output by the mixer.

It can be seen from the above equation that the distance between the target and the radar is proportional to the frequency of the intermediate frequency signal. Due to the different distances in different lanes between target and radar, the corresponding intermediate frequency range is different. Moreover, the vehicle needs a certain time to pass through the radar detection area. Therefore, the traffic flow, vehicle speed, lane occupancy, vehicle type, and other parameters can be counted by analyzing the power spectrum of intermediate frequency signal with time.

The operation process of microwave ranging and velocity measuring devices is affected by many factors. Therefore, there are many interference factors that affect the measurement effect, especially the isolated single pulse interference. This interference is prone to produce some wrong data, i.e., invalid data. The input of these data to the filtering algorithm will increase the time complexity of the algorithm or destroy its filtering performance, which will result in low accuracy. Therefore, it is necessary to optimize the logical process of data acquisition and filter the single pulse interference before data filtering to ensure the accuracy of the collected data.

A remote traffic microwave sensor is a kind of detector that uses radar to monitor microwave transmission to detect traffic data. It has the characteristics of advanced technology, low cost, and convenient use. It can provide real-time online traffic information on the road, which has up to eight lanes of traffic flow, road utilization,

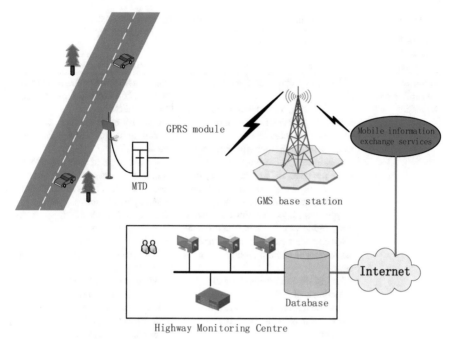

Fig. 2.17 Principal diagram of remote traffic microwave sensor

vehicle speed, driving direction, and vehicle model. More importantly, with the help of public network transmission detection data, it avoids the transmission line laying operation and reduces the amount of engineering.

As shown in Fig. 2.17, the detected data are transmitted to the control center by GPRS module, GSM base station, and the Internet for real-time processing [11].

2.2.5   Comparison

Each detector has its own advantages and disadvantages. When selecting the detector, it can be selected according to the actual situation. The advantages and disadvantages of various information collection technologies are summarized in Table 2.1.

Table 2.1 Comparison of advantages and disadvantages of various traffic detectors

Information collection technology	Advantages	Disadvantages
Geomagnetic coil information collection	• Mature and understandable technologies • Flexible design can meet the needs of various implementation situations • Extensive practical basis • Compared with nonburied detectors, the equipment is cheaper • Provide basic traffic parameters, such as flow, appearance, occupancy, speed, headway, and vehicle clearance • Types with high-frequency excitation can provide vehicle classification data • Some models can be installed under the pavement without cutting the pavement • The installation time is shorter than the induction coil. • It can be used in areas where induction coil is not applicable, such as bridge deck, etc. • Sensitivity to road vehicle pressure is lower than that of induction coil • Some models can transmit data by radio	• Installation and maintenance need to close lanes, causing interference to traffic flow • It is easy to be damaged when installing on roads with poor pavement quality • Reinstallation of detectors may be required in pavement renovation and maintenance of road facilities • Multiple detectors are often needed to detect traffic flow conditions in specific areas • Reducing road life • Sensitive to road vehicle pressure and temperature • When the vehicle type changes greatly, the accuracy will decrease • Regular maintenance of the detector is required • Installation needs to open the road or excavate pipelines under the road • Installation and maintenance need to close lanes, causing interference to traffic flow • In order to detect stationary vehicles, special sensors should be used to design or use special signal processing software • For models with small detection area, multiple detectors are needed to detect all lanes
Microwave radar information collection	• In the short wavelength range for traffic management, microwave radar is not sensitive to bad weather • Direct detection of vehicle speed can be realized • Realization of multilane detection	• The beam width and transmitted waveform of the antenna must be suitable for specific application requirements • Doppler microwave radar cannot detect stationary vehicles • The vehicle counting effect of Doppler microwave radar at intersections is not good
Ultrasonic information collection	• Realization of multilane detection • Easy installation	• Environmental factors such as temperature change and strong airflow disturbance will affect the detection performance of the sensor. To this end, some models have designed temperature compensation devices

(continued)

Table 2.1 (continued)

Information collection technology	Advantages	Disadvantages
		• When the vehicle on the highway runs at medium or high speed, the detector adopts large pulse repetition period, which will affect the detection of occupancy rate
Video image information collection	• Multiple detection areas can detect multiple lanes • Easy to increase and change the detection area • A large amount of data can be obtained • When multiple cameras are connected to a video processing unit, a wider range of detection can be provided	• Vulnerable to bad weather, such as fog, rain and snow, and shadow • Shadows projected by vehicles on adjacent lanes • Change of light level • Change of contrast between vehicles and roads • Water mark, salt, frost, and spider web on camera lens may affect detection performance • In order to obtain the best effect of vehicle appearance and speed detection, the camera needs to be installed at a higher height

2.3 Mobile Traffic Information Collection Technology

With the increasing demand of the advanced traveler information system (ATIS) for real-time dynamic traffic information, the traditional fixed traffic information collection method is relatively mature and the information processing means and methods are basically complete. But there are also shortcomings, which are mainly manifested in the following aspects.

(a) The coverage in the road network of fixed traffic information collection method is relatively low; the collected traffic information cannot fully reflect the traffic state of road network. At present, when laying fixed traffic acquisition equipment at home and abroad, it is necessary to consider road grade, traffic monitoring objectives, equipment investment, etc. Therefore, it will only be laid in important sections and key sections, which will cause a lot of information to blind areas.

(b) Due to the limitation of its own technical characteristics, different acquisition methods have different acquisition characteristics and environmental adaptability. Thus, the reliability of information sources is not high.

(c) Fixed traffic information collection needs to destroy the road or affect the normal traffic flow in the process of installation and maintenance. The maintenance needs to spend a lot of manpower and material resources every year.

To sum up, the fixed traffic information collection method cannot fully meet the massive, comprehensive, and real-time information needs of the intelligent transportation system. Therefore, the traffic management departments and researchers all over the world are conducting the selection and experiment of the traffic mobile collection technology, hoping to make up for the shortcomings of the fixed collection technology with the help of the characteristics of the mobile collection technology and improve the entire traffic information collection system so as to better serve the subsystems of the intelligent transportation system.

2.3.1 Information Collection Using Floating Car

2.3.1.1 Mobile Traffic Information Collection Technology

In the early 1990s, foreign countries used model simulation or field tests to verify the feasibility of using mobile vehicles to collect data. Among them, the VERDI system in Germany, ADVANCE experimental project T31, and AMI-C system in the United States are the most typical. The VERDI mobile detection system realizes real-time mobile transmission of road traffic information through vehicle mobile communication unit and monitoring center. The mobile transmission network adopts the GSM network. The real-time data collected by the monitoring center includes not only vehicle location information and vehicle speed information but also vehicle running status information, weather information, road information, etc.

The AMI-C mobile vehicle detection system in the United States emphasizes the transmission of multimedia information between the vehicle and the monitoring center through advanced communication technology. The vehicle installs a variety of multimedia display and acquisition equipment, so the traffic information collected by the floating car is more abundant and comprehensive. As these multimedia devices become the future standard configuration of American cars, the AMI-C system will gradually transform to the IP-Car floating vehicle detection system. The IP-Car system believes that each vehicle on the road can be used as a detection vehicle, and each vehicle has a unique IP number. A large amount of information collected corresponds to the IP number of the vehicle so that the monitoring center can grasp the road traffic information more comprehensively in real time according to these information and IP numbers.

By studying and analyzing the research status of mobile traffic information collection technologies at home and abroad, the technologies can be divided into active test vehicle technology and passive detecting vehicle technology. Active test vehicle technology is usually called floating car technology. This method was proposed by Wardrop and Charlesworth of the British Road Research and Test Institute in 1954, which can obtain the traffic volume, driving time, and driving speed of a certain road at the same time. At present, it is mainly used for comprehensive traffic investigation.

2.3.1.2   Active Test Vehicle Technology

This is the test method of the active test vehicle technology: in a specific test vehicle, traffic data collector records the vehicle speed, travel time, or travel distance information at any time through manual, Distance Measuring Instrument (DMI) or (Global Position System) GPS and other equipment, through the DMI device. It can collect and record information such as vehicle speed and driving distance every half second or even in smaller time intervals. GPS equipment can also record the vehicle position and speed information of the test vehicle every second.

Using active test vehicle technology usually requires selecting a specific vehicle as the test vehicle. The main purpose of this vehicle traveling in normal traffic flow is traffic data collection, which is usually called active test vehicle technology. The main advantages of this mobile collection technology are the following: (1) It can provide real-time traffic information under specific driving behavior conditions. (2) The detailed data of the entire driving process of the vehicle can be recorded in detail through DMI or GPS equipment. (3) The initial investment of the equipment is relatively small. However, there are some shortcomings: (1) The reliability of the information source is affected by both the data collector and the recording instrument. (2) The large amount of data collected by DMI or GPS equipment will bring storage problems. (3) The travel time of the entire road network is estimated only by the data from a particular test vehicle, which will bring about large errors.

2.3.1.3   Passive Detecting Vehicle Technology

Passive detecting vehicle technology refers to the process of installing auxiliary instruments or other remote sensing equipment on vehicles traveling in normal traffic to complete the collection of traffic information. The detecting vehicles can be personal vehicles, taxis, buses, or other commercial vehicles. The purpose of these vehicles to travel is not to collect traffic information but only to use the instruments and equipment on these vehicles to collect road traffic flow information in real time without obstructing the purpose of the vehicle itself. These vehicles and the traffic management or monitoring center communicate in real time through various wireless transmission technologies.

According to the different equipment installed on the vehicle, passive detecting vehicle technology can be divided into the following types: detecting vehicle technology based on beacon technology (Signpost), detecting vehicle technology based on automatic vehicle identification technology (AVI), radio station positioning (Radio) detecting vehicle technology, mobile phone positioning (GSM)-based detecting vehicle technology and GPS-based detecting vehicle technology, etc. The mobile acquisition technology using the above-mentioned passive detecting vehicle technology has the following advantages:

1. Low cost of data collection

Once the relevant hardware is installed, the system data acquisition is quite easy and inexpensive. And it is no need to frequently install or maintain the relevant equipment.

2. Ability to obtain continuous data

Through the onboard equipment, real-time road traffic conditions can be continuously obtained 24 h a day. Although some commercial vehicles or buses have a timetable, as long as the vehicle is running, there will be real-time data collection.

3. Directly reflecting the characteristics of actual traffic flow

Because the detecting vehicles are driving directly in the traffic flow, the collected data is not affected by the outside or subjective. Also, the drivers of these vehicles are randomly selected, so the collected data can more fully reflect the characteristics of the actual traffic flow.

Of course, the passive detecting vehicle technology also needs further improvement in the following areas:

1. Large initial investment

No matter what kind of technology is used to establish a mobile collection system for traffic information, it is necessary to purchase the necessary onboard equipment and roadside facilities and also to train relevant technical personnel to monitor and operate.

2. It is difficult to change once the system is established

Because the system needs to build related base stations and antennas when it is initially established. Once these facilities are built, they cannot be changed easily. Therefore, the system coverage needs to be well considered during system construction to ensure that the data collected by the detecting vehicle can be uploaded in real time.

3. System construction brings personal privacy issues

The traffic information collection equipment will be installed on social vehicles or private vehicles so that the driver's driving habits and travel locations will be monitored by the monitoring center, so this is also a problem that needs to be considered during the construction of the system.

4. The system is only suitable for large-scale traffic data collection

Since the initial investment in detecting vehicle technology is large and the travel area of the detecting vehicles is relatively free, the detecting vehicle technology is not suitable for small-scale traffic data collection.

The number of passive detecting vehicles, the selected sample size, is mainly affected by the urban road network, urban traffic composition, and other variable factors in the city, such as weather, urban infrastructure construction, large-scale activities, etc. At present, relevant research at home and abroad has not figured out a fixed standard and formula to calculate this number, so each city needs to study according to its own characteristics so as to choose the appropriate number of detecting vehicles.

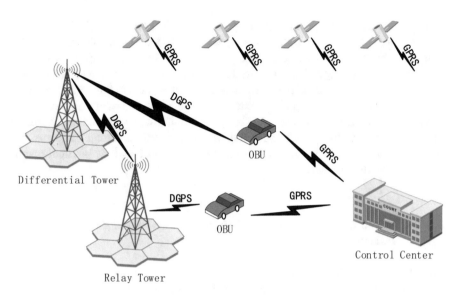

Fig. 2.18 Floating car detection system composition

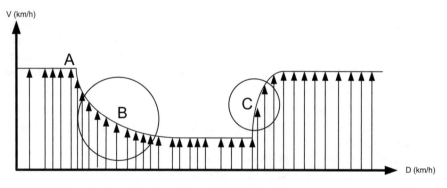

Fig. 2.19 Basic principles of traffic mobile collection technology

2.3.1.4   Basic Principles of Traffic Mobile Collection Technology

Regardless of whether it is active test vehicle technology or passive detecting vehicle technology, the basic principles of data collection are basically the same. The following takes the detection of a certain road traffic status as an example to analyze the data collection and analysis process of the traffic mobile collection technology.

As shown in Fig. 2.18, the real-time vehicle speed data is processed and compressed by a specific algorithm, packaged with vehicle location information and driving road information, and sent to the monitoring center using the onboard wireless transmitter. The monitoring center obtains road traffic information by processing these data. It can be seen from Fig. 2.19 that the vehicle is in a normal driving state at point A. Then at point B, the vehicle speed drops to a certain threshold and continues to drive at low speed for a period of time, thereby judging that the vehicle is in a congested state. At point C, the vehicle speed gradually increases and gradually returns to the normal driving speed. Thereby it can be judged that the vehicle leaves the congestion and the road traffic starts to flow smoothly. The driving vehicle and the monitoring center realize real-time communication of information through GPRS and GPS networks. The monitoring center obtains real-time traffic information and vehicle operating status information through information processing and releases information to the driving vehicles and the public through wired and wireless methods.

2.3.1.5 Introduction to Nagoya's P-DRGS Based on Floating Car Information

PRONAVI car navigation system is a dynamic navigation demonstration system based on floating car information developed by P-DRGS. The system uses real-time probe information obtained from 1500 taxis operating in Nagoya and real-time information obtained from the Japan Road Traffic Information Center (JARTIC) to predict current traffic conditions and provide users with traffic information reference at the same time. The basic operation mode of the P-DRGS system is to collect various vehicle information to the DRGS center and then send traffic prediction information to the car navigator through wireless signal transmission. The car navigator is also used to collect traffic information at the same time and realize the traffic forecast. The system structure of PRONAVI is shown in Fig. 2.20. The functions currently provided by the system mainly include the following:

1. **Map display function**
   Map representation of the surrounding area of Nagoya, with the function of zooming in and out.
2. **Real-time traffic information release function**
   The P-DRGS center updates the collected construction, accident, control, and other traffic information every 5 minutes, and the user can freely set the information to be displayed or not displayed on the map.
3. **Detailed traffic information reference**
   Click on the construction, accident, control, and other signs on the map to immediately provide detailed traffic information for reference. It also provides system information, local area information, and traffic traveling permit-related information.
4. **Automatic display function of the user's current location**
   According to GPS information, automatically display the user's current location.
5. **Route search function**
   Within the scope of the service, based on the user's departure time, provide up to 5 multimode traffic shortest routes between any departure place and

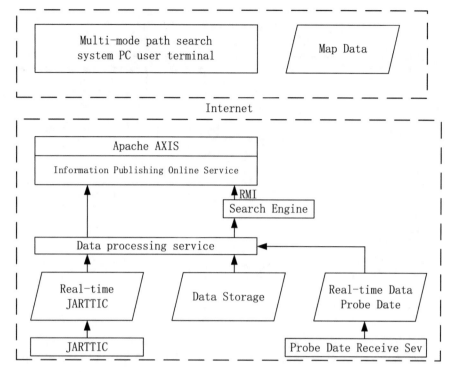

Fig. 2.20 PRONAVI system composition

destination, and provide transfer-related detailed information for users who need to transfer. In addition, take the World Expo. The field is the destination, providing traffic information that prioritizes surrounding traffic control and park-and-ride.

6. **Management of usage history and login points, etc.**

It can display the history of use and the management of login points on a separate screen to improve the convenience of the service.

7. **Print preview function**

The searched path and other results can be previewed on the map and printed directly.

8. **User authentication**

In order to prevent the use of unauthorized users, each time the system is used, user authentication is required to enter the service through the Internet. The system was demonstrated to the world at the 2004 World ITS Conference, which attracted the attention of participants from various countries, and major Japanese newspapers also rushed to report. Later, during the 2005 Nagoya World Expo, a 1000-person experiment was conducted to test the stability of the central system and the validity of PRONAVI information. According to the plan of the

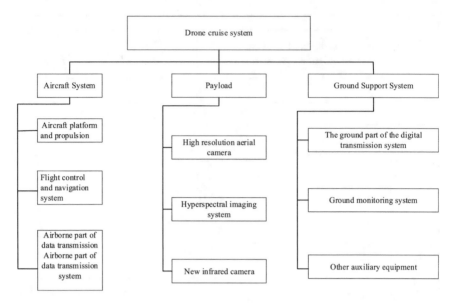

Fig. 2.21 Drone system composition [12]

P-DRGS cooperative group, the system will continue to be improved, including the improvement of system service functions and the expansion of service scope.

2.3.2 *Information Collection Using UAV*

With the continuous development of drone technology, it has unique advantages such as simple structure, low cost, low risk, flexible maneuverability, and strong real-time performance. This makes drones more widely used, especially in the field of traffic information collection. The use of drones has further expanded the mobile traffic information collection technology in space (Fig. 2.21).

In the field of traffic information collection, drones are usually equipped with video recording equipment, using integrated GPS positioning system, wireless communication system, and high-resolution video camera system to capture and photograph ground targets and then extract images and video data from them. The traffic information collection-oriented detection technology by drones is using self-driving aircraft as the flight platform, high-resolution digital cameras as sensors, obtaining high-resolution image data as the direct target, and obtaining road traffic flow information and the purpose of video Image processing technology as data processing method. Through GPS positioning and navigating technology, and with the help of advanced drone flight control systems and wireless communication systems, real-time data transmission between drones and ground traffic monitoring centers can be realized. Also, real-time monitoring and information collection of road traffic flow from various angles and directions, different altitudes, and multiple locations.

2.3.2.1 Characteristics of Drone Traffic Information Collection Technology

Drones also have obvious advantages in the application of traffic information collection due to their inherent characteristics [13]:

1. **Wide detection range**

   Due to low-altitude flight, adjustable cruise height, flexible change of viewing angle, and no obstruction between vehicles, it can realize the detection of the point and line and surface traffic detection from local to wide area, which is conducive to the rapid and efficient control of the situation by the traffic management department.

2. **Diversified collection of information**

   Through continuous investigation of specific areas, it is possible to track and detect the spatial position and operating status of a single vehicle and also to collect macro traffic information such as traffic density, traffic flow, average speed, and distribution of traffic facilities.

3. **Flexible**

   Drones can fly on roads and bridges and even enter tunnels to conduct accident site surveys and evidence collection. The flying height ranges from tens of meters to hundreds of meters and is not affected by road traffic, showing unique flexibility and motility.

4. **Emergency rescue**

   In the event of natural disasters such as earthquakes, floods, tsunamis, and blizzards, all ground transportation is paralyzed, and drones can be dispatched immediately to observe the actual situation in-depth on-site, searching for personnel, and establishing communication relays.

5. **Low risk**

   Without considering the risk of the driver, it can perform high-risk tasks in severe weather or polluted environments.

6. **High efficiency**

   The drone has a short preparation time and can be dispatched at any time, which has the characteristics of low investment and high efficiency.

7. **Low cost**

   The market price of small drones is relatively low, ranging from several thousand yuan to tens of thousands of yuan.

At the same time, drone information collection technology also has some shortcomings:

1. **Limited load**

Fig. 2.22 Small drone

Due to the limited volume and mass of small civilian drones, their effective load is also limited. Their cameras, pan-tilts, and other communication equipment must be controlled within several kilograms.

2. **Battery life**

Affected by the weight of the load, the endurance of the drone is generally controlled within two hours.

3. **Weatherrequirements**

The drone's requirements for weather, especially wind, generally need to be below level 6 wind, and the temperature is within a certain range, and it is in a clear weather state to be able to fly safely and collect traffic information.

4. **Platform vibration**

During the cruise of the drone, the camera is always in a state of vibration; coupled with the influence of operation and weather, it is very easy to cause the collected video to appear jitter or blur, which will increase the difficulty of video detection.

5. **Sight blocked**

In the process of traffic monitoring, drones are often disturbed by various vegetation, construction equipment, pedestrians, buildings, and other factors. Therefore, in the process of traffic monitoring, if drones want to extract traffic parameters more accurately, they must overcome the influence of these series of factors and better serve the traffic monitoring work (Fig. 2.22).

2.3.2.2 Application of Drones Traffic Information Collecting Technology

1. **Daily road condition monitoring**

   Larger fixed-wing drones are usually used in daily road condition monitoring, which have many advantages such as fast speed, long range, high flying altitude, etc. and can usually be equipped with high-definition digital cameras, which can clearly capture and record real-time ground traffic conditions. Also, it is very suitable for daily patrols and road condition collection tasks in the air. If the fixed-wing drone is operated in cruise mode, without manual intervention, it will automatically go back and forth on the planned road section every day and send traffic video information to the road command center through the mobile Internet. The road command center can be intuitive and fast, the locality has a clear understanding of the traffic conditions of the road, and it can publish the real-time obtained road conditions to the information platform so that the drivers on the road and the people who are about to travel can obtain the latest and most accurate road conditions information independently and then choose the route and time of travel to lay the foundation for safe and convenient travel [14].

2. **Auxiliary traffic control**

   When there are bad behaviors in the road network, such as escaping after a traffic accident, breaking the card, etc., the drones equipped with the video transmission system and the positioning system can track and locate the vehicle in the accident for a long time and transmit the relevant information. It also guides law enforcement officers to intercept and dispose of them, which contributes to the fight against road crimes, while also avoiding the harm caused by law enforcement vehicles directly intercepting or even chasing the vehicles involved in the incident on the road, and guarantees the personal safety of road law enforcement officers.

3. **Emergency response and on-site command**

   When traffic jams, accidents, or even natural disasters occur on the road, which seriously affect the safety of traffic operation, making decision-makers understand the situation on the spot faster and more intuitively is a difficult problem. The coverage of fixed cameras is limited and cannot meet the needs of full video coverage. Road administration and traffic police vehicles and personnel carrying onboard surveillance systems may be isolated away from the scene by the long traffic flow. At this time, a fast-response video transmission, especially on-site command platform, is particularly important. In developed countries in Europe and the United States, this kind of rapid response platform is usually based on helicopters. However, due to the high price of helicopters, high operation and maintenance costs, and scarcity of quantities, it is currently not suitable for batch applications in China. In this case, it is more suitable to use drones that are inexpensive and relatively easy to maintain. The flying speed of fixed-wing drones is much higher than that of ordinary vehicles, and its travel is not restricted by terrain and interference. It can fly directly to the location of the

Fig. 2.23 Vehicle detection technology route of drones

incident, rush to the scene as soon as possible, and then transmit live video from the air scene to the commander dispatch center. Drones can have a bird's eye view of the actual traffic flow on the ground, which is conducive to the road management department to grasp the overall situation, conduct overall command and correct guidance, and solve the problem of road congestion [15].

After an emergency occurred, the rotary wing drone can be equipped with different task modules such as megaphones, sirens, and traffic command lights, so that it can reach the scene faster and perform dispatching and command tasks before the road administration and law enforcement personnel arrive on the scene. Traffic management personnel can directly direct and clear the traffic on the spot by means of the traffic control equipment carried by the rotor drone. The use of drones greatly improves the on-site information collection and transmission capabilities after emergencies.

4. **Assistance in rescue**

When a serious traffic accident occurs in the road network, a large area of traffic congestion may ensue, especially when rescuers arrive at the scene often only to find that some rescue equipment and medicines cannot arrive at the scene in time and the injured cannot get it. When a serious traffic accident occurs in the road network, a large area of traffic congestion may ensue, especially when rescuers arrive at the scene and then often only find out that some rescue equipment and medicines cannot arrive at the scene in time and the injured cannot get timely transferring and treatment. The addition of drones can effectively solve these problems. Rotor-wing drones can avoid congested traffic and crowds, deliver the required materials and equipment to the scene, and return the information of the injured so that outfield rescuers can obtain enough information to prepare for follow-up treatment [16] (Figs. 2.23 and 2.24).

Fig. 2.24 Assistance in rescue with drones

2.3.2.3   Traffic Detection Based on Drone Video

The use of drones to collect traffic information is mainly to extract traffic information with the help of videos and photos taken by the high-definition video equipment carried by the drone platform [17]. The traffic information usually includes three categories: vehicle detection, traffic density and flow, and vehicle trajectory and speed (Figs. 2.25 and 2.26).

1. **Vehicle detection based on drone video**

     For the video traffic information collection system, the detection of moving vehicles is the basis and prerequisite for the function of the system. Only when vehicles are detected quickly, accurately, and reliably can the next traffic parameter extraction be carried out. Vehicle detection based on drone video can be roughly divided into image preprocessing, background modeling, vehicle detection, and other processes.

     When using drone video to detect vehicles, several influencing factors should also be considered, which have a greater impact on the accuracy of traffic parameter extraction [19].

     The first is the traffic state. The impact of traffic conditions on vehicle detection during morning and evening rush hours and normal peak hours is mainly reflected in the impact of vehicle speed and traffic density on the detection algorithm. For example, the inter-frame difference method has a higher recognition rate for higher speed vehicles and a very low recognition rate for lower speed or stationary vehicles. Higher traffic density requires higher image segmentation algorithms, and lower density makes vehicle detection easier. It is difficult to model the background when the traffic density is high.

Fig. 2.25 Traffic detection based on drone video

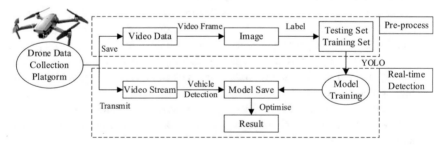

Fig. 2.26 Technical route of vehicle detection by drone video [18]

Then comes the lighting conditions. The lighting conditions are mainly different in different time periods (morning and evening peaks, flat peaks) and weather conditions (sunny, cloudy, etc.). Lighting conditions are measured by illuminance and visibility. There are two main effects of lighting conditions: target clarity and target shadow.

The flying height of the drone also has a certain impact. Due to the different shooting heights of drones and the different video equipment used, the pixel size of the vehicle in the video image is also different, which will affect the selection of the vehicle detection algorithm. At the same time, the model composition also has a certain impact.

2. **Traffic density and flow extraction based on drone video**

Traffic density is one of the three basic parameters in traffic flow theory, which is of great significance to the study of traffic flow. Limited by the current traffic information collection technology or equipment, the automated acquisition of traffic density has not yet been achieved. Drone combines the advantages of its flying height and angle of view with vehicle detection technology to automatically and accurately extract this important traffic parameter.

The traffic density extraction process is roughly as follows: obtaining the video taken by the drone, selecting the area where the traffic density needs to be extracted, calculating the ratio of the actual length to the image pixel density for calibration, using the vehicle detection algorithm to detect the vehicle, and finally calculating the traffic density.

3. **Vehicle trajectory and speed extraction based on drone video**

Vehicle trajectory is of great significance for extracting traffic parameters, studying driving behavior, and traffic safety analysis. At present, vehicle speed detection is limited by detection technology and equipment, and there are problems such as mutual occlusion between vehicles. This article gives full play to the advantages of drones collecting traffic videos and proposes a vehicle trajectory and vehicle speed extraction method based on drone video.

At present, the extraction methods of vehicle trajectory mainly include GPS positioning and tracking and ground fixed video camera detection. The GPS positioning and tracking error is about 15 m, which is unacceptable for traffic parameter extraction and traffic micro-level research. In the case of ground-fixed video camera detection, the detection range is small and the observation angle is not ideal. Taking advantage of the drone's ability to collect large-scale and multiview traffic videos above the road can make vehicle trajectory extraction easier to implement.

The main process of vehicle trajectory extraction is as follows: converting the drone video image into a certain type of chromaticity image (HSV/HSL), using the corresponding algorithm to process the image, finding the vehicle coordinates, processing the next frame of image to determine the new coordinates, and then recording the vehicle coordinates in each frame of image and displaying it in the current frame of image to get the vehicle trajectory.

2.3.3   Crowdsourcing Information Collection Technology

UGC (user-generated content) is the use of the Internet to distribute work, discover ideas, or solve technical problems [20]. Traffic information user-generated content is the feedback of traffic information by travelers through mobile terminals. It is an important means of traffic information collection. It can be divided into active generated content and implicit generated content.

Holders of smart terminal devices (such as smartphones, Pads, etc.) download and install specific applications and use the information reporting function provided by the application to actively upload traffic information on the traffic scene to the application in the form of text, voice, pictures, videos, etc. This way of actively reporting by users is called active generated content. Users start specific applications installed on smart terminal devices (such as smartphones, Pads, etc.) and grant the application the right to obtain location information. The application operation background uses the real-time location obtained. Information uses a certain algorithm to calculate the road traffic flow conditions at the traffic scene. This information collection method of obtaining user real-time location information is called implicit generated content. The hidden generated content traffic information collection method is similar to the floating car principle and can be regarded as a generalized floating car. In the past, taxis were used as the main data source for traffic information calculations on floating cars because taxi floating cars accounted for urban traffic flow. The proportion is relatively low and has some of its own business characteristics, which is quite different from the operation law of urban traffic flow. The penetration rate of smart terminals is relatively high, and users in the traffic flow are generally equipped with smart terminals. Now the information of a large number of various drivers is added to effectively supplement the limit on the number of taxis, which can effectively reflect the state of urban traffic flow.

The rise and practicality of user-generated content are mainly due to the development of the mobile Internet and the popularization of smart user terminals, especially the popularization of smartphones. A good application can quickly expand the number of users and guide it with extensive user participation. With more and more users joining in, the information continues to be enriched and improved. It also makes the information more and more accurate. Users post road conditions through Weibo, WeChat, etc. It is also manifestation of user crowdsourcing, and data accuracy can be improved through mobile phones for this information (Fig. 2.27).

As traffic participants and travelers, they need real-time road condition information, and they are also direct experiencers of road condition information. If a traffic accident occurs, travelers passing by the scene can share the accident information in a certain way, and then they can travel for other trips. The more people who participate, the more comprehensive, accurate, and timely information will be. This is user-generated content, and users of information are also providers of information. Using this method can make up for other collection methods that cannot cover a large area and the fact that the actual road conditions cannot be determined, such as the impact of traffic incidents, traffic control, etc. If there are users on-site, these conditions are shared with other users. Other users can then make decisions about their driving route. The biggest advantage of user-generated content lies in the characteristics of large coverage area, low cost, and timely response. The disadvantage is that when the number of users or the number of feedbacks is small,

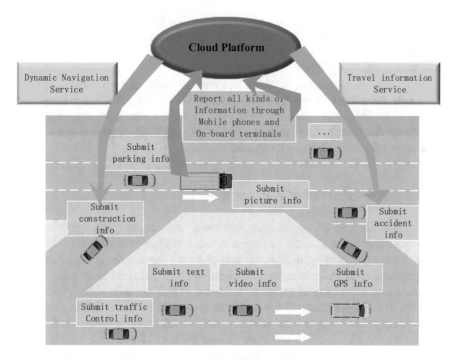

Fig. 2.27 Traffic travel service platform based on generated content

comprehensive and timely road condition information cannot be obtained, and false information is not easy to judge and reported data may be expired, etc.

2.3.3.1   Crowdsourcing of Traffic Information Users

The following are the common ways of crowdsourcing real-time traffic conditions users [21]:

1. **Open the location interface of the mobile device**

   Mobile terminal devices such as smartphones, PADs, and car terminals can send information about their position, speed, and direction to data processing centers at regular intervals during movement through pre-installed software such as road condition and navigation software. In this way, each terminal can be seen as a floating vehicle, and when the volume of users reaches a certain level, the road conditions of an area or road can be calculated.

2. **Direct user feedback**

   When traffic congestion, incidents, or traffic equipment (signal lights) fails on the road, users can reflect the relevant situation to the data center via mobile phones, landline phones, vehicle-mounted communication devices, etc. They can also use SMS, voice, pictures, text, video, etc. Of course, calling the data center directly is the most direct and effective way. Traffic radio stations can also use road condition information reported by informants as their main source of information.

3. **Users publish via Weibo, WeChat, etc.**

At present, the amount of Weibo and WeChat users has been very large. These people have formed the habit of sending Weibo and WeChat anytime and anywhere. Many people also send out the traffic conditions they encounter through the search and attention of Weibo and WeChat. A large number of the latest road information can be obtained.

2.3.3.2 Generated Content Platform Workflow

The main players in space-time generated content are the requesters of crowdsourcing tasks and the generated content participants, who are connected through the space-time generated content platform. As shown in Fig. 2.28 [22], the platform is at the center of the workflow. The platform is responsible for the integrated processing of requested tasks and participant information. Generally, the platform first preprocesses the task/participant information and then hands it to the task assignment engine. Subsequently, the task allocation engine allocates tasks based on task characteristics and optimization goals and feeds back corresponding information to requesters and participants. According to different task requirements, the platform can directly feedback the task execution results to the requester or integrate the execution results (as shown in the dashed box on the left in Fig. 2.28)

Fig. 2.28 The workflow of the space-time generated content platform

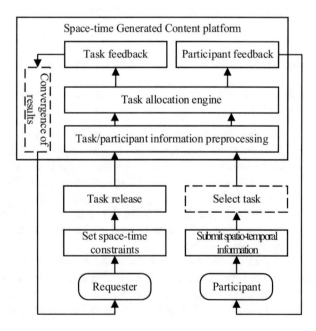

and then feedback to the requester. The workflow is described from the perspective of the requester and the participant.

Task requester workflow: When the requester intends to use the space-time generated content platform to complete the task, they need to perform the following steps in sequence. First, the requester needs to set the space-time constraints of the task. For example, dispatch tasks usually need to set the dispatch time and location. After the setup is completed, the requester can submit the task to the platform. Subsequently, the requester waits for platform feedback [23].

Generated content participant workflow: To complete the task, participants first need to submit their space-time information (current location, etc.) for the platform to determine whether it meets the relevant space-time constraints. On some platforms, participants can browse and choose tasks independently, as shown in the dashed box on the right in Fig. 2.28. Subsequently, participants waited for platform feedback.

The above is the general workflow of the space-time generated content platform. The uploaded data can be divided into two categories. One category is positioning information, including vehicle location, driving direction, and speed information. The other is the voice, text, image, video, etc. reported by the user. These data are processed in different ways and then released in multiple forms.

Depending on the real-time nature of tasks, tasks can be divided into static offline tasks and dynamic online tasks. The main difference between them is that the situation of the task and the participant in the dynamic online tasks is unknown, and participants and tasks arrive randomly (or according to some unknown distribution). The platform can only make allocation decisions based on the temporal and spatial distribution of current participants and tasks but cannot predict the temporal and spatial distribution of future participants and tasks. For static offline tasks, all tasks and participant information are known, so the global optimal goal can often be easily obtained.

Take the logistics delivery application as an example. This application is a typical static offline task. Usually, before delivery, the delivery person (can be regarded as a crowdsourcing participant) already knows all the delivery locations. Therefore, the platform can arrange the best delivery route for the dispatcher. But in practical applications, it is usually faced with dynamic online tasks, and the tasks are dynamically changing in real time. For example, in the Didi Travel platform, taxi drivers are unpredictable for users who may request taxi services in the future. Therefore, the platform can only match reasonable orders based on the information of the known driver and passenger at the current moment. The current matching strategy will have an impact on the matching results of future orders, even if the current matching strategy is optimal. When considering the order distribution at the next moment, the current match is probably not very reasonable. Therefore, in various spatiotemporal crowdsourcing applications, the real-time characteristics of crowdsourcing tasks will affect the work strategy of crowdsourcing platforms to a large extent.

Depending on the demand of participants for tasks, tasks can be divided into single-participant demand tasks and multiparticipant demand tasks. The difference lies in the number of participants required to complete the task. Some tasks have a single requirement for crowdsourcing participants. Such tasks are often limited to only one participant. For example, special vehicle services such as Didi Travel and the recent emergence of driving services are typical single-participant demand tasks. On the other hand, some tasks require multiple participants to complete the task in coordination, either due to the variety of required abilities or due to the large workload of the task or the more complex requirements for work skills. For example, it is a multiparticipant demand task that requires participants with multiple abilities to hold a party together. For this kind of work involving multiple people, how to control the quality of the collaboratively completed work is not simple. Therefore, in all kinds of different spatiotemporal crowdsourcing applications, the different demands of participants for crowdsourcing tasks will affect the way of controlling the quality of the completion of crowdsourcing tasks.

Depending on whether crowdsourcing participants have the right to choose tasks, space-time generated content tasks can be divided into two forms: participants actively choose tasks and platform active assignment tasks. For tasks that participants actively choose, crowdsourcing participants have the right to choose tasks, and they can choose appropriate tasks based on their own preferences.

However, in many application scenarios, crowdsourcing participants share the same preferences, which results in a small number of cost-effective tasks being in short supply, while many less cost-effective tasks go unrequested, resulting in a lower overall utility of the space-time generated content platform. Therefore, this has also led to the emergence of another type of platform active distribution tasks. Such tasks are only dispatched by the space-time generated content platform according to the task allocation algorithm. The generated content participants who are assigned to the task should perform the task; otherwise, they will be punished by the platform. Taking real-time private car applications as an example, the Didi Travel platform uses the participant's active selection task, and each of its private car tasks (crowdsourcing tasks) can be selected by the private car driver (crowdsourcing participant). When multiple drivers choose a task, the platform adopts the strategy of "grab the order" to ensure the uniqueness of the task assignment. But the nature of the task is the participant's active selection task. In contrast, Shenzhou Special Vehicle and Uber platforms use active platform assignment tasks, and each driver can only passively wait for the platform to assign tasks. When the task is assigned, the driver must perform this task. Otherwise, you will be punished. Therefore, in various spatiotemporal generated content applications, generated content participants' right to choose tasks will affect the platform's allocation mechanism for generated content tasks.

In terms of data processing, the mobile terminal regularly sends data such as position and speed to the center. Unlike monitoring equipment, the mobile terminal sends driving data upwards for a certain period of time; e.g., it can send data to the center once every minute. These data contain 12 position data with a time interval of 5 seconds so that more detailed data can be obtained without increasing the frequency of sending and the amount of data. Subsequently, the road data can be obtained more accurately by matching the roads in the background to achieve a higher level of accuracy. User-reported data includes location and other multimedia data, and some data that cannot be directly quantified can be added to the traffic information release using manual review and processing. The data collected in different forms are calculated by the data processing center through certain models, using automatic or manual judgment, and these user crowdsourced data are processed on time and data fused with data from other sources to produce the final real-time traffic information. However, there may be a large amount of false data in the user crowdsourcing data, which needs to be compared and manually judged by certain algorithms to remove the false data and ensure the accuracy of the data.

2.3.3.3 Problems with Current Generated Content Technology

Whether it is a crowdsourcing map service such as Kay Lade or Tencent Lobo, the data sensed by traffic detection equipment (in this article, roadside, vehicle, floating car, and other data are collectively referred to as sensor data) and vehicle data are processed separately, and the two types of data are not truly integrated. Due to the huge user base, most of the users have basically had an awareness of "generated content." The map basically relies entirely on data to analyze and display road conditions. The data is visually displayed on the map to guide users in traffic. The problem with this is when there are only a few users sending information on certain road sections and they are bad users or users whose credibility cannot be determined for the first time. The service platform has to choose these "possibly bad data" because there is no other data. At this time, the information displayed on the map to the user has a greater risk.

The difference is that for domestic crowdsourcing map service providers, the road condition information marked on the map mainly relies on various sensor data such as roadside sensors, vehicle terminals, and floating car data. Most of the data used as auxiliary means are only displayed on the map in the form of simple icons. The platform has not really audited its effectiveness. Therefore, the data is not actually integrated into the real-time update of traffic conditions. The reasons for this situation are many and varied and may be because domestic users have not yet developed a strong enough awareness of crowdsourcing and that the proportion of users who actually report traffic information is not very high among all users. This paper, as a research topic, will not consider these background factors but will focus on how to fuse data and sensor data to obtain real-time traffic status to improve and enhance the real-time and accuracy.

Task allocation and quality control are the two core issues in generated content data management, which have more research significance in the generated content environment. The optimization goals and constraints of the two issues of task allocation and quality control have undergone major changes in the generated content environment. As far as the task assignment problem is concerned, the task assignment problem in generated content research is usually only described by the offline static bipartite graph matching model. However, this method is difficult to adapt to the real-time constraints of tasks in the spatiotemporal generated content environment and the needs of path planning when participants complete multiple tasks. As far as the quality control problem is concerned, the quality control problem in generated content research usually takes as the optimization goal to maximize the expected correct rate of a single participant or group of participants to complete the task. In some real-time spatiotemporal generated content applications, the optimization goal of quality control is to minimize the spatiotemporal cost of participants to complete tasks.

The privacy protection issue in generated content is also a great challenge because the spatiotemporal generated content platform needs to allocate tasks based on the location of the generated content task (or the location of the generated content task requester) and the location of the generated content participants. Therefore, any spatiotemporal generated content platform has the potential risk of leaking the privacy of task requesters and participants. Take the space-time generated content platform for special vehicles as an example. Once the platform is attacked and private information is leaked, the accurate daily travel information of each car order requester and car driver in the past will be published to the public. Therefore, the privacy protection of generated content is its unique core research problem. This question aims at how to design a privacy protection strategy. It not only protects the spatiotemporal information of task requesters and generated content participants but also guides the platform to carry out effective task assignments based on the protected spatiotemporal information.

2.3.3.4   Generated Content Application Case

In the past 10 years, generated content technology has been closely related to people's daily life. For example, the early generated content platform usually refers to the "question and answer system" platforms such as Wikipedia, Yahoo Answers, Baidu Know, etc. It has become a necessity for modern people to acquire knowledge. In recent years, due to the single task type supported by the early generated content platform, it can no longer meet the current needs of Web applications with diversified data types and complex tasks. This led to the birth of a new generation of "online generated content platforms." It is a large-scale online job recruitment and task subcontracting management platform, for example, Amazon Mechanical Turks (AMT), Crowd Flower, o Desk, etc. This kind of generated content platform not only brings a new technological revolution but also creates a huge market economy value. According to the American Amazon's annual report, as of 2010, the company's annual profit on the AMT-generated content platform has exceeded 520 million US dollars. Therefore, generated content technology has brought huge potential to the technological revolution in today's Internet age. As the "People's Daily" report on generated content in 2014 stated, "The generated content model is the general trend" [24] (Fig. 2.29).

Real-time traffic monitoring affects people's daily travel and lifestyle at all times. In recent years, with the popularity of portable mobile computing devices, mobile navigation software developed by location-based service providers, such as

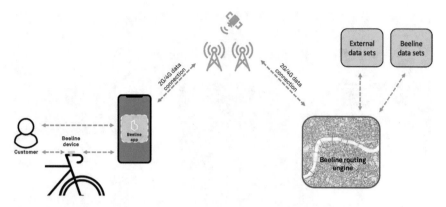

Fig. 2.29 Generated content transportation service concept "beeline"

domestic Baidu map and AutoNavi map, or foreign Waze, can already provide real-time road condition monitoring information more accurately. The precise traffic monitoring information obtained by this type of software is mainly derived from the acquisition and analysis of sensor data in a large number of users' mobile devices. By obtaining the spatial distribution information of a large number of users at different times and the corresponding various sensor data, this type of software can analyze and infer real-time traffic conditions. In other words, mobile navigation software releases a potential generated content task while the user is using its software. That is to share the user's spatiotemporal information and sensor data. And its users have passively become generated content participants. Such scenarios are also called "participatory sensing" in mobile Internet research (Fig. 2.30).

1. Personalized traffic information service

From the research status at home and abroad, it can be seen that whether it is the government, enterprises, or travelers, they are all seeking transportation service methods with more specific information, richer content, and more diverse forms. Personalized traffic information service refers to the provision of more accurate, more specific, and more valuable traffic information services to meet individual needs based on the traffic needs of individual travelers.

Personalized traffic information services obtain data from multisource heterogeneous data sources. Through processing and fusion technology, it is integrated into complete traffic information. It is not just pure red, yellow, and green traffic flow data, but is released in the form of information such as videos, pictures, and voices. This form makes the display of road conditions more intuitive and convenient to use. Compared with most current written expressions, we can see the existing road conditions at a glance. Drivers on the way can listen to the voice broadcast of the road conditions ahead, which can greatly improve the safety and convenience of driving.

Fig. 2.30 Waze interface of Google's crowdsourcing map application

Through the development of application systems such as the Internet, mobile phones, car machines, etc., when users use navigation or road conditions applications, they can not only receive information provided by others but also publish their own road conditions information through spatial positioning, photographing, video, text, and recording to realize the sharing of traffic information and ensure timely, safe, and accurate information through strict review and comparison mechanisms.

Users can decide which information to filter and strengthen according to their current needs. Users can customize key sections and hot spots on their travel routes according to their daily travel conditions. Users can set the time and view their customized information within the set time. For example, at work hours at 7: 30 in the morning, customized messages can be sent to the mobile phone via SMS, message reminders, etc. This way makes the road conditions on your travel route clear at a glance and provides a time and route basis for the upcoming travel.

Present information as vividly as possible through text, pictures, voice, and video can provide users with a convenient and efficient way to obtain information.

2. Push traffic information based on driving position and direction

When the vehicle is driving, the traffic information application system will automatically broadcast real-time voice reports on the road conditions ahead and surrounding roads according to the user's driving route and direction. Provide users with personalized push services, including providing dynamic navigation, road conditions along the way, surrounding information, etc. Help users choose the best route to avoid congestion and improve travel efficiency.

Of course, it can be used not only by motorists but also by people who travel by public transportation. If it shows serious congestion ahead, you can change to the subway or other transportation to travel. This function is to tailor real-time personal traffic conditions for each user. Realize one-to-one traffic and road condition reporting, making the road condition application more pertinent. It will bring good news to fellow travelers.

3. Destination-based traffic information service

   Under the condition of setting the destination, the service center plans multiple driving routes to the destination according to the user's current location and driving direction and continuously provides traffic information on possible routes. When there is a traffic accident or serious congestion along the way, it can provide users with detour options.

   Before you travel, you can learn about the road conditions by checking the comments of road friends. During the trip, you can make complaints and interactions with road friends about the traffic situation or learn about traffic information. Or strive for a sense of identity, excrete traffic jams and irritability, and kill time to make good relationships.

4. Smart travel traffic information service

   According to the user's daily driving route, analyze the user's driving habits. Using artificial intelligence technology, the system automatically provides users with traffic information along the way, road condition broadcast, and travel time predictions for common routes without the user having to enter a customized route. Eliminate the user's tedious setting work, and intelligently realize the tailor-made route function. Use GIS and statistics-related technology to analyze the user's travel route and area. Find the routes and areas that users often take. Thereby, traffic information can be pushed in a targeted manner to achieve intelligent traffic information services. The application of this function will greatly improve user stickiness. And through the frequent use of users, the accuracy of information push is further improved. Let people feel the true wisdom of travel.

References

1. Yan L (2010) A study on the city traffic flow detection techniques and applications. Hunan University. (Published in Chinese)
2. http://www.hbgs.com.cn/infon.asp?ID=457380
3. Shu K, Ren J, Wang B et al (2011) Design of real-time traffic information collection system based on ultrasonic measuring. Res Explor Lab 30(8):70–73. (Published in Chinese)
4. Cheng XZ, Zhang Z, Wang W (2004) Study of long-distance pulse echo system with ultrasonic ware. Chinese J Sci Instrum S2:179–182. (Published in Chinese)
5. Li CL, Shu HS (2013) An ultrasonic positioning system. Res Explor Lab 32(2):39–44. (Published in Chinese)
6. Zhao WD, Zhao Y, Cheng SS (2018) Analysis of the overall architecture of the public security video image information application system. China Secur Protect Technol Appl 1:9–16. (Published in Chinese)

 7. Zhang G (2014) Research on the key technics of processing the video road traffic information. Hunan University. (Published in Chinese)
 8. Huang JS (2010) Design and realization of traffic information acquisition system based on microwave. University of Electronic Science and Technology. (Published in Chinese)
 9. Xie YS (2011) Study on distance and velocity measurement system for freeway based on microwave radar. Zhejiang University. (Published in Chinese)
10. Cheng WM, Li CL (2015) Radar-based displacement/distance measuring techniques. J Electron Measur Instrum 29(9):1251–1265. (Published in Chinese)
11. Deng C (2010) Microwave radar and video sensor fusion for vehicle detection and classification technology research. Wuhan University of Technology, (Published in Chinese)
12. Yin H, Pei NS, Yu L (2018) Application of UAV aerial survey technology in highway engineering. J China Foreign Highway 38(2):1–5. (Published in Chinese)
13. Zhao JH, Xu BK, Wang BX (2017) Current status and prospects of research on UAV-based traffic monitoring. J Comput Prod Circulat 9:125. (Published in Chinese)
14. Song SW (2014) Mini unmanned aerial vehicle application and research for expressway monitoring system. Shijiazhuang Railway University. (Published in Chinese)
15. Chegn L (2014) Application of UAV in the field of traffic emergency command. China ITS J 4: 128–130. (Published in Chinese)
16. Reng T, Zhao SJ, Cheng R et al (2015) Design of navigation system based on unmanned helicopter and intelligent vehicle. J Shenyang Univ (Nat Sci) 27(5):385–389. (Published in Chinese)
17. Study on traffic parameters extraction in unmanned aerial vehicle videos. Chongqing Jiaotong University (2017). (Published in Chinese)
18. Jiang SJ, Luo B, Liu J et al (2017) Real-time vehicle detection based on UAV. Bull Survey Map (S1):164–168. (Published in Chinese)
19. Sheng LX, Liu J (2017) Research on highway emergency response based on UAV. Electron World 12:131–133. (Published in Chinese)
20. Tong YX, Yuan Y, Cheng YR et al (2017) Survey on spatiotemporal crowdsourced data management techniques. J Software 28(1):35–58. (Published in Chinese)
21. Zhai ZQ, He L (2014) Exploration of the traffic information service pattern based on user generated content. In: The 9th China intelligent transportation annual conference. (Published in Chinese)
22. Cheng X (ed) (2017) Research on the parking information service pattern based on crowdsourcing. J Lingnan Normal Univ 38(3):100–104. (Published in Chinese)
23. Chui QQ (2015) A study on the data model and scheme of intelligent traffic guidance based on crowdsourcing map. Beijing University of Posts and Telecommunications. (Published in Chinese)
24. Lin L (2014) Crowdsourcing model, the general trend. People's Daily (2014-4-4). (Published in Chinese)

Chapter 3
Traffic Data Management Technology in ITS

MA Xiaolei, LUAN Sen, and YU Xiaofei

The basic purpose of data processing is to extract and deduce some specific valuable and meaningful data from many disordered and incomprehensible data. This chapter will focus on traffic data acquisition, cleaning, storage, and data mining technology.

3.1 Data Cleaning Technology

3.1.1 Importance of Data Cleaning

With the development of information processing technology, all walks of life have established a lot of computer information systems, accumulating a lot of data. In order to effectively support the daily operation and decision-making of organizations, data is required to be reliable and accurate and can accurately reflect the real-world situation. Data is the basis of information, and good quality of data is the basic condition for the effective application of all kinds of data analysis (e.g., OLAP, data mining, etc.). People often complain that data is rich, but information is poor. The reasons are as follows: one is the lack of effective data analysis technology; the other is the poor quality of data, such as the wrong data input, the different representation methods caused by different sources of data and the inconsistency between data, etc. These reasons lead to many "dirty data" existing in the raw data. The main manifestations include spelling problems, printing errors, illegal values, null values, inconsistent values, abbreviations, multiple representations of the same entity (repetition), noncompliance with referential integrity, etc.

MA Xiaolei (✉) · LUAN Sen · YU Xiaofei
School of Transportation Science and Engineering, Beihang University, Beijing, China
e-mail: xiaolei@buaa.edu.cn

© Tsinghua University Press 2022
W. Yunpeng et al. (eds.), *Intelligent Road Transport Systems*,
https://doi.org/10.1007/978-981-16-5776-4_3

The purpose of data cleaning (data cleansing, data scrubbing) is to detect the error and inconsistency in data and then eliminate or correct them so as to improve the quality of data. Data cleaning can also be seen from its name as "washing away the dirty data." It is the last procedure to find and correct the identifiable errors in the data, including checking the data consistency, dealing with invalid values and missing values, etc. Because the data in the data warehouse is a set of data for a certain topic, and these data are extracted from multiple business systems and contain historical data, so we cannot avoid the problem that some data are wrong and some data conflict with each other. These wrong or conflicting data are obviously unwanted, which is called "dirty data." To wash out "dirty data" according to certain rules is data cleaning. The task of data cleaning is to filter those data that do not meet the requirements and submit the filtering results to the business department to confirm whether to filter them out or to extract them after correction by the business department. There are mainly three types of data that do not meet the requirements: incomplete data, wrong data, and duplicate data. Data cleaning is different from questionnaire audit. Data cleaning after input is usually completed by computer.

Data is the soul of the real world. Whether it is the increasingly popular risk investment in the market or the accurate advertisement on the Internet, both of them depend on every piece of data. Especially in the field of transportation, data is one of the core competitiveness. There are all kinds of data in the real world. The traffic managers and passengers have their own data, which represent the department of transportation management and the individual travel themselves to some degree. Correct data play an important role in improving travel efficiency, making long-term strategies, making correct decisions, and maintaining efficient development of transportation. In the research of data cleaning, a variety of data quality management models are proposed, which implement data quality control from different levels. At the same time, many effective cleaning algorithms are also proposed to solve the problems of data quality.

In practice, there is no good solution to deal with the data quality problems existing in business data. We can only check and process the "dirty data" by writing a complex database language. This method is not only difficult but also inconvenient for later maintenance, which is unimaginable for users who are not familiar with the database language. Data processing must start from data cleaning, according to the characteristics of the existing decision-making system, develop the methods to improve the quality of data, then put forward the steps to implement data cleaning, and finally realize a data cleaning framework that can automatically realize the cleaning of duplicate data, association error data, and dictionary data. This realizes the effective control of data quality and provides data quality guarantee for system decision-making, scheme implementation, and further data mining in the future.

Theoretically, data does not exist in isolation. There are various constraints among data, which describe the relationship among data. Data must be able to meet the association among them, but not contradictory. The authenticity, completeness, and self-consistency of data are the attributes of data itself, which is the basis of ensuring data quality.

This constraint can be found in many examples. There is always an intelligence room for war-themed films and TV works to collect information gathered from all sides, and then the staff will analyze and summarize it to identify whether the information is correct and can be adopted and finally make reasonable assumptions and analysis on the basis of the information to support the war decision-making. If the source of information is wrong, the impact will be huge and even directly lead to the failure of the war. Fundamentally, it is also a matter of data quality. Thus, people have paid great attention to the problem of data quality. Nowadays, this problem is more important; whether business or investment analysis, data quality is crucial. The same is true in the field of transportation, from the construction plans of roads to the statistics of traffic flow and passenger flow of each road, which are inseparable from the data quality. It can be said that high-quality data can guide the correct decision-making and promote the decision-making behavior of people in a good direction.

At present, the historical data in the field of transportation does not consider the demand of integration, analysis, and decision-making for future work, so a large number of the data cannot meet the quality requirements. High-quality traffic data is not only the basis of ITS (intelligent transportation system) its effective function, but also the basis of road planning and design, traffic signal optimization, and traffic information release and also the premise of big data analysis. Without high-quality data, the results of big data analysis cannot reflect reality, so there is no significance of analysis. The big data platform has different data sources involving many departments such as road traffic, public transportation, external transportation, and special event transportation. Besides, the data generation standards are not unified, so the first problem faced by the construction of transportation big data platform is the data quality. In order to solve the problem of data quality, it is necessary to unite multiple departments associated with transportation and formulate a unified strategy in order to achieve the basic requirements of data analysis.

3.1.2 The Main Content of Data Cleaning

The concept of data quality needs to be discussed from multiple dimensions. Generally speaking, the dimensions of data quality include integrity, consistency, accuracy, and timeliness. The quality of data is also evaluated from the four dimensions. From a certain point of view, data, as a trusted source of specific applications, needs to be provided to the right people at right time and right place to make the right decisions. The research on data cleaning is orthogonal to the work on view integration. Our research also assumes that the conflict at the concept/pattern level has been resolved and that a global reconciliation architecture has been obtained. However, data reconciliation at the instance level is faced with completely different challenges and difficulties. When it is necessary to integrate data from multiple sources, such as in data warehouse, federated database system, or global network information system, data cleaning becomes more meaningful. This is because data sources often contain different forms of redundant data. In order to obtain accurate and consistent data, it is necessary to merge different forms of data and eliminate duplicate data. However, it is very challenging to get high-quality data. Only meeting several evaluation criteria of data quality alone cannot guarantee satisfactory results.

In general, data missing can be mainly divided into the following types:

1. incomplete data

    This kind of data mainly refers to the data that lack some information, such as the name of the supplier, the name of the branch company, and the region information of the customer. For this kind of data, it can be written into different Excel files according to the missing content and submitted to the customer. It is required to complete it within the specific time and then write it into the data warehouse.

2. Error data

    The reason for this kind of error is that the business system is not robust enough. It is directly written into the database without judgment after receiving the input. For example, the numerical data is input into full-width numeric characters, there is a "enter" operation after the string data, the date format is not correct, the date is out of bound, and so on. This kind of data also needs to be classified. For problems similar to full-width characters and invisible characters before and after the data, we can only find them by writing SQL statements and then ask the customer to extract them after the business system corrects them. Errors such as incorrect date format or date out of bounds will lead to ETL running failure. These errors need to be picked out in the business system database in the form of SQL statements and submitted to the business department for correction within a time limit. After correction, they can be extracted.

3. Duplicated data

    For this kind of data, especially in dimension tables, all the duplicate data can be exported for customers to confirm and sort out.

    Data cleaning is an iterative process, which cannot be completed in a few days but can only continuously find and solve problems. Customers are generally required to confirm whether to filter or correct the data. For filtered data, write it to an Excel file or write it to a data table. In the early stage of ETL development, an email of filtered data can be sent to the business department every day to urge them to correct errors as soon as possible. At the same time, it can also be used as the basis for data verification in the future. Data cleaning needs to pay attention not to filter out the useful data, for each filter rule should be carefully verified and require customer confirmation.

    The problem of data quality can be traced back to the census of the United States in the 1950s.

Abuse of abbreviations

    The abuse of abbreviations will lead to ambiguity, confusion, and reduced quality of data.

Data input error

The data input error is completely human error, which can be improved to a certain extent under certain specifications.

Embed control information into data

For example, embedding printer control commands that format the output in the data domain is difficult to identify.

Different phrases with the same meaning

For example, the acronym ASAP is similar to the phrase "as soon as possible" in meaning. Such problems are similar to abbreviation abuse.

Similar or duplicate records

Similar or duplicate records will not only increase the load of database storage data but also greatly reduce the quality of data. Especially in data mining, similar or duplicate records can easily lead to the failure of establishing the mining model. Therefore, how to accurately and efficiently identify the similar/duplicate data in the data source is considered to be one of the problems to be solved in the research of data cleaning.

Missing value

There are many reasons for missing values, which can be divided into subjective and objective reasons. Subjective reasons are mainly human errors or deliberate concealment, such as refusing to disclose privacy or deliberately omitting to fill in key information. The objective reasons are mainly mechanical reasons, such as the failure of data storage or loss caused by the failure of the storage device or the failure of data collection for a certain period of time caused by the failure of the collection device.

The changes in spelling

Different units

Different units lead to different meanings of the same data in different unit scales.

Invalid code

Invalid coding will lead to meaningless data and even cause data anomalies, such as association failure.

The above listed are the main reasons for data quality problems, but in actual operation, there are many reasons that cause "dirty data." Due to the limit of length, this section will not list them. In short, to solve the problem of data quality, we should start by analyzing these main reasons.

According to different task requirements and characteristics of the environment, the process of data cleaning is different. According to the summary of general data cleaning tools, the general process of data cleaning can be divided into four steps, as shown in Fig. 3.1.

1. Analyze the characteristics of the data. To solve the problem of data quality, we should first analyze the causes of the data quality problem and the characteristics of data sources. The main task of this step is to summarize the characteristics of the data in order to prepare for the formulation of cleaning rules. In addition to domain knowledge, sample data can also be analyzed by manual analysis or data

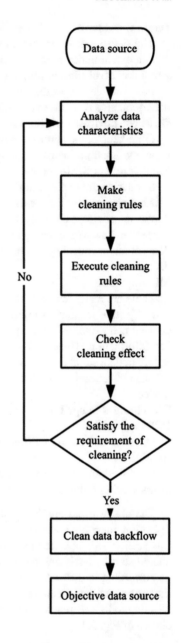

Fig. 3.1 The steps of data cleaning

analysis programs. Through this step, we can know which specific data quality problems may exist in the data source and provide the basis for the next step of making cleaning rules.

2. Make cleaning rules. After summing up the characteristics of data sources, the corresponding cleaning rules are made according to the existing cleaning algorithms. Generally speaking, there are four cleaning rules: inconsistent data

Table 3.1 The common methods of data cleaning

Cleaning methods	Description
Format	According to the standard format of data definition, format some inconsistent or nonstandard data
Merge/delate	Merge duplicate records according to business rules and delete duplicate data
Replace	Replace noncompliant values with compliant values
Split	Single attributes can be divided into multiple attributes or combined into one attribute, with the main purpose of eliminating pattern conflicts

detection and processing, null value detection and processing, similar/duplicate record detection and processing, and illegal value detection and processing, as shown in Table 3.1.

3. Execute cleaning rules. The most important step in data cleaning is to execute cleaning rules. The implementation of cleaning rules generally has a sequence. Due to the strong relevance of domain knowledge and environmental dependence of data cleaning, it is difficult to form a unified general standard, and the data quality problems are difficult to summarize because they are scattered and complex, so different cleaning rules can only be formulated according to different problems.

4. Check the cleaning effect. This is a reviewing step in cleaning. According to the generated cleaning report, check the data cleaning results, find the problems in the cleaning process, manually deal with the problems that cannot be handled by the program, evaluate the cleaning effect, and improve and optimize the rules and algorithms that do not meet the cleaning requirements. Then clean again as needed until the requirements are met. Data cleaning is a process that needs to be iterated and repeated many times. Only through continuous comparison and improvement can we get the ideal results.

3.1.3 The Main Method of Data Cleaning

The principle of data cleaning must be based on the analysis of characteristics of data source, and the reasons for the data quality problems must be analyzed deeply by applying the idea of backtrack. After analyzing every link of data flow, the corresponding methods and schemes are summarized continuously, and a theoretical cleaning model is established. Gradually, the cleaning algorithms and schemes can be applied to the actual cleaning work. These algorithms, strategies, and schemes are applied to the identification and processing of data, and the control of data quality is realized. Data cleaning is generally divided into four cleaning methods: full manual, full computer, human–computer synchronous combination, and human–computer asynchronous combination.

1. Manual cleaning. The characteristic of this cleaning method is slow with high accuracy. It is generally used in data sets with a small amount of data. In large

data sets, due to the limitation of the human, the speed and accuracy of cleaning will be significantly reduced. Therefore, this cleaning method is generally used in some business systems of a small company.

2. Computer cleaning. The advantage of this cleaning method is that the cleaning is completely automated, which frees people from complicated logical tasks to complete more important things. In this way, the cleaning program is written according to the specific cleaning algorithm and cleaning scheme so that it can automatically execute the cleaning process. The disadvantage is that the implementation is difficult, and the later maintenance is difficult.

3. Human–computer synchronous cleaning. Some special cleaning requirements cannot be well realized only by cleaning program, which requires the synchronous cooperation of humans and computers. By designing an interface for human–computer interaction, when the cleaning program cannot deal with the problems, the human intervention is used to deal with it. This method not only reduces the complexity and difficulty of programming but also does not require a lot of manual operation. However, the disadvantage is that people must participate in the cleaning process in real time.

4. Human–computer asynchronous cleaning. The principle of this kind of cleaning is basically similar to the human–computer synchronous cleaning. The only difference is that when encountering problems that cannot be handled by the program, it does not directly require human participation but records the abnormal situation in the form of a report and then continues the cleaning work. According to the cleaning report, the corresponding treatment can be carried out by humans in the later stage. This is a very feasible cleaning method, which can not only save human resources but also improve the cleaning effect.

Generally speaking, data cleaning is the process of condensing the database to remove duplicate records and converting the rest into a standard acceptable format. The standard model of data cleaning is to input the data to the data cleaning processor, clean up the data through a series of steps, and then output the cleaned data in the desired format (as shown in Fig. 3.1). Data cleaning from the aspects of accuracy, integrity, consistency, uniqueness, timeliness, effectiveness of data to deal with the missing value, value out of range, inconsistent code, duplicate data, and other issues.

Data cleaning is generally for specific applications, so it is difficult to summarize the unified methods and steps. But according to the different types of data, the corresponding data cleaning methods can be given.

1. Method to clean incomplete data (i.e., missing value)

In most cases, missing values must be filled in manually (i.e., cleaned up manually). Of course, some missing values can be derived from the own data sources or other data sources, which can replace the missing values with average, maximum, minimum, or more complex probability estimates so as to achieve the purpose of cleaning up.

For example, the recovery method is based on the trend of historical data obtained by smoothing the collected historical data of multiple days. Through this method, the trend of historical data of any day can be obtained directly, and the obtained data changes smoothly with small fluctuation after recovering.

The recovery method based on weighted estimation of measured data and historical data not only considers the influence of the previous traffic flow in the actual environment on the later period but also considers the change characteristics of the historical traffic flow in the current period so as to further reduce the random fluctuation in the actual road traffic environment.

The data recovery method based on adjacent time does not need to rely on a large number of historical data, the calculation method is simple, and the recovery effect is satisfactory.

2. Error value detection and solution

Statistical analysis can be used to identify possible error values or outliers, such as deviation analysis and identification of values that do not comply with distribution or regression function; simple rule base (common sense rules, business specific rules, etc.) can also be used to check data values, and constraints between different attributes and external data can be used to detect and clean up data.

3. Detection and elimination of duplicate records

Records with the same attribute value in the database are considered duplicate records. By judging whether the attribute values between records are equal, the equality of records is detected. The equal records will be merged into one record (merge/clear). Merging/clearing is the basic method of eliminating duplication.

4. Detection and solution of inconsistency

Data integrated from multiple data sources may have semantic conflicts. The data inconsistency is inside or between data sources. Constraints of integrity can be defined to detect inconsistencies, and relationships can be found by analyzing data to keep data consistent. Currently, the data cleaning tools can be divided into the following three categories.

The data migration tool allows you to specify simple conversion rules, such as replacing the string "gender" with "sex." The Prism Warehouse of Sex Company is a popular tool, which belongs to this category.

Data cleaning tools use specific domain knowledge (such as postal address) to clean data. They usually use syntax analysis and fuzzy matching technology to clean multisource data. Some tools can indicate the relative cleanliness of the source. Tools "Integrity" and "Trillum" fall into this category.

Data audit tools can find rules and relationships by scanning data, so this kind of tool can be regarded as a deformation of data mining tools.

With the continuous development of public transport big data mining, data cleaning has gradually become a research hotspot. Its main task is to detect and recover "dirty data" (eliminate errors or inconsistent data) and solve data quality problems. There are many kinds of traffic big data. Different kinds of data have different structures and problems. This section will introduce the data cleaning process of common bus big data.

Example 3.1 The data cleaning of bus smart card.

The goal of data mining for the bus smart card database is specific. One is to assist the manager to make a decision, and the other one is to provide data for the bus plan.

Bus data preprocessing is to filter and correct the data in the established data warehouse and get the available data information for subsequent data mining through simple statistics; data mining is to analyze the preprocessed with data mining algorithm. Using public transport data to do preprocessing needs to focus on the following data issues.

1. Data missing

   The occurrence of data missing will have a bad impact on the later data processing; in particular, the missing of key data may require a different algorithm design, even leading to the analysis that cannot be carried out. Generally speaking, the lack of bus smart card data is rare. For the absence of data, it is important to consider whether other similar data can be used instead. If there is no alternative data, it is possible to calculate based on experience or other similar data.

2. Data error

   There may be some error data on the bus smart card, such as the input of time format as 24:30:00, etc. In order to ensure that data analysis does not produce wrong results, it is necessary to check the validity of data, delete redundant data, and find data that may generate noise to the analysis results by classification and filter. If the transaction time is 2:35:46 am, this is obviously wrong. According to the investigation, it is due to the hardware or technical problems of the smart card and ticket gate that makes individual smart card records errors. But the probability of such an error is small, so deleting these records will not have a great impact on the analysis results, so these error records can be deleted.

3. Redundant data

   The redundant data is duplicate data, in fact. The existence of redundant data actually improves the accuracy of data. When some data is wrong, it can recover information through redundant data. But redundant data will bring some trouble in later processing, even leading to wrong analysis. For example, repeated smart card data records will lead to a high statistical passenger flow. For redundant data, it should be treated according to the specific analysis purpose of data, and the data storage consumption should be considered.

4. Consistency of data

   The data of the public transportation system involves multiple equipment and departments. However, the goals of data collection of different equipment manufacturers and departments are different, which makes the data with the same meaning are different among different equipment manufacturers and departments. Such inconsistency may be due to the difference in data precision, data unit, data storage format, or inconsistency of data definition.

5. Obsolete data

As time goes by, some data with a long time may be invalid. Whether the data is out of date is a problem relative to the analysis goal. For obsolete data, it can be processed separately and stored separately.

6. Delete useless fields

There are many fields in the data table, some of which are meaningless to the data analysis of this book. Therefore, these redundant fields can be deleted to make the data analysis faster. The smart card number, transaction date, transaction time, smart card type, vehicle number, line number, and record number are significant for data analysis. So, card number, city number, industry number, card physical type, monthly ticket type, receivable amount, driver number, and ticket supervisor number make the analysis easier and faster.

Example 3.2 The data cleaning of GPS data.

GPS data is mainly collected by the GPS in vehicle, and the data format may not be consistent with the data format in the actual demand for research, so it is necessary to transform the GPS data format.

GPS data preprocessing mainly includes the following aspects.

1. If the analysis of line number is not correct, the SIM card number and vehicle information table should be used to determine the correct line number.
2. The marks of the direction of upward and downward are not correct, so it is impossible to judge the moving direction of the bus.
3. The direction information should be multiplied by 10. For example, 21 corresponds to $210°$.
4. Coordinate translation is needed to match the GPS data of the bus line; longitude = longitude + 0.006, and latitude = latitude + 0.001.
5. The number of information recorded in each GPS file is limited, which cannot match the information of the IC card.
6. GPS time is GMT, which needs to add 8 h to the original time.

Taking the GPS data on April 7, 2010 as an example, 5567 GPS data files were received, including 4274 vehicles equipped with GPS (excluding redundant files and error files). On these buses equipped with GPS, 3725 vehicles can identify the line number, and the line coverage rate of GPS vehicles is 87.2%. Among these vehicles with line number identification, 2125 vehicles with fixed price account for 57% of all GPS vehicles with line number identification, 1515 vehicles with step price account for 40.7% of all GPS vehicles with line number identification, and 85 vehicles cannot identify the pricing method, accounting for 2.3% of all GPS vehicles with line number identification. Through the GPS vehicle number and line number, we can match with the records in the smart card database. We find that about 2548 buses equipped with GPS equipment can match the records in the smart card database, and the matching rate is 68.4%.

7. When the vehicle speed is less than 3 km/h, the GPS positioning signal will drift, generally waiting for a red light near the intersection, road congestion, or temporary parking. At this time, GPS navigation data will cause an error, which will affect the normal navigation. Theoretically speaking, the positioning point of the vehicle at this moment should basically remain stationary, but the actually

measured positioning point does not stop at a point but falls in a circle centered on the actual position of the vehicle. In addition to detecting the instantaneous speed of the receiver, we can also calculate the distance between the current positioning point and the previous positioning points, supplemented by the heading angle as the basis of judgment for this error. If they continue to be less than the upper limit of GPS normal positioning error, it means that the vehicle is basically in the stop or at a very low speed. At this time, in order to prevent the wrong matching caused by signal drift, this positioning data will be regarded as invalid data, and road matching processing will not be carried out. The matching point corresponding to the current positioning point can be regarded as the current real position of the vehicle, and the subsequent positioning points will not be matched until three distance values in front of it are more than the upper limit of the error when GPS normal positioning.

8. When the vehicle moves to the position under the bridge or tunnel that affects the signal reception, the received signal will be temporarily interrupted. In this case, interpolation should be used for data compensation in the current driving direction of the vehicle until the normal GPS positioning data is received. The so-called "jump point" generally means that the distance between a certain point or some points and the previous point exceeds the maximum possible movement distance. When GPS signal is blocked by high buildings or other interferences, this situation often occurs. In order to prevent the wrong map matching, the method to deal with this problem is to discard the positioning value of the point, filter it out, and then use the interpolation derivation method to simulate the GPS data reception in the current driving direction. If the GPS data is still moving far away, and the relative distance between the front and back points is within the allowable range, there is no abnormal speed, the interpolation should be stopped, and the current GPS data should be treated as the correct data for normal processing. This book uses the average speed to judge whether it is a jump point. When the average speed between this point and the previous point is greater than 200 km/h, it is considered that this point is a jump point. The judgment of jump point mainly considers the effective range of the speed of the vehicle. When the speed of the vehicle is greater than the maximum theoretical value, the point is removed for interpolation. When the velocity is greater than a certain value, the angle between two segments composed of three adjacent locating points with the middle point as the vertex is greater than 60°, but the middle point is not a node; the point should be removed for interpolation.

3.2 Data Storage Technology

3.2.1 Data Format

Data format is the format of data stored in files or records. It can be in the form of numerical value, character, or binary number. It is described by data type and numerical length. The data format should meet the following conditions.

1. Ensure that all information required for recording is stored.
2. Improve the storage efficiency and ensure the full utilization of storage space.
3. The format is standardized to ensure the data exchange between relevant data processing systems.

According to the characteristics of data record length, it is generally divided into fixed-length format and variable-length format. The records in the former file have the same length, while the length in the latter file is determined by the length of the record value.

The data format in the computer field is not a new concept, but the data format in the transportation field has more abundant meaning. Before the concept of big data appeared, dynamic traffic data was mainly based on relational data tables, relying on Oracle, IBM DB2, SQL server, and other relational databases to convert the data of fixed detection equipment and mobile detection equipment on the road into table files with standard structure. However, with the increasing means of collecting traffic information, especially video images, voice records, traffic websites, smart phones, and other means of obtaining traffic information, the stored traffic data add text files, video files, audio files, pictures, websites, and other semistructured and unstructured data besides the traditional table.

The complexity of data format has brought about changes in the way of data organization, management, and usage. Relying on a relational database alone cannot meet the requirements of traffic data analysis of big data. Therefore, it is necessary to introduce a distributed file system and nonrelational database as beneficial supplements. The format of urban traffic big data is an eternal problem that data analysis needs to face. With the rapid development of the Internet of things, cloud computing, and mobile Internet, a large number of intelligent terminal devices will have the ability to produce data. Various types of traffic data will be integrated and refined, finally processed into information products to serve the whole society.

At present, the complete cleaning processing of the whole traffic big data is roughly divided into three steps, as shown in Fig. 3.2. First of all, the basic data collected by the transportation system is stored in the database, such as smart card data, GPS data, etc., then the data needs to be exported to the intermediate environment, and different intermediate tools are selected according to the size of the file, and then some professional data cleaning work is carried out by using the data processing software. Of course, this process also has great differences in different data cleaning processes. Not all cleaning processes need all the steps. It needs to be analyzed according to the data quality, data structure, and other aspects. For

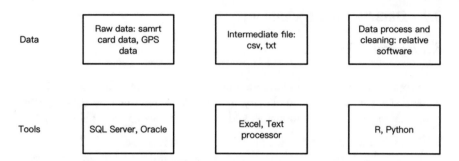

Fig. 3.2 The processing of data cleaning

example, when the lack of data is less or the rule is more obvious, you can directly use the database query for cleaning. When the lack of data is serious, we need to use data cleaning tools for professional processing. Therefore, this process needs specific analysis. The following is a brief introduction of the common data structures in the tools of each department.

1. **Data in Original Database**

Database is a general data processing system for a department or an application area. It stores the collection of relevant data belonging to enterprises, institutions, groups, and individuals. The data in the database is established from the global view and is organized, described, and stored according to a certain data model. Its structure is based on the natural relationship between data, so it can provide all the necessary access paths, and the data is no longer for a certain application, but for the whole organization, with the overall structural characteristics.

The data in the database is established for the purpose of many users to share their information, which has gotten rid of the restrictions and constraints of specific procedures. Different users can use the data in the database according to their own demand, and multiple users can share the data resources in the database at the same time; that is, different users can access the same data in the database at the same time. Data sharing not only meets the requirements of users for information content but also meets the requirements of information communication between users. Strictly speaking, the database is a long-term storage in the computer, organized, and shared data set. The data in the database is organized, described, and stored together with a certain data model, which has the characteristics of less redundancy, high data independence, and easy extension, and can be shared by multiple users in a certain range.

This kind of data set has the following characteristics: it is not repeated as much as possible and serves a variety of applications of a specific organization in an optimal way. Its data structure is independent of the application program that uses it. In addition, deletion, modification, and query of data are managed and controlled by unified software. From the history of development, database is the advanced stage of data management, which is developed from the file management system.

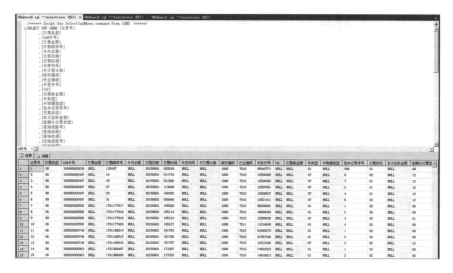

Fig. 3.3   The data of smart card of Beijing subway

(a) The real client/server architecture.

(b) Graphical user interface makes system management and database manage-ment more intuitive and simpler.

(c) Rich programming interface tools provide users with more choices for programming.

(d) SQL Server is fully integrated with Windows NT, which makes use of many functions of NT, such as sending and receiving messages, managing login security, etc. SQL Server can also be well integrated with Microsoft BackOffice products.

(e) It has good scalability and can be used in laptops running Windows 95/98 or large multiprocessors running Windows 2000.

(f) With the support of web technology, users can easily publish the data in the database to the web.

(g) SQL Server provides data warehouse, which is only available in Oracle and other more expensive DBMS.

Common basic traffic data are stored in the database, and different data formats have obvious differences. For the same type of data, different statistical companies will have obvious differences. Figures 3.3, 3.4, 3.5, and 3.6 show several common data formats.

2. **Intermediate Basic Data**

When the data structure is complex or data cleaning is difficult, we need to use professional data processing software. Usually, it is difficult to connect the database with professional software directly. Therefore, when the data environ-ment permits, Excel or other software is usually used for data transfer. At this time, the data storage format is usually csv or txt. A delimited file is a text-based file that contains a separator predefined by delimited table data. The separator can

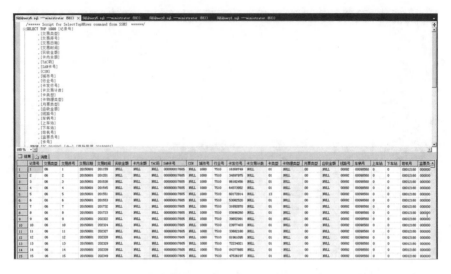

Fig. 3.4 The data of smart card of Beijing bus

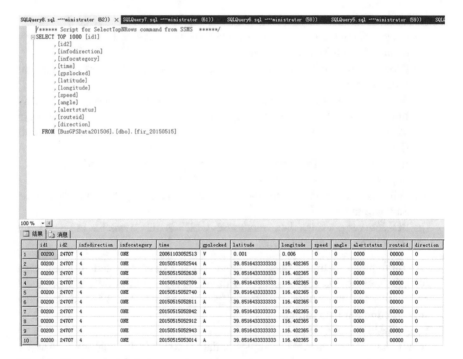

Fig. 3.5 The format of GPS data

be a tab, comma, semicolon, or any non-alphanumeric character. In datasets, there are two types of files with a separator, the comma separator separated by an extended header line. The CSV file may or may not have the first row as the header row, which contains the names of the data columns.

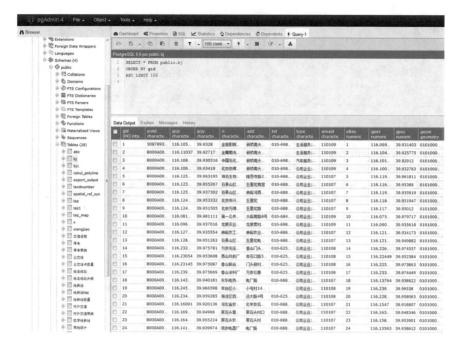

Fig. 3.6 GIS data in PostgreSQL

The CSV file separated by comma is as follows:

```
2.01109E 113, 200011501, 0, 0, 0, FALSE, FALSE, FALSE
2.01109E 113, 200011504, 0, 0, 0, FALSE, FALSE, FALSE
2.01109E 113, 200011506, 0, 0, 0, FALSE, FALSE, FALSE
2.01109E 113, 200028101, 0, 0, 0, FALSE, FALSE, FALSE
2.01109E 113, 200028103, 0, 0, 0, FALSE, FALSE, FALSE
```

The file separated by vertical line is as follows:

```
C5AA|157|01/25/201118:38|02/06/201119:01|02/29/201206:01|LA|LA|
5|SB|34.65|EB/WB|Tuxford St||34.65|SB|Golden State Frwy, Rte
5||On Ramp|Full|BridgeConstruction||All|2|Y|03/01/201122:46|
N||N|
C5TA|304|03/07/201111:21|03/14/201120:01|01/13/201206:01|LA|LA|
5|SB|36.86||Brandford St||36.86|SB|Golden State Frwy, Rte 5||On
Ramp|Full|SlabReplacement||All|1|Y|04/11/201120:01|N|08/10/
201112:33|N|
C5TA|312|03/07/201111:24|03/14/201120:01|01/13/201206:01|LA|LA|
5|NB|37.41|EB|Osborne St||37.41|NB|Golden State Frwy||On Ramp|
Full|SlabReplacement||All|1|Y|03/08/201120:01|N|08/10/201112:
34|N|
C5TA|308|03/07/201111:22|03/14/201121:01|01/13/201206:01|LA|LA|
5|NB|37.41|NB|GoldenStateFrwyRte5||37.41||OsborneSt||OffRamp|
Full|SlabReplacement||All|1|Y|03/07/201120:01|N||N|
```

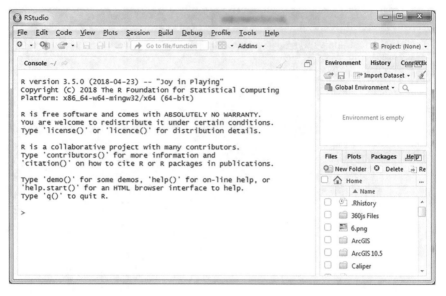

Fig. 3.7 The IDE of R-Studio

3. **Data in Professional Statistics Software**

Usually, when the error data is complex, we need to use professional data processing software to clean the data through a professional algorithm. The commonly used software includes R-Studio, Python, and so on. R-Studio (hereinafter referred to as R) as a statistical analysis software (as shown in Fig. 3.7), it is a statistical analysis and graphical display in one, can run on UNIX, Windows, and MacOS operating systems, and embeds a very convenient and practical help system. Compared with other statistical analysis software, R has the following characteristics:

R is free software, which means that it is completely free and open source. You can download any relevant installation program, source code, package, and documentation from its official website and its image. The standard installation file itself has many modules and embedded statistical functions, which can directly realize many common statistical functions after installation.

R is a kind of programmable language. As an open statistical programming environment, its grammar is easy to understand, and it is easy to learn and master the grammar of the language. After understanding, you can compile your own functions to expand the existing language. That is why its update speed is much faster than that of general statistical software (such as SPSS, SAS, etc.). Most of the latest statistical methods and techniques can be obtained directly in R.

All R functions and data sets are stored in packages. Only when a package is loaded can its contents be accessed. Some commonly used and basic packages have been included in the standard installation files. With the emergence of new statistical analysis methods, the packages included in the standard installation files are also changing with the version update. In the other version of the installation file, the packages include base—R basic module, mle—maximum likelihood estimation module, ts—time series analysis module, mva—multivariate statistical analysis module, survival—survival analysis module, etc.

R is highly interactive. As shown in Fig. 3.7, except that the graphic output is in another window, its input and output windows are both in the same window. If there is an error in the input syntax, a prompt will appear in the window immediately. It has the memory function for the previously entered commands and can be reproduced, edited, and modified at any time to meet the needs of users. The output graphics can be directly saved as JPG, BMP, PNG, and other image formats and can also be directly saved as PDF files. In addition, R has a good interface with other programming languages and databases.

The basic data type of R is not a single scalar, but a vector, defined as a set of index values with the same type. The types of these values define the class vector. There are five typical types of data in R: number, integer, complex, logic, and characteristic. Given the statistical property of R, the number is the most relevant type, and it is a precise type of double precision.

In order to support the numerical algorithm, R uses atomic vector, matrix, and array, which are respectively expressed as 1, 2, and n-dimensional data structures. These data structures require isomorphic data; although statistical analysis can be inferred from a data format using the mathematical expression of these structures, it is not intuitive, especially when users need to include the purpose of nondigital data analysis. To support heterogeneous and complex data, R uses two additional data formats: data frame and list.

Data frame is a data structure similar to matrix. Unlike the matrix structure, the columns of a data frame can contain different types of data.

Example 3.3 Construct a basic matrix by R.

```
> A < - matrix (c(c(2,4,3), c(1,5,9)), nrow= 3,ncol= 2)
> A
     [, 1]   [, 2]
[1, ]   2    1
[2, ]   4    5
[3, ]   3    9
```

Example 3.4 Construct a data frame without specific column name by integrating multiple vectors; the data frame uses the contents of the vector as the default column name.

```
> A < - data.frame(c(2,4,3), c('one', 'five', 'seven'))
> A
c,2..4..3, c..1....5....7..
1    2    "one"
2    4    "five"
3    3    "seven"
```

Example 3.5 Data frame with command columns.

```
> A =data.frame (x=c(2,4,3)), y=c('one', 'five', 'seven'))
> A
x   y
1   2    one
2   4    five
3   3    seven
```

List is a vector structure. Unlike vectors, the elements in a list do not have to follow a generic type, but can be any object. To access an element of a list, use double bracket notation ([[x]]).

Example 3.6 Using R to build a list of multiple data structures: vector, matrix, and data frame.

```
> A < - c(1,2,3,4,5)
> B < - matrix (c(2,4,3,1,5,7), nrow=3, ncol=2)
> C < - data.frame (x=c(1,2,3), y=c("two", "four","six"))
> D < - list (A, B, C)
> D
[[1]]
[1] 1 2 3 4 5
[[2]]
     [,1]   [,2]
[1,]  2     1
[2,]  4     5
[3,]  3     7
[[3]]
x   y
1   1    two
2   2    four
3   3    six
```

Example 3.7 R has a support function called read.table, which can read any delimited file. To see the complete syntax of this function, the user can enter "? Read." The table in R-Studio console will display the complete document of the function, including most of the parameters and examples and will be displayed in the R-Studio window in the lower right corner. The table does not need to be changed. In most cases, the table can be called with a few parameters.

```
> A < - read.table(file= "PasadenaDet_20110930_Sample.csv", sep=",")
> A
V1          V2         V3   V4   V5   V6      V7      V8
2.01109e+13  200011501  0    0    0    FALSE   FALSE   FALSE
2.01109e+13  200011504  0    0    0    FALSE   FALSE   FALSE
2.01109e+13  200011506  0    0    0    FALSE   FALSE   FALSE
2.01109e+13  200028101  0    0    0    FALSE   FALSE   FALSE
2.01109e+13  200028103  0    0    0    FALSE   FALSE   FALSE
2.01109e+13  200028303  0    0    0    FALSE   FALSE   FALSE
```

3.2.2   Data Storage Method

Big data is a concept with rich connotation and denotation, which is not limited to a specific technology, method, or system. However, when people actually accumulate data, organize data, query data, and analyze application data, they need to have a real method, mode, or system, which is what the engineering and technical personnel in the area of information technology and even the transportation field most want to see. Engineers and data analysts are far less interested in ideas than in real systems and tools.

The data storage structure can be obtained by the following four basic storage methods.

1. **Sequential Storage Method**
   The method stores the logical adjacent nodes in the adjacent storage units in physical location, and the logical relationship between nodes is reflected by the adjacent relationship of the storage units. This storage method is called sequential storage structure, which is usually described by the array in a program language. This method is mainly used for the linear data structure. The nonlinear data structure can also be stored by some linearization method.

2. **Linked Storage Method**
   This method does not require that the logically adjacent nodes are adjacent in physical location, and the logical relationship between nodes is represented by pointer fields. This storage method is called linked storage structure, which is usually described by the pointer in the programming language.

3. **Index Storage Method**
   In this method, the node information is stored with an additional index table, which is composed of several index entries. If each node has an index entry in the index table, the index table is called a dense index. If a group of nodes only correspond to one index in an index table, the index table is called a spare index. The general form of an index is keyword or address. Keywords are the data items that can uniquely identify a node. The address of the index in dense index indicates the storage location of the node. The address of the index in sparse index indicates the starting storage location of a group of nodes.

4. **Hash Storage Method**
   The basic idea of this method is to determine the storage address of the node directly according to the keywords of the node.

The aforementioned four basic storage methods can be used alone or combined to store and image the data structure.

Different storage structures can be obtained by using different storage methods for the same logical structure. The choice of the storage structure to represent the corresponding logical structure depends on the specific requirements, mainly considering the convenience of operation and the space-time requirements of the algorithm.

The logical structure, storage structure, and operation of data are a whole, so it is not advisable to understand one aspect in isolation without paying attention to the relationship between them. Storage structure is an indispensable aspect of data structure: different storage structures of the same logical structure can be labeled with different data structure names.

The logical structure, storage structure, and operation of data are a whole, so it is not advisable to understand one aspect in isolation without paying attention to the relationship between them. Storage structure is an indispensable aspect of data structure: different storage structures of the same logical structure can be labeled with different data structure names.

In essence, urban transportation big data is the application of big data concepts and technologies in the transportation area, with more emphasis on user-oriented application services and product generation. However, the basic principles and methods of the underlying database system and operating system are "borrowlism." At present, data management systems for big data applications emerge endlessly in the market and are warmly sought after. Both the old IT famous companies and some cutting-edge companies are advancing bravely in the wave of big data and have developed a large number of system products, but their basic principles and modes are similar, mainly based on MapReduce distributed data file storage and calculation.

3.2.3 Distributed Storage

In order to ensure high availability, high reliability, and economic, big data generally uses distributed storage to store data and uses redundant storage to further ensure the reliability of data. The information storage mode of Hadoop Distributed File System (HDFS) based on Hadoop is a popular data storage structure, as shown in Fig. 3.8. The construction of a cloud storage service system based on HDFS can solve the problem of mass data storage in intelligent transportation and reduce the cost of implementing distributed file system. Hadoop distributed file system is the bottom implementation part of the Hadoop framework, which is an open-source cloud computing software platform. It has the characteristics of a high transmission rate and high fault tolerance. It can access the data in the file system in the form of stream so as to solve the access speed and security problems. In the final analysis, the solution to the data storage problem in the distributed file system is to divide big problems into small ones. When a large number of files are evenly distributed to multiple data servers, the number of files stored in each data server is less. In

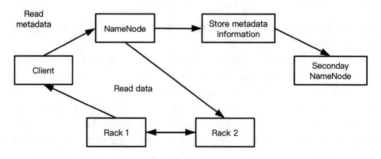

Fig. 3.8 The structure of HDFS

addition, by using large files to store multiple small files, the number of files stored in a single data server can always be reduced to the size that can be solved by a single computer. Large files can be divided into several relatively small fragments and stored on multiple data servers (at present, many local file systems have no problem in supporting large files; for example, ext3 file system uses 4K blocks, and the maximum file size can reach 4 TB, while ext4 can support larger files, which is limited by the storage space of the disk).

In theory, a distributed file system can be composed of a client and multiple data servers. The client decides which data server to store the files to according to the file name. However, once a data server fails, the problem will become complicated. Because the client does not know the message of the down of data server and still connects it for data access, the reliability of the whole system will be greatly reduced, and the client may be very inflexible when deciding the data allocation, because it cannot make different distribution strategies according to the file characteristics.

There are mainly three types of distributed storage:

1. Distributed file system. A large number of unstructured data such as files, pictures, audio, and video are stored. These data are organized in the form of objects. There is no relationship between objects. These data are binary data, such as GFS, HDFS, and so on.
2. Distributed key-value system. It is used to store semistructured data with simple relationship. It provides key-based addition, deletion, modification and query operations, caching, and solidification storage, such as Memached, Redis, DynamoDB, etc.
3. Distributed database system. It stores structured data, provides SQL relationship query language, and supports multiple table association and nested subquery, such as My SQL Sharding cluster, MongoDB, etc.

The main characteristics of distributed storage are as follows:

High availability It refers to the ability of the distributed storage system to provide normal service in the face of various exceptions. The availability of the system can be measured by the ratio of the time when the system stops service and the normal service time. For example, the availability of 99.99% requires that the downtime in a year should not exceed $365 \times 24 \times 60/10,000 = 53$ min.

High reliability It mainly refers to the data security index of the distributed system, data reliability without loss, mainly using multicomputer redundancy, single disk RAID, and other measures.

High scalability It refers to the ability of distributed storage system to improve the storage capacity, computing, and performance of the system by expanding the size of cluster servers. With the increase of business volume, the performance requirements of the underlying distributed storage system are higher and higher. Automatically adding servers can improve the service capability, which is divided into scale up and scale out. The former refers to adding and upgrading server hardware, and the latter refers to increasing the number of servers. To measure the scalability, the cluster should have linear scalability, and the overall performance of the system has a linear relationship with the number of servers.

Data consistency It refers to the data consistency among multiple copies of the distributed storage system, including strong consistency, weak consistency, final consistency, causal consistency, and sequential consistency.

High security The distributed storage system is protected from malicious access and attack, and the stored data is protected from theft. The Internet is open. Anyone can visit the website at any time, any place, and by any means. There should be corresponding solutions to all kinds of existing and potential attacks and theft means.

High performance The common indicators to measure the performance of the distributed storage system are system throughput and system response delay. System throughput is the total number of requests that can be processed in a period of time, which can be measured by QPS (query per second) and TPS (transaction per second). The response delay of a system refers to the time consumed by a request from sending to receiving, which is usually measured by the average delay. These two indicators are often contradictory. It is difficult to achieve low latency in pursuit of high throughput. If low latency is pursued, the throughput will be affected.

High stability This is a comprehensive index to evaluate the overall robustness of the distributed storage system. For any exception, the system can face it calmly. The higher the system stability, the better.

The key technologies of distributed storage are as follows.

1. Metadata management. In the big data environment, the volume of metadata is also very large, and the access performance of metadata is the key to the performance of the whole distributed file system. Common metadata management can be divided into centralized and distributed metadata management architecture. Centralized metadata management architecture uses a single metadata server, which is easy to implement, but there are some problems such as single point of failure. The distributed metadata management architecture disperses the metadata on multiple nodes, which solves the performance bottleneck of the metadata server and improves the scalability of the metadata management architecture. However, the implementation is complex and will cause the problem of metadata consistency. Besides, there is another distributed architecture without a metadata server, which organizes data through online algorithms and does not need a dedicated metadata server. However, the architecture is difficult to ensure data consistency, and its implementation is more complex. Its file directory traversal is inefficient and lacks the global monitoring and management function of the file system.

2. System elastic extension technology. In the big data environment, the increase of data size and complexity is pretty rapid, which requires the high performance of system extension. To realize the high scalability of the storage system, we must first solve two important problems, that is, the allocation of metadata and the transparent migration of data. The allocation of metadata is mainly realized by

static sub-tree partition technology, while the latter focuses on the optimization of the data migration algorithm. In addition, big data storage system has a large scale and high failure rate of nodes, so it also needs to complete certain adaptive management function. The system must be able to estimate the number of nodes according to the amount of data and the workload of calculation and dynamically transfer the data between nodes to achieve load balancing. At the same time, when the node fails, the data must be able to be recovered through mechanisms such as copy, which cannot affect the upper application.

3. Optimization technology within the storage hierarchy. When building a storage system, it needs to be considered based on cost and performance. Therefore, the storage system usually uses multilayer and different cost-effective memory components to form the storage hierarchy. Big data is large in scale, so building an efficient and reasonable storage hierarchy can reduce the system energy consumption on the premise of ensuring system performance. The principle of data locality of reference can optimize the storage hierarchy from two aspects. From the angle of improving performance, hot data can be identified and cached or prefetched by analyzing application characteristics. The access performance is improved by an efficient cache prefetching algorithm and a reasonable cache capacity ratio. From the perspective of cost reduction, the cold data with low access frequency is transferred to low-speed and cheap storage devices by using the information lifecycle management method. It can greatly reduce the construction cost and energy consumption of the system on the basis of small sacrifice of the overall performance of the system.

4. Storage optimization technology for application and load. The traditional data storage model needs to support as many applications as possible, so it needs to be universal. Big data has the characteristics of large-scale, high dynamic, and fast processing. The general data storage model is usually not the model that can improve the application performance most, and the attention of big data storage system to the upper application performance is far more than the pursuit of generality. Aiming at application and storage optimization is to couple data storage with application, simplify or expand the function of distributed file system, and customize and deeply optimize file system according to specific application, specific load, and specific computing model so as to achieve the best performance of application. This kind of optimization technology manages more than 10 billion bytes of big data on the internal storage systems of Internet companies such as Google and Facebook and can achieve very high performance.

The urban traffic big data platform needs powerful computing power to realize analyzing and processing large, complex, and disordered traffic data. Traffic data modeling and spatiotemporal indexing based on big data platform, historical data mining, distributed processing and fusion of traffic data, and traffic flow dynamic prediction all need the distributed computing capability of big data platform, namely high-performance parallel computing model MapReduce. MapReduce is a programming model for massive data processing, which simplifies the complex data

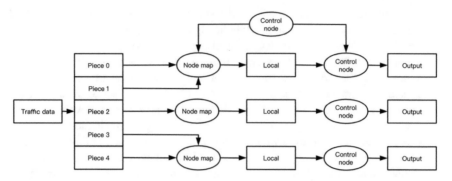

Fig. 3.9 Distributed storage of traffic data

processing and calculation process. The data processing process is divided into "map" stage and "reduce" stage, and its execution logic model is shown in Fig. 3.9.

MapReduce achieves reliability by distributing large-scale operations on datasets to network nodes. Each node will periodically send back the completed work and status updates. If a node keeps silent for more than a preset time interval, the master node records the node status as dead and then sends the tasks belonging to this node before to other nodes. MapReduce is a parallel computing model based on the data partition, which has good fault tolerance.

If distributed computing is to use the network to complete data sharing and computing, then local computing is the traditional database-centric computing mode; distributed computing undoubtedly has great potential and advantages.

The local computing mode of the database center is to concentrate the processing power and load of the software system on one or two database servers. If we want to improve the computing capacity, we can only continuously improve the hardware level of database servers, from ordinary dual-core and multicore PCs to minicomputers, to medium-sized computers and supercomputers. With the improvement of processing capacity, the construction cost of the system is also higher and higher.

There is a sharp contrast between the two computing modes. Distributed computing manages all data and computing tasks through software, and resources are shared through the network. Local computing transfers all the computing resources to the computer in the computing center for processing. Through comparison, we can find that both of them are multitask management, but one is centralized multitask management, the other is distributed multitask management. In the case of huge amount of data, each has its own advantages and disadvantages, which can form complementary advantages. It is necessary to select the appropriate technology according to the needs of practical application.

3.3   Data Mining and Visualization

3.3.1   Data Query

The direct purpose of data query is to obtain the data set that users expect to see from the massive data by the method of querying. Generally, the data will be managed with the help of database software (such as SQL server of Microsoft and DB2 of IBM). This section takes SQL Server as an example to briefly introduce data query. Given a sample set "taxi_traject," it describes the taxi trajectory, including vehicle ID, event, number of passengers, time, latitude and longitude, speed, direction, and status.

1. **Simple Query**
   The purpose of simple query is mainly to check the variables in the database.

Example 3.8 "Select*from taxi_traject" to find all the samples

Example 3.9 "Select top 10*from taxi_traject" to find the top 10 rows in the sample table, as shown in Fig. 3.10.

2. **Condition Query**
   Condition query is a targeted data query, usually with the help of the keyword "where" to judge the condition. Common condition judgment operators are shown in Table 3.2.
   On the basis of simple query, add "where" to condition query.

Example 3.10 "Select top 10*from taxi_traject where passenger=1" indicates that the data record with the number of passengers is 1, as shown in Fig. 3.11.

Example 3.11 "Select top 10*from taxi_traject where speed between 20 and 30" indicates the data record with low speed, as shown in Fig. 3.12.

Example 3.12 "Select top 10*from taxi_traject where passenger=1 and speed between 20 and 30" indicates that the data record with the number of passengers is 1 and low speed, as shown in Fig. 3.13.

	vehicleid	event	passenger	time	lat	lon	speed	direction	status
1	096224	4	0	20121001001326	116.0789490	39.9822617	0	236	1
2	488153	4	0	20121001001326	116.4261322	39.9230347	47	264	1
3	156349	4	0	20121001001131	116.0093155	39.6825676	0	86	1
4	426486	4	0	20121001001326	116.2067871	40.1832962	0	208	1
5	189283	4	0	20121001001326	116.0409164	39.7553825	0	118	1
6	189846	4	2	20121001001325	116.3153305	39.5613785	0	152	1
7	453356	4	0	20121001001329	116.2772217	39.8641624	39	236	1
8	214184	4	2	20121001001327	116.0241699	39.8751259	0	0	1
9	191439	4	2	20121001001325	116.4979858	40.0048637	0	138	1
10	214889	4	2	20121001001327	116.5000153	39.9956131	0	340	1

Fig. 3.10 The top rows data

Table 3.2 Common operators in SQL Server

Types of operators	Symbol or keywords	Meaning
Comparison operators	>, <, =, <=, >=	Larger, less, equal, not greater than, not less than
	between... and...	The value in an interval (including boundary)
	in (x1, x2, xn)	Value in the list
	like	Fuzzy query, % means any character
	is null	Judge whether null value
Logical operators	and	Multiple conditions hold simultaneously
	or	Any one of the conditions holds
	not	The condition does not hold

	vehicleid	event	passenger	time	lat	lon	speed	direction	status
1	489664	4	1	20121001001328	116.4271164	39.9359322	38	176	1
2	162451	4	1	20121001001328	116.4107132	40.0141106	67	8	1
3	194736	4	1	20121001001328	116.3911743	39.8896599	56	72	1
4	174598	4	1	20121001001329	116.3835983	39.9319534	41	86	1
5	100184	4	1	20121001001329	116.3091049	40.0014534	65	0	1
6	164920	4	1	20121001001327	116.3513641	39.9858131	75	268	1
7	174812	4	1	20121001001329	116.4836273	39.8957481	71	178	1
8	194305	4	1	20121001001329	116.5574112	39.8676186	26	272	1
9	093453	4	1	20121001001329	116.3593597	40.0620804	45	6	1
10	077435	4	1	20121001001329	116.4859467	39.9054832	60	52	1

Fig. 3.11 Query record based on the number of passengers

	vehicleid	event	passenger	time	lat	lon	speed	direction	status
1	155474	4	0	20121001081131	116.3739243	39.9530029	23	210	1
2	194305	4	1	20121001001329	116.5574112	39.8676186	26	272	1
3	194373	4	1	20121001001329	116.3825302	39.9423981	25	134	1
4	194725	4	0	20121001001329	116.3030548	39.8949547	26	90	1
5	162480	4	1	20121001001329	116.4549255	39.8566246	30	76	1
6	194665	4	1	20121001001329	116.3640213	39.8335457	23	272	1
7	154882	4	2	20121001001325	116.3966827	39.8339424	21	12	1
8	566754	4	0	20121001001326	116.3684311	39.8915482	23	4	1
9	495146	4	1	20121001001327	116..3430481	39.9126549	28	264	1
10	431530	4	0	20121001081132	116.4550476	39.8823242	30	2	1

Fig. 3.12 Query record based on the speed of vehicles

Example 3.13 "Select top 10*from taxi_traject where vehicleid like '10%'" indicates the data record beginning with 10, as shown in Fig. 3.14.

3. **Sort query**

Sort query is to sort data by one or several variables. Generally, the keyword "order by (x)" is used to complete the sorting from small to large, that is, order by (x) in desc. If there is a "where" condition query, "order by" needs to be placed after the where statement.

	vehicleid	event	passenger	time	lat	lon	speed	direction	status
1	194305	4	1	20121001001329	116.5574112	39.8676186	26	272	1
2	194373	4	1	20121001001329	116.3825302	39.9423981	25	134	1
3	162480	4	1	20121001001329	116.4549255	39.8566246	30	76	1
4	194665	4	1	20121001001329	116.3640213	39.8335457	23	272	1
5	495146	4	1	20121001001327	116.3430481	39.9126549	28	264	1
6	453209	4	1	20121001001329	116.6331329	39.9009552	23	350	1
7	470477	4	1	20121001001330	116.0998230	39.9339867	28	264	1
8	174971	4	1	20121001001330	116.3545532	39.8488884	25	86	1
9	164358	4	1	20121001001330	116.3197861	39.9222069	30	104	1
10	492224	4	1	20121001001328	116.4446869	39.8750763	26	88	1

Fig. 3.13 Query record based on the number of passengers and the speed of vehicles

	vehicleid	event	passenger	time	lat	lon	speed	direction	status
1	100184	4	1	20121001001329	116.3091049	40.0014534	65	0	1
2	101684	4	0	20121001001329	116.4544449	39.6780624	0	126	1
3	101664	4	0	20121001001329	116.4191055	39.7937775	0	0	1
4	101794	4	0	20121001001329	116.5837021	40.0794907	0	0	1
5	101624	4	0	20121001001330	116.4551849	39.9025536	0	168	1
6	104133	4	0	20121001001327	116.4580841	40.1738472	0	6	1
7	104080	4	0	20121001001350	116.1098022	39.9236717	0	12	1
8	102314	4	0	20121001001356	116.6012726	40.3431816	0	238	1
9	102504	4	0	20121001001356	116.3284149	39.8962517	0	138	1
10	102934	0	0	20121001001356	116.2679062	39.9061317	2	124	1

Fig. 3.14 Query record based on the vehicle ID

	vehicleid	event	passenger	time	lat	lon	speed	direction	status
1	470463	4	0	20010101000000	0.0000000	0.0000000	0	0	1
2	470463	4	0	20121001001339	116.5268936	39.9401970	0	54	1
3	470463	4	0	20121001001351	116.5268860	39.9401932	0	54	1
4	470463	4	0	20121001001353	116.5268860	39.9401932	0	60	1
5	470463	4	0	20121001001403	116.5268784	39.9401932	0	62	1
6	470463	4	0	20121001001415	116.5268784	39.9401932	0	68	1
7	470463	4	0	20121001001427	116.5272446	39.9402161	28	84	1
8	470463	4	0	20121001001439	116.5285950	39.9402275	32	88	1
9	470463	4	0	20121001001446	116.5293884	39.9402351	36	88	1
10	470463	4	0	20121001001451	116.5299149	39.9402313	21	88	1

Fig. 3.15 Query record based on time

Example 3.14 "Select top 10*from taxi_traject where vehicleid='470463' order by time", and the result is shown in Fig. 3.15.

4. **Subquery**

Subquery makes the result of one day's select statement a part of another day's select statement. Subquery must be enclosed in parentheses.

Example 3.15 "Select count (distinct vehicleid) from taxi_traject where vehicleid in (SELECT distinct vehicleid FROM taxi_Traject2)" indicates that from the table taxi_traject, the number of different vehicles counted, which also appears in table taxis_traject2, is 12348.

3.3.2 Data Classification

Classification analysis is simply to divide a large data packet into several categories or groups according to its inherent data characteristics. In the area of transportation, data classification is a problem that needs to be faced almost every day. Many continuous and compound data need to be classified according to their region, traffic flow characteristics, and attributes when entering the database. There are many kinds of methods of data classification, such as k-nearest neighbor (k-NN), support vector machine, neural network, and so on. Data classification and data clustering are twin technology. This section only introduces the decision tree algorithm as a case.

1. **Introduction of Decision Tree Algorithm**

    The origin of the decision tree algorithm is the concept learning system (CLS algorithm), then develops to the n-ary tree (e.g., ID3) method, and finally evolves to C4.5 algorithm, which can deal with continuous values. There are also some famous decision tree algorithms such as classification and regression tree (CART) and assistant algorithm.

    In general, the decision tree algorithm is a process that uses the information gain of information theory to find the attribute field with the largest amount of information in the sample database to establish a node of the decision tree. Then, it establishes branches of the tree according to the different values of the attribute field and repeats the next node and branch of the tree in each branch set. The root node of the decision tree is the whole data set space, each subnode is a test of a single variable, which divides the data set space into two or more blocks, and each leaf node is a record belonging to a single category.

    The decision tree is divided into classification tree and regression tree. The classification tree makes decision tree for discrete variables and the regression tree makes decision tree for continuous variables. The quality of the tree depends on the classification accuracy and the size of the tree. The construction of decision tree mainly consists of the following two stages.

    (a) The stage of establishment. Training data is selected to establish a decision tree. The decision tree is established according to breadth first until each leaf node contains the same class tag.
    (b) The stage of adjustment. Test the decision tree with the remaining data (except training data). If the established decision tree cannot answer the question correctly, the user should adjust the decision tree (prune and add nodes) until a correct decision tree is established. In this way, the attribute

values are compared at each internal node of the decision tree, and the conclusion is obtained at the leaf node. A path from the root node to leaf node corresponds to a rule, and the whole decision tree corresponds to a set of disjunctive expression rules.

The reason why decision tree technology can be widely used is mainly due to the following points.

(a) Decision tree can generate understandable rules. Compared with some data mining algorithms, the rules generated by the decision tree algorithm are easier to understand, and the establishment process of the decision tree model is also very intuitive.
(b) The computation of decision tree is small.
(c) Decision tree supports both discrete data and continuous data.
(d) The output of the decision tree contains the ordering of attributes. When generating the decision tree, the test attributes are selected according to the maximum information gain, so the relative importance of the attributes can be roughly judged in the decision tree. However, decision tree technology also has some shortcomings; for example, the cost of training a decision tree is very large, and classification is easy to make mistakes in the case of too many class labels.

2. **Description of the Construction Method of Decision Tree**

The input of decision tree construction is a group of examples with class labels, and the result of construction is a binary-ary tree or n-ary tree. The inner node (nonleaf node) of a binary-ary tree is generally expressed as a logical judgment in the form of $a_i = v_f$, where a_i is an attribute, a_f is an attribute value of the attribute, and the edge of the tree is the branch result of the logical judgment. The inner node of the n-ary tree is an attribute, and the edge is all the values of the attribute. The number of attribute values is equal to the number of edges under the node. The leaf nodes of the tree are class tags.

The decision tree is constructed recursively from top to bottom. Taking the n-ary tree as an example, the construction idea is as follows: if all the examples in the training example set are of the same kind, they are regarded as the leaf nodes of the decision tree, and the content of the nodes is the category label. Otherwise, an attribute is selected according to a certain strategy, and the example set is divided into several subsets according to each value of the attribute so that all the examples in each subset have the same attribute value on the attribute. Then, each subset is recursively processed in turn. This idea is the principle of divide and conquer. Binary-ary tree is the same, and the only difference is to choose a good logical judgment.

The general steps of constructing decision tree include data preparation, data preprocessing, constructing decision tree, and decision tree checking.

The process of constructing a decision tree by recursive segmentation is as follows.

(a) Find the initial split. The whole training set is regarded as the set to generate the decision tree, and each record of the training set must be classified. To find the initial split is to decide which field is the best classification index at present. The general approach is to try all the attribute fields, quantify the quality of each attribute field split, calculate the best split, and repeat until the records in each leaf node belong to the same class.

(b) Data pruning. Pruning is a technology to overcome noise, and it can simplify the decision tree and make it easier to understand. It can be divided into two types: forward pruning and backward pruning. Forward pruning is to decide whether to continue to divide or stop the impure training subset while generating the decision tree. Backward pruning is a two-stage method: fitting and simplifying. First, a decision tree that completely fits the training data is generated, and then pruning is started from the leaves of the tree, and a tuning set or adjusting set is used when pruning to the root. If there is a leaf that does not reduce the accuracy or other measures (not getting worse) on the tuning set after pruning, the leaf will be pruned; otherwise, it will stop. Theoretically speaking, backward pruning is better than forward pruning, but the computational complexity is large. Pruning usually involves some statistical parameters or thresholds, such as stop threshold. It is worth noting that pruning is not good for all datasets, just as the smallest tree is not the best (with the largest prediction rate). When data is sparse, over-pruning should be avoided. In a sense, pruning is also a bias, which is good for some data and bad for some data.

The key to constructing a good decision tree is how to choose good logical judgment or attribute. For the same set of examples, there can be many decision trees that can match this set of examples. In general, from the perspective of probability, the smaller the tree, the stronger the prediction ability of the tree. To construct a decision tree as small as possible, the key is to select appropriate logical judgment or attribute. Since it is a nondeterministic polynomial (NP) problem to construct the smallest tree, we can only use heuristic strategy to select good logical judgments or attributes. There are many classification technologies like decision tree; these algorithms can play a very good role in intelligent transportation accident detection, anomaly recognition, and traffic state prediction. Especially in the context of big data, many research topics that need data sampling and small sample analysis will be able to process and calculate in the scale of full sample and massive data, which further improves the expression ability of classification algorithm for real problems.

3.3.3 Data Clustering

Clustering analysis is another kind of data mining technology commonly used in urban traffic big data analysis. Traditional cluster analysis methods include system cluster, decomposition, addition, dynamic cluster, ordered sample cluster, overlapping cluster, and fuzzy cluster. Clustering analysis tools based on k-means and k-medoids have been added to many famous statistical analysis software packages, such as SPSS, SAS, etc.

1. **Introduction of Cluster Analysis**

Clustering is to divide the whole data into different groups and make the gap between groups as large as possible while the gap within groups as small as possible. Different from classification, users do not know how to divide the data into several groups, nor do they know the specific criteria of grouping before clustering. Therefore, the characteristics of the data set are unknown in clustering analysis. It is also called unsupervised learning, in which data with the same certain characteristics are clustered. Classification means that the user knows that the data can be divided into several categories and divides the data to be processed into different categories according to the rules, which is also called supervised learning.

From the perspective of machine learning, clusters are equivalent to hidden patterns. Clustering is an unsupervised learning process of searching clusters. Unlike classification, unsupervised learning does not rely on predefined classes or training instances with class tags. The tags of unsupervised learning need to be automatically determined by a clustering learning algorithm, while the instances or data of classification learning have class tags. Clustering is observation learning, not example learning.

From the perspective of practical application, clustering analysis is one of the main tasks of data mining, and clustering can be used as an independent tool to obtain the distribution of data, observe the characteristics of each cluster data, and focus on further analysis of specific clusters. Clustering analysis can also be used as a preprocessing step for other algorithms, such as classification and qualitative induction.

The commonly used clustering methods mainly include partition method (k-means, k-medoids, etc.), hierarchical clustering method (BIRCH, CURE, etc.), density-based method, grid-based method, and model-based method.

2. **The Impact Level Determination Model of Expressway Traffic Events Based on k-Means Method**

k-means algorithm is a typical clustering algorithm based on distance. Distance is used as the evaluation index of similarity; that is, the closer the two objects are, the greater the similarity is. The algorithm considers that the cluster is composed of close objects, so the final goal is to get a compact and independent cluster.

The process of the k-means algorithm is described as follows. First, k objects from n data objects are randomly selected as the initial cluster centers. For the remaining objects, they are assigned to the most similar clusters (represented by cluster centers) according to their similarity (distance) with these cluster centers. Then, the centers of each new cluster (the average of all objects in the cluster) are calculated. This process is repeated until the standard measure function begins to converge. Generally, the mean square error is used as the standard measure function, and the specific definition is as follows.

$$E = \sum_{i=1}^{k} \sum_{p \in c_i} |p - m_i|^2 \tag{3.1}$$

where E is the sum of mean square deviation of all objects in the cluster; p represents a point in the cluster, which can be multidimensional; and the mean value of cluster c_i is m_i, which can be multidimensional.

The clustering shown in the formulation aims to make the obtained k clusters as compact as possible in each of them and separate as much as possible between them. For example, the time and space influence range of traffic events are grouped into four categories, namely $k = 4$, $p = [\text{time}_p, \text{space}_p]$, $m_i = [\text{time}_i, \text{space}_i]$.

Clustering analysis is an indispensable and important technology in the field of transportation, especially for massive discrete time series data sets, such as traffic accidents, and long-time and large-area congestion. It is necessary to cluster in the discrete samples of many years of historical data to obtain the feature set and then define the event or congestion type. The twin application of data classification and cluster analysis in the era of urban traffic big data can not only bring more detailed and diverse single data interval fields for continuous data sets and discrete data sets but also realize the multiple integration and analysis of multisource and multidimensional data. This brings vitality to the analysis and mining of full sample data.

3. **Spatial Cluster**

As a research direction of clustering analysis, spatial clustering is to divide the concentrated spatial data objects into classes composed of similar objects. The objects in the same class have high similarity, while the objects in different classes have great differences. As an unsupervised learning method, spatial clustering does not need any prior knowledge, such as predefined classes or labels with classes. Because spatial clustering methods can classify spatial objects according to their attributes, they have been widely used in urban planning, environmental monitoring, traffic management, and other fields.

GIS spatial clustering analysis technology provides a new idea for urban traffic big data analysis [1]. In the GIS data of urban traffic, information can be divided into two categories: attribute information reflecting the nonspatial attributes of spatial objects and spatial information reflecting the spatial position of spatial objects (also known as coordinate information). Therefore, according to the information of clustering objects, GIS spatial clustering can be divided into the following three types.

Attribute Clustering

There is no essential difference between the attribute information of GIS objects and that of general objects. However, in GIS, attribute information relates to spatial information through entity relationship, forming spatial entity. So, GIS attribute clustering is basically the same as the multidimensional clustering method of general objects.

Coordinate Clustering

Spatial coordinate information describes the spatial position of the object. The similarity of spatial coordinate data has the following three main characteristics.

(a) The low dimension and format consistency of coordinate information make the clustering operation easier, and the clustering is more obvious, which reflects the simplicity and effectiveness of spatial coordinate clustering.
(b) The essence of spatial coordinate information clustering is to find the dense areas of object distribution in space, such as the measurement of dense passenger flow, the distribution of urban bus stops, and so on. From the abstract "category" to the concrete and intuitive "region," spatial coordinate clustering has a different meaning from the general clustering operation. At the same time, the diversity and low dimension of spatial regions cause the density of spatial clusters, which increases the complexity of clustering.
(c) Spatial information is the basic information or the first information of GIS when processing objects. Attribute information is based on spatial information and depends on spatial information. Therefore, the processing of spatial coordinate information is not only the data mining of GIS objects but also the first problem to be dealt with in clustering operation.

Spatial-Attribute Information Hybrid Clustering

GIS object is a spatial entity that connects spatial information and attribute information. Spatial entity is the basic unit of storage and processing in GIS. Therefore, all kinds of operations in GIS, including data mining and visualization, should be able to operate at the level of spatial entity. In other words, attribute information and spatial information should be connected for processing. As a part of spatial data mining, clustering should also be able to operate on mixed high-dimensional vectors containing both spatial information and attribute information. However, due to the different information formats and meanings expressed by spatial information and attribute information, we cannot simply treat the spatial information and attribute information in mixed vector as equivalent. So, the definition of distance and the interpretation of clustering results by mixed vector are the problems to be solved in mixed clustering. Spatial-attribute information hybrid clustering is still a frontier problem in the field of spatial clustering.

4. **k-Means Algorithm Application**

Example 3.16 Taking the detection data of Beijing road network as the model input, including the traffic flow, speed, and occupancy, this example clusters 418 sample detection stations within the Fifth Ring Road. The data chart is shown in Fig. 3.16 (showing the first 10 rows).

First, the centroids of four initial points are randomly determined. Then, each point in the dataset is assigned to a cluster; that is, the centroid nearest to each point is found, and it is assigned to the cluster corresponding to the centroid. After

rowid	routename	flow	speed	occupancy	location
HI8012d	宋庄路北口->东铁营桥	3	85	2	南三环东段
HI8043d	安华桥->安贞里	57	62.8	3	北三环东段
HI9426a	北七家桥南->定泗路口北	12	64.7	1	京承高速
HI8027c	莲花桥->新兴桥	47	72.7	4	西三环南段
HI9358b	志远西桥->团河桥东	64	60	43	南五环
HI7060d	北小街桥->小街桥东	61	67.8	5	北二环东段
HI3009b	小街桥东->北小街桥	86	66.4	10	北二环东段
HI7065c	白纸坊桥南外环指示牌->白纸坊桥南外环指示牌北	47	70.7	4	西二环南段
HI9333a	观音堂桥北->观音堂桥南	66	70	24	东五环
HI5056a	广安门桥->白纸坊桥	66	65.7	5	西二环南段

Fig. 3.16 Data sample in detection station

this step, the centroid of each cluster is updated to the average value of all points in the cluster, and the pseudo-code is as follows.

```
Create k points as the starting centroid, which can be randomly
selected (within the data boundary)
When the cluster allocation result of any point changes
    For each point in the dataset
        For each centroid
            Calculate the distance between the centroid and the data point
        Assign data points to the nearest cluster
    For each cluster, the mean value of all the points in the cluster is
calculated and taken as the centroid
```

It is easy to construct the cluster model based on the KMeans in the sklearn module based on Python. The code is as follows (Fig. 3.17).

```python
from sklearn.cluster import KMeans
import pandas as pd
inputfile='test.xlsx'
data=pd.read_excel(inputfile)
X=data.loc[:, ['flow', 'speed', 'occupancy']] # choose 3
columns to train
kmodel=KMeans(n_clusters=4).fit(X) # constrcut the model
wtih 4 clusters
category=kmodel.predict(X) #classification
print(category) #show the results, as shown in Figure 3-17
```

We can show the result of cluster further based on Python. Regarding the flow and speed as the two-dimension coordinate, the different types of detection station are shown with different color to show the cluster results. The code is as follows.

```
In [2]: print(category)
[3 0 0 0 3 3 0 0 3 0 0 0 0 3 0 0 0 0 0 3 2 0 0 0 0 0 0 3 0 0 0 0 0 0 1 0 0
 0 0 0 0 3 3 0 3 0 0 0 0 0 0 0 0 3 3 3 0 0 3 3 3 0 0 3 0 3 0 3 0 0 0 0 0 0 0
 0 0 2 1 0 1 0 0 3 0 3 1 0 0 3 0 0 1 3 0 3 0 0 0 3 0 0 0 0 0 3 1 3 3 0 0 0
 0 1 1 3 0 0 0 3 0 3 3 0 0 0 0 3 1 0 0 0 0 3 0 2 3 0 0 0 1 0 0 3 3 2 1 0 0
 0 0 0 0 0 0 0 0 0 0 3 3 0 0 3 0 0 0 0 0 0 3 0 0 0 2 1 0 3 0 0 0 0 3 3 3 0 3
 0 0 1 0 3 0 0 0 3 0 0 0 3 3 3 0 3 0 0 0 2 1 0 0 0 0 0 0 0 1 3 0 0 3 3 1 0 0
 0 0 0 0 0 3 0 0 0 3 0 0 0 3 3 0 0 3 0 0 0 3 0 0 0 3 2 3 0 0 0 3 0 0 0 3 3
 0 3 0 0 0 0 0 3 3 0 3 0 0 0 1 1 0 0 3 0 3 0 0 0 0 2 0 0 3 1 1 3 3 0 0 3 0
 0 3 1 0 0 3 3 0 3 0 3 0 0 3 0 3 0 3 2 0 3 3 0 0 0 0 3 3 0 0 0 0 0 0 0 0 0 0
 0 0 3 0 0 1 3 3 0 0 0 1 0 0 0 3 0 0 0 3 0 0 0 3 0 0 0 3 0 0 0 0 3 3 0 0 0 0 0 3
 0 0 0 0 3 0 0 0 0 3 3 0 0 0 0 0 0 0 0 0 0 0 0 3 3 0 0 0 0 0 0 0 0 0 0 0
 0 0 0 0 0 3 0 3 3 3 0]
```

Fig. 3.17 Results of cluster

```
import matplotlib.pyplot as plt
data['category']=category
plt.figure(figsize=(7, 5), dpi=500)
color=['r', 'b', 'g', 'c']
for i in range(0, 4):
subdata=data[data['category']==i]
plt.scatter(subdata['flow'], subdata['speed'], c=color[i])
# scatter
    plt.title('GCZ_category') #title
    plt.legend(('category 1', 'category 2', 'category 3',
'category 4'))
    plt.savefig('GCZ_category') # save images
    plt.show() # show images
```

The result of the above code is shown in Fig. 3.18. From the figure, we can see the clear boundary between the four clusters of detection stations. It means that the cluster is effective. However, the choice of the k value is set based on experience or the distribution of the data; appropriate adjustment of it may get a better cluster result. The cluster result of three clusters of detection stations is shown in Fig. 3.19.

3.3.4 Data Association

Considering the big data thinking, when many cross-industry data are accumulated at the same temporal and spatial context, it is often more important whether there is a correlation between them. This kind of correlation analysis plays an important role in the in-depth analysis of the traffic characteristics of the cross-industry. This section takes gray relational analysis (GRA) as an example [2] to explain its basic principles and methods and introduce its application in the field of transportation.

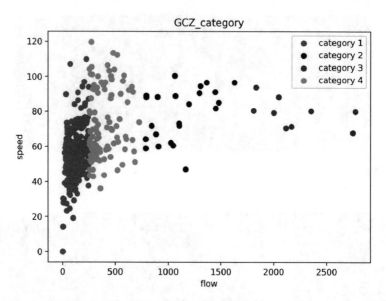

Fig. 3.18 The cluster results of four clusters in the detection station

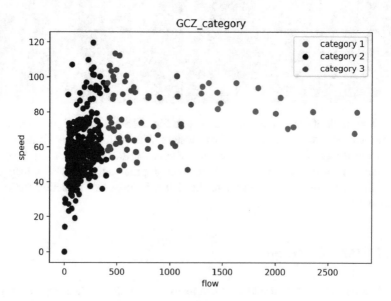

Fig. 3.19 The cluster results of three clusters in the detection station

1. Correlation Analysis Method

Gray relational analysis is an analysis method based on gray system theory. The research object is "small sample" and "poor information" uncertain system with part information known and part information unknown. The basic idea of gray correlation analysis is to judge whether the relationship is close or not according to the similarity degree of the geometric shape of the sequence curve. The closer the curve is, the greater the correlation degree between the corresponding sequences is, and vice versa.

The specific calculation step of gray correlation analysis is as follows.

We set $X_0 = (X_0(1), X_0(2), \ldots, X_0(k), \ldots, X_0(m))$ and $X_i = (X_i(1), X_i(2), \ldots, X_i(k), \ldots, X_i(m))$ as the reference sequence and comparison sequence, respectively, where $i = 1, 2, \ldots n$.

The commonly used dimensionless processing method includes initial value transformation, mean value transformation, range transformation, and effect measure transformation. The initial value transformation is often used in the correlation analysis of the dynamic series of the relatively stable socioeconomic system. The specific calculation formulation is as follows.

$$X_i = \left(\frac{X'_{i(1)}}{X'_{i(1)}}, \frac{X'_{i(2)}}{X'_{i(1)}}, \ldots, \frac{X'_{i(m)}}{X'_{i(1)}} \right) \tag{3.2}$$

The formulation to calculate the gray correlation coefficient $\gamma(X_0(k), \gamma(X_i(k))$ is as follows.

$$\gamma\left(X_0(k), (X_i(k))\right) = \frac{X(\min) + aX(\max)}{\Delta_{0i}(k) + aX(\max)} \tag{3.3}$$

where $X(\min) = \min_i, \min_k |X_0(k), X_i(k)|$; $X(\max) = \max_i, \max_k |X_0(k), X_i(k)|$; $\Delta_{0i}(k) = |X_0(k), X_i(k)|$, $a \in [0,1]$ The resolution coefficient is equal to 0.5 based on the principle of minimum information.

The gray correlation $\gamma(X_0, X_i)$ is calculated by aggregating the gray correlation coefficient $\gamma(X_0(k), \gamma(X_i(k))$ at each point $k = 1, 2, \ldots, m$. The calculation formulation of gray correlation is as follows:

$$\gamma(X_0, X_i) = \frac{1}{m} \sum_{k}^{m} \gamma(X_0(k), \gamma(X_i(k)) \tag{3.4}$$

We can get the gray correlation $R[R = \gamma(X_0, X_i)]$ by the above steps. According to the correlation between the comparison sequence and the reference sequence, we can judge the effect of each factor on the traffic noise. The larger the correlation is, the greater effect the factor has. We call it a major impact factor, and vice versa, we call it a minor impact factor.

2. **Application in Transportation Field**

Example 3.17 Based on the traffic noise of a southern city in China from 2002 to 2009, we discuss the application of gray correlation analysis in the influence factor of urban traffic noise.

Traffic noise data comes from the average value of four traffic noise fixed monitoring stations in the city. In order to truly reflect the traffic noise situation of the city, the average value method is used to optimize the noise monitoring data on the basis of noise census. The steps are as follows.

Establish a Data Sequence

The number of motor vehicles, the length of traffic road, and the density of motor vehicles will directly affect the urban traffic noise. The GDP, permanent population, and other factors as the main indicators of urban characteristics reflect the urban traffic noise level to a certain extent. This book selects five factors, including resident population, GDP, number of motor vehicles, length of traffic road, and density of motor vehicles to establish data sequence. Then, the relationship between five factors and urban traffic noise is analyzed by gray correlation analysis. Among them, the five factors of urban resident population, GDP, motor vehicles, road traffic line length, and driving motor density are the comparison sequence of influencing traffic noise, that is, $X_i = (X_i(1), X_i(2), \ldots, X_i(k), \ldots, X_i(m))$. The noise of urban traffic is reference sequence, that is, $X_0 = (X_0(1), X_0(2), \ldots, X_0(k), \ldots, X_0(m))$. The basic data of the five factors is as follows (Table 3.3).

The aforementioned equation is used to do the processing of initialization on the basic data of these five factors. The dimensionless processing result is shown in Table 3.4.

Calculation of the gray correlation coefficient and correlation

Let $a = 0.5$, and using DPS statistical analysis software to process the data after initialization, we get $X(\min) = 0$, $X(\max) = 2.161$. Then, the correlation coefficient is

$$\gamma\left(X_0(k), (X_i(k))\right) = \frac{0 + 0.5 \times 2.161}{\Delta_{0i}(k) + 0.5 \times 2.161} \tag{3.5}$$

By substituting each correlation coefficient into the equation, the correlation can be obtained as follows: $\gamma_1 = 0.902$, $\gamma_2 = 0.586$, $\gamma_3 = 0.810$, $\gamma_4 = 0.915$, $\gamma_5 = 0.882$. The order of correlation is $\gamma_4 > \gamma_1 > \gamma_5 > \gamma_3 > \gamma_2$. The correlation of γ_1, γ_4, and γ_5 are all greater than 0.880.

The correlation sequence shows that the correlation of urban traffic noise with lane length and resident population is the largest, which is 0.915 and 0.902, respectively. It indicates that lane length and resident population have a great correlation with urban traffic noise. Generally speaking, the longer the road lane, the lower the traffic flow and the lower the traffic noise, which has a positive impact on the urban traffic noise. The increase of permanent population will lead to the increase of traffic noise, showing an obvious negative effect.

The correlation of the density and the total number of motor vehicles to the urban traffic noise is 0.882 and 0.810, respectively, which indicates that the density and the total number of motor vehicles have a great correlation with the urban traffic noise. In general, the higher the traffic density and the total number of motor vehicles, the higher the traffic noise.

Table 3.3 Five types of related traffic data in 2002–2009

Year	2002	2003	2004	2005	2006	2007	2008	2009
Traffic noise/dB(A)	70.42	71.46	71.07	70.23	69.91	70.08	70.78	70.55
Permanent resident population	441,637	448,495	465,333	484,300	513,400	538,100	549,200	542,200
GDP (100 million)	548	636	822	922	1137	1502	1735	1693
Number of motor vehicles	122,345	130,472	141,258	152,542	162,874	174,520	182,765	189,863
Lane length/km	341	345.2	362.1	368.2	383.6	400.8	404.4	413.1
vehicle density (vehicles/km)	358.8	378	390.1	414.3	424.6	435	452	460

Table 3.4 The initialization of the five types of data

Year	2002	2003	2004	2005	2006	2007	2008	2009
Traffic noise/dB(A)	1	1.015	1.009	0.997	0.993	0.995	1.005	1.002
Permanent resident population	1	1.016	1.054	1.097	1.162	1.218	1.244	1.228
GDP (100 million)	1	1.161	1.5	1.682	2.075	2.741	3.166	3.089
Number of motor vehicles	1	1.066	1.155	1.247	1.331	1.426	1.494	1.558
Lane length/km	1	1.012	1.062	1.080	1.125	1.075	1.186	1.211
Vehicle density (vehicles/km)	1	1.054	1.087	1.155	1.183	1.212	1.260	1.282

Compared with the other four factors, the correlation between GDP and urban traffic noise is small, only 0.586. The impact of GDP on urban traffic noise has no obvious positive or negative effect. Generally speaking, with the increase of GDP, the government's investment in urban road construction will also be relatively increased, which is bound to improve the urban traffic conditions and reduce urban traffic noise pollution. However, the increase of GDP will also lead to the increase of urban motor vehicles, which will lead to the aggravation of traffic noise pollution. Various factors will lead to a smaller impact of GDP on urban traffic noise.

3.3.5   Spatiotemporal Data Analysis

Most of the existing geographic information systems use two-dimensional or three-dimensional spatial data organization and management, which is based on layers and the map processing mode. It only describes the instantaneous data and cannot process the temporal data. When the data changes, the new data will replace the old data, and the system becomes another transient. The old data no longer exists, so it is impossible to analyze the history of data and even unable to predict the future trend. This kind of GIS is called static GIS. At present, there is no complete and definite framework for the understanding and expression of spatiotemporal relationship. The research on the integrity and consistency of spatiotemporal information needs to be deepened and improved. The complexity of spatiotemporal data and the limitations of existing computer technology hindered the development of GIS. The research of spatiotemporal data model can provide more powerful support for the application of GIS. If time, which is as important as space, is introduced into GIS, or time dimension or time variable is added on the basis of two-dimensional GIS, temporal GIS is formed [3, 4].

1. **Temporal GIS**

   Temporal GIS is a system that can express the temporal behavior of geographical phenomena. When GIS can manage and express the spatial and temporal information of geographical objects, it can reproduce and analyze the dynamic real world through vivid storage, from the real world to people's conceptual world, then to the digital world, and finally to the real world through the feedback

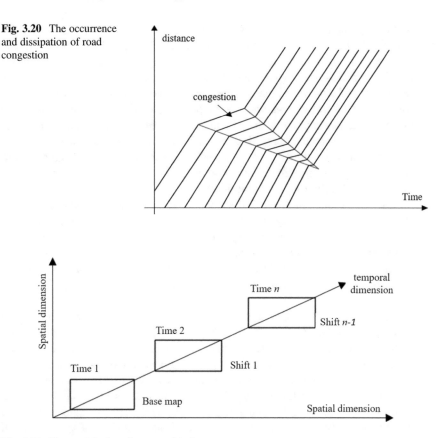

Fig. 3.20 The occurrence and dissipation of road congestion

Fig. 3.21 The model of spatiotemporal index

of the activities modified by users. The three stages correspond to the three research fields of geographic information science, namely, the research of geographic cognitive model, the research of geographic concept computing method, and the research of the relationship between geographic information science and society.

With the development of GIS, more and more application fields require GIS to provide perfect time series analysis function and fully display the application of GIS system in time and space, such as the occurrence and dissipation of road congestion caused by accidents (see Fig. 3.20). The gathering and dissipation of regional commuter passenger flow is a spatiotemporal dynamic process, and most of its spatiotemporal data have strong time sensitivity. Based on this feature, applying the basic principle of temporal GIS and visualization technology of scientific computing, using spatiotemporal index model (see Fig. 3.21), and using object-oriented programming language, spatiotemporal index object can be constructed, and then a professional application model that simulates and displays spatiotemporal dynamics of urban traffic from the perspective of GIS spatiotemporal is constructed. The expression of time dimension is not the ultimate goal of spatiotemporal research. The spatiotemporal data model emphasizes the use of spatiotemporal analysis tools and techniques to simulate the dynamic process and explore and mine the information and rules hidden in spatiotemporal data on the basis of the reasonable organization of spatiotemporal data. Under the current conditions, an important direction of spatiotemporal data model research is to propose an operational spatiotemporal data model oriented to the professional application field on the basis of the existing functions, expand the space-time analysis function of the application system, and answer the questions caused by time changes in the professional field.

2. Research of Spatiotemporal Database

The core problem of the spatiotemporal database is to study how to effectively express, record, and manage the real-world entities and their relationships changing with time. At present, the main problems of research include the data model of expressing spatiotemporal changes, the organization and access method of spatiotemporal data, the version of the spatiotemporal database, the quality control of the spatiotemporal database, the visualization of the spatiotemporal data, and so on.

According to the different ways of spatiotemporal expression, the current spatiotemporal data model can be divided into four types: space-based spatiotemporal data model, time-based spatiotemporal data model, spatiotemporal integration data model, and spatiotemporal thematic composite integration data model. Generally speaking, the historical data stored in traditional models is relatively simple, but the ability of temporal analysis is poor. The base state with amendments model is suitable for vector and easy to implement in the current model, but the ability of temporal analysis is weak. The spatiotemporal composite model includes topological information needed by temporal analysis, but it is difficult to combine with current models. The integrated model provides a feasible scheme, but the operation is complex and data redundancy is prominent. Due to the complexity and particularity of temporal, there are few instances of the spatiotemporal database based on the spatiotemporal data model running in the application system.

3. Research of Temporal GIS Realization Method

At present, there are two ways to realize the time dimension: one is to extend the traditional relational model, the other is to use the object-oriented method. Because the traditional relational model is rich in semantics, perfect in theory, and has many efficient and flexible implementation mechanisms, people can try to add a time dimension to the traditional relational model, expand the relational model, and process temporal data with relational algebra and query language, so as to realize the storage, representation, and processing of spatiotemporal data supported by relational model directly or indirectly. However, the data types of traditional relational models are relatively simple and lack expression ability. Many entities and structures in reality are difficult to map to relational models. Therefore, in recent years, many researchers have begun to explore how to express complex geographic information in a more natural way, and the object-oriented method has become the research focus. In addition, there are event-based research methods and feature-based research methods, which are also the research direction of temporal implementation methods.

4. **The Application of Temporal GIS**

The purpose of theoretical research is to be applied. Therefore, the application research of time dimension is also an important content. At present, it is widely used in the field of land and resources, especially in the cadastral information system. The content related to time-dimension is more common, while the research in the field of transportation is not much, but the time-dependent problem in the field of transportation is also a common problem. Therefore, it is of great significance to explore the temporal application in the field of transportation to solve the problems that cannot be solved in the current traffic information system.

3.3.6   Geocoding

Geocoding refers to the process of mapping the data in the database to the corresponding graphic elements on the map according to the geographical coordinates or spatial address of each data point (such as province, city, block, floor, etc.), that is, assigning X and Y axis coordinate values to each data, so as to determine the location of the data on the map. With the help of GIS geocoding technology, the original information system and spatial information can be integrated to realize the information spatial visualization in daily urban traffic so as to facilitate the spatial analysis and decision-making application under the support of spatial information, thus becoming an important function in urban traffic GIS data.

It is not difficult to find that nonspatial resources have specific places of occurrence, which is also a key link in the association between nonspatial data resources and spatial data by analyzing the existing urban traffic data resources. By using the coding technology of geographical location, the location of data resources can be determined in the geospatial reference range, the connection between spatial information and nonspatial information can be established, and the integration of various information resources can be realized. Through the geocoding of traffic objects, the basic coding principles and methods of traffic objects can be formulated to facilitate the communication of traffic information. Besides, it can accurately express the coding standard of traffic information. The traffic information released after coding can meet the needs of most users so that users can accurately understand the traffic conditions of each road network, the specific location of traffic accidents, road construction, and its impact on the road so as to organically combine various traffic information application systems and play the comprehensive role of traffic information platform.

The whole coding system is divided into four levels: basic application, professional application, extensible application, and open application.

1. Basic application. Road, section, node code, and corresponding graphic elements directly applied in the Standard of Shanghai Road, section and node coding, and the main basic application information including ground center line, elevated center line, etc.
2. Professional application. Inherited from basic object code, the road section, node graphic elements, and codes are extended to two-way sections with direction and corresponding node and code. It can also be divided and defined manually. For expressway, freeway, and other level highways, the on and off ramps and bridge sections are also considered for professional application. The main professional application information includes ground release section, expressway release section, freeway release section, national and provincial highway release section, etc.
3. Extensible application inherits from professional object coding. For the actual traffic infrastructure and equipment, according to their specific physical location, it can be extended coding from the actual directed road section, subnode, ramp, etc., mainly including the following aspects.

    Road traffic: camera, information board, loop detector, toll station, traffic signal, public security checkpoint, visibility meter, weather station, etc.

    Public transportation: parking lot, bus line, bus stop, urban railway line, urban railway station, ferry terminal, etc.

    External transportation: airport, railway station, wharf, coach station, etc.
4. Open application. Only the coding principle is defined, and no coding method is defined. It mainly includes nongeographic traffic objects, such as meteorological information, GPS information, vehicle information, event information, various statistical information, etc.
5. Demonstration of examples.

Example 3.18 Regional code

Beijing can be divided into regions according to road network or regional functional attributes. The results are shown in Fig. 3.22.

Example 3.19 Network code

The road network is coded according to different road levels or attributes, such as national highway, urban expressway, provincial highway, and freeway, as shown in Fig. 3.23.

Example 3.20 Road code

The essence of road code is to divide a road into several road sections, as shown in Fig. 3.24. Select the sample road section from the geocoding information table on the right side of Fig. 3.24, and the corresponding road section can be shown in the block diagram on the left side of Fig. 3.24 in bright colors.

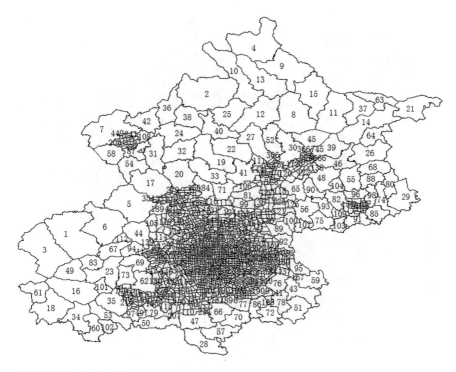

Fig. 3.22 Regional code

Example 3.21 Node code

It can be understood as the positioning of things, such as the positioning of subway station, a university, entertainment and catering point, etc. The following is an example of the detection station on the Beijing urban expressway.

First, address matching is performed for the detector location on Beijing Ring Road in Fig. 3.25; that is, longitude and latitude coordinates are matched for each point (see the next section for address matching). This step can be done with the help of map API, and different coordinate systems can be selected. The matching result is shown in Fig. 3.25 (black box on the right).

After obtaining the longitude and latitude coordinates, the spatial distribution of the detection points can be displayed with the help of the GIS system. Here, a simple operation example is given with the help of ArcGIS.

(a) Open ArcGIS and import the data of detection points and shp files of road network.
(b) In the left window, right-click the "data" layer and click the "show XY data" button.
(c) Select longitude and latitude data in the pop-up dialog box to get the graphical interface, as shown in Fig. 3.26.

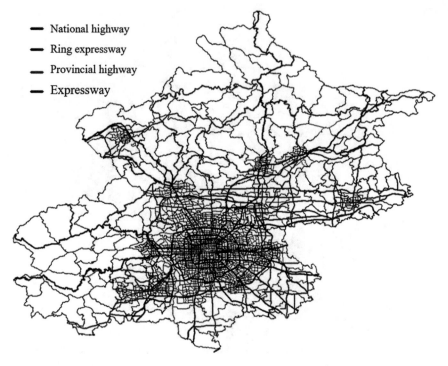

— National highway
— Ring expressway
— Provincial highway
— Expressway

Fig. 3.23 Network code

3.3.7 *Location Matching*

1. Introduction of Location Matching

Location matching is the process of establishing the corresponding relationship between the literal description location and its spatial geographic coordinates. Location matching has been described the "nuclear power" to promote the application of spatial geographic information. The location matching service follows specific steps to find matching objects for addresses.

There are three steps to realize the location matching. The first step is to split and standardize the location string to be matched. The second step is to associate the standardized key location value with the geographical entity in the database to find the potential location. The third step is to specify a score for each candidate location according to the proximity to the location and finally match the location with the highest score, that is, to update the geographical entity coordinates to the corresponding records in the attribute data. In this way, the placename location of the record is matched [5].

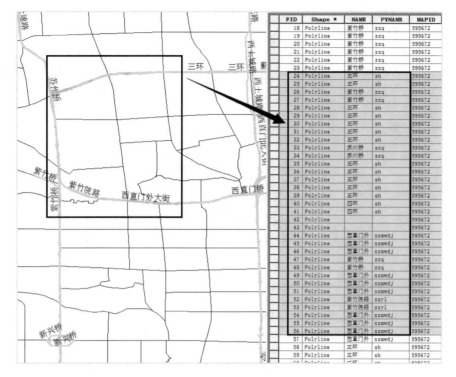

FID	Shape *	NAME	PYNAME	MAPID
18	Polyline	紫竹桥	zzq	595672
19	Polyline	紫竹桥	zzq	595672
20	Polyline	紫竹桥	zzq	595672
21	Polyline	紫竹桥	zzq	595672
22	Polyline	紫竹桥	zzq	595672
23	Polyline	紫竹桥	zzq	595672
24	Polyline	三环	sh	595672
25	Polyline	三环	sh	595672
26	Polyline	紫竹桥	zzq	595672
27	Polyline	紫竹桥	zzq	595672
28	Polyline	三环	sh	595672
29	Polyline	三环	sh	595672
30	Polyline	三环	sh	595672
31	Polyline	三环	sh	595672
32	Polyline	三环	sh	595672
33	Polyline	苏州桥	szq	595672
34	Polyline	苏州桥	szq	595672
35	Polyline	三环	sh	595672
36	Polyline	三环	sh	595672
37	Polyline	三环	sh	595672
38	Polyline	三环	sh	595672
39	Polyline	三环	sh	595672
40	Polyline	四环	sh	595672
41	Polyline	四环	sh	595672
42	Polyline			595672
43	Polyline			595672
44	Polyline	西直门外	xzmwdj	595672
45	Polyline	西直门外	xzmwdj	595672
46	Polyline	西直门外	xzmwdj	595672
47	Polyline	紫竹桥	zzq	595672
48	Polyline	紫竹桥	zzq	595672
49	Polyline	西直门外	xzmwdj	595672
50	Polyline	西直门外	xzmwdj	595672
51	Polyline	西直门外	xzmwdj	595672
52	Polyline	紫竹院路	zzyl	595672
53	Polyline	紫竹院路	zzyl	595672
54	Polyline	西直门外	xzmwdj	595672
55	Polyline	西直门外	xzmwdj	595672
56	Polyline	西直门外	xzmwdj	595672
57	Polyline	西直门外	xzmwdj	595672
58	Polyline	三环	sh	595672
59	Polyline	三环	sh	595672

Fig. 3.24 Road code

Fig. 3.25 Sample
information of detection
point after location code

   In the process of matching, two types of data are needed, one is the geographical location entity information, such as street location, postcode, house number,
landmark name, etc., and the other is the map positioning information (spatial
coordinates), which plays a role of spatial reference in the matching process, such
as street map data, postcode map data, house number map data, landmark name
map data, etc. After the completion of the matching, the former is given
geospatial coordinates, which is also the core part of the location matching model.
   The realization of location matching of place names should have three elements: first, to identify the geographical objects that need to be matched because
different geographical objects need different processing methods; second, to have
a certain reference system, which can be based on coordinates or geographical

Fig. 3.26 The special
distribution of detection
points

indications; and third, location standardization model must be established, and the model determines the structure of location database, the standard of location splitting, and the accuracy of location matching.

2. **Development Requirements of Location Matching Technology**

The traditional method is to split the location one by one and extract the spatial information matching the location from the basic geographic database. But there are some problems with this method:

(a) Most of the work in the process of application in practice of the location data needs to be handled manually.
(b) The location data is not standardized, there are a lot of garbage data, and manual processing is difficult.
(c) Massive location data processing needs a lot of time and energy, and the processing efficiency is low.
(d) There are many human errors in the process of data processing.

There are many massive location data in urban traffic big data. In order to obtain the spatial location information of these massive location data, the traditional data processing methods cannot meet the relevant requirements of data processing. In order to solve the above technical problems, we can use GIS-based automatic location analysis and matching tool. Using location automatic analysis and matching algorithms, the efficiency and accuracy of data processing are greatly improved and the cost of data processing is saved.

3. **Case Introduction**

This paper mainly introduces the theoretical framework of location matching technology based on thesaurus. It converts location information into location thesaurus by constructing a hierarchical location library and realizes location matching by using Chinese word segmentation technology based on location dictionary [6]. The system architecture is shown in Fig. 3.27.

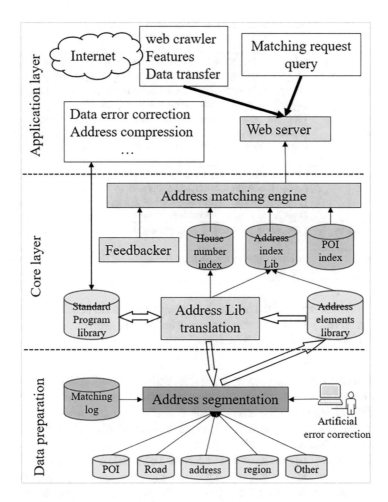

Fig. 3.27 The system architecture of location matching

(a) Application layer. The application layer uses the software functions provided by the core layer to provide various external applications. At present, location matching mainly has the following applications:

- Providing external location matching engine service.
- Location matching engine is used to realize duplicate checking and error correction of location data.
- Batch processing of location data without longitude and latitude indication, that is, geocoding.
- On the existing location data, a location matching engine is used to realize large-scale expansion of geographic information data. At present, because location matching technology is mainly used for location matching

Table 3.5 The introduction of functions of each module in the core layer

Module	Function
Location matching manager	Responsible for managing and coordinating the six modules of search and realizing the function of encapsulating basic user requirements.
Parameter parser	Responsible for parsing external parameters, including parsing strings containing multiple parameters and parameter values.
Location segmentation	It is responsible for splitting the user input location into more detailed location element units. Each location element unit has address element name, address level, parent address information, longitude and latitude, etc.
House number retriever	Realize the given road, find the corresponding house number in the house index file. After the user address is split by the location segmentation, if the address element contains the house number, the corresponding house number is searched in the house index file through its house ID.
Combined sorter	Responsible for sorting multiple matching results and returning the final result.
Test module	Responsible for testing of each module and the stability of the other modules.

services, many of the related descriptions of location matching in this book refer to location matching services. The location matching engine service is implemented by Apache + DSO.

(b) The core layer. The core layer provides software function implementation for location matching, which is the most important layer in location matching and directly related to the effect of location matching. At present, the location matching core layer is mainly composed of six modules, and the functions of each module are shown in Table 3.5.

(c) Data layer. The data layer undertakes the tasks of data extraction, processing, and production (conversion) in the whole location matching architecture and provides the necessary data support for the upper layer. At present, according to the function, the data layer can be divided into three tools, namely, the creation tool of the location element library, the location conversion tool, and the location segmentation interactive tool. These three tools bear different roles and tasks in the location matching data layer and are indispensable.

References

1. Veronika V, Hrube P (2015) Traffic accident, system model and cluster analysis in GIS. Acta Informatica Pragensia 4:64
2. Liu H, Wang W, Zhang Q (2011) Grey systems: theory and application. Grey Syst Theory Appl 4883(4):44–45

3. Liu G, Zhou B, An M, Yang G (2007) A preliminary probe into temporal GIS and its data model. Beijing Surveying and Mapping, Beijing
4. Wu X, Cao Z (2002) Basic conception, function and implementation of temporal GIS. J Earth Sci (03):241–245
5. Hong Y (2008) Research and experiment on matching method of city address and non-spatial information. Liaoning Technical University, Fuxin
6. Sun Y, Chen W (2007) Address matching technique based on word segmentation. In: China Association of Geographic Information Systems Annual Meeting

Chapter 4
Intelligent Transportation Information Interaction Technology

WU Xinkai and WANG Pengcheng

4.1 Overview

4.1.1 Concept

Intelligent transportation refers to a technology of enhancing the connections among vehicles, roads, and the environment by making full use of advanced science and technology, and thereby improving safety and efficiency [1]. Traveling comes with information collection and analysis. For example, during driving, the driver needs to adopt different driving strategies according to different traffic scenes, including velocities of surrounding vehicles, space headway, state of signal lights, pedestrian motion state, etc.

The realization of intelligent transportation is based on information collection and interaction; i.e., driver–vehicle–road–cloud collaboration could be achieved by utilizing various sensors and communication facilities. In the traditional sense, information interaction means the process of signal sending and receiving. It is the transmission and exchange of varied intelligence, data, and technical knowledge of natural and social life. Usually, the process consists of information source, information, information transmission channel, receiver, feedback, and noise.

In the field of intelligent transportation, information interaction technology is the core of the entire intelligent transportation system, and most functions of intelligent transportation rely on information interaction [2]. With the rapid development of wireless communication technology, more traffic information such as vehicle driving state and road environment can be obtained through the in-vehicle network, Internet of vehicles (IoV), cooperative vehicle infrastructure systems (CVIS), and other technologies; the efficient and deep integration of these information and the

WU Xinkai (✉) · WANG Pengcheng
School of Transportation Science and Engineering, Beihang University, Beijing, China
e-mail: xinkaiwu@buaa.edu.cn

© Tsinghua University Press 2022   151
W. Yunpeng et al. (eds.), *Intelligent Road Transport Systems*,
https://doi.org/10.1007/978-981-16-5776-4_4

establishment of a driver–vehicle–road–environment integrated intelligent transportation system can significantly improve the operating efficiency of the transportation system and the transportation safety.

With the development of information technology, vehicles are not just a way of transportation but have penetrated into our daily lives and entertainment. The growing vehicle ownership, however, has also led to traffic congestion. In cities, a large amount of vehicle exhaust emissions, caused by high-density vehicles' frequent brakes and stops, have also led to environmental pollution. As an important part of the intelligent transportation system, the Internet of vehicles technology is regarded as an important way to alleviate this situation.

The Internet of vehicles is a technology that deeply integrates automobiles, communications, and the Internet [3, 4]. It interacts information among people, vehicles, roads, environment, and infrastructure according to certain communication protocols and communication standards. In the Internet of vehicles, communications are available between vehicles and vehicles, road testing facilities, and the cloud, transmitting information like vehicle's position, driving velocity, acceleration, driving intention, and driving route, and receiving information from other vehicles, roadside units, and the cloud at the same time, which could optimize driving efficiency and driving experience. At the macro level, making an organic combination between all kinds of detailed information of in-use vehicles, collected by sensors such as radio-frequency identification (RFID) and cameras, and infrastructure information, traffic condition, and other information could also optimize traffic control and improve traffic efficiency [5]. As a result, energy could be considerably saved, and driving safety could be greatly improved. The technology of the Internet of vehicles is considered to be a significant share of the future intelligent transportation and is widely believed to be a necessary technology to make autonomous driving and unmanned driving coming true.

In general, the network architecture of the Internet of vehicles can be divided into three layers: perception layer, network layer, and application layer, as shown in Fig. 4.1:

1. Perception layer: Perception is the source of all information processing and a prerequisite for Internet of vehicles information processing. Through onboard sensors such as millimeter-wavelength radar, lidar, camera, and positioning technologies like GPS, the driving vehicle's information like in-vehicle and out-of-vehicle state, traffic conditions, and road environment are transmitted to the driver, and the driver will then make driving decisions based on the information received. There are currently many driver assistant systems such as automatic cruise technology and front collision avoidance warning technology, which can perceive the conditions of the road and vehicles in front of the vehicle to help the driver to make more reasonable and complete decisions.

2. Network layer: The network layer is the physical structure of information transmission. It delivers information through in-vehicle networks, Internet, WiFi, and wireless communication technologies, processes various collected information, and has functions like the Internet of vehicles network access and vehicle

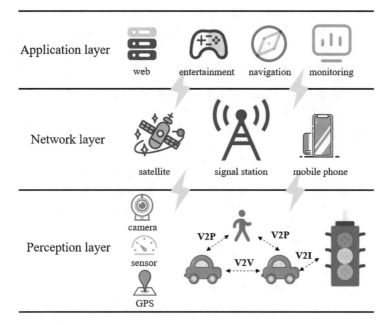

Application layer
web   entertainment   navigation   monitoring

Network layer
satellite   signal station   mobile phone

Perception layer
camera
sensor
GPS
V2P   V2P   V2V   V2I

Fig. 4.1 Internet of vehicles architecture design

management. The network layer could also provide certain information interaction and communication resource allocation for each vehicle to ensure the balance between communication stability and communication load.

3. Application layer: The application layer is the highest layer of the Internet of vehicles architecture, which can provide differentiated services on the basis of different users. It can also provide in-vehicle entertainment, emergency rescue, and other functions according to the different needs of users in various scenarios.

Generally, the Internet of vehicles system involves the following five key technologies [6]:

1. Sensor technology: In driving, human drivers need to rely on their sight, hearing, and touch to judge the traffic conditions, weather and environment, and the state of the vehicle itself. For a vehicle, however, to determine the number of lanes, the spacing from the vehicle in front, the road infrastructure ahead, traffic lights and signs, etc., the stable and accurate work of the sensors are required. Now, more and more researches are focused on data fusion. The perception results collected by all sensors are fused and analyzed to increase the accuracy of recognition and driving safety. Then, the driving instructions are transmitted to the corresponding part of the vehicle to be executed.

2. Positioning technology: Location is a critical and fundamental dynamic parameter in driving, and a series of other functions can only be implemented on this basis. Today's intelligent vehicles mainly use a GPS positioning system or

Beidou navigation system, combined with the environmental information provided by local maps to locate the vehicle. In addition, the Internet of vehicles can also use other positioning technologies, such as the microwave radar, which is able to maintain more stable measurement accuracy in severe weather. Microwave radar can also measure the distance headway between the vehicle and its front one, sense the position and velocity of the vehicle in front, and then ensure the stability and safety of driving. With the improvement of positioning technology, positioning at the decimeter or even centimeter level can be guaranteed, which provides the technical support to achieve automatic driving and unmanned driving.

3. Wireless communication technology: After receiving the information collected by sensors, vehicles need to transmit that information to other vehicles or roadside and cloud devices. The vehicle also needs to obtain information from other vehicles and the network. All these require the implementation of wireless transmission technology. For example, vehicles could communicate within a certain range through a single hop or multiple hops, complete information interaction using specific frequency radios, and exchange information with vehicles and the cloud through roadside units.

4. Communication security of vehicles: if not protected, communication between vehicles in the Internet of vehicles could easily cause privacy leakage. For example, an open network is easy for hackers to actively carry out replay attacks, Sybil attacks, and so on. Therefore, in order to ensure data security and driving safety, the privacy and anonymity of communication, and the integrity and reliability of data, it is necessary to design supporting algorithms.

5. Cloud computing technology: After collecting various driving data of a large amount of vehicles, it is necessary to quickly calculate the collected data and feed it back to the traffic control and the driving vehicles in time. The integration of these data requires high computing capacity, so it needs to be carried out in the cloud. With the growth of edge computing technology, some data could also be processed in the edge.

In today's development of intelligent vehicles, intelligent networking is booming, and the technology of intelligent vehicles themselves is also enjoying rapid improvement and change. Automatic parking system, adaptive cruise system, forward anti-collision system, etc., have all been widely applied in real driving. In the future unmanned driving and even automated driving, the networking technology needs to widely connect the information exchanges among vehicles, with the data obtained by the sensor technology of each vehicle served as an auxiliary, to form a real-time communication network. Therefore, to reach the deep and organic integration of cooperative vehicle infrastructure systems control and driving assistance systems, technology development of intelligent vehicle must be closely linked, rationally connected, and deeply integrated with the vehicle-to-vehicle collaboration and cooperative vehicle infrastructure systems technologies in the Internet of vehicles.

4.1.2   Development Status

In 2015, the US Department of Transportation issued the Intelligent Transportation System Strategic Plan, and the Internet of vehicles technology made great progress with the theme of intelligentization and informatization. In 2016, the US Department of Transportation launched the "Connected Vehicle Pilot Deployment" program in Wyoming, New York City, and Tampa, with an investment of more than 45 million dollars, as presented in Table 4.1. In the development of the Internet of vehicles technology, the United States only authorized the dedicated short-range communication technology standards. Under the new standards of the present day, the United States has launched a number of V2X technology pilots in succession and pushed forward the cellular Internet of vehicles communication technology with all efforts, stimulating the development of the Internet of vehicles technology.

In December 2019, China released the White Paper on Internet of Vehicles Intelligent Property Rights, which sorted out and summarized the current status and development trends of intelligent property rights, patents, and intelligent property litigation in the Internet of vehicles. In January 2021, China's Ministry of Industry and Information Technology acknowledged in an announcement that C-V2X technology is the only standard for China's Internet of vehicles technology.

In the development of the Internet of vehicles, wireless communication technologies are mainly the following three types [7].

1. DSRC

   DSRC (dedicated short range communication) works in the 5.9-GHz frequency band, with a bandwidth of 75 MHz and a communication distance of about 1 km. DSRC consists of the physical layer standard IEEE802.11p and the

Table 4.1 Pilot deployment projects of connected vehicles

Location	Application focus	Content
Wyoming	Cargo transportation	Wyoming plays a key role as a very important cargo transportation node within the United States and between the United States and Canada and Mexico; focuses on the development of commercial transportation and has developed V2I (vehicle-to-infrastructure) and V2V (vehicle-to-vehicle) technology and related APPs, supporting services such as dynamic navigation, parking warning, information warning, etc.
New York	Traffic safety	It is hoped to build V2V and V2I facilities to improve traffic safety; assist urban development goals, and reduce losses due to traffic collisions; and to evaluate and apply the connected automated vehicle technology in dense intersection areas of typical metropolitan traffic management systems
Tampa	Traffic congestion	Tampa Hillsboro has established a reversible fast lane to address the urban congestion problem; deployed V2V and V2I applications to reduce traffic congestion and improve safety at the same time; improved the efficiency of public transportation vehicles; and reduced traffic conflict of mixed traffic regions

network layer standard IEEE 1609. Using MAC (multiple access control), it can achieve distributed operation and apply to point-to-point communication mode. DSRC can use each frequency band with a single channel or multiple channels according to different application types and different communication standards.

Because of its technical characteristics, DSRC is mainly used in the V2V environment. Most messages in the V2V scene are about emergency accidents. For example, if a vehicle accident occurs, the rear vehicles need to be notified in time to avoid secondary collisions, thereof the extremely strict requirements for the delay and reliability of information transmission. The short-range communication characteristics of DSRC technology undoubtedly satisfy these two important requirements. In actual applications, the vehicle must be within the coverage of the roadside unit to be able to access the network smoothly. However, in the high-speed moving scene, the connection time between the vehicle and the roadside unit is very short, and the vehicle needs to switch networks constantly, which is prone to communication interruption. If the vehicle is in a high-density scene, the communication range can be expanded by using the surrounding vehicles as signal relay stations, but even so, it is difficult to ensure stable communication between the vehicle and the roadside unit. Besides, under high-speed moving and high-density vehicle scenarios, the high transmission rate and high delay characteristic of data will cause the competition between vehicles for channels to become extremely fierce, ultimately leading to the weakening of communication performance and much more poor communication quality between V2V.

2. C-V2X

C-V2X technology is a V2X technology based on cellular networks, including LTE-V2X (long-term evolution–vehicle to everything) and 5G-V2X technology that is still in development. LTE-V2X is a V2X Internet of vehicle wireless communication technology that gradually evolved based on LTE (long-term evolution) mobile communication technology, which includes two communication modes of cellular communication and direct communication. Since cellular communication relies on the existing LTE cellular network, it mainly supports communication of wide network coverage and high bandwidth requirements. However, direct communication does not need to pass through a base station, so two user nodes can directly communicate, i.e., point-to-point communication.

Compared with DSRC technology, C-V2X technology has several advantages. First, C-V2X has a higher network capacity and can meet the requirements of high bandwidth and large data volume. Second, the communication range of the cellular network is much larger than the DSRC communication range. The vehicle has plenty of time to apply for connection with the base station with no need to switch networks frequently. This not only improves the communication quality but also reduces the network control signaling overhead. Moreover, C-V2X can directly transform and upgrade the existing base station facilities and equipment without having to redeploy gateway facilities, lowering the cost. Although technically speaking, C-V2X is more reliable and stable in communication, in commercial applications, DSRC's industrial chain is more matured.

3. 5G communication technology

In 2019, China moved into the 5G age, and the Internet of vehicles has also seen further development due to the boosting of 5G communication technology. At present, V2X (vehicle-to-everything) technology occupies a major position in the development of the Internet of vehicles, while C-V2X (cellular vehicle-to-everything) technology is not yet mature. The standard is still under discussion and revision, and few projects have been implemented. The 5G network's excellent technical features of high transmission and low latency satisfy the diverse needs and information transmission requirements of users in the development of the Internet of vehicles and tackle one of the core issues of the Internet of vehicles. And 5G-related technologies, such as mobile edge computing (MEC), software-defined networking (SDN), etc., can also be deeply integrated with the development of the Internet of vehicles to resolve the problem of version incompatibility. With the application of 5G technology, the Internet of vehicles will become more intelligent.

According to the definition of 5G made by the International Telecommunication Union (ITU), 5G network is the next-generation cellular wireless communication that can provide 20 Gb/s speed, 1 μs delay, 1 million connections per square kilometer, and 99.999% of network stability. It is generally accepted in this industry that 5G will be commercially available in 2020. The communication industry divides 5G applications into enhanced mobile broadband (eMBB), massive machine type communications (mMTC), and ultra-reliable low-latency communication (uRLLC). Among them, eMBB is equivalent to the change from 3G network rate to 4G network rate, providing users with a better application experience, while mMTC and uRLLC are brand new scenarios launched for the industry. At present, new applications such as autonomous driving and virtual reality have an incredible need for 5G since autonomous driving requires millisecond-level delay and absolute reliability. The latency of 5G is usually less than 5 ms, which is unattainable for DSRC and LTE-V. The formulation of 5G communication standards is mainly completed by the ITU and 3GPP (The 3rd Generation Partnership Project), mostly including technical conditions and standards of things like frequency spectrum. The World Radio communication Conference 2015 (WRC-15) gave 6 low-frequency frequency bands (below 6 GHz) and 11 high-frequency frequency bands (above 6 GHz) as candidate frequency bands.

The current global consensus is that frequency bands below 6 GHz are used to meet requirements for 5G network coverage and network capacity; frequency bands above 6 GHz are used to meet the requirements for 5G network capacity and are also used as signal tunnels (backhaul). US spectrum planning is dominated by the Federal Communications Commission (FCC) and will be focused on the high-frequency spectrum. The EU spectrum plan is formulated by the Radio Spectrum Policy Group (RSPG) of the European Commission, and the mid-and-low-frequency spectrum is the development priority for their frequency band. In August 2016, China issued the National Radio Management Plan (2016–2020), which pointed out that spectrum resources no less than 500 MHz should be reserved for 5G. In June 2017, the Ministry of Industry and Information Technology successively announced the 5G spectrum plan: the low frequency is 3.3–3.4 GHz, 3.4–3.6 GHz, and 4.8–5 GHz, and the high frequency is clearly defined as the 24.75–27.5 GHz frequency band and the 37–42.5 GHz frequency band. Countries around the world are accelerating 5G trials and commercial plans and are striving for the leading position in the development of 5G standards and industries. China conducted large-scale networking trials in 2018 and will officially commercialize 5G networks in 2020. In the 187th meeting of 3GPPRAN in November 2016, the PolarCode (polarization code) promoted by Huawei became the eMBB scene coding scheme of 5G control channel and has continuously increased the influence of Chinese manufacturers in the world. The overall 5G technology can be divided into wireless technology (radio technology) and network technology (network architecture).

Technologies in the wireless field include large-scale antenna arrays, ultradense networking, new multiple access and full-frequency access, etc., and technologies in the network field include software-defined networking (SDN), network function virtualization (NFV), edge computing, etc. Large-scale antennas are the core technology of Pre5G, the essence of which is a technology that improves system capacity and spectrum efficiency through quantitative changes in antennas. The new multicarrier technology improves spectrum efficiency by reducing spectrum leakage. The ultradense networking is essentially a technology that could significantly promote frequency reuse efficiency through the quantitative changes of the density of microcomputer stations deployed in a unit area, which is one of the main means to meet the demand for 5G capacity that grows by a thousand times. Ultradense networking and high-frequency communication will significantly increase the demand for small base stations. Mobile edge computing is characterized by ultralow delay, ultralarge bandwidth, localization, and high real-time analysis and processing, which reduces the occupation of the core network and backbone transmission network, lowers the end-to-end delay, and promotes the research and development of network slicing technology. It is predicted that 5G will bring about considerable changes in the economic market and communication [8].

In development, there are also many challenges, which require new technology to overcome [9].

1. Multiversion compatibility issues. The current Internet of vehicle C-V2X standard has two versions: the LTE-V2X version based on R14 and the enhanced R15 version. The R16 version based on the 5G-V2X standard has also been perfected. It supports direct communication between V2V and V2I and can be applied in scenarios such as vehicle formation and semi-autonomous driving under the V2X environment. In subsequent developments, there will be a long period of time in which the two versions of LTE-V2X and 5G-V2X coexist. Just like the long-term coexistence of 3G and 4G, 4G and 5G, LTE-V2X needs to be compatible with the 5G version, and the 5G version also needs to be compatible with the LTE-V2X version. In addition to solving the compatibility problems in communication

caused by different versions, LTE-V2V also needs corresponding hardware facilities to support the compatible upgrades.

2. The huge investment. According to the plan, China will achieve a higher degree of autonomous driving in 2035. In order to achieve this goal, in addition to major technological development and innovation, huge capital investment is also needed. As of September 2020, the holdings of motor vehicles in China have reached 365 million, and the highway mileage has reached 142,600 kilometers. Such a large number requires a great deal of installation of onboard terminals and roadside units regardless of the commercial technology type, and a lot of sensing equipment, such as cameras and radars, none of which could be completed without considerable capital investment.

3. Difficulties of deep analysis of massive data. The future Internet of vehicles will be an ecosystem with highly coordinated and unified perception. All data in the ecosystem will be intertwined, bringing a challenge to not only storage space but to data processing. In the existing data processing, only a small part of the data has been paid attention to, while many valuable data, more hidden information of which has not been dug out, has not been used on what they are really worth it. With the development of the Internet of vehicles technology and the application of a variety of services, not only will there be more requirements for information and communication, but also high requirements for data processing capabilities. How to allocate computing resources, how to adjust the priority of different applications, and how to optimize the performance of the system are all issues that need urgent solutions.

4.2   Architecture and Key Technologies

4.2.1   Architecture

4.2.1.1   System Composition

The intelligent transportation information interaction system is designed to connect all kinds of traffic participants together and serve the traffic participants through communication technology, cloud computing technology, big data analysis technology, and cooperative vehicle infrastructure systems technology. For example, with the CVIS technology, the roadside facility can release the traffic condition information of the road ahead and notify the driver in advance to take a reasonable route planning. Figure 4.2 shows an intelligent traffic information interaction scenario in a V2X environment.

V2X-based traffic scenarios include the following four typical scenarios.

Vehicle-to-vehicle communication (V2V): Information interaction between vehicles is achieved through wireless communication technology. The content of information interaction could be dynamics information, steering intent information, etc. In the case of poor traffic conditions, each intelligent vehicle acts as a

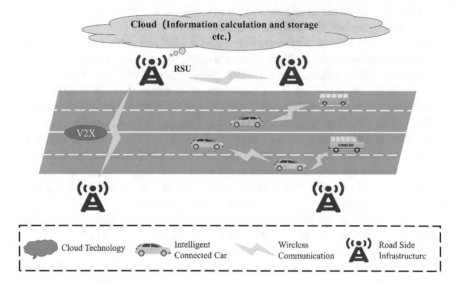

Fig. 4.2 Intelligent traffic information interaction scenario based on V2X

communication node and quickly broadcasts congestion information to remind vehicles behind to plan their routes in advance.

Vehicle-to-infrastructure communication (V2I): This is information interaction between vehicles and road side units (RSUs), traffic lights, etc. For example, during communicating with RSU, RSU is responsible for the storage and intermediate carrier of information interaction because it can store the information of the intelligent vehicle and can also forward and broadcast the necessary information. From this perspective, it also indirectly expands the communication range of vehicles and makes up for the lack of communication capabilities of a single vehicle. In the case of the limited computing power of a single vehicle, RSU can also serve as a computing center to provide support for vehicle's decision-making.

Vehicle to person (V2P): In the foreseeable future, intelligent vehicles can also interact with pedestrians. For example, intelligent vehicles send location information to smart devices worn by pedestrians (such as electronic bracelets) to warn pedestrians that there are vehicles coming in ahead so that the safety of the pedestrians is enhanced.

Vehicle-to-cloud communication (V2C): Vehicles can be connected to cloud servers. For example, vehicles could obtain the latest software patches and bug review information from the traffic management department. In this regard, V2C communication can also be combined with the intrusion detection system (IDS) of information security protection of the Internet of vehicle to strengthen the security of intelligent vehicles and build a sharing platform of cloud-sharing network security of the Internet of vehicles.

4.2.1.2   Technical Realization of the Internet of Vehicles Communication System

The nature of the Internet of vehicles system is the networking of automotive electronic systems. After networking, data can be exchanged with other communication terminals. The common automotive telematics processor is T-Box (Telematics Box), and information can be sent and received through the T-Box intelligent vehicle. Now, dozens or even hundreds of electronic control units (ECUs) have been integrated into most vehicles, which can be divided into power train domain, high-speed information service domain, body control domain, and safety domain, according to the functional areas. Among them, the power train domain includes the control module of the vehicle, the high-speed information service domain refers to vehicle audio and video, Wi-Fi, Bluetooth, and digital broadcasting, the body control domain includes doors, sunroofs, air conditioning, etc., and the safety domain has in-vehicle network intrusion detection systems, firewall technology, etc.

A common distributed architecture of a gateway-based in-vehicle network is shown in Fig. 4.3.

The automotive internal network system mainly refers to CAN bus (controller area network, CAN) and CAN-FD bus (controller area network flexible data-rate), LIN bus (local interconnect network, LIN), media oriented system transport (MOST), and FlexRay (FlexRay Consortium). Among them, the CAN bus is mainly applied to automobile power train domain and is currently the most popular automobile bus system; the LIN bus is mainly used in the body domain; the MOST bus is used in the high-speed information service domain for its fast speed and high transmission rate; and FlexRay is a high-speed, fault-tolerant, deterministic onboard

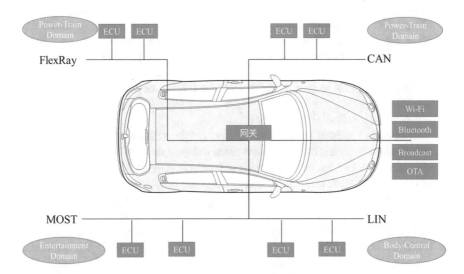

Fig. 4.3 A common distributed automotive internal network system

bus system specially designed for in-vehicle local area networks. It has high real-time performance and strong reliability and is considered to be the next generation bus system.

Intelligent vehicles can use the vehicle-mounted communication unit T-Box to interact with surrounding traffic participants, such as V2I, V2P, V2C, and V2V. At this stage, the research of the Internet of vehicles is mainly focused on the V2V process, that is, to design a system or communication protocol for safe and efficient information communication between intelligent vehicles. Compared with the previous wired fixed connections in the past, intelligent vehicles have characteristics like high-speed mobility and dynamic networking. The network used in the V2V scenarios is a vehicular ad hoc network (VANET), which is a special network form based on the mobile ad hoc network (MANET).

Compared with ordinary MANET, VANET has the following obvious characteristics [10, 11].

1. Instability

When an intelligent vehicle moves at a high speed on a structured road, the communication channel will become complicated and changeable, and it will be easily affected by the surrounding buildings and trees, resulting in poor and unstable Internet of vehicles communication like data loss and signal interruption. In addition, in high-speed movement, the V2V process can also produce Doppler frequency shift effects. For example, when a high-speed moving vehicle passes by a low-speed vehicle, the Doppler frequency shift effect is likely to change significantly, causing big deviations in the results of information interaction.

2. High dynamics

Each vehicle in the VANET system can be regarded as a separate network node. For example, a vehicle driving on a highway will have a relative velocity difference in most cases. When the relative velocity is large, it is likely to cause changes in the communication topology structure of the VANET, which will make the communication connection time between vehicles become very short. To link longer, each vehicle needs to increase its own signal power. Moreover, the changeable network communication structure will also increase the burden on intelligent vehicles to create a network node list, thereby affecting communication efficiency.

3. Single node positioning technology

Real-time location information can be obtained through the onboard GPS intelligent vehicle and combined with other onboard sensors, the vehicle's motion state model can be formulated, which will help improve the speed and reliability of the establishment of the vehicular ad hoc network.

Due to the fast mobility of the intelligent vehicle, it requires its decision-making speed to be fast enough to respond in the shortest time. However, because of VANET's instability, high dynamics, single-node positioning technology, and other features, there will be delay and Doppler frequency shift in the communication of V2V, V2I, etc. These characteristics will greatly affect communication quality. Eventually, they will directly affect the correctness and reliability of the decisions made by intelligent vehicles. Here is a quote from the speech of Liu Yunjie, an academician of the Chinese Academy of Engineering, at the 2020 World Connected Automated Vehicles Conference: "Lidar requires 150 ms. If the delay and jitter of in-vehicle communication are greater than 150 ms, then the instructions may go wrong." It can be seen that problems such as communication delay are one of the factors that limit the development of the Internet of vehicles.

4.2.2   Key Technologies Based on the Physical Layer

The Internet of vehicles belongs to the network communication, so it can learn from the layer deployment and structural design of computer networks. When designing, it is necessary to consider its characteristics and build its own unique network transmission method. The physical layer (PHY) is the bottom layer of the open system interconnection (OSI) model of a computer network. It refers to the physical medium (i.e., the communication channel) for actual data transmission. Note that the physical layer does not refer to a specific physical device, nor the physical medium for signal transmission, but the physical connection, which, on top of the physical medium, provides the original bit stream for the upper layer (data link layer).

Unlike the previous wired and fixed connection communication methods, the physical layer of the Internet of vehicles is more complicated. When a vehicle drives on structured roads similar to urban roads, its communication in the Internet of vehicles will be affected in many ways [12–14]. The striking feature of the wireless channel is its multiple paths, meaning that the wireless channel will eventually produce a signal copy with time delay and frame loss for a single signal (or the same signal) after reflection and scattering. For example, due to the covering of buildings, trees, or other obstacles, wireless signals will reflect and scatter during transmission. Signals that contain the same information but are in different transmission paths will undergo vector superposition after reaching the signal receiving devices. Then, they become time-varying signals, which will cause signal attenuation, thereby also reducing the quality of communication and the reliability of the content. This is also one of the difficulties in applying traditional wireless communication methods to the Internet of vehicles. Therefore, for the development of the network communication mode of the Internet of vehicles, how to establish a high-speed mobile and high-dynamic physical layer specification with stable transmission, strong security, and low communication delay is an important research topic.

To deal with the defects of network transmission delay and instability caused by the features of the Internet of vehicles like high dynamics, scholars and front-line workers have proposed some feasible technologies applied to the physical layer of the Internet of vehicles and verified the reliability and feasibility of these technologies by simulation or testing.

4.2.2.1   Communication Channel Modeling Analysis

To make the model closer to the wireless signal transmission in real space, it is necessary to be clear about the characteristics of wireless signal transmission and the intermediate process, including the possible impact, loss in real space, etc. The essence of the Internet of vehicles is radio transmission. Combined with the theory of radio wave propagation, the steps for modeling and analysis of the Internet of vehicles communication channels are as follows.

1. Deterministic multipath channel modeling

    The Internet of vehicles environment can be formulated by a continuous input and output linear time-varying (LTV) model. The deterministic modeling method generally adopts the ray-tracing method and the finite-difference time-domain (FDTD) method. Among them, $\hat{y}(t)$ is the continuous output value and $x(t)$ is the continuous input value.

$$\hat{y} = \sum_i y_i(t)x(t - \tau_i(t)) \tag{4.1}$$

    Equation (4.1) shows that the output of the Internet of vehicles communication channel $\hat{y}(t)$ is the vector superposition of all signal copies produced by reflection (or scattering) of $x(t)$. The delay in the communication process is expressed by $\tau_i(t)$, and the actual signal value (source signal) calculates each time input $x(t)$. The reason for adopting the linear time-varying model is that as the intelligent vehicle moves, the traffic scene may change. For example, the different densities of the surrounding buildings may cause different communication delays.

2. Multipath channel statistical model

    The modeling of the deterministic multipath channel establishes the propagation mode of the wireless channel, taking into account the environmental information of the wireless channel, such as the impact of traffic scenes and surrounding obstacles, but the multipath propagation of the wireless channel is difficult to make comprehensive consideration, so it is certain that there will be a gap between the modeling of the new multipath channel and the actual space. Therefore, it is also necessary to build a statistical model of the wireless channel to simulate the relationship between the propagation environment and the communication quality (delay and frame loss) when the intelligent vehicle is moving fast.

3. Path loss model of the multipath channel

    In the Internet of vehicles environment, after the wireless signal sent by the intelligent vehicle is reflected by the multipath of different objects, the vector superposition of the multipath signal will cause the multipath attenuation, and the signal received at the receiving end will experience loss. This is because the signal power between the signal receiving terminal and the signal sending terminal is inversely proportional to the distance. The attenuation caused by the reflection of the wireless channel or the change with the distance is expressed by the Path Loss, which is a positive value in dB (decibel).

Considering generalized intelligent transportation scenarios, the loss model can be calculated by the following equation:

$$PL(d) = PL(d_0) + 10\alpha \log\left(\frac{d}{d_0}\right) + N\left(0, \sigma_p^2\right) \tag{4.2}$$

where $PL(d_0)$ represents the basic outdoor propagation loss, α represents the path loss index, d_0 represents the reference distance, which is the expected transmission distance in a specific, non-dynamic environment, $N\left(0, \sigma_p^2\right)$ represents the normal distribution, and σ_p represents the standard deviation of the penetration loss. The normal distribution term can describe the influence of different obstacles and scattering objects on the transmission signal.

4.2.2.2 Evaluation of Wireless Channel Communication Performance

The above describes how to simulate the propagation of wireless channels, the time delay and frame loss of intelligent vehicles moving at high speed, and the loss model formulated during the propagation. The traffic scenes will vary greatly with the urban development degree and terrain changes. Therefore, the wireless channel transmission performance under different traffic situations needs to be evaluated, and the indicators could be Doppler spread, time delay, coherent bandwidth, etc.

4.2.2.3 Communication Channel Estimation Technology

In the intelligent transportation scenario, when the wireless signal sent by the intelligent vehicle is transmitted on the channel, due to the characteristics of the channel transmission itself (reflection, scattering, etc.) and the influence of obstacles, the signal received by the receiving end will be less credible, because the signal may have dropped frames. The idea is to restore the data sent, so the receiving end needs to thoroughly check and correct the received data according to the characteristics of the wireless channel. This situation is similar to when the two parties still choose to exchange information even if they are aware of the open environment, but because the information is incomplete, the receiving end needs to infer the original data based on the incomplete information.

Channel estimation can be understood as a mathematical formulation of the channel's influence on the input signal, and a high-performance channel estimation as an algorithm or method that can minimize transmission errors. Academic and industrial circles have proposed some theoretically feasible channel estimation schemes.

1. Channel estimation method based on the intermediate training series

In the deterministic multipath channel modeling, it is known that the communication channel of intelligent vehicles changes over time. In order to tackle the problems caused by high-speed change characteristics of intelligent vehicle channels, intermediate training sequences could be inserted periodically during data transmission. As a result, it is ensured that the data could be any length.

2. Channel estimation method based on the Wiener filtering

The Wiener filtering is essentially something to minimize the mean square value of the estimation error (the difference between the expected response and the actual output of the filter). In the Network of vehicles system, the Wiener filtering could be arranged between the channel estimation module and the equalization module, and it is needed to solve the optimal coefficients to minimize the mean square error (MSE) of the front and rear channel responses. Understanding the corresponding parameters is necessary when designing the Wiener filtering. Its disadvantage is that the maximum delay extension is approximately calculated, so the result obtained is not the optimal solution.

In general, the difficulty of communication at the physical layer of the Internet of vehicles is that the communication environment of the intelligent vehicle is more complex and changeable. In addition, the correctness and integrity of the communication data directly affect the rationality of the intelligent vehicle's decision-making, so the physical layer design needs full consideration of the actual situation of each part. The existing physical layer technologies include wireless channel physical modeling and simulation, channel estimation, interference cancellation strategies, etc., which have certain practical significance for improving the communication quality of wireless channels. Fundamentally speaking, the characteristics of the wireless channel do not change easily, so a more reasonable technical route could be one to make the most of the characteristics of the wireless channel, such as selecting a more stable multicarrier transmission scheme.

4.2.3 Key Technologies Based on the MAC Layer

The optimized design of the physical layer of the Internet of vehicles aims to solve problems such as the poorer communication quality caused by the impact of the traffic environment. Now the physical layer is assumed to have improved the communication quality and ensured the correctness and reliability of the data. But the essence of VANET is broadcasting; i.e., after the vehicle sends information, most of the surrounding vehicles can receive the data. This is similar to the broadcast mechanism of the CAN bus; that is, the information sent by one ECU can be sent to any other ECUs. Because this assumption is important, this feature has also become the main target for the development of the MAC layer.

There are several key problems in VANET, and each problem needs a corresponding MAC layer solution [15, 16]. Here are a few typical problems.

1. Hidden nodes

   Because of the wireless transmission characteristics of VANET, two intelligent vehicles may not perceive each other's existence. If they send data packets simultaneously, it will create grouping conflicts of the two intelligent vehicles in the effective communication range. This is the well-known hidden node problem. The hidden node phenomenon may seriously affect the security of the Internet of vehicles system under certain circumstances. The high-speed moving intelligent vehicle will also bring system dynamics problems; that is, due to the rapid changes in the position of the vehicle, the frequency-variable channel will be weakened, which in turn lead to other issues.

2. Scalability

   At present, the bandwidth of the frequency channel allocated to VANET is 5–20 MHz. When the intelligent vehicle operates in an intense traffic area, the wireless communication channel is prone to blockage.

3. Communication topology

   When the intelligent vehicle is operating normally, it can be assumed that it will periodically transmit important information such as dynamics to the surrounding vehicles. The communication topology can be a single-hop unicast or single-hop broadcast. Assuming a vehicle spots an accident, it needs to disseminate the accident as soon as possible. The communication at this time should be selected as a single-hop broadcast. Different security levels and emergency solutions require different communication topological structures.

The main purpose of MAC layer optimization is to solve the above problems. According to different technical points, the corresponding solutions can be subdivided as follows:

1. Message priority division

   Different types of messages (such as vehicle dynamics, road conditions, signal lights) may be transmitted in the same channel, making it impossible to estimate the importance and urgency of the message. So, a message priority mechanism could be added to the MAC layer to transmit the urgent and important information first after determining the message type. This mechanism helps to sort the situation and could also use the channel access time to determine the information delay.

2. Information-distance mapping relationship

   In most instances, the intelligent vehicle closer to the current vehicle should obtain information faster and could establish a distance-related network communication topology based on this information.

3. Robustness of communication performance

   In the VANET environment, wireless communication weaknesses may easily occur, including slow weakness and fast weakness. So, it is feasible to make a model concerning the degree of weakness and search the objective factors that are most related to the weakness degree and the conditions that could enhance communication robustness.

Next, take D2D (device-to-device) as an example to introduce the D2D central MAC solution. D2D is a communication method in which two equivalent user nodes communicate directly. In the network structure composed of D2D communication nodes, each user node can send and receive signals and has the function of automatic routing (forwarding messages). It is worth noting that in the D2D network, the communication node also assumes the roles of server and client. The communication node can sense the existence of the surrounding intelligent vehicles, allowing the intelligent vehicles to freely form a traffic flow or special queue. Its technical advantages are summarized as follows.

1. It meets a variety of intelligent transportation needs, such as road condition information push, large-scale network security upgrade patch transmission, V2X information interaction, etc.
2. Its working frequency band is approved. It is a derivative technology of LTE communication technology using the frequency band of the cellular system that could guarantee a good user experience for the two communicators, regardless of the communication distance.
3. D2D allows a large amount of information interaction between intelligent vehicles and transportation infrastructure. Compared with Bluetooth transmission, D2D does not require matching and the transmission speed is faster.

D2D can be applied to V2V communication in the Internet of vehicles. For highway driving, the sudden braking, lane change, overtaking, and other behaviors of the vehicle will threaten the surrounding vehicles. D2D can be applied to the highway scene. For example, when the intelligent vehicle in front suddenly decelerates, you may send out early warning information through D2D. At this time, the surrounding vehicles receive the early warning information and adopt appropriate driving strategies. In some special cases, such as when the driver's inattention or bad weather like thick fog blocks the driver's sight, the intelligent vehicle can actively control the vehicle to avoid dangers, thereby effectively reducing the accident rate (Fig. 4.4).

In addition, D2D can also be applied to vehicle identity recognition. For example, when an ordinary intelligent vehicle is close to a police vehicle or an ambulance performing a task, D2D technology can identify vehicles with special characteristics, thereby attracting the attention of surrounding vehicles. From this perspective, D2D communication could be well used in future emergency medical assistance or special tasks (as shown in Fig. 4.5).

On the whole, the communication topology of the Internet of vehicles will become more and more complex, and the heterogeneous network composed of different types of communication nodes will also become larger and larger. There will be new requirements for the service level of the Internet of vehicles in various industries, such as the transportation industry, urban travel services, rescue services, etc. The MAC layer of the Internet of vehicles will also face new challenges, requiring continuous optimization and improvement to meet security requirements.

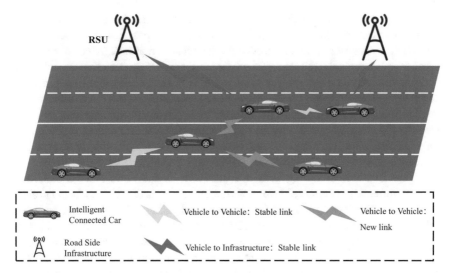

Fig. 4.4 D2D-based intelligent transportation communication architecture

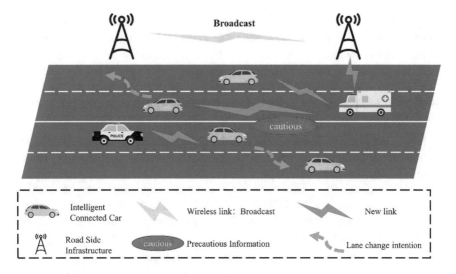

Fig. 4.5 Schematic diagram of D2D-based intelligent transportation application

4.3   Technical Standards

4.3.1   Overview of the Development of Technical Standards

With the increasing application of Internet technology in cities, traffic information interaction has become more intelligent and convenient. It is necessary to normalize the transportation industry information to address the existing problems of uncertainty and low efficiency so as to ensure standardized and orderly informatization in the transportation industry, improve the efficiency of informatization services, and to accelerate the construction of a powerful transportation country, meet the demand of the time, and promote the effective development of the transportation industry. The Ministry of Transport Science Research Institute and Beijing Traffic Operation Monitoring and Dispatching Center and other institutions responded by formulating the Urban Public Transport Management and Service Data Exchange Specification in 2017, which stipulates the overall requirements for data exchange of urban public transportation. In 2019, the Transportation Informationization Standard System (2019) was promulgated by the Ministry of Transport in 2019, including industrial standards in the basic and general field, infrastructure, data resources, information application, network security, and engineering specifications. And in the information-application aspect, the Transportation Informationization Standard System (2019) established specifications for the data exchange message, exchange index and format, exchange interface, and other technologies. As a long-term research topic, information interaction has already had a relatively solid foundation. The existing intelligent transportation information interaction technology also refers to the information interaction content in the communication field. The later part of this chapter will introduce various standards for the application of intelligent transportation information interaction technology on vehicles, which have a high penetration rate in the traffic, by taking the Internet of vehicles as an example.

In the Internet of vehicles technology standard system, there are two levels of wireless and application, which are distinguished by the underlying wireless communication technology. At present, there is two mainstream Internet of vehicles communication technology standard routes around the world [17]:

1. DSRC technology: Based on IEEE 802.11p, it provides short-distance wireless transmission technology, which is widely used in vehicle-to-vehicle and vehicle-to-infrastructure communication.
2. C-V2X technology: V2X inter-vehicle communication technology carried by cellular networks includes LTE-V2X and NR-V2X (new radio vehicle-to-everything), supporting vehicle–vehicle, vehicle–infrastructure, vehicle–network, and vehicle–person applications.

By reasons of the coupling between the upper-level application standards and the deployment areas of the Internet of vehicles technology and the differences in technology accumulation in different countries and regions, the technical standards for the Internet of vehicles at home and abroad are various.

The research on V2X communication in the United States is mainly about DSRC. In 1999, the Federal Communications Commission of the United States used the 5.9 GHz of 75 MGz bandwidth as DSRC. The American Society for Testing and Materials (ASTM) issued the DSRC standard (ASTM E2213-02 2003) in 2002, which further promoted the study and application of DSRC technology in the Internet of vehicles. In 2004, the American Institute of Electrical and Electronics Engineers revised the 802.11p (i.e., wireless access in the vehicular environment,

WAVE) and gradually began to formulate the WAVE standard. In the same year, the American Computer Association firstly created the term "VANET" for the international standard seminar on the Internet of vehicles. Additionally, the United States has actively pushed forward the legislation. In December 2016, the United States promulgated the NPRM (Notices of Proposed Rule Making) on V2V communication in the Federal Motor Vehicle Safety Standard, which gave suggestions on operating frequency, communication capability, market penetration, etc., of the V2V communication devices. In June 2017, SAE (Society of Automotive Engineers) prepared and established a C-V2X working group to conduct researches on enhanced applications and direct communication, etc.

Japan is a rising star. The big three of the automobile industry of Toyota, Honda, and Nissan have promoted the development of the Internet of vehicles. With the popularization of 4G and the development of big data cloud, IT companies such as NEC and Hitachi have also begun to enter the Internet of vehicles market in Japan. As early as 2003, Japan developed a mature road traffic information communication system of Internet of vehicles information system, which is a system that sends instant road traffic information such as road congestion and traffic restrictions to the vehicle navigator and has covered the whole country. In February 2012, Japan published the ARIB STD-T109 standard of the safety applications, which is for the prevention of V2V or V2I collisions and uses the 10 MHz frequency in the 700 MHz frequency band. Field tests were conducted in Hiroshima and Tokyo, and large-scale tests have begun.

The term "Internet of vehicles" was first mentioned at the 2010 China International Internet of Things (Sensor Network) Expo and China Internet of Things Conference. The comfortable and safe driving experience and the convenience this term described gave the people then a good idea of what to expect. But at that time, the Internet of vehicles was only a concept with no actual technology and product. In October of the same year, the State Council proposed in the "863" plan the research on key technologies of intelligent vehicles and cooperative vehicle infrastructure systems, as well as key technologies of collaborative control of regional transportation in large cities. During the "Twelfth Five-Year Plan" period, the Ministry of Industry and Information Technology started from various aspects such as industrial planning and technical standards to increase support for in-vehicle information services and accelerate the full development of the automotive Internet of things industry. In 2013, Datang first publicly proposed the LTE-V technology based on the LTE system, which has become the LTE-V2X standard of 3GPP. LTE-V2X, as a comprehensive communication solution for CVIS, can provide low-latency, high-reliability, high-speed, and secure communication in a high-speed mobile environment. LTE-V2X meets the needs of multiple applications of the Internet of vehicles and is based on the TD-LTE communication technology. Therefore, it can make the most resources such as TD-LTE deployed networks and terminal chip platforms, saving network investment and reducing chip costs. In July 2014, Alibaba and SAIC signed a cooperation agreement to carry out research and development related to connected vehicles and to create a connected vehicles ecosystem. The automobiles mentioned above will be equipped with Alibaba YunOS operating system, integrating resources like Alibaba Cloud, Amap, and Alibaba communications. In general, the potential scale of China's Internet of Vehicles is huge, but there is currently no overall policy, and the development of the entire industry is disordered.

From a vertical point of view, researches on the Internet of vehicles began to spring up in the era of 4G communications. The 3GPP of the International Organization for Standards clarified the basic concepts, network architecture, and application scenarios of C-V2X in the R14 version. The key technologies required for the Internet of vehicles include wireless communication technology, high-precision positioning technology, high-precision map technology, intrusion detection technology, privacy security technology, vehicle-person interaction technology, traffic signal optimization technology, etc.

The current 3GPP R16 standard was frozen in August 2020, and R16 has improved in the Internet of vehicles communication ways, side-link physical layer enhancements, HARQ retransmission, synchronization, resource allocation, etc. R16 has introduced the channel state information reference signal and the side-line HARQ mechanism, which supports the measurement and feedback of the side-line link and the feedback of the side-line HARQ information to the base station separately. R16 has also improved the structure of the synchronization signal block, with better synchronization performance in low-latency and high-reliability scenarios.

At the end of 2018, the Ministry of Industry and Information Technology of the People's Republic of China officially promulgated the 5.905–5.925 GHz dedicated frequency band, which supports the evolution to the 5G standard for the industrial application of connected automated vehicles. After that, the United States also allocated a 5.9 GHz 20 MHz bandwidth for C-V2X in December 2019, and European countries have also turned from the IEEE 802.11p supporters to the technology neutrals. China has provided a series of policy supports for C-V2X and has gradually formed a complete industrial ecology in the productization. Chinese companies such as Datang and Huawei have taken a leading position in the Internet of vehicles application standards, industrial cooperation, and application development. In 2019, the C-V2X "Four Layers" Interoperability Application Demonstration was held, which added communication security demonstration scenarios and effectively verified the C-V2X communication security technology plan. In the future, China is expected to truly realize the ultralarge bandwidth and low-latency and reliable communication of mobile facilities, supporting the massive connection of mobile terminals, promoting the research and large-scale application of the Internet of Vehicles, and moving to fully automated driving.

4.3.2   DSRC Technology and Standards

The architecture of the DSRC technology stack is shown in Fig. 4.6. The underlying physical layer (PHY) and media access control layer (MAC) are defined by the IEEE 802.11p protocol. In the upper network service, DSRC adopts the IEEE1609 series

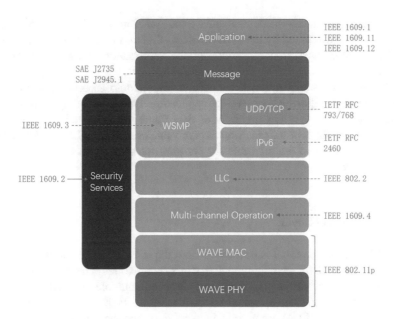

Fig. 4.6 DSRC protocol stack structure diagram

of standards, which defines the multichannel operation and the security of each communication entity.

The IEEE 802.11 series of standards define the wireless environment of the shared medium for WiFi services, which are suitable for fixed devices or low-speed scenarios. In order to support the high-speed mobile scenes such as the Internet of vehicles, technical improvements have been made based on the 802.11 series of standards, and the stricter 802.11p protocol of adjacent channel indicators has been formulated. The 802.11p protocol was released together with 802.11 as the supplementary protocol of the later one. The data link layer is divided into a MAC layer and an LLC layer. The MAC layer of 802.11p uses enhanced distributed channel access (EDCA) to support differentiated QoS demands and to replace the DCF in 802.11. IEEE 1609.4 defines the MAC extension layer on the MAC layer, allowing multichannel operation [18].

The logical link control (LLC) layer, as a part of the data link layer, provides the upper layer with a method to process any type of MAC layer.

The network layer and transmission layer above the link layer provide transmission modes of different message types. IEEE 1609.3 functions in the network layer and the transmission layer and defines WAVE short message (WSM) and the corresponding WSMP (WAVE short message protocol) as methods of information transmission. Among them, the WSMP with a smaller protocol header is often used for anti-collision single-hop information, and IPv6 is usually used for multihop information.

On the basis of the network layer and the transmission layer, SAE J2735 and SAE J2945.1 standards issued by the SAE (Society of Automotive Engineers) have been applied. These two standards regulated the information content and structure of DSRC.

The uppermost application layer can be divided into secure applications and nonsecure applications according to application types, including two message types. SAE J2735 provides a variety of data frame formats, realizing road safety applications by the sent status information of vehicles on the road. The SAE J2945/1 defines the minimum requirements for performance such as message sending rate and accuracy.

4.3.2.1   Physical Layer

The IEE 802.11p protocol is developed from the IEEE 802.11a and is a dedicated protocol standard for vehicle-mounted wireless networks. The physical layer of 802.11p still uses orthogonal frequency division multiplexing (OFDM) as the modulation method. The specified bandwidth of IEEE 802.11a is 20 MHz. The 48 subcarriers of OFDM are used to transmit data, and the four subcarriers are for tracking frequency position and phase deviation, which are only suitable for the indoor static environment. For the purpose of adapting to the dynamic time-varying of the fast fading channel in the high-speed mobile environment, reducing the interference of Doppler frequency shift and multipath propagation to the signal, ensuring real-time and robust road message transmission, and reinforcing the safety of vehicle driving in the Internet of vehicles environment, 802.11p has made adjustments to the specific parameters of the physical layer, which are reflected in the halved channel bandwidth, transmission rate reduced by half, and increased time-domain parameters. The specific parameters of IEEE 802.11p are shown in Table 4.2.

Table 4.2   Physical layer parameters of IEEE 802.11a and IEEE 802.11p

Physical layer parameters	IEEE 802.11a	IEEE 802.11p
Standard bandwidth (GHz)	5.15–5.875	5.85–5.925
Physical layer rate (Mbps)	6–54	3–27
Modulation method	BPSK, QPSK, 16QAM, 64QAM	Constant
Encoding rate	1/2, 2/3, 3/4	Constant
Number of subcarriers	52	Constant
Symbol period (µs)	4	8
Protection period	0.8	1.6
Fast Fourier transform period	3.2	6.4
Detection time	16	32
Subcarrier interval	0.3125	0.15625
Communication range (m)	30–45	300–1000

The physical layer of IEEE 802.11p can be divided into a physical medium dependent (PMD) layer and a PLCP (physical layer convergence procedure) layer [19]. The PMD layer is used as a physical medium connection layer to convert the signal to a specific medium or vice versa. The PLCP layer maps the ATM cells to the physical media and defines the specific management information.

When the signal is sent, PLCP requests PDM to transmit frame and provide data transmission rate, transmission power, and coded bit information. The PMD layer performs OFDM modulation, and the flowchart is presented in Fig. 4.7. OFDM converts high-speed data signals into parallel low-speed subdata streams and then modulates them to each subchannel for transmission. The process includes fast Fourier transform (FFT), wave filtering, and radiofrequency modulation.

When receiving, the PMD receiving end processes and demodulates the message of each channel and transmits the received frame to the PLCP layer. In contrast to the sending process, the receiving steps include fast Fourier transform, clock recovery, radiofrequency demodulation, etc. When receiving frame information, the RSSI (received signal strength indication) can be provided simultaneously.

During signal transmission, if the signal has a time delay, the OFDM symbols of a certain time slot will be overlapped on the time slot of another path. If the overlap is too long, the real symbols of the adjacent time slots will be disordered, forming intersymbol interference (ISI). Therefore, it is necessary to insert a guard interval between the code elements before sending the modulated signal when the modulated signal reaches the receiving end through the wireless channel, because the subcarriers cannot maintain a good orthogonal state due to the severe intersymbol interference caused by multipath propagation.

On the other hand, if the multipath delay occurs, the idle guard interval will be included in the integration time when calculating the fast Fourier transform, which leads to the incapability of the integration to include the entire waveform. As a result, the orthogonal state between the subcarriers will be destroyed, thereby causing the intercarrier interference (ICI). Therefore, it is necessary to fill in the cyclic prefix signal in the guard interval, sample the post-time of each OFDM, and copy the samples to the front of the OFDM symbol to ensure the waveform of the OFDM symbol within the integration time.

When sending, the PLCP sublayer adds the PLCP preamble, PLCP header, PSDU, Tail bits, and PAD bits to the MAC frame sent by the MAC layer to form the IEEE 802.11p physical layer protocol data unit (PPDU). The MAC frame length, transmission data rate, and transmission power are included in the transmission. When receiving, the PLCP layer parses out the MAC frame from the PPDU through the reverse operation during transmission and transmits it to the MAC layer. The PPDU frame structure is presented in Fig. 4.8.

The PLCP preamble consists of a short training sequence that repeats ten cycles and a long training sequence that repeats two cycles, which is 32 μs in total. The 12 training sequences are used for receiver's timing synchronization, carrier frequency offset estimation and channel estimation, etc.

The PLCP header contains six fields: rate, reserved, length, parity, tail, and service. The rate field is responsible for the modulation and code rate information

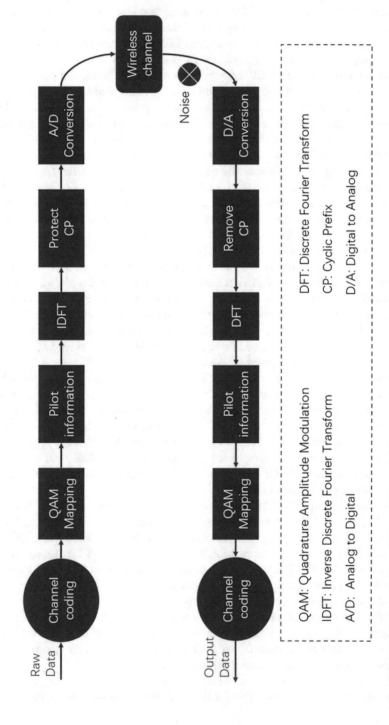

Fig. 4.7 OFDM system basic modulation and demodulation structure diagram

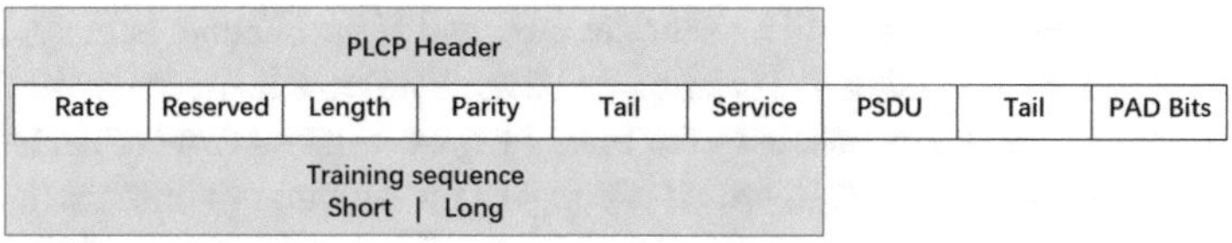

| Rate | Reserved | Length | Parity | Tail | Service | PSDU | Tail | PAD Bits |

PLCP Header

Training sequence
Short | Long

Fig. 4.8 IEEE 802.11p PPDU architecture

of the subsequent transmission data symbols; the length field indicates the number of bytes of the PLCP service data unit (PSDU) of the PLCP sublayer that is requested from the MAC layer to send by the physical layer. Parity is used for odd–even checking. The service field and PSDU are used as data to form multiple OFDM symbols, which are sent at the rate described in rate. The TAIL field is used for bit scrambling, and the PAD field is used to fill the number of bits required for matching the modulation codes.

4.3.2.2   MAC Layer

The MAC layer of IEEE 802.11p continues to use the relevant content of IEEE 802.11a. It adopts the DCF (distribution coordinate function) mechanism and the EDCA (enhanced distribution channel access) mechanism and forms the three main channel access mechanisms of the IEEE 802.11p MAC layer algorithm with the distribution coordinate function, point coordinated function, and hybrid coordinated function. On a certain level, it has made the improvement in simplifying the communication process between nodes, access control methods, and access priority processing.

1. DCF mechanism

   The DCF mechanism provides distributed multiple access control, using the carrier sense multiple access/collision avoidance (CSMA/CA) protocols as the basis to make each node equally getting access to the channel. CSMA/CA uses energy detection carrier detection and energy carrier hybrid detection to perform carrier monitoring and decides whether there are other nodes sending data by testing the strength of the signal [20].

   Before the node sends data, it first monitors if the channel is busy or not in the arbitration interframe space (AIFS). When the channel is idle and continues in the distributed interframe spacing (DIFS), the node starts to send frames; when the channel is occupied, it will continue to monitor until the channel is idle in the DIFS time. Then the node starts to perform a random backoff; i.e., the number of random backoffs is used as the initial delayed transmission time. If the wireless channel becomes idle during the process, the backoff count of the counter is decremented by 1; if the channel is busy, the counter hangs up until the channel becomes idle, and the backoff process continues based on the previous backoff. When the count is decremented to 0, the backoff process ends. If the channel is free at this time, the data will be sent directly, and if the channel is busy, the next backoff process is executed. The specific backoff process is presented in Fig. 4.9.

The CSMA/CA protocol uses the binary exponential backoff (BEB) to determine the duration of backoff, which is equal to the set random number multiplied by the duration of each time slot. In the backoff process, the random number range selected by the node is called the contention window (CW), and the backoff time is determined by the following equation:

$$t_{\text{back-off}} = \text{Random}(0, \text{CW}_i) \times \text{Slottime} \qquad (4.3)$$

where CW_i represents the maximum contention window for the ith backoff, Random() represents a random number, Slottime represents the length of a time slot, and i represents the number of backoffs, also known as the backoff order.

2. EDCA mechanism

The Internet of vehicles needs to ensure the accuracy and low latency of information transmission when driving a vehicle and the priority access of high-priority information. The IEEE 802.11p protocol continues to use the EDCA mechanism in IEEE 802.11e, which defines the priorities by setting different EDCA access parameters for different access categories (AC) to support different access priorities.

The EDCA algorithm defines the eight service types based on IEEE 802.11d and the four access types of this layer [21]. The four access types are arranged from low to high according to the priorities: nontimely flow (AC0), background flow (AC1), video flow (AC2), and voice flow (AC3), and the priorities are shown in Table 4.3.

EDCA is based on strict priority. According to the QoS parameters of the service, it is mapped to the corresponding AC, and fixed MAC competition parameters are set so that different access types could correspond to different competition parameters to distinguish priorities. When the channel competition is fierce, channels of low priorities will actively back off.

Though DSRC technology has gone through decades of research and development, has established detailed protocol standards, and has been verified and applied to a certain extent, it still has certain limitations due to reasons such as communication performance and deployment costs. First, in complex scenarios with many obstacles, the line-of-sight transmission of DSRC will be greatly affected; second, DSRC has higher requirements for infrastructure deployment; moreover, the long-term-development automated driving has higher requirements in terms of communication range, delay, and robustness. Although the DSRC technology is strongly supported and promoted by the United States, its commercialization process is not smooth. The Federal Communications Commission recently passed a proposal to reallocate the 75 MHz spectrum in the 5.9 GHz frequency band, part of which will be used for C-V2X technology.

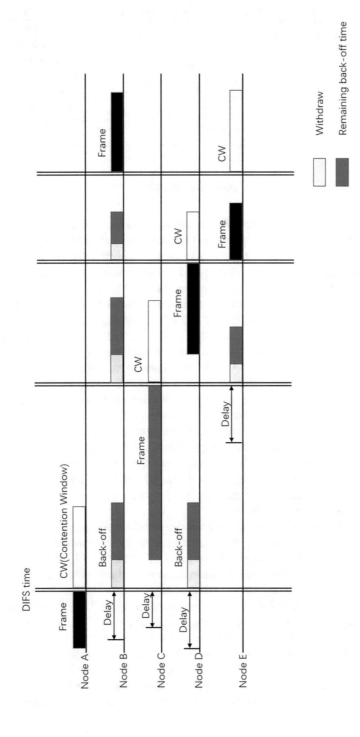

Fig. 4.9 Node backoff process

Table 4.3 Mapping relationship between EDCA priorities and access categories

Priority	Access category (AC)	Direction information
0	0	Nontime flow
1	0	Nontime flow
2	0	Nontime flow
3	1	Background flow
4	2	Video flow
5	2	Video flow
6	3	Video flow
7	3	Voice

4.3.3   C-V2X Technology and Standards

The cellular communication network arranges multiple base stations neatly and has been used to support remote information transmission and infotainment services, having advantages like large capacity and high reliability [22]. By utilizing the technological and industrial advantages of cellular mobile communication and combining with the short-distance direct communication technology, the complex scenes of high-speed movement of nodes between vehicles could be adapted to and the low-latency and high-reliability V2V and V2I communications could be realized.

LTE-V2X is the first cellular Internet of vehicles (C-V2X) wireless communication technology and has defined two communication methods for vehicle applications: centralized method (LTE-V-Cell) and distributed method (LTE-V-Direct) [23]. The LTE-V-Cell method is also called the cellular type. This method uses the base station as the control center for centralized control and data forwarding. It has functions like centralized allocation, congestion control, and interference coordination and supports large-bandwidth and large-coverage communications, significantly improving LTE-V2X networking efficiency. LTE-V-Direct communication is independent of the cellular network and does not require base stations as support to achieve direct communication between the vehicles and surrounding nodes. At the same time, LTE has established the basic system architecture and technical principles of C-V2X, as shown in Fig. 4.10.

C-V2X is led by 3GPP and standardized into two development stages: LTE-V2X and NR-V2X, while considering backward compatibility and forward compatibility [24]. 3GPP completed the standardization of LTE-V2X in R14. This standard is oriented to the basic communication requirements of road safety service, focusing on vehicle-to-vehicle communication and gradually supporting the environmental needs of other vehicle-to-infrastructure communications. LTE-V2X introduces a PC5 interface communication method that works in the 5.9 GHz frequency band in cellular communication. This interface supports V2X short-distance direct communication and is used to realize assisted driving and semi-autonomous driving. 3GPP continues to perfect the LTE-V2X standard in R15 to satisfy the enhanced needs of some Internet of vehicles. 3GPP started the research on NR-V2X in R16, which is mainly based on 5GNR's PC5 interface and Uu interface enhancement.

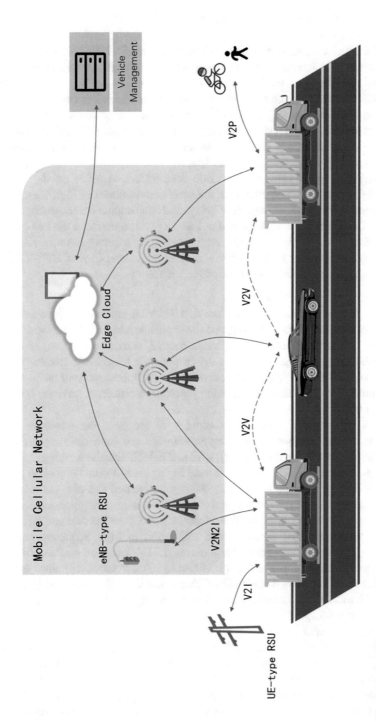

Fig. 4.10 C-V2X system architecture

4.3.3.1 Key Technologies of LTE-V2X

LTE-V2X provides four direct communication modes of V2V, V2P, V2I, and V2N based on the PC5 interface and also supports the network communication mode based on the Uu interface. To meet the needs of low-latency V2X services, LTE-V2X has carried out research on direct communication between devices, which can not only centrally allocate PC5 interface resources through the network but also distribute PC5 interface resources in a resource pool. LTE-C2X transmits the basic safety messages (BSM) including velocity, location, time, allocation, steering, braking, etc., on the PC5 interface.

In order to support multi-to-multinode communication mode on the basis of mobile cellular network and cope with the severe environment of high communication frequency, fast time change, low delay, and multisource heterogeneous in the Internet of vehicles, LTE-V2X carried out the targeted research on the key technologies of frequency bands, synchronized management, wireless resource allocation methods, etc. According to the evolution of the standard versions of LTE-V2X, the key technologies of LTE-V2X are introduced as follows.

1. Physical layer structure design

    The standard protocol architecture of LTE-V2X consists of three parts: the physical layer, the data link layer, and the application layer. The physical layer is mainly in charge of frame transmission control, channel activation, and failed services; the data link layer is mainly responsible for reliable transmission of information, providing error and flow control; the application layer mainly fulfills communication initialization, releasing procedures, broadcast services, and other related tasks.

    In a typical LTE scenario, a stationary device or a medium-to-low-speed mobile device terminal communicates with a fixed base station with a low working frequency band (2.6 GHz), while LTE-V2X sets the working frequency band at a higher 5.9 GHz, geared toward higher mobile communication parties. LTE-V2X adopts the pilot frequency encryption method and adds to the number of the demodulation reference signal columns to increase the pilot density in the time domain, thereby reducing the reference signal time interval and effectively implementing channel detection and estimation in high frequency bands in typical high speed scenarios. In addition, LTE-V2X uses single-carrier frequency-division multiple access (SC-FDMA) in modulation to reduce peak-to-average ratio, supports flexible modulation and coding schemes, and is compatible with multiple message types, meeting various application requirements of the Internet of vehicles.

2. Resource allocation method

    Internet of Vehicles service of high communication frequency, quick topology changes, and unstable service group size could cause resource conflicts easily. LTE-V2X uses a fixed-size physical sidelink control channel (PSCCH) to indicate the number of resources of the physical sidelink share channel (PSSCH). In order to further reduce the PSCCH overhead, the PSCCH's resource granularity is divided into subchannels of the same size, transmitting data with one or more subchannels.

Combining with the characteristics of the road safety service cycle, LTE-V2X adopts a semi-persistent allocating mechanism based on perception channels. Through sending the transmission resources of node reservation periodicity, LTE-V2X can take into account the needs of other nodes, which is good for the receiving node to sense the resource status and avoid resource conflicts. This not only reduces the signaling overhead but also improves the reliability of message transmission.

3. Synchronization mechanism

In the LTE-V2X cellular system, with the base station used as the synchronization source, the nodes covered by the cellular are synchronized with the fixed base station, and the communication between the terminals needs to synchronize the uplink and downlink data. In the LTE-V2X system, since the communication node supports the global navigation satellite system (GNSS) module with high-frequency accuracy, the communication node can directly obtain accurate synchronization information. Considering that LTE-V2X and LTE cellular systems share the same carrier and the signal of LTE-V2X direct communication will interfere with the uplink synchronization of the LTE cellular network, the base station is still used as the synchronization source and broadcasts the time deviation between the base station and the GNSS to make adjustments and compensations in nods. Therefore, there are three synchronization sources in LTE-V2X: GNSS, base station, and node. LTE-V2X uses the base station as the synchronization source within the cellular network coverage and is configured with the synchronization method, determining the synchronization source outside the cellular network by configuration method to achieve unified synchronization timing.

4.3.3.2   NR-V2X Key Technology

The NR-V2X based on the 5G New Radio has been improved upon the LTE-V2X version, which basically follows the LTE-V2X communication architecture [25]. It adopts the new 5G NR technology on the PC5 interface to support more flexible V2X services, which supports the millimeter-wave and has introduced edge computing and communication slicing and other characteristics as well. Therefore, it can provide a greater reliability, a lower latency, and a higher data transmission rate. The 5G NR-based NR-V2X technology is in smooth evolution under the cellular communication system [26], supporting the future vehicle communication system in terms of technological development and industrial promotion. NR-V2X will support vehicles in adjacent areas to exchange real-time information through collective perception of environment (CPE) and locally realize the collective perception of the surrounding environment by the vehicle network, thereby reducing the probability of accidents.

To support applications such as vehicle–vehicle driving intention coordination, sensor sharing, and formation driving, NR-V2X conducts further research on direct communication. With reference to 3GPP R16, the key technologies of NR-V2X are summarized as follows.

NR Sidelink is composed of PSCCH, PSSCH, PSBCH, and PSFCH (physical sidelink feedback channel). The first three channels follow the arrangement of LTE-V2X. PSFCH is introduced by NR-V2X to support HARQ transmission; on the basis of supporting broadcast methods, unicast and multicast mechanisms are used to support complex services requirements, and a feedback mechanism is introduced to achieve higher reliability compared to broadcast mechanisms; it supports link adaptive operations including link path loss measurement-based opening loop power control mechanism, HARQ feedback mechanism of direct link, channel state measurement, and feedback mechanism of direct link; NR-V2X supports a centralized control scheme based on base station control (mode1) and distributed nod self-selection scheme based on sensing (Mode2); in synchronization, the synchronization sources of NR-V2X include the sub-synchronization mechanisms of GNSS, gNB, eNB, and NR UE, and through the live sidelink synchronization signal block (S-SSB), synchronization messages can be delivered; in the future, NR-V2X will also conduct research on the power saving mechanism of V2X terminals, the enhancement mechanism of resource allocation methods, and the coordinated allocation of resources between UEs.

4.4   Typical Applications and Future Trends

4.4.1   Typical Applications

One of the purposes of intelligent transportation is to ensure the safety of drivers and passengers and reduce possible traffic accidents and casualties. Therefore, many technologies are implemented focusing on safety. In addition to safety, efficient driving of a vehicle also requires the knowledge of the specific location of the vehicle on the road, without which all applications are just an illusion.

4.4.1.1   Cooperative Positioning of Vehicles Based on Information Interaction

The current Internet of vehicles positioning technology mainly relies on GPS or the Beidou navigation system, but GPS or the Beidou navigation system cannot provide stable positioning service under all conditions [6]. Generally speaking, GPS positioning requires communication with at least three satellites because we are in three-dimensional space. And the approximate position could be obtained through the mathematical calculations of satellite positions and communication delays among different satellites. In some specific scenes, such as dense landscapes, cities with tall

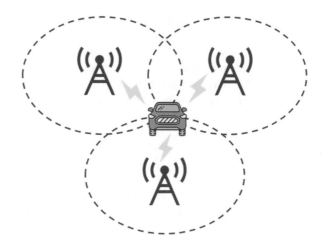

Fig. 4.11 Principle of triangulation location

buildings, or tunnels and basements with direct obstructions, it is difficult to use GPS to locate. Even if the vehicle is in a relatively open place, the positioning accuracy is between 5 m and 10 m. Considering the general road width is 3.5 m, it is not possible to distinguish the main and auxiliary roads under this positioning accuracy. Therefore, the Internet of vehicles generally requires road-width-level positioning accuracy, that is, no more than 3 m.

With the rapid development of GPS positioning technology, the accuracy of GPS is getting higher and the price of GPS receivers lower, so the application is more and more extensive. In view of the GPS positioning accuracy and the occlusion in memorable scenes, GPS is far from being able to meet the requirements of autonomous driving.

Alternative methods of GPS positioning can be divided into two categories of noncooperative positioning and cooperative positioning.

In the noncooperative positioning, also known as the triangulation location, each node needs to communicate with at least three anchor nodes. The principle of the triangulation location is shown in Fig. 4.11. Measure the distance from the three nodes to the current node, and then draw circles with the three nodes as the centers, and get the position of the current node, which is the intersection of the three circles. However, in reality, three anchor nodes are difficult to find, and if a large number of roadside units are pre-arranged to serve as anchor nodes, it will significantly increase the cost.

Cooperative positioning, which uses multiple vehicles to confirm the distance to each other and transmit ranging information, is a low-cost and high-efficiency method for vehicles to achieve precise positioning. By installing a radio transceiver on the vehicle and measuring the time difference or signal strength between different vehicles, it can estimate the distance between the vehicles with the filtering algorithm. Then, it can calculate the approximate position of the vehicle, greatly reducing errors. There are two types of collaborative positioning: central collaborative positioning and distributed collaborative positioning. In centrally coordinated positioning, the positioning node transmits the position information of the vehicle to the central processor in a multihop manner, and the central processor calculates the precise position of each node. Since the central system is an indispensable part of this positioning, its scalability is poor, making it difficult to be applied to the Internet of Vehicles. Distributed collaborative positioning does not rely on the central processor, and each node of which can complete the final precise positioning through multiple iterations, so it is more and more favored by researchers. In each iteration, the node obtains the position information from other nodes, and through a certain algorithm, the position of each vehicle can be estimated.

Cooperative positioning algorithms are mainly divided into two types: Bayes-based type and non-Bayesian type. The most classic non-Bayesian cooperative positioning algorithms are the minimum mean square estimator and the maximum likelihood estimator. They use the loop iteration method to minimize the cost function by building a cost function that includes the measured value so as to get an accurate estimation of the final position. As for the Bayesian collaborative positioning algorithm, it referred to the classic insight of Bayesian probability and can estimate more accurately by introducing a priori information. A large number of existing research studies are about Bayesian probability. By analyzing the existing research, it can be found that the current research has the following shortcomings. First, with the increase in the nodes participating in the positioning, the accuracy could be improved step by step. But at the same time, there will also be an increase in the overhead of communication and calculation. Second, in the traditional distributed collaborative algorithm, all nodes need to send their own location information to other neighboring nodes, causing extremely severe data packet collisions and network congestion for the vast amount of data. Third, there will be some abnormal nodes in the network, and the positioning of abnormal nodes will have large deviations or even errors, which leads to serious problems in their own positioning. Then, the positioning information will lose its value as a reference and could even affect the position calculations of the surrounding nodes. As a result, the positioning accuracy of the entire network could be lost. Fourth, in the distributed collaborative positioning algorithm, the nodes are through loop iterations to complete the final position estimation. The current common method is to set a maximum round of loop iterations and stop the iterations as long as the preset value is reached. In actual experiments, it is found that in many situations, most nodes do not need such a large number of iterations, which causes a serious waste of computing power resources.

In order to improve the accuracy of GPS positioning, researchers have proposed many improved algorithms in recent years, such as differential global positioning system, real-time dynamic positioning, assisted global satellite positioning system, satellite-based augmentation system, ground-based augmentation system, etc. The differential global positioning system requires a reference station that knows its precise position and corrects the error through the GPS information it receives to improve the GPS positioning accuracy. However, in the urban environment, there is the influence of multipath, so the accuracy will be significantly reduced, and the difference algorithm cannot be effectively used. For real-time dynamic positioning, the positioning accuracy can reach centimeter level. It uses the carrier phase dynamic real-time differential technology in the positioning and the reference station for correction. But unlike the previous algorithm, in order to ensure positioning accuracy, it needs to communicate with five satellites at the same time, which is also difficult to achieve in urban scenarios. As for the assisted global satellite positioning system, when the GPS signal of the mobile phone is poor, it could use the reference station to provide certain information to the mobile phone to reduce the satellite acquisition time. But this algorithm is incapable of improving the positioning accuracy, so it can only be regarded as compensation for a poor GPS signal environment. In the satellite-based augmentation system, the geosynchronous satellite can transmit error correction information to the ground receiver to improve positioning accuracy, but the signal penetration is just as poor as the GPS signals in an urban scene. Ground-based augmentation systems are commonly used in aircraft navigation. This system calculates the correction value by arranging GPS ground facilities near the airport and feedback the value to the landing aircraft to improve the positioning accuracy.

Although the above algorithms can be applied in the Internet of vehicles, there are many difficulties in the Internet of vehicles, especially in urban scenarios. Therefore, it is a better choice to use wireless communication to realize the precise positioning of the Internet of vehicles. Compared with the map-based positioning method, the wireless communication-based positioning method does not require more onboard equipment nor the deployment of new drive test facilities and has a better performance in cost control. Parker et al. proposed a positioning algorithm based on received signal strength ranging. On the basis of GPS positioning, each vehicle shares its own location information with neighboring vehicles, and the precise positioning of the vehicle is achieved through the Kalman filter algorithm. However, in the follow-up research, it was found that the accuracy of this algorithm is very poor and cannot be used for independent ranging. Alam proposed a vehicle positioning method based on carrier frequency offset. After the information sent by the driving vehicle is received by the roadside unit, the signal carrier frequency will be shifted due to the Doppler frequency shift and more precise positioning of the vehicle can be obtained through the design of a reasonable filter and the information of the road itself. However, this algorithm relies heavily on the roadside unit, and requires two roadside units to detect carrier frequency offset at the same time to complete the algorithm. Therefore, it is needed to conduct further research on cooperative positioning.

4.4.1.2 Cooperative Perception Based on Information Interaction

The perception of the environment is one of the keys to unmanned driving. The perception system detects the environment in real-time through a variety of sensors installed on the vehicle, provides necessary information for the driving decision of the vehicle, gives early warning of possible dangers, assists the vehicle to make correct driving decisions and path planning, and executes the driving action.

In complex environments, such as the shielding of high-rise buildings and the mutual shielding of high-density vehicles, the sensors of a single vehicle will have certain blind areas. And every sensor will also be affected by various environments. For example, if the camera is under strong light or night environment, it cannot collect effective information, and in rainy and snowy days, Lidar will generate a lot of noise and so on. How to overcome these problems is also a challenge in the current research on vehicle environment perception technology. With the progress of communication technology and 5G technology, the idea of the Internet of vehicles has emerged. Through the V2V and V2I communications in the Internet of vehicles, a vehicle can interact with the surrounding environment and integrate the perceived information of multiple vehicles, thereby expanding its perception domain and ensuring a safe and reliable driving environment. Multivehicle cooperative perception can significantly improve detection accuracy, expand the scope of detection and perception, and dramatically compensate for the lack of perception of a single vehicle in solving problems like occlusion and blind spots, which are of great significance.

Commonly used onboard sensors include lidar, millimeter-wave radar, and cameras. Lidar is the most widely used sensor in the development of unmanned vehicles. It has a long detection range and high detection accuracy. It can perform well under strong light and dark night environments and operate in different weather. The radar, however, performs poorly in the conditions like atmosphere, smoke, and dust, and its perceiving ability is significantly reduced in extreme environments such as haze, rain, and snow and cannot operate around the clock. The millimeter-wave radar has a farther detection range and a lower cost. There is a significant Doppler effect in its detection. And it requires higher speed measurement accuracy and strong penetrating power, which can pass through rain and snow, etc., and can operate all the time. But, it cannot identify the details of the detection target, so it cannot distinguish pedestrians or certain obstacles. The camera is cheaper. It is able to get richer visual information, such as lane lines, signal lights, pedestrians, etc., and carry out deep recognition. But, it has a limited range and is easily affected by strong light and weak light, rain, and snow. It cannot obtain depth-related information, and its performance in ranging is far poorer than lidar.

In the perception of a single vehicle, improving the accuracy, adapting to complex scenes and target movements, and overcoming interference from the environment and other objects have always been in the center of the research. However, sensors have their own limitations, and each type of sensor cannot always ensure the reliability of the information. Naturally, multisensor data fusion has become a reliable research direction. Fusing the detected data of multiple sensors could analyze and obtain more accurate results because it learns from the strengths of every sensor and overcomes the shortcomings of the single-sensor perception. In the actual driving, the vehicle will be affected by the surrounding buildings, trees, and other vehicles. And when the vehicle is in a bad position, the information it can perceive is very limited, which could cause possible safety problems. Solving this problem through the sensors of a single vehicle is difficult, because it also needs information exchanges with other vehicles.

With the V2X technology, vehicles on the road can be equipped with sensing equipment to collect traffic status information, gather the information to the roadside in real-time through V2X technology, and integrate the information with the information directly sensed by the roadside. In addition, as the penetration rate of V2X in-vehicle devices continues to increase, the state information of vehicles installed with onboard devices can also be accurately obtained through V2X, which is mutually corroborated by the information collected by sensors. With the use of multiple sensing devices, the synchronization and de-duplication between the sensing results of different devices have become a new research topic. This has promoted the birth of a global sensing fusion technology based on edge computing, which will eventually lead to the realization of the digital twin of traffic physics systems and provide support for local traffic coordination and global traffic control.

4.4.2   Future Development Trends

With the rapid development of technologies such as the Internet of vehicles, the Internet of things, artificial intelligence, and wireless communications, the entire field of intelligent transportation is conducting exploration and related research on the basis of information interaction.

As a brand-new technology, unmanned driving is attracting more and more attention, especially the attention of vehicle companies and governments. Autonomous driving is considered to be an important technology to liberate human labor, and it can significantly improve traffic safety, alleviate traffic congestion, and optimize traffic efficiency. Perceiving the environment is one of the most basic requirements of autonomous driving. Many accidents resulting from unmanned driving are also caused by failures in recognition of a single vehicle and the inability to make correct decisions. To overcome the limitation of a single vehicle's perception, perceptual fusion and decision-making through the V2V and V2I information interactions will be the significant trend in the future. Vehicles exchange and share information frequently and in large quantities, causing urgent requirements to formulate the physical structure and standards for information interaction. With the intelligentization of vehicles, each vehicle in the Internet of vehicles becomes an independent intelligent node that can communicate, store information, and compute. How to effectively manage these communications and rationally use computing resources is of great significance to future development.

The current research of the physical layer technology of vehicle communication is conducted around the strict vehicle channel conditions, by calculating effective channel estimation, eliminating inter-carrier interference, etc., to ensure the stability and the performance of the communication technology in the static conditions and the high-speed and complex Internet of vehicles, respectively. However, with the development and innovation of wireless communication technology, researchers have turned to relying on wireless channels from confrontation with wireless channels. Fundamentally, under the high dynamic Doppler effect of the Internet of vehicles communication channel, the widely used OFDM technology is far from the best multicarrier transmission scheme in the current high-speed time-varying situation. Therefore, it is significant to explore a more stable and reasonable multicarrier transmission scheme. In such a scheme, we not only require better time-frequency positioning capabilities but also require the ability to better adjust the selected scheme according to the characteristics. In addition, the time-frequency dual-selection characteristics of the vehicle communication channel also lead to double diversity in the time-frequency domain, and the flexible index modulation in a two-dimensional time-frequency resource grid is also a possible research direction.

5G can greatly increase the data rate of multiple users. However, considering the complexity and high speed of IoV communications, the application of large-scale antenna technology in IoV communications still has a long way to go. Taking advantage of large-scale antenna technology to reduce the impact of the imperfect channel conditions in IoV communications is a difficult problem. In the progress of the Internet of vehicles, the complexity of the network topology of the Internet of vehicles will increase greatly. The increase in communication modes and large-scale distributed connections in heterogeneous networks will also cause more challenges. In the future, users will also put forward higher requirements for service quality. The enormous amount of network throughput requires lower access and waiting time. Therefore, the future Internet of vehicles needs better design and architecture for the MAC layer.

With the large-scale commercialization of 5G on a global scale, 6G research and exploration have gradually attracted attention. With the help of innovative research in basic disciplines such as mathematics and physics, 6G is deeply integrated with big data analysis and artificial intelligence to achieve a deep coupling of communication, perception, and computing. In the field of intelligent transportation, 6G can use communication signals to detect, locate, and recognize the target objects, and the wireless communication system can obtain the surrounding road conditions and environmental information. 6G can also acquire and generate ultrahigh resolution images through higher computing power and edge computing technologies, realize centimeter-level or even millimeter-level positioning, and complete more accurate identification of traffic participating entities. And it can connect the intelligent transportation system to the Internet of things, form a deep coupling with other terminals, and use massive data for driving analysis.

In addition, the application of communication technology has made the traditional closed vehicles gradually become the connected automated vehicle with open wireless communication, and the security issues of traditional networks have also extended to the field of intelligent transportation. Different from the traditional computer network information security issues, automobile information security issues not only bring about personal privacy leakage and property and economic losses to users but may even lead to public safety issues such as mass deaths and injuries. In recent years, world-renowned auto companies such as BMW, Toyota, Tesla, JEEP, etc., have discovered varying degrees of in-vehicle network information security vulnerabilities in their related models, which has aroused great concern in the industry. According to statistics, 56% of consumers said that one of the main factors they will consider when buying a vehicle in the future is whether they can guarantee the security of personal information and protect their privacy. It can be seen that the information security of connected automated vehicles has become the focus of social attention and is an important subject that the automotive industry and academic research institutions need to consider in the future.

References

1. Lixin M, Faping W (2020) Overview of the research and application of key technologies V2X of internet of vehicles. J Automot Eng 10(1):1–12. (in Chinese)
2. Wenke W (2017) Overview of the development and application of internet of vehicles technology. Automot Pract Technol 3:88–91. (in Chinese)
3. Cunzhi Z (2011) Overview of vehicle self-organizing network (VANET). J Hubei Radio Television Univ 31(11):157–158. (in Chinese)
4. Jialang C, Wei N, Weigang W et al (2014) Overview of research on application of vehicle self-organizing network in intelligent transportation. Comput Sci 41(S1):1–10. (in Chinese)
5. Libo Z, Bingbing L, Wang X (2017) Overview of Internet of Things information perception and information interaction technology. Mod Comp (Professional Edition) 20:34–39. (in Chinese)
6. Xiang C, Rongqing Z, Chen C (2020) 5G internet of vehicles technology and application. Science Press, Beijing. (in Chinese)
7. Hannes H, Laberteaux KP (2013) VANET: vehicular applications and inter-networking technologies. In: Limin S, Yunhua H, Xinyun Z, Hongliang L, Maohua Z (eds) VANET in-vehicle network technology and application. Tsinghua University Press, Beijing. (in Chinese)
8. Bing W, Dudu G (2016) Transportation information technology and application. Machinery Industry Press, Beijing. (in Chinese)
9. Daxin T, Xuting D, Jianshan Z (2020) Vehicle network technology. Tsinghua University Press, Beijing. (in Chinese)
10. Sassi A, Charfi F, Kamoun L et al (2011) The impact of mobility on the performance of V2X communication. In: 2011 4th International conference on logistics, Hammamet, Tunisia
11. Zhang H, Lu X (2020) Vehicle communication network in intelligent transportation system based on Internet of Things. Comput Commun 160:799–806
12. So S, Petit J, Starobinski D (2019) Physical layer plausibility checks for misbehavior detection in V2X networks. In: Proceedings of the 12th conference on security and privacy in wireless and mobile networks, pp 84–93
13. Wang C, Li Z, Xia X et al (2020) Physical layer security enhancement using artificial noise in cellular vehicle-to-everything (C-V2X) networks. IEEE Trans Veh Technol 69(12): 15253–15268
14. Elhalawany BM, EL-Banna AAA, Wu K (2019) Physical-layer security and privacy for vehicle-to-everything. IEEE Commun Mag 57(10):84–90
15. Li S, Liu Y, Wang J (2019) Astsmac: application suitable time-slot sharing MAC protocol for vehicular ad hoc networks. IEEE Access 7:118077–118087
16. Johari S, Bala KM (2021) TDMA based contention-free MAC protocols for vehicular ad hoc networks: a survey. Veh Commun 28:100308
17. Jin Bo H, Yanming. (2020) Overview and prospects of the development of C-V2X internet of vehicles industry. Telecommun Sci 36(3):97–103. (in Chinese)
18. Lixin M, Faping W (2020) Overview of the research and application of key technologies of V2X internet of vehicles. J Automot Eng 54(1):4–15. (in Chinese)
19. Lin Z, Shiping C (2013) Overview of VANET-oriented 802.11 research. J Shanghai Univ Electric Power 29(6):568–573. (in Chinese)

20. Huiming D (2018) Research and improvement of physical layer and MAC layer protocol of vehicle communication system. Beijing University of Posts and Telecommunications, Beijing. (in Chinese)
21. Qi T (2020) Research on the optimization of Internet of Vehicles channel access mechanism based on DSRC technology. Beijing University of Posts and Telecommunications, Beijing. (in Chinese)
22. Shanzhi C, Shi Yan H, Jinling. (2020) Overview of cellular Internet of Vehicles (C-V2X). Natl Sci Found China 34(2):179–185. (in Chinese)
23. Chen Shanzhi H, Jinling SY et al (2018) LTE-V2X Internet of Vehicles technology, standards and applications. Telecommun Sci 4:1–11. (in Chinese)
24. Rongbin G, Yongdong Z, Kainan Z et al (2020) The evolution of key technologies of C-V2X Cooperative Vehicle Infrastructure Systems. In: The 15th China intelligent transportation conference. (in Chinese)
25. IMT-2020 (5G) Promotion Group (2019) C-V2X service evolution white paper. IMT-2020 (5G) Promotion Group, Beijing. (in Chinese)
26. Chen S, Hu J, Shi Y, et. (2017) Vehicle-to-everything (V2X) services supported by LTE-based systems and 5G. IEEE Commun Standards Mag 1(2):70–76

Chapter 5
Traffic State Analysis and Prediction Technology

MA Xiaolei, LUAN Sen, TANG Erlong, YAN Haoyang, and LI Yujie

The complete acquisition, accurate real-time evaluation, and prediction of urban road traffic state are the basis for accurately grasping the behavior of urban road traffic system, making traffic management decisions scientifically and giving full play to the potential of traffic facilities. This chapter focuses on the analysis and prediction of traffic status to meet the actual needs of urban traffic management.

5.1   Concept and Connotation of Traffic State

5.1.1   Concept of Traffic Status

Urban road traffic state means the general operation of traffic flow, with multi-scale, multi-variable, random, and time-varying characteristics. With regard to different traffic conditions, various traffic control and management schemes and diverse travel schemes are needed. The traffic state analysis is the foundation of traffic control and guidance, especially under the abnormal events. The acquisition of traffic state information is necessary to determine the emergency evacuation plan of government departments, rescue and relief of public security departments, and avoid or travel of the public, which can buy valuable time to save lives, assets, and prevent the situation from deteriorating. Therefore, it is crucial to define traffic state, set up the evaluation index system, and design a reasonable calculation method of traffic state evaluation index.

Although there are many related researches and some beneficial results on road traffic status at home and abroad, most of the studies have been used to describe the

MA Xiaolei (✉) · LUAN Sen · TANG Erlong · YAN Haoyang · LI Yujie
School of Transportation Science and Engineering, Beihang University, Beijing, China
e-mail: xiaolei@buaa.edu.cn

© Tsinghua University Press 2022
W. Yunpeng et al. (eds.), *Intelligent Road Transport Systems*,
https://doi.org/10.1007/978-981-16-5776-4_5

traffic flow of intersections. The evaluation indexes are mainly extracted directly from the data collected by subgrade detectors at intersections, including flow, speed, and occupancy, and so on, which usually appears in traffic signal control system. Although the traffic operation state of a specific intersection can be explained, it is difficult to describe the traffic operation of the whole road network comprehensively. Especially after the occurrence of abnormal events, the uncertainty factors increase, causing an increase in demand for travel information. Therefore, it is urgent to establish a systematic evaluation index system of road traffic status based on multi-source information to meet the needs of different user subjects at the same time.

Road traffic state is the general operation of traffic flow, and could be analyzed from micro, meso, and macro perspectives.

From the micro point of view, the operation state of traffic flow can be described by some basic vehicle operation parameters, performing as the quantity of traffic flow, the vehicle speed, the queue length, the duration of delay time, and other quantitative indicators.

From the meso point of view, the operation state of traffic flow is generally described as the comprehensive level of traffic condition of a certain section or intersection, and has normal and abnormal states. Normal state refers to that all vehicles can run orderly, safely, and smoothly; abnormal state refers to abnormal state and traffic congestion state. In most cases, traffic congestion is divided into slight, normal, and severe congestion according to its severity degree. To sum up, the meso level of road traffic state has four categories: smooth, slight congestion, congestion, and severe congestion.

Among them, smooth refers to the traffic condition that the average travel speed is not significantly lower than the maximum speed limit of the section where the road is located, including the traffic condition which can be driven according to the maximum speed limit and the traffic condition in the stable flow state and the middle part. In the first case, the traffic flow is minimal, and the driver is not affected by or is basically not affected by other vehicles in the traffic flow, and has a very high degree of freedom to choose the desired speed. But in the second case, the traffic flow is in the better and middle part of the stable flow. The vehicle speed is affected by other vehicles, and gradually increases as traffic flow grows, and the traffic service level has decreased obviously.

Congestion refers to the traffic condition in which the average travel speed is obviously lower than the maximum speed limit of the road section. In terms of the severity degree, it can be divided into slight, normal, and severe congestion. Among them, slight congestion refers to the traffic condition that the average travel speed is still within the acceptable range. At this point, the traffic is in the inferior status of the stable traffic flow, the speed and driving freedom is strictly restricted, and the travel comfort and convenience are weak. And at this time, a slight increase in traffic volume might cause traffic problems.

Severe congestion refers to the traffic condition when the average travel speed is lower than the accepted range. Among them, the acceptable range varies with the city size, road, and intersection level. At this time, the traffic may be in the state of unstable flow, the slight increase of traffic volume or the small disturbance in the traffic flow will produce large traffic operation problems, and even lead to traffic interruption. Or, the more serious situation is that the traffic is in the state of forced flow, and the vehicles follow the car in front to stop and go, often in line. At this time, the traffic volume and speed change from large to small or even zero, and the traffic density increases as traffic volume decreases.

Normal congestion is a transition between slight and severe congestion. It is worse than slight congestion and better than severe congestion.

From the macro point of view, the operation state of traffic flow can be described as the traffic congestion degree of road network or local road network, and is represented by traffic congestion index.

At present, the most widely used traffic state is the level of service mentioned by HCM. The definition of the level of service in HCM2000 is: the level of service is to describe the operation of traffic flow, and is usually characterized by speed, travel time, driving freedom, traffic interruption, comfort, and convenience.

Traffic state should contain two meanings. One is reflecting the factual state of traffic flow operation; that is, along with the traffic flow variation, the traffic state is constantly changing. The other is reflecting the psychological feelings of traffic travelers to the traffic flow condition; that is, under different traffic states, drivers have different feelings to the traffic flow operation. At the same time, traffic state should be a dynamic state or network mode generated in the process of traffic flow operation, which can be analyzed from different levels according to different traffic management needs.

According to system theory, the state of a system is the state of a material system and the aggregate of a group of state variables. The state variable of a system is the smallest subset of the system variables that can represent the complete state of the system at any time.

The traffic state is produced with the operation of the traffic system. It is a system state first, but the traffic system is complex with human participation, and has different characteristics from other systems. Referring to the general definition of system state, and the features of road traffic system, traffic state is defined as an objective reflection of the overall operation of traffic flow, which can be represented by a group of index variables reflecting traffic flow behavior as different aspects and different granularity.

Through analyzing the traffic state, we can get the elements of traffic state.

1. *Time element*: The traffic state changes with time, and the traffic state should be limited by time.
2. *Space elements*: The traffic state changes with the change of space, and the traffic state should be limited by space.
3. *Object elements*: The description object of traffic state is the overall operation condition of traffic flow in different time and space, not the operation condition of single vehicle.
4. *Condition elements*: Traffic flow behavior to produce traffic state. Traffic flow that does not change behavior will not produce new traffic state.

5. *Result elements*: The traffic state is measurable and can be represented by a set of index variables which reflect the traffic flow behavior as different aspects and different granularity.

The concept of traffic state is analyzed and its characteristics are summarized as follows.

1. *Objectivity*: The objectivity of traffic state means that in the road traffic system, due to the objective existence of traffic flow, the traffic state reflecting the overall operation of traffic flow is also objective, which is not influenced by human will.

2. *Dynamic randomness*: One of the biggest characteristics of traffic flow is dynamic, which is because the urban road traffic system is closely related to the activities of travelers, with many random factors: on the one hand, the unexpected change of traffic demand, such as travel purpose and personal preference; on the other hand, the unexpected change of traffic supply capacity, for example, traffic accidents, road expansion, and weather will affect the evolution of traffic flow. The dynamic random change of traffic flow determines that the traffic state must be a dynamic random change.

3. *Continuity*: Traffic flow behavior is to produce traffic state. This book thinks that traffic flow behavior is a continuous changing process. Therefore, traffic state is also a continuous changing state.

4. *Hierarchy*: The traffic state changes with spatial variation. Different spaces make different traffic states. The urban road network has a hierarchical structure, and the traffic state of different levels is distinct.

5. *Correlation*: Road traffic network is composed of interconnected road sections and intersections. The change of traffic state of a road section or intersection may lead to the change of traffic state of adjacent road sections or intersections, or even the evolution of traffic state of the whole road network.

6. *Periodicity*: Due to the relative stability of road network structure in a certain period of time, combined with the experience of travelers and managers on travel habits and control measures, and the adaptability of road network itself to the environment, although the traffic state will change dynamically and randomly, it still presents periodic regularity on the whole.

7. *Testability*: The purpose of traffic state research is to better serve traffic management decision-making. Therefore, traffic state could be quantified, that is, the index variables representing traffic state are measurable.

Traffic behavior on the road can be divided into two categories according to the interference between vehicles: free traffic flow driving and non-free traffic flow driving. In the case of free traffic flow driving, drivers can choose higher speed and better lane according to road conditions, vehicle conditions, and their conditions. However, in the case of non-free traffic flow driving, vehicles are vulnerable to the interference of other vehicles, and the choice of speed and lane is affected to a certain extent, even there is a stop and go behavior.

The purpose of traffic condition evaluation and prediction is to serve traffic management decision. Different traffic states reflect different traffic flow operation conditions, so corresponding traffic control measures can be formulated according to different traffic states, but the traffic state is continuous, and changes with time and space, so many traffic control measures need to be formulated, which is not conform with the reality. Therefore, the traffic state can be divided into several categories (levels) according to the similarity of traffic flow, reflecting the qualitative evolution of traffic flow in different stages, and then the targeted traffic control measures could be formulated.

Before the classification of traffic status categories (levels), it should be based on the following consensus.

1. The traffic state category is a collection of traffic states which reflect similar traffic flow behavior. Traffic state is an objective reflection of traffic flow operation, and it is a quantity that changes with time and space. However, there is some similarity in traffic flow at different times or spaces. Therefore, the category of traffic state can be understood as the collection of traffic states which reflect similar traffic flow behavior.
2. There are differences between different traffic status categories. Different traffic state categories describe different traffic flow operation conditions, reflect distinct traffic flow behavior, and have different impacts on traffic management decision-making.
3. The differences between different traffic status categories can be evaluated. By dividing the traffic state into different categories, describing the qualitative change behavior of traffic flow in different stages, and scientifically evaluating the differences between them, traffic managers can formulate targeted traffic control measures. Based on the concept of traffic state, the traffic state can be represented by a series of indicators variables which reflect traffic flow behavior to different sides and different granularity. Therefore, the traffic state can be quantified by designing scientific and reasonable index variables, thus realizing the evaluation of the difference between the traffic status categories, and laying a foundation for further distinguishing the different traffic states.
4. The traffic state classification is subjective. The category of traffic state is a collection of traffic states that reflect similar traffic flow behavior. How to define the similarity of traffic flow behavior is different between people, especially among participants in road traffic system. These participants have different ages, gender, educational degree, and psychological features. Therefore, the understanding of similarity is different, then the cognition of traffic state category is diverse. The classification of the categories is subjective and there must be differences. Traffic status can be classified according to the statistical results or decision-making needs of traffic participants' subjective feelings.

   The traffic state is objective and can be represented by a group of index variables which reflect the traffic flow behavior in different aspects and different granularity. The category of traffic state is subjective, which is generally divided according to the statistical results of subjective feelings of traffic participants or decision-making needs. Traffic state and traffic state category describe the operation of traffic flow from different angles, forming a relatively complete concept system of traffic state.

5.1.2 Index System of Traffic State

A large number of pedestrians and vehicles passing on the road have the character-istics of similar fluid on the whole. In traffic engineering, the flow of people and vehicles passing on the road is collectively referred to as traffic flow (Traffic Stream or Traffic Flow). Generally, traffic flow discussed in traffic engineering mainly refers to car flow.

Traffic flow is a whole and macro concept. Through the analysis of a large number of observation data, it is found that traffic flow has a certain characteristic tendency. Therefore, traffic flow characteristics are conceptualized. Traffic flow characteristics refer to the qualitative and quantitative characteristics of traffic flow operation state. The physical quantities used to describe and reflect the characteris-tics of traffic flow are called traffic flow parameters.

The traffic flow parameters used to describe the traffic state are divided into macro parameters and micro parameters. Among them, macro parameters are used to describe the operating state characteristics of traffic flow as a whole, mainly includ-ing traffic volume, speed, traffic density, occupancy rate, and queue length; micro parameters are used to describe the operating state characteristics of interrelated vehicles in traffic flow, including time headway and space headway.

5.1.2.1 Volume

Volume is also called traffic volume, which refers to the number of vehicles passing through a road or a section within a time unit. Traffic volume is not a static and constant volume, it has spatiotemporal characteristics. One method to measure the traffic characteristics of a city is to observe the change of traffic volume in time and space in a series of positions in the road system, and draw its isograms. When the traffic volume exceeds a certain level, it is considered that congestion occurs. However, the problem of this judgment is that the same flow level can correspond to two different traffic states, so this parameter should be combined with other methods, rather than used alone.

5.1.2.2 Speed

Speed is the second basic parameter characterizing traffic state, which refers to the distance the vehicle passes in unit time. There are many different concepts of speed for different research purposes.

From the micro point of view, each vehicle has instantaneous speed and average travel speed, and average running speed in a specific period of time. Instantaneous speed (also known as instant speed and location speed) is the instantaneous speed of vehicles passing through a certain place (or at a certain time), which provides support for road design, traffic control, and planning. The average running speed of a single vehicle refers to the ratio of the length of the road section to the time used when the vehicle passes through a specific length of road. The driving time does not include the stopping time of the vehicle due to various reasons. Because the average speed of the vehicle does not consider the parking delay in the process of operation, it cannot accurately reflect the running characteristics of the vehicle, so the concept is rarely applied in practice. The average travel speed of a vehicle is the ratio of the length of the road section to the total time used when the vehicle passes through a specific length of road. Considering the possible parking delay, this concept of speed can better reflect the running state of the vehicle in a specific section and a specific period of time.

From the macro point of view, the average speed of traffic flow is divided into the time mean speed (average location speed) at a specific location and the space mean speed (average travel speed) on a specific road section. The time mean speed is the arithmetic average of the velocity of all vehicles passing through a certain section during the observation time, and the space mean speed is the quotient of the observation distance and the average travel time used by the vehicle to pass through the distance. The former reflects the traffic flow operation at a specific observation site, while the latter reflects the traffic flow in the specific road section space. When the two speed values are obviously lower than the normal value, it indicates that the traffic at the observation site or observation section is in a crowded state.

5.1.2.3   Traffic Flow Density

Traffic flow density refers to the number of vehicles per unit road length at a certain time snapshot, i.e.,

$$K = N/L \qquad (5.1)$$

where K is the traffic density (vehicle/km· Lane); N is the number of vehicles (vehicle); L is the length of observation section (km).

Under normal circumstances, a large traffic flow corresponds to a large traffic density. However, when the road traffic is very congested and the traffic flow is stagnant, the traffic flow is approximately equal to zero, while the traffic density is close to the maximum. Therefore, it is difficult to express the actual state of traffic flow only by using traffic flow index, while traffic density index can make a better evaluation. Although traffic density can directly indicate the nature of traffic state, this parameter is rarely applied in reality due to the difficulty in data collection.

5.1.2.4   Time Headway and Space Headway

In the same direction of traffic flow, the space distance between two adjacent vehicles is called space headway. Because it is very difficult to measure the space headway in the process of traffic flow, this index is generally not used.

In the same direction traffic flow, the time interval between two adjacent vehicles passing a section of the road is called time headway. In a specific period of time, the average time headway of all vehicles on the observation section is called average time headway.

Time headway is a very important micro traffic characteristic parameter. Its value is closely associated with the characteristics of driving behavior, vehicle performance, and the specific situation of the road. At the same time, it is also affected by traffic volume, traffic control mode, intersection geometry, and other factors. Similar to traffic flow, the same time headway also corresponds to two different traffic states, so it can't be used to distinguish traffic state degree alone.

5.1.2.5 Occupancy

Occupancy includes space occupancy and time occupancy.

In a certain section of the road, the ratio of the total length of the vehicles to the total length of the road section is called space occupancy, which is usually represented as a percentage. Space occupancy directly reflects the level of traffic density, and can also indicate the actual occupation of the road. Similarly, the difficulty of collection data makes it also rarely used in practice.

Time occupancy refers to the ratio of the total time occupied by the vehicles to the length of the observation time in a certain observation time T. The calculation formula is

$$\text{occupy} = \sum \Delta t_i / T \qquad (5.2)$$

where occupy is the time occupancy; Δt_i is the time occupied by the i-th vehicle (s); T is the length of the observation period (s).

The size of time occupancy can reflect the state of traffic operation. In the case of small traffic volume, the time occupancy is relatively low due to the high speed. With the increase of traffic volume, the vehicle speed decreases. Therefore, the time occupied by vehicles increases and the time occupancy rate increases significantly. When traffic congestion occurs, although the traffic volume passing through the detector may be reduced, the time occupancy rate is still at a high level due to the obvious decrease of vehicle speed.

5.1.2.6 Queue Length

Queue length refers to the number of vehicles queuing at traffic discontinuities (intersections, accident points, etc.). Queue length can be used as a measure of traffic congestion. In general, heavy congestion usually leads to long queues. Therefore, queue length can intuitively reflect the degree of traffic congestion. In other side, vehicles arriving at the intersection during the red light period must queue in front of the stop line due to the existence of traffic signal control. It is generally considered that the queue that can pass in 1–2 signal cycles does not belong to the category of traffic congestion.

The evolving traffic flow data can reflect the dynamic spatiotemporal characteristics of traffic flow. If there are a certain number of traffic detectors on the road, and the traffic parameter data are sampled according to a certain time interval, the monitoring of traffic flow attitude can be realized by analyzing the change law of these traffic data. Which traffic parameter data should be used as the basis of congestion discrimination should be considered from the efficiency, effect, economy, and reliability of traffic state discrimination.

Urban road traffic state refers to the overall operation status of traffic flow, and lays the foundation for flexible traffic management and control. The traffic state on urban roads can be divided into three states: smooth, congested, and severely congested.

Smoothness generally refers to the traffic condition in which the average travel speed is not significantly lower than the maximum speed limit specified in the road section, including the traffic condition in which the maximum speed limit can be used, and the condition in which the traffic is in the better and middle part of the stable flow. In the first case, the traffic volume is very small, the user is not affected or basically not affected by other vehicles in the traffic flow, and has a very high degree of freedom to choose the desired speed (provided that it is not greater than the maximum speed limit). In the second case, the vehicle speed will be influenced by nearby vehicles. This influence will gradually increase with the increase of traffic flow, and the comfort and convenience of vehicle driving will decrease significantly.

Congestion refers to the traffic condition in which the average travel speed is significantly lower than the maximum speed limit of the road section. Generally, congestion can be divided into slight congestion, congestion, and severe congestion. Here, we think that slight congestion refers to the traffic state in which the average travel speed is still within an acceptable range, when the traffic is in the poor part of the stable traffic flow range. The speed and driving freedom are strictly restricted, and the comfort and convenience are low. At this time, a small increase in traffic volume will cause problems in operation. Severe congestion refers to the traffic state when the average travel speed is lower than the acceptable range. This generally accepted speed standard varies with the size of the city, the level of roads and intersections, and the time of occurrence. At this time, the traffic is in the range of unstable flow, and a small increase in traffic volume or a small disturbance in the traffic flow will cause large operation problems, or even traffic interruption. What's more serious is that the traffic is in the state of forced flow, and the vehicles often line up and stop with the vehicles in front. In this case, the traffic volume and speed change from a large value to zero gradually at the same time, while the traffic density increases with the decrease of traffic volume. Representing a transitional state between slight congestion and severe congestion, congestion is not worse than that of slight congestion, and better than that of severe congestion.

Severe congestion refers to a phenomenon of too many cars and slow speed, and usually occurs in the metropolitan areas of the world, the expressways connecting the two cities, and the areas with high automobile utilization rate during rush hours. In addition, people often call the roads that are easy to jam as traffic bottlenecks. The reasons are generally due to the increase of car utilization, insufficient capacity or improper design of roads, and intersections. The existence of congestion will cause a variety of effects, such as increasing the commuting time, reducing the time available for work, causing economic losses to drivers and the region; causing drivers to feel angry and irritable, increasing their pressure, and further damaging their health; wasting fuel and pollution as the engine is still running in traffic jam. Traffic congestion not only wastes energy, but also causes air pollution, which reduced the quality of life in metropolis. In turn, a large number of residents moved to the suburbs (the so-called suburbanization). It is difficult to react to emergencies, when there is an urgent need, it is also difficult to reach the destination due to traffic congestions, etc. Therefore, traffic congestion is an urgent problem to be solved.

Congestion can be divided into two categories according to the causes of it. One is frequent congestion, which usually occurs in a fixed place and time, such as the congestion in the morning rush hour and evening rush hour. At this time, the traffic volume on the road is nearly saturated or oversaturated, and the driving speed of vehicles on the road is relatively slow, the degree of freedom is relatively low, and even there is a stop and go situation when it is serious. The other is non-recurrent congestion, also known as incident congestion, which is generally caused by some random traffic incidents, such as traffic accidents, vehicle breakdowns, bad weather, large-scale activities, road construction and maintenance, etc. Incident congestion is irregular and unpredictable, and may last for a long time. It can only be managed by means of on-site organization, coordination, and command. However, for the congestion caused by large-scale activities, the time and place of occurrence can be estimated to a certain extent, so the countermeasures are similar to the frequent traffic congestion management can be adopted.

5.2   Discrimination and Analysis of Traffic States

By observing the actual operation of urban traffic and analyzing the collected traffic information, we can find that the traffic state of road section and region has its own rules. In the past, these rules need the long-term experience of traffic managers to find out. Now, on the basis of the massive information of intelligent transportation system, pattern recognition, artificial intelligence, etc., we can find out the traffic patterns hidden in the traffic information, and provide decision support for the traffic flow control and guidance of regional road network. In essence, traffic congestion also reflects one of traffic states, and the related research such as traffic congestion state discrimination is also a subset of traffic state analysis. Therefore, this section will focus on the related methods of traffic state discrimination and analysis in detail.

5.2.1 Road Traffic States Discrimination

The traffic state of a road network is generally discriminated by macro, meso, and micro traffic flow parameters. Among them, the evolution processes of network characteristics and overall states of traffic networks are captured by macro traffic parameters; meso parameters mainly refer to the traffic state of intersection and road section; micro parameters mainly refer to the vehicle operation state and interaction relationship. Therefore, road network traffic state identification is a complex system analysis problem involving multi-scale, multi-variable, highly random, and time-varying.

The existing traffic state identification methods can be divided into two categories: manual identification methods and automatic identification methods.

Artificial discrimination is the earliest, easiest, and the most widespread method, which is used to report traffic congestion and traffic incident information to traffic management center in daily life. Manual discrimination methods include citizen report, full-time personnel report, civil radio, closed-circuit television monitoring, aviation monitoring, etc. On the whole, the main advantages of this non-automatic discrimination method are convenient, straightforward, economical, and efficient. The disadvantages are that there are local witnesses at that time, the obtained crowded and incident locations are usually not accurate, and special personnel are needed to screen and confirm the report, so the workload and intensity of personnel are relatively large. Notably, motived by the popularity of mobile phones, it has gradually become an important method to identify congestion and events in most urban areas. Highway patrol team can not only find events, but also quickly carry out event response and clearance activities. It is still an attractive method in practical application from the perspective of event management. Fixed observers are more suitable for short-term needs (such as special activities or during highway construction and maintenance). Closed-circuit television can be used as a method to judge the state of artificial traffic, and it can also be used as a method to confirm the alarm of telephone and ACI algorithm. The former requires continuous observation by the operator. Generally speaking, the artificial identification method is generally suitable for the identification of traffic state in urban roads, while the application in expressway is greatly affected by time and weather, with longer detection time and lower detection rate. Therefore, expressway and urban expressway are more suitable to use the automatic identification method of traffic state.

For automatic identification of ACI, most of the ACI methods indirectly judge the existence of traffic congestion and traffic events by identifying the abnormal changes of traffic flow parameters obtained by traffic detector. We call this method indirect ACI method. To identify traffic congestion, the direct ACI method is to use image processing to identify whether there are slow or stopped vehicles. This kind of method actually "sees" the occurrence of traffic congestion and traffic events, rather than through the analysis of traffic flow characteristic parameter data to detect their existence. The direct ACI method is far better than the indirect ACI method in the speed of discrimination in terms of detection effects, especially in the case of low traffic volume, it can also distinguish traffic emergencies well, but it needs more intensive traffic monitors (cameras), higher capital investment to ensure the reliability of discrimination, and the meteorological conditions also have a greater impact on it.

Here are some classic traffic state discrimination methods:

5.2.1.1 California Algorithm

California Department of transportation obtained road occupancy data by using loop coil technology from 1965 to 1970, and eventually developed the California algorithm, referred to as California algorithm. The main principle of the algorithm to judge the traffic state is to judge the traffic state according to the correlation between the change of the upstream and downstream occupancy (difference change, change rate, relative difference) and the corresponding threshold detected by the detectors set up in the upstream and downstream of the road at different times. The basic identification logic of California algorithm is shown in Fig. 5.1.

where

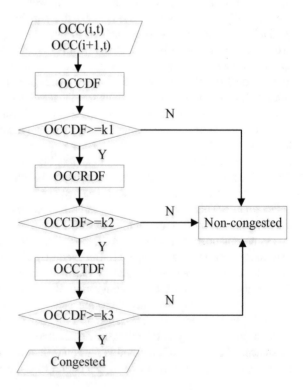

Fig. 5.1 California algorithm flow

$$\begin{cases} \text{OCCDF} = \text{OCC}(i,t) - \text{OCC}(i+1,t) \geq K_1 \\ \text{OCCRDF} = \dfrac{\text{OCCDF}}{\text{OCC}(i,t)} \geq K_2 \\ \text{OCCTDF} = \dfrac{\text{OCC}(i+1,t-2) - \text{OCC}(i+1,t)}{\text{OCC}(i+1,t-2)} \geq K_3 \end{cases} \quad (5.3)$$

where OCC (i, t) is the occupancy measured at t time of the i detection station, OCCDF is the difference of upstream and downstream occupancy of the road section, and OCCRDF is the relative difference of upstream and downstream occupancy of the road section, and the OCCTDF is the relative difference of downstream occupancy at the beginning of congestion, K_1, K_2, and K_3 are corresponding thresholds.

Later, scholars have made a series of improvements to the above California algorithm, and developed more than ten improved congestion automatic identification algorithms based on the original California algorithm, among which the California #7 algorithm and California #8 algorithm have the best congestion recognition effect. In California #7 algorithm, the current measured downstream occupancy replaces the relative difference of occupancy, which eliminates the error alarm of common compression waves in the case of large traffic volume. It is also found that when the downstream occupancy data is less than the set threshold (usually 20%), there is an event. California #8 algorithm is the most complex but best performance algorithm of all improved California algorithms [1]. California #8 algorithm repeatedly detects traffic volume compression waves, which are the main reasons for the decrease of upstream traffic movement speed, and these compressed waves are likely to break traffic in a large traffic situation. It is found that the compressed wave can be detected in time and can delay the alarm upstream.

5.2.1.2   Exponential Smoothing Method

Usually, the traffic parameter data collected by the detector contains a lot of noise. If we directly use it to identify traffic congestion states, it will lead to a high rate of misjudgment. Firstly, the exponential smoothing method smooths the original traffic data to remove short-term traffic interference, such as random fluctuation, traffic pulse, and compression wave, and then compares the processed data with the preset threshold to judge whether there is congestion. The exponential smoothing method for traffic parameters is introduced as follows.

$$\text{ST}_i(t) = \alpha T_i(t) + (1 - \alpha)\text{ST}_i(t - 1) \quad (5.4)$$

where α is the smoothing coefficient, $0 < \alpha < 1$, and the general value range is 0.01–0.3; $T_i(t)$ is the traffic flow parameter value of the i-th detection station at t time; $\text{ST}_i(t)$ is the traffic flow parameter smoothing value of the i-th detection station at t time.

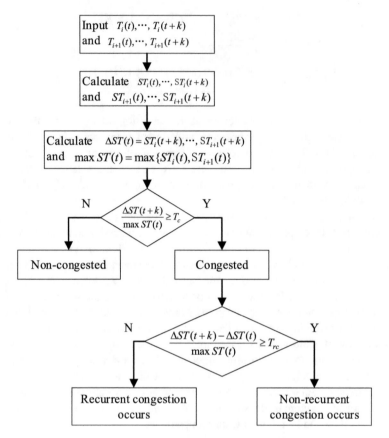

Fig. 5.2 Flow chart of exponential smoothing discriminant algorithm

The discrimination flow of the algorithm is shown in Fig. 5.2.

5.2.1.3 McMaster Algorithm

McMaster algorithm is developed by the Department of Civil Engineering of McMaster University in Canada based on catastrophe theory. The two-stage process is developed to identify the traffic congestion: (1) identifying the existence of congestion; (2)
. As shown in Fig. 5.3, this algorithm represents the traffic flow and occupancy data in two-dimensional space, and divides the flow-occupancy two-dimensional graph into four regions. Therefore, each region represents a different traffic state.

Region 1 represents the normal (non-congested) traffic state; region 2 represents the traffic state upstream of the occasionally congested location; region 3 represents the slow traffic flow blocking state, which generally means the congestion downstream of the detection station; region 4 represents the traffic state upstream of the

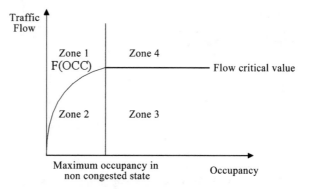

Fig. 5.3 The state classification of McMaster algorithm

frequently congested location. By checking the distribution of the measured data points in the four areas, the start, duration, and end time of the traffic congestion at the downstream of a detection station can be distinguished. The algorithm provides that in three consecutive sampling periods, if the vehicle speed falls below the threshold, or the occupancy exceeds the threshold, or the flow and occupancy are outside the non-congested area, congestion can be determined. When the speed, flow, and occupancy exceed their respective thresholds in two consecutive sampling periods, the traffic congestion of the road section is also be identified.

5.2.1.4   Standard Deviation Method

The standard deviation algorithm uses the arithmetic mean of traffic parameter data (flow or occupancy) of n sampling periods before time t as the prediction value of traffic parameters at time t, and then uses the standard normal deviation to measure the change degree of traffic parameter data relative to its previous mean value. When it exceeds the preset threshold, it is considered that there is occasional traffic congestion.

Let the actual value of traffic parameters at time t be $x(t)$. The actual values of traffic parameters in the n sampling periods before time t are $x(t-n)$, $x(t-n+1)$, ..., $x(t-1)$. The discriminant formula is

$$\text{SND}(t) = \frac{x(t) - \widehat{x}(t)}{S} \geq K \tag{5.5}$$

where $\widehat{x}(t)$ is the current predicted value of traffic parameters, S is the standard deviation of traffic parameters in the previous n sampling periods, K is the decision threshold, and SND is the normal deviation.

Figure 5.4 shows the process of congestion identification using standard deviation method. The parameters of traffic flow, the occupancy, and the speed can be applicable for this method. In this algorithm, the time window width n of moving average has a great influence on the effect of congestion detection.

Fig. 5.4 Congestion
discrimination logic flow of
normal deviation method

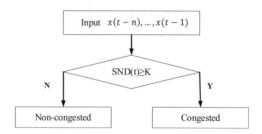

5.2.1.5 Double Section ACI Algorithm Based on ANN

ANN is one of the artificial intelligence technologies which appear and develop rapidly in recent years, and has been applied effectively in many fields. Because ANN can simulate the memory and processing ability of human brain to information, and is good at extracting useful knowledge from massive data, a double cross section ACI algorithm based on ANN is proposed. The basic idea of using ANN to distinguish traffic state is to train ANN with a large number of traffic data including congested and non-congested states to determine its optimal structure and weight. For a specific group of traffic data, the traffic state is determined by comparing the output result of ANN with the decision threshold.

1. *Data preparation*: The traffic data is preprocessed to remove the random components in the data; the data sets are divided into training and testing subsets of ANN; the predicted values of all data are estimated in time order to calculate the input variables.
2. *ANN model calibration*: The number of hidden nodes and the connection weights of ANN model are determined by using the input variable data of the upstream and downstream of the target road section.
3. *The determination of decision threshold and the discrimination of traffic state*: According to the traffic state corresponding to each group of data in the data set, under the requirements of ensuring the traffic congestion discrimination rate, error rate, and average discrimination time, the decision and threshold of ANN model upstream and downstream are optimized. Based on this, the traffic index is discriminated, and the decision results are fused by "or" operation, and the traffic status of the target road section is given.

The DS-ANN algorithm is proposed based on two ANN models, which are associated with the detection stations in the upstream and downstream of the target road, respectively. The ANN models of upstream and downstream stations have two input variables, and each ANN model will give the traffic congestion index data corresponding to its own location. Therefore, two congestion indices can be obtained at the same time. In order to make full use of the traffic state information provided by the two congestion indexes, the two results can be fused by or operation. This fusion method plays an important role in improving the efficiency of ACI algorithm.

5.2.2   Traffic States Analysis Method

As an important part of traffic state analysis, the discrimination result of traffic state is mainly used to measure urban road traffic operation. Therefore, in order to realize the real-time response to the traffic flow state of urban road network, it is necessary to extract and analyze the traffic information, so as to obtain the useful information that can scientifically represent the state of urban traffic flow, and then realize the mutual cooperation between the traffic control system and the guidance system. With the use of a large number of traffic control equipment and the deepening of intelligent transportation system research, traffic management and control center not only accumulates a large amount of historical traffic information, but also provides rich and accurate real-time traffic information, which creates conditions for traffic state analysis through traffic information processing. And the regional traffic flow feature extraction, traffic mode division, congestion time, and space analysis methods are gradually applied to the traffic state analysis.

In order to provide decision support for traffic flow control and guidance of regional road network, according to the differences of analysis objects and traffic management needs, urban road traffic state analysis can be divided into three levels: micro, meso, and macro traffic state analysis.

5.2.2.1   Analysis of Micro Traffic State

The micro traffic state analysis mainly focuses on the basic research of traffic flow, and estimates the traffic state parameters by establishing the micro traffic flow model. At present, two typical traffic flow models are cellular automata model and car following model.

Cellular automata are essentially a dynamic system that is defined in a cellular space composed of discrete and finite state cells, which is evolved in discrete time dimension according to certain local rules. At present, the cellular automata model describing the traffic flow operation characteristics is divided into two categories: one is one lane model, two lane model, and multi lane model with one-dimensional road traffic flow as the research object; the second is two-dimensional network traffic model. Traffic flow can be divided into small units based on cellular automata theory. Therefore, traffic effects caused by traffic lights, driver overreaction, bottle-neck of road section, and ramp connection point can be described better. Therefore, many nonlinear complex phenomena in the actual traffic operation can be described and simulated by cellular automata model, and the essential characteristics of traffic flow can be presented more accurately.

The car following model takes a single vehicle as the research object, and the traffic flow is equivalent to the dispersed particles. By studying the interaction between the adjacent front and rear vehicles on the road, the running characteristics and laws of vehicles are revealed. In the car following model, the motion law of each car is described as a coupled ordinary differential equation with discrete state variables and time and space variables. In the car following model, due to the change of vehicle speed in front, the distance between vehicles changes, and the driver has delayed reaction. The resulting traffic flow micro disturbance propagates along the vehicle flow direction. In the case of unstable traffic state, the micro disturbance is amplified to a certain extent, and the vehicle density in local area increases, resulting in traffic congestion. Therefore, the study of car following model should be carried out by numerical simulation, which can analyze the stability characteristics of vehicle operation, reveal the evolution mechanism of traffic flow operation, and determine and divide the road traffic state.

5.2.2.2 Analysis of Meso Traffic State

Meso traffic state analysis mainly includes the classification and discrimination of traffic state, and the analysis objects are road sections, intersections, or weaving areas.

At present, the classification of traffic state is mainly divided into two categories: one is represented by the U.S. Capacity Manual, which characterizes the traffic state as the level of service and divides it into six levels A–F; the other is based on different research objectives, traffic flow change characteristics, and traffic measurement characteristics. Generally speaking, the choice of traffic state parameters is determined by many qualitative or hierarchical parameters. The basic traffic parameters such as flow, average speed, and occupancy can be obtained directly through fixed detectors. After further calculation or conversion of the basic parameters, specific traffic characteristics parameters such as road saturation or congestion, travel time, driving delay, and traffic density can be obtained. Vehicle queue length, vehicle delay, parking time ratio, congestion coefficient, acceleration noise, and other indicators used to divide the traffic state level can be obtained through the mobile traffic flow detection technology.

The methods of traffic state identification can be divided into manual identification and automatic identification. Among them, the manual identification method is time-consuming and labor-consuming, and the detection accuracy is low, and the time span is short, so it has great limitations in the application of the actual traffic state identification. The automatic discrimination method has better mobility and flexibility, which can make up for the shortage of manual discrimination method. With the expansion of the layout range and the increase of density of all kinds of traffic detectors, automatic identification method has become the main means of traffic state identification. The specific discrimination method will be presented in Sect. 5.2.1.

5.2.2.3   Analysis of Macro Traffic State

Macro traffic state analysis is mainly to analyze and evaluate the overall operation state of macro road network, and the research object is traffic network. The analysis method is mainly through the analysis of intersection accessibility matrix and link connectivity, combined with the appropriate traffic parameters for reasonable road network traffic state modeling, based on the traffic state of the road network for spatial-temporal characteristics analysis. With the rapid development of communication technology, floating car data has become a widely available and highly applicable traffic data source in urban road network. The traffic flow parameter data obtained by the floating vehicle GPS device can cover a wider space including those roads without traffic monitoring infrastructure, so it provides good data support for the traffic state research at the road network level. However, the traffic state of the road network level is always changing with time and space, showing a high degree of dynamic, random, and complex, and its traffic flow parameters also have the characteristics of massive, high-dimensional, and multi-temporal. It is difficult to accurately model it by traditional analysis methods based on mathematical formulas and statistical models. At present, some data-driven visualization methods or data mining algorithms are mostly used in the traffic state analysis at the road network level. By analyzing the temporal and spatial characteristics of traffic state, it helps traffic managers to carry out traffic control at the macro level. There are several typical analysis methods.

1. *Cluster analysis*: Cluster analysis is a statistical method which divides a group of samples into different categories according to a certain similarity principle between objects, so that the similarity between the data in the same category is as small as possible, and the difference between the data in different categories is as large as possible. Compared with the supervised classification algorithm, clustering algorithm is an unsupervised machine learning method, which can effectively find out the internal structure characteristics of different samples from a large number of data and divide the data into a certain number of subsets without much prior information about the data structure and distribution of samples, and then provide important help for the next step of data analysis. With the rapid development of clustering technology, clustering analysis has been widely used in mathematics, statistics, biology, computer science, economics, and other disciplines. It is not only an important part of data mining and statistical analysis, but also a basic research problem in pattern recognition, image processing, and other fields. In the field of intelligent transportation, the cluster analysis technology can divide the massive traffic data into several traffic states with obvious characteristics, so as to dig out the traffic flow change rules under different traffic states. It is of great scientific significance to analyze the causes of traffic congestion, improve the traffic flow theory and provide decision support for the formulation of traffic management measures. With the in-depth study of different clustering problems by scholars at home and abroad, clustering technology has formed a systematic method system, and put forward a large number

of clustering algorithms suitable for different problems. At present, according to the principle of clustering, common clustering algorithms can be divided into the following: partition clustering method, such as K-Means algorithm; density-based clustering methods, such as DBSCAN, OPTICS; fuzzy clustering method, such as FCM algorithm.

2. *Visualization method*: The visualization method uses the corresponding two-dimensional geographic spatial expression or three-dimensional spatial visualization means, combined with GIS map to represent the spatio-temporal variation of traffic flow parameters, in order to explore the spatio-temporal congestion characteristics. This method can intuitively and clearly observe the temporal and spatial variation characteristics of traffic flow, which is helpful for managers to take scientific and reasonable traffic control measures. Visualization method is often used as a way to express the traffic state at the road network level, which needs to be combined with specific clustering methods or statistical methods for in-depth analysis of traffic state.

3. *Macro basic graph method*: In order to describe the relationship among speed, density, and flow, the predecessors built a traffic flow model based on historical data and statistical knowledge, and presented it in the form of basic relationship graph, so it is also called basic relationship graph of traffic flow. According to the different research objects, the same form can be used to get the macro basic chart. The macro basic map takes the road or macro road network as the research object, and analyzes the variation law and relationship between the average traffic flow parameters of all vehicles in multiple roads or multiple sections. The common model structures of macro basic graph are as follows: the relationship model of average density, average speed,and average flow of road network; the relationship model of vehicle mileage and cumulative number of vehicles in road network; the relationship model of completed traffic volume and vehicle mileage in road network.

5.2.2.4 Case Study

Example 5.1 Taking the road network of the Third Ring Road in Beijing as an example, and analyzing the traffic state of the road network at the macro level. The road network is shown in Fig. 5.5, with a total of 14,990 road sections. Taking the road section as the unit, the speed index with 2 min as the update frequency is selected as the traffic flow parameters to analyze the temporal and spatial variation of the traffic state of the road network in one day.

The specific methods are as follows.

Firstly, the traffic state at the network level is defined as an n-dimensional vector composed of m timestamps and n road sections. Let A_{ij} represent the traffic state of road section i in j period, then the global traffic state of the road network can be defined as the matrix X of $m \times n$.

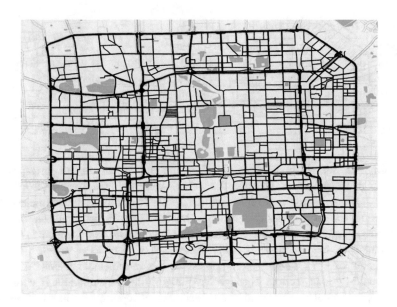

Fig. 5.5 Road network in the third ring road

$$X_{m \times n} = \begin{bmatrix} A_{11} & \cdots & A_{1n} \\ \vdots & A_{ij} & \vdots \\ A_{m1} & \cdots & A_{mn} \end{bmatrix} \tag{5.6}$$

In order to more accurately characterize the traffic state, A_{ij} is defined as the section traffic accessibility index, which is expressed by the ratio of the actual average speed and free flow speed of section i in period j. The closer the value is to 1, the more smooth the section is, the smaller the value is, the more serious the congestion situation is. The formula is as follows.

$$A_{ij} = \frac{V_{ij}}{V_{fi}} \tag{5.7}$$

The non-negative matrix factorization (NMF) algorithm is used to extract the features of the matrix composed of the high-dimensional vector. As a semi-supervised learning algorithm, the core idea of NMF is to decompose an objective matrix X into two non-negative submatrices W and H. That is to say, given a non-negative matrix X of order $m \times n$, which is composed of n samples and m-dimensional eigenvectors of each sample, and taking it as the input, the input matrix is decomposed into two non-negative matrices W and H of order $m \times s$ and order $s \times n$ through iterative calculation, so as to approximate the input matrix X to the maximum. Through matrix decomposition, the m-dimensional eigenvectors of each sample in the input matrix X can be approximated to the linear combination of

Table 5.1 Time clustering results

Cluster	Average free flow index	Time	Cluster name
Cluster 1 (red)	0.7645	9:50 am—17:10 pm	Daytime transition phase
Cluster 2 (blue)	0.9840	22:48 pm—6:38 am	Free flow
Cluster 3 (green)	0.6921	17:12 pm—18:50 pm	Evening peak
Cluster 4 (purple)	0.7546	6:40 am—9:48 am	Morning peak
Cluster 5 (yellow)	0.8331	18:52 pm—22:46 pm	Night transition phase

s-columns in the matrix W after weighting the components in the corresponding columns of the matrix H. Usually, the variable s should take a value far less than n and m, so that the matrix H represents a simplified low dimensional representation of the input data for further data analysis. In this example, the input matrix corresponds to the global traffic state X of the road network, and its dimension is equal to the number of road sections m. The time series is taken as n samples, and the eigenvector of each sample is composed of the m-dimensional column vector represented by road section. Through NMF algorithm, the submatrix H composed of n s-dimensional eigenvectors is obtained, and H represents a low dimensional representation of the global traffic state of the road network.

Based on the low dimensional expression of the global traffic state of the road network, the K-Means algorithm is used to cluster the two submatrices in time and space to mine the temporal evolution law and spatial structure characteristics of the road network traffic state. At present, K-Means is a widely used clustering analysis method. By determining an objective function, it is used to evaluate the advantages and disadvantages of clustering results, and the data sample partition results that meet the minimum objective function are regarded as the final clustering results. For the data sets with n samples, the basic idea of clustering method is: firstly, determine the number of categories k to be divided into data sets, create an initial partition result, so that each category contains at least one sample, and meet the requirements that each sample belongs to and belongs to only one category; then, according to some measurement method, data samples in different categories are continuously exchanged through cyclic iteration until the result of the partition of the minimum value is obtained.

Combined with GIS map of 3D space and road network, the spatiotemporal clustering results are analyzed visually.

The results of cluster analysis are as follows.

The K-Means algorithm is used to cluster, and the clustering classification is obtained, as shown in Table 5.1. Free flow clustering has the highest average traffic index, corresponding to the most smooth traffic state. In contrast, the morning and evening peak hours show the lowest average traffic free flow index, representing the most congested cluster.

In order to show the time evolution law of road network traffic state more intuitively and clearly, three dimensions of s-dimensional clustering results are selected for visualization, and the results are shown in Fig. 5.6.

According to the above example, this method based on clustering algorithm and visualization can effectively show the time variation law of traffic state in a day, and

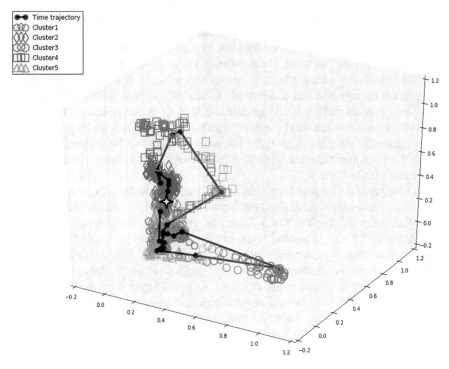

Fig 5.6 Time clustering results and evolution process of road network traffic state (*star* pattern represents early morning, black track shows the evolution process of traffic state in one day in hours)

has the ability to accurately divide the traffic state according to the traffic flow parameters. The results can provide decision support for the traffic flow control and guidance of regional road network.

5.3   Basic Theory and Method of Prediction

5.3.1   Overview of Traffic State Prediction

Traffic state prediction is based on the historical and existing traffic operation data, using intelligent methods to determine the future operation of traffic targets. As a branch of forecasting, traffic state prediction is an important part of modern intelligent transportation system, and it is also one of the ways to solve the problem of traffic congestion recognized by the international community. Under the background of information age, the city puts forward new requirements for the traffic management department to scientifically judge and predict the changes of urban road network and expressway traffic state, and actively adopt electronic information release unit, traffic broadcast, and other guidance methods to ensure the smooth flow of urban traffic. Based on the current or past changes in the road network traffic state, we can reasonably infer the traffic state information in the future and release reliable information. On the one hand, it can provide guidance for traffic management and control strategy by taking effective measures in advance to avoid possible traffic congestion or jam. On the other hand, it can also provide travel reference for travelers, conduct benign traffic navigation, actively guide residents' travel, and improve the use efficiency and travel efficiency of urban roads by making reasonable travel planning, which is of great significance to alleviate traffic congestion.

Traffic state prediction can be divided into many categories according to different classification standards.

According to the length of prediction cycle, it can be divided into three categories: short-term traffic state prediction, medium-term prediction, and long-term prediction. Generally, the prediction of 5–30 min is classified as short-term prediction, the prediction of 30 min to several hours is called medium-term prediction, and the prediction of time span not less than 1 day is called long-term prediction. Among them, the results of short-term traffic state prediction can be directly applied to advanced traffic information systems (ATIS) and advanced traffic management systems (ATMS) to provide travelers with real-time and effective travel information to help them better choose the route, realize route guidance and to save travel time, alleviate road congestion, reduce pollution and save energy.

According to the content of prediction, it can be divided into continuous traffic state prediction and discrete traffic state prediction: one is that the traffic state is continuous and mainly forecasts the traffic flow parameters according to the traffic state index variables; the other is that the traffic state is discrete and mainly forecasts the category or level of traffic state (mainly for congestion state).

According to different scenarios, traffic state prediction can be divided into three categories: micro traffic state prediction, meso traffic state prediction, and macro traffic prediction. Among them, the microscopic traffic state prediction mainly focuses on the road intersection, the meso level on the urban single trunk road or expressway, and the macro level on the urban road network.

Because the change of traffic state is not only related to the traffic state of this section in the past several periods, but also affected by the traffic state of upstream and downstream, traffic environment and other factors, especially for large-scale road network, there are many kinds of traffic roads, such as intersections, main roads, expressways, overpasses and so on. Its traffic state always changes with time and space, showing a high degree of dynamic, random, and spatiotemporal complexity. The problem to be solved in traffic state prediction is to find out the law from the traffic state changes with randomness and uncertainty, according to the traffic state parameters obtained by vehicle detector, combined with other influencing factors, and establish the prediction method and model to predict the traffic state changes in the future. An excellent traffic state prediction method must have the following points.

1. *High accuracy*: The prediction results must be able to accurately reflect the future traffic flow operation, and the prediction error is within the acceptable range.
2. *High efficiency*: The prediction results directly serve the current traffic control and guidance system, so the prediction model must have fast computing ability to ensure the real-time results.
3. *Strong adaptability*: Because the traffic flow in the road network is changing all the time, the prediction model should have strong adaptability, which can not only have high prediction accuracy in the unblocked state, but also have high accuracy in the traffic congestion state.
4. Considering the spatiotemporal characteristics of traffic flow comprehensively. The temporal variability and spatial mobility of traffic flow make the prediction model not only need to consider the correlation of time series of traffic flow data, but also need to consider the correlation characteristics of adjacent road traffic flow data.

   Therefore, scholars and traffic engineers at home and abroad have invested a lot of manpower and material resources to carry out corresponding research in order to alleviate traffic congestion. At first, the research on urban road traffic state mainly focused on single intersection, trunk road, or some road sections. By means of manual data collection, induction coil data, license plate recognition data, radar data, and other methods, through statistical model construction, simulation experiment analysis, and other methods, the traffic state of intersections, expressways, or trunk roads was predicted from the micro and meso level. In recent years, with the development of intelligent transportation system and the continuous improvement of road traffic data acquisition equipment in China, it is possible to provide real-time dynamic traffic data through information collection, processing, and analysis. At present, many urban roads in our country have established traffic information collection system based on embedded sensors, TV monitoring system based on cameras and traffic violation monitoring system, etc., initially realizing the real-time monitoring and surveillance of traffic status. Taking Beijing as an example, the urban road traffic flow information collection methods mainly include the detection coil set in the signal control system of more than 1600 intersections, which is used to collect the traffic flow, speed, and occupancy of motor vehicles and store them in the plug-in database; nearly 2000 traffic flow section detectors are set on urban expressways and main roads, It can obtain all kinds of traffic flow data such as vehicle speed, flow, and occupancy rate in real time; it has more than 600 travel time detection sections; it has set up vehicle violation detection equipment at more than 1500 intersections to provide vehicle violation information; it has installed more than 240 traffic information boards on urban expressways, connecting lines and some main roads. With the more reasonable cost performance of acquisition equipment, the continuous improvement of acquisition quality and accuracy, the increasingly rich acquisition means and sources, and the continuous strengthening of the integration of different acquisition means and information, the information acquisition, processing, and utilization ability of transportation system has been upgraded to a new level, in order to realize the deep mining, integration, and application of information, it not only lays the foundation for improving the informatization level of traffic command, management, and service, but also provides massive data support for traffic research, which also urges a large number of scholars at home and abroad to establish a variety of data-driven algorithms, which are committed to mining useful information for traffic management and control from massive travel data, so as to achieve the purpose of easing traffic congestion. As the accuracy of traffic state prediction is greatly affected by the quality of data, it benefits a lot from the data-centric traffic change. With the continuous progress of GPS technology, the emergence of floating car data also makes the traffic state prediction break through the bottleneck of data collection by infrastructure. Its controllable large coverage and real-time performance lay a data foundation for the prediction of large-scale urban road network traffic state, and it is also more conducive for traffic managers to control the current situation of urban congestion from the macro level.

Summarizing the relevant research, the methods of traffic state prediction can be roughly divided into the following four categories.

5.3.1.1   Prediction Based on Parameter Model

The basic idea of this kind of method is to estimate the future traffic flow parameters by establishing an accurate prediction model. These methods include not only linear models such as multiple linear regression model, adaptive model, time series method, filtering method, and exponential smoothing method, but also nonlinear models such as wavelet analysis and chaos theory.

Multiple linear regression model is a statistical method to study the correlation between a random variable and several controllable variables. Through the statistical analysis of historical data, the mathematical functional relationship model between the independent variable and the dependent variable is found out, and then the measured value of the independent variable is substituted to output the predicted value. Multiple linear regression model is a simple and practical model, which is easy to be applied to large-scale road network traffic flow prediction. However, the prediction accuracy of multiple linear regression model is low for the road sections with no or scarce historical data.

The adaptive model selects road condition indicators that can be detected in real time, such as forecast interval, emergencies, weather-related factors, road occupancy, average road network travel time, etc., and dynamically changes the proportion of each forecast factor in the regression model, so as to strengthen the adaptive ability of the model in order to overcome the shortcomings that the ordinary linear regression model cannot reflect the nonlinear change and uncertainty of traffic flow. This method is based on the linear regression model, which is easy to calculate, easy to implement, easy to large-scale application, and it is also easy to collect road condition indicators to change the weight, and has good real-time performance. However, in the parameter estimation of the model, there is no more scientific selection mechanism for the influence of road condition indicators on the weights theoretically or empirically.

Time series method is to predict traffic flow data using the theory and method of time series. The principle of this method is to process the collected traffic flow data series, and fit it into a parameter model according to the inherent statistical characteristics of the data series, and use this model to realize the prediction of future traffic flow data. The common models in time series method include autoregressive model (AR), moving average model (MA), autoregressive moving average model (ARMA), and autoregressive integral moving average model (ARIMA).

Kalman filtering is a linear filtering method proposed by RE Kalman in 1960. In this method, a state-space model composed of state equation and observation equation is used to describe the filter, and the parameters in the model are estimated and predicted iteratively. In the aspect of traffic state prediction, the state-space model composed of state equation and observation equation is used to describe the traffic system, and the state equation and observation equation and Kalman filter recursive algorithm are used to predict the traffic flow respectively. Kalman filter recursive algorithm is a matrix iterative parameter estimation method for linear regression analysis model, which has the advantages of flexible selection of prediction factors and high accuracy, and the prediction accuracy of the model does not depend on the prediction time interval. However, because the model is based on a linear estimation model, when the prediction interval is less than 5 min, the randomness and nonlinearity of traffic flow change are strong, the performance of the model will decline.

Exponential smoothing method uses exponential smoothing model to predict traffic flow parameters, and then uses least square principle or genetic algorithm to realize parameter estimation. The feature of the model is that the correction of the latest observation value to the prediction is constantly considered in the prediction process, so as to synthesize the error of the previous prediction and add it to the next prediction. After repeated iteration, a linear combination of all the previous observations is obtained, and finally the prediction result is formed. Its weight is an exponential weight which is declining continuously, and relatively speaking, the closer the observation value is, the greater the weight value is in the correction of prediction deviation, which is the origin of the name of exponential smoothing. Exponential smoothing model uses recursive calculation, which requires less data storage. It only needs to store the previous estimated value and filter parameters. Moreover, the calculation is relatively simple and does not need training, so it is widely used in the early prediction.

The prediction method based on wavelet analysis uses wavelet analysis theory to decompose the time series data of traffic flow to get the decomposed signals with different resolutions, and forecasts each decomposed signal respectively. Finally, the final result is obtained by synthesizing the prediction results.

The prediction method based on chaos theory applies chaos theory to traffic state analysis, and uses chaos theory to judge whether the traffic system is chaotic or not. If the traffic system is identified as a chaotic system with some chaotic characteristics, it can be predicted in the short term rather than in the long term. The general process of using phase space reconstruction technology to predict traffic state is to reconstruct the phase space of time series data of traffic state, determine the embedding dimension and time delay parameters, then find out the last known point in the phase space, take the known point as the center, find out the nearest several related points in the phase space, and use these related points as the basic fitting function. And it is used to predict the next point, and finally the predicted value is separated according to the predicted point.

5.3.1.2 Prediction Based on Nonparametric Model

This kind of method does not need to establish accurate model expression, its basic idea is to find out the change rule of traffic flow data through statistical analysis of a large number of historical or survey data, and take it as the basis of prediction, such as nonparametric regression algorithm, k nearest neighbor (KNN), decision tree, and so on. Among them, the nonparametric regression prediction method establishes the case database according to the relationship between the dependent variable and the independent variable in the historical data. In the prediction, the current traffic state to be predicted is regarded as the neighbor state of the past state. According to the principle of pattern recognition, the neighbor states similar to the current input state in the case database are found, and the traffic state is predicted according to these neighbor states. It does not need prior knowledge, only needs enough historical data, and with the increase of cases in the case database, it can consider more traffic state change trends.

5.3.1.3 Prediction Based on Artificial Intelligence

The basic idea of this method is to realize the prediction of traffic flow by means of machine learning algorithms and artificial intelligence technology. Common methods include neural network, support vector machine, deep learning algorithm, etc.

Neural network prediction method is to simulate the structure and function of human brain nervous system by engineering technology, send the information obtained through many paths to the neural unit, which processes the acquired signal, then transmits the information to more neural units, and also gives different weights according to the correlation between the processed data information and neurons. By training neural network model with a large number of historical data, a mapping relationship between output and input is obtained. If the corresponding input is given, the relevant prediction results can be obtained by using the mapping relationship. The characteristics of this complex nonlinear system are identified by neural network. The traffic state of urban roads can be predicted by taking the local road network or a line with strong correlation in the road network as the research object. According to the different neural network models, the current algorithms used in traffic state prediction include BP neural network, radial basis function recursive neural network, generalized regression neural network, fuzzy neural network, wavelet neural network, etc.

Support vector machine is a new machine learning algorithm. It can get the global optimal solution by solving convex optimization problem, so there is no local extreme value problem of general neural network. So it is widely used in pattern recognition, regression estimation, probability density function estimation, and other aspects once proposed. It has become a powerful method to overcome the difficulties encountered by traditional machine learning algorithms, such as small sample, dimension disaster problem, overfitting problem, and local optimal problem.

The concept of deep learning is derived from the research of artificial neural network, which simulates the multi-layer perceptual structure of human brain to recognize the data pattern. It combines the low-level features to form a more abstract high-level representation (attribute category or feature) to discover the distributed characteristic representation of data. Traditional machine learning and signal processing technology explore the shallow learning structure with only single layer nonlinear transformation. One common feature of shallow model is only a single simple structure which transforms the original input signal to the spatial characteristics of a specific problem, while deep learning has stronger representation ability than shallow learning. In recent years, as a new field of data mining, deep learning has shown excellent performance in processing unstructured data such as image, text, voice, etc., and has also made some achievements in traffic state prediction. There are two commonly used structures, convolutional neural network and recursive neural network.

5.3.1.4 Prediction Based on Combination Model

This kind of prediction method is to combine two or more different types of prediction models for final prediction. The purpose of introduction to intelligent transportation technology is to give full play to the advantages of each prediction model, overcome their own defects, and achieve the purpose of improving the prediction accuracy. For example, the combination of wavelet analysis and time series analysis method, fuzzy reasoning combined with neural network prediction method, a variety of prediction methods combined with artificial intelligence technology intelligent prediction method.

5.3.2 Analysis of Traffic State Prediction Methods

This section will focus on some mainstream advanced algorithms in the field of traffic state prediction at present to introduce the theoretical system of the algorithm in detail, and explain the examples, including typical time series prediction algorithm, neural network-related algorithm, support vector machine, depth learning algorithm, etc.

5.3.2.1 Time Series Prediction Method

There are three common time series models: autoregressive model (AR model), moving average model (MA model), and autoregressive moving average model (ARMA model). The formula of p-order autoregressive model AR (p) is

$$y_t = \phi_1 y_{t-1} + \phi_2 y_{t-2} + \cdots + \phi_p y_{t-p} + e_t \qquad (5.8)$$

where p is the order of autoregressive model, y_t is the observation value of time series in t period, that is, dependent variable or explained variable; $\{e_t\}$ is white noise sequence, that is, random error; ϕ_1, ϕ_2, ϕ_p are autoregressive parameters to be estimated. The formula of q-order moving average model MA (q) is

$$y_t = e_t - \theta_1 e_{t-1} - \theta_2 e_{t-2} - \cdots - \theta_q e_{t-q} \qquad (5.9)$$

where q is the order of moving average model, y_t is the observation value of time series in t period, and $\{e_t\}$ is white noise sequence, which indicates the error or deviation of time series model in t period.

ARMA (p, q) model is a combination of the above two models, in the form of

$$y_t = \phi_1 y_{t-1} + \phi_2 y_{t-2} + \cdots + \phi_p y_{t-p} + e_t - \theta_1 e_{t-1} - \theta_2 e_{t-2} - \cdots$$
$$- \theta_q e_{t-q} \qquad (5.10)$$

The basic principle of autoregressive sliding average model method is to treat the data series formed by the predicted object with time as a random time series. The future development of the sequence has the dependence and continuity on the past development and change of the predicted object. According to this, the appropriate mathematical model can be established, and the disturbance errors can be considered, so as to get the future change state of the predicted object.

Comparing the above three models, we can see that when $q = 0$, ARMA $(p, 0)$ model is AR (p) model; when $p = 0$, ARMA $(0, q)$ model is MA (q) model.

It is noted that the prediction object of ARMA model method is a zero mean stationary random time series. From the line diagram of stationary random time series, there is no obvious upward or downward trend, and the fluctuation is up and down near the horizontal axis of each observation value. However, many traffic phenomena always show a certain upward or downward trend, which constitutes a non-zero mean nonstationary time series. Therefore, before the ARMA model is established, the time series should be treated with zero mean and stable.

The zero mean processing is to use the original sequence $\{y_t\}$ $(t = 1,2,\ldots, n)$ minus the mean value of the sequence to obtain a new time series with zero mean.

The stationarity treatment usually refers to the difference stabilization treatment. In practice, many nonstationary sequences can be transformed into stationary sequences only when one or more differential is performed. This time series is a homogeneous nonstationary sequence. And the order of homogeneous means the number of differences.

The processed zero mean stationary time series can be predicted by ARMA model following the following steps.

1. *Model recognition*: By using autocorrelation and partial correlation analysis methods, the randomness and stationarity of the given sample sequence are analyzed, and the sample sequence should belong to AR (p), MA (q), and ARMA (p, q).

   If the partial correlation function of sequence $\{y_t\}$ is truncated after step p, AR (p) model can be selected; if the autocorrelation function of sequence $\{y_t\}$ is truncated after step q, MA (q) model can be selected; if the autocorrelation function and partial correlation function of sequence $\{y_t\}$ are tail supporting, ARMA (p, q) model can be selected.
2. *Parameter estimation*: Based on the determined model and order, the undetermined parameters of the model are estimated.
3. *Model validation*: The rationality of the initially established model is tested. If the test results do not meet the requirements, return to the first step to select the model again. However, with the increase of model order and time series length, model checking and repeated correction will become very complicated. At this time, we can use computer to complete the cumbersome calculation process.
4. *Forecast*: The future value of time series of traffic state data is predicted by the model that has passed the test.

   In addition, there is a more commonly used improved ARMA model— ARIMA (p, d, q) model, which is a famous time series prediction method [2] proposed by Box and Jenkins in the early 1970s, so it is also called Box-Jenkins model or Box-Jenkins method.

A stationary time series can always find its stationary random process or model; a non-stationary random time series can usually be transformed into stationary by difference method, and the corresponding stationary random process or model can also be found for the stationary time series after difference. Therefore, if a non-stationary time series is made stationary by d-difference, and then a stationary ARMA (p, q) model is used as its generating model, then the original time series is said to be a time series satisfying the autoregressive moving average model, which is ARIMA (p, d, q), where ARIMA (p, d, q) is called differential autoregressive moving average model, AR is autoregressive, and p is autoregressive term; MA is moving average, q is the number of moving average terms, and d is the difference times when the time series becomes stationary. Generally speaking, the value of d is usually 0,1,2.

For example, an ARIMA (2,1,2) time series must be differentiated once before it becomes a stationary series, and then an ARMA (2,2) model is used as its generation model. Of course, an ARIMA $(p,0,0)$ process represents a pure AR (p) stationary process, and an ARIMA $(0,0,q)$ process represents a pure MA (q) stationary process.

The basic idea of ARIMA model is to regard the data series formed by the prediction object with the passage of time as random series, and describe this series approximately with a certain mathematical model. Once the model is identified, future value can be predicted by the past and present value of time series. General process is given as follows.

1. *Data preprocessing*: Including the above-mentioned zero mean processing and stationary processing.
2. *Parameter estimation*: Parameter estimation is the core part of modeling, usually using recursive algorithm for estimation. Parameter estimation needs a given order before it is carried out. Usually, a upper limit of order is given. Then, according to certain estimation methods, parameter estimation is carried out from low order to high order one by one, and the estimated value of parameters is finally selected by combining the fixed order criterion.
3. *The order of the model is determined*: For each group of given order and the corresponding parameter estimation, the order of the model is compared and judged according to the order determination criteria, and the ideal model for data fitting is finally determined.

ARIMA model is especially suitable for stable traffic flow. When the traffic condition changes rapidly, the model will show obvious deficiencies in delay. ARIMA model can be used to predict future observations according to the following steps: (a) to identify the stationarity of time series; (b) to stabilize the data; (c) according to the law of time series, establish the corresponding model; (d) to estimate the parameters of the model and test whether it has statistical significance; (e) to test the hypothesis and diagnose white noise.

5.3.2.2 Neural Network Algorithm

The basic unit of neural network is neuron, which is a multi-input and single output information processing unit. The input can be compared to the dendrite of neuron in bioengineering, the output can be compared to the axon of neuron, and the calculation can be compared to the nucleus. The artificial neuron on the model is shown in Fig. 5.7.

Among them, $a_1, a_2, \ldots, a_n$ are n input signals of neurons, $w_1, w_2, \ldots, w_n$ are the weight of the corresponding input, which represents the connection strength between

Fig. 5.7 Structure of neurons

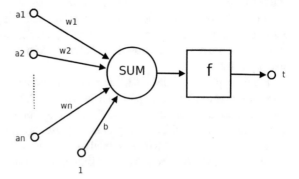

each signal source neuron and the neuron; SUM is the total input of the neuron, f corresponds to the membrane potential of the biological nerve cell, which is called the activation function; y is the output of the neuron.

Therefore, the input-output relationship of artificial neuron can be described as follows

$$y = f(\mathrm{SUM}) \tag{5.11}$$

$$\mathrm{SUM} = \sum_{i=1}^{n} w_i x_i \tag{5.12}$$

The function $y = f(\mathrm{SUM})$ is called the characteristic function, also known as the action function or the transfer function. The characteristic function can be regarded as the mathematical model of neurons, which choose different activation function f, and the range of output y is different. The commonly used functions are linear function, symbol function, Sigmoid function, and hyperbolic function.

The relationship between neuron and neural network is the relationship between elements and the whole. A large number of the same neurons are connected together to form the neural network. Neural network is a highly nonlinear dynamic system. Neural network has the ability of self-learning and self-adaptive. It can analyze and master the potential laws of the two by a large number of corresponding input-output data prepared in advance. Finally, according to these rules, the output results are calculated with new input data. The process is called training.

Neural network has the following characteristics:

1. Neural network is an effective method to approximate any nonlinear mapping.
2. Neural networks have learning ability, that is, they can be trained with sample data.
3. Neural network has generalization ability, that is, a trained neural network can produce accurate response to any input.
4. Neural network can run qualitative and quantitative data simultaneously.
5. Neural networks can be easily applied to multivariable systems.
6. Neural network has high parallel implementation ability, and has higher fault tolerance than traditional methods.

   In the network of reticular structure, there may be bidirectional connection between any two neurons. Feedforward network, feedback network, and self-organizing network are the topological structures of three representative neural networks.
1. *Feedforward network*, which consists of input, hidden, and output layers. Hidden layer can be multi-layer. The output of the input layer is the input of the hidden layer, and the output of the hidden layer is the input of the output layer. There is no connection between the neurons in the same layer, so the feedforward network is also called forward network.

   Feedforward network is the most common network structure. Its information processing ability mainly comes from the multiple compound action of nonlinear output function, and the network structure is relatively simple and easy to realize.

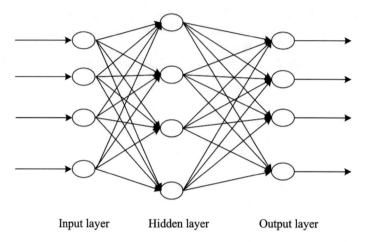

Input layer   Hidden layer   Output layer

Fig. 5.8 Structure of BP neural network

BP network is a typical network. In addition, perceptron network, linear neural network, RBF network, and GMDH network also belong to feedforward network.

2. *Feedback network*: Feedback network is also called recursive network or regression network. There is feedback between output and input layer, and there is information flow in neurons both in same layer and different layers. In the feedback network, the input signal determines the initial state of the feedback system, and the system needs to be stable after a series of state transitions. Elman and Hopfield neural network are typically used; CG network model, regression BP network and Boltzmann machine network are also feedback networks.

3. *Self-organizing network*: Self-organizing network is an unsupervised learning network, which simulates human beings to adapt to unpredictable environmental changes automatically according to past experience. Because there is no supervision signal, this kind of network usually uses the competition principle for network learning, including self-organizing competition network and self-organizing feature mapping neural network.

Neural network has developed rapidly, and there are dozens of neural network models. Back propagation (BP) neural network is the most mature and widely used neural network model.

BP neural network is a multilayer feedforward network trained by error inverse propagation algorithm. It was originally proposed by Paul Werboss in 1974, but it did not spread. Until the mid-1980s, Rumelhart, Hinton and Williams [3], Yann Le Cun [4] rediscovered BP algorithm, and it was not known until the algorithm was included in parallel distribution processing. At present, BP algorithm has become the most widely used neural network learning algorithm. According to statistics, nearly 90% of the neural network applications are based on BP algorithm.

The basic structure of BP network is shown in Fig. 5.8. The model structure includes input layer, hidden layer, and output layer. In input layer, each neuron is receiving input information from the outside and input it to the middle layer; the middle layer is the layer of information transformation. According to the demand of information change ability, the intermediate layer can be designed as single hidden layer or multi hidden layer structure; the information transmitted to each neuron in the output layer can be further processed. Then the process of forward propagation of learning is completed, and the information processing results are output from the output layer to the outside world. Sigmoid function is commonly used in hidden layers. The input and output nodes can use Sigmoid function or linear function.

BP algorithm includes forward propagation and back propagation. Input information is transmitted from input to output layer in forward propagation. Only the next layer's state of neurons will be affected by current layer's state of neurons. When the actual output is not in line with the expected output, the error will be returned along the connecting path through the output layer. The weight of each neuron is updated by the error gradient descent. The process of forward and backward propagation of information is a process of adjusting weights of each layer. It is also a process of neural network learning and training by modifying the connection weights between neurons at each layer. This process is carried out until the error signal output of the network is reduced to an acceptable level or preset learning times.

BP model transforms the function problem of a set of input/output samples into a nonlinear optimization problem, and uses the most common gradient descent method in optimization technology. If the neural network is regarded as the input-output mapping, the map is a highly nonlinear mapping. The program diagram of BP algorithm is shown in Fig. 5.9.

Because BP network has the function of fitting any non-linear function, it can be used to replace the complex non-linear mathematical relationship, and achieve more accurate description of the real traffic system. Of course, BP network has a high demand for training samples, and the required sample size is large. It is most commonly used in urban road and expressway section traffic flow and other microscopic traffic information prediction.

The traffic information prediction algorithm based on BP network can be summarized as follows:

1. Raw data acquisition and preprocessing.
2. The network structure is designed to determine the number of layers, input layer neurons, hidden layer neurons, and output layer neurons.
3. The original data is used to construct training sample pairs.
4. Design learning algorithm to train neural network.
5. The performance of the network is evaluated by the test sample pair and the preset error. If the current cumulative error is greater than the preset error, return to (4) adjust the learning algorithm and retrain the network; if the cumulative error is less than the preset error, enter (6), or enter (6) when the number of training times reaches the maximum number of preset training times.
6. For the trained BP network, a set of known data is input to get the corresponding traffic information prediction value.

Fig. 5.9 Block diagram of
BP algorithm

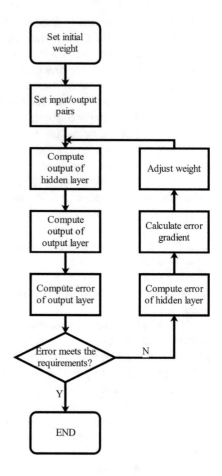

Specifically, let X be the input data set. For a single sample $(x^{(i)}, y^{(i)})$, the input part is $x^{(i)} = [x^i_1, x^i_2, \ldots, x^i_n]$. n means that each sample has n features, that is, the input layer has n neurons. In the regression problem, $y^{(i)}$ is usually a specific value. Firstly, the input layer and the hidden layer are linearly summated, and the value of each neuron in the hidden layer is obtained through the activation function. After the value of the hidden layer is obtained, the feedforward calculation is continued to get the value of the output layer. The relationship between the hidden layer and the input layer is similar to that between the input layer and the hidden layer. The loss function can be defined as the mean square error function, and the parameters in the network model can be updated according to the gradient descent principle.

Among them, the value of each neuron in the hidden layer can be obtained through the activation function, as shown in Eq. (5.14).

$$Z_j = \sum_{i=1}^{n} W_{1i}^{(1)} x_i + b_i \tag{5.13}$$

$$h_j = f(Z_j) \tag{5.14}$$

where $W_{1i}^{(1)}$ is the weight matrix connecting the input layer and the hidden layer; b_j is the bias; the value of the j-th neuron in the hidden layer of h_j; f is the activation function. The activation function of hidden layer and input layer is similar to the above formula and will not be described in detail.

Through the effective learning of historical traffic data, the gradient descent method is used to update the parameters. Before overfitting, every learning of the model is toward the direction of error reduction. Based on enough samples and enough training times, the model can learn the evolution trend of road network traffic flow, and make prediction based on the trend. The prediction results of traffic flow parameters are obtained.

5.3.2.3   Traffic State Prediction Based on Support Vector Machine Regression

Support vector machine (SVM) was first proposed by Cortes and Vapnik in 1995. It shows many unique advantages in solving small sample, nonlinear and high-dimensional pattern recognition, and can be extended to other machine learning problems such as function fitting. According to the basic principle of road traffic flow prediction, support vector machine regression prediction model is adopted.

Assuming that $x_i \in Z^n$ is the variable that affects the road traffic flow parameters, and y_i is the predicted value of the road traffic flow parameters, the support vector machine regression prediction model is to find the correlation between the two.

$$f : Z^n \rightarrow Z \tag{5.15}$$

$$y_i = f(x_i) \tag{5.16}$$

As mentioned above, the traffic flow operation state of the previous period in the actual traffic environment will affect the latter period. The change of traffic flow is time series. It is advisable to predict the traffic flow in future according to the traffic flow data in adjacent period.

The traffic flow data of the first n sampling time adjacent to the prediction time is selected as input, namely x_i, and the estimated results of the predicted time traffic flow parameters are taken as the output, i.e., y_i. The specific process of the model is as follows.

Select the original data, filter and repair the data, and build the training set. If the current traffic flow parameter to be predicted is k_i, the training set data of $t + 1$ in the next period is $x_i = [k_i(t), k_i(t-1), \ldots, k_i(t-n)]$.

According to the characteristics of known data sets, the kernel function is selected to initialize the penalty factors and other parameters. By comparing the training effect, the radial basis kernel function (RBF) with better performance and less deviation is selected. The function form is as follows:

$$K(x_i, x) = \exp\left(-\frac{\|x_i - x\|^2}{\sigma^2}\right) \tag{5.17}$$

The optimization problem is constructed and solved. The training samples and the corresponding traffic flow parameters (i.e., sample labels) of the samples are taken as input to solve the problem.

$$\min\left\{\frac{1}{2}\sum_{i=1}^{N}\sum_{j=1}^{N}(a_i^* - a_i)(a_j^* - a_j)K(x_i, x) - \sum_{i=1}^{N}(a_i^* - a_i)y_i + \sum_{i=1}^{N}(a_i^* + a_i)\varepsilon\right\}$$

$$s.t. \ \sum_{i=1}^{N}(a_i - a_i^*) = 0$$

$$0 \le a_i \le C$$

$$0 \le a_i^* \le C$$

$$i = 1, 2, \cdots, N$$

$$\tag{5.18}$$

The optimal solution is substituted into the decision function, and the test set is used to predict the future traffic flow parameters.

$$f(x) = \sum_{i=1}^{N}(a_i^* - a_i)K(x_i, x) + b \tag{5.19}$$

For traffic prediction, the specific steps of the prediction algorithm based on support vector and regression are as follows:

1. The historical traffic volume data is normalized to generate data sets.
2. The kernel function is selected to determine the parameters of SVM. After obtaining the sample data set, the radial basis function (RBF) is selected as the kernel function, and the optimization parameters ε and C of quadratic programming are determined. ε and C have great influence on the learning ability and generalization ability of SVM algorithm. If the penalty coefficient C is small, the training error will increase, and if the penalty coefficient C is large, the learning accuracy will improve, but the generalization ability of the model will become worse. Therefore, it is very important to select an appropriate penalty coefficient C for the prediction model. An appropriate value of C can reduce the interference of outlier samples and improve the stability of the model. The prediction ability of the model is controlled by the insensitive loss function ε. When the value of ε is

Fig. 5.10 Schematic
diagram of a local road
network in Beijing

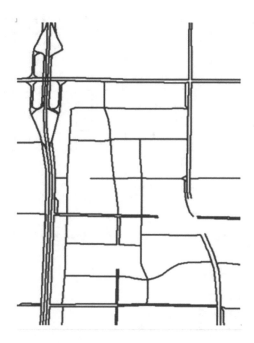

large, the learning accuracy will be low and the generalization ability will be
reduced. When the value of ε is small, the prediction model will be too complex
and the training time will be longer.

Therefore, we can use the method of dynamic adjustment to determine the
parameters, first fix one of the parameters with a priori knowledge method, and
then use the enumeration method to determine the other parameter. Finally, the
optimized parameters are fixed to determine the non-optimized parameters, so as
to determine the final ε and C.

3. Input data set to generate prediction function.
4. Prediction and error analysis. According to the generated prediction function, the
   traffic flow information in the future period is predicted, and the prediction results
   are evaluated and analyzed. If it is found that the relative error is large, it is
   necessary to return to (2) and readjust the parameters of SVM.

Example 5.2 This part will take the prediction speed as an example to analyze the
above two prediction algorithms (BP neural network and support vector machine),
select an area in Beijing as the research object, and give the comparison results and
conclusions.

The road network is shown in Fig. 5.10, including 278 road sections, covering an
area of 2.4124 square kilometers (1.63 km × 1.48 km). The vehicle data update
frequency is 2 min, and the time range is from June 1, 2015 to July 31, 2015, a total
of two months.

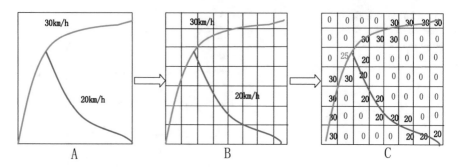

Fig. 5.11 Grid process

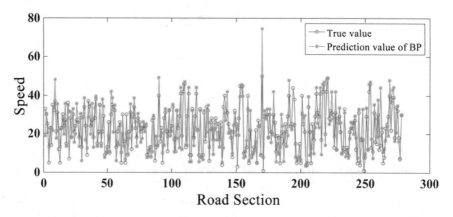

Fig. 5.12 Comparison of BP neural network prediction results and actual values at 08:00 in the morning peak

Datasets include DataInput dataset and DataInput_Lables dataset.

The DataInput dataset is the state-space matrix of road network every 2 min, and the size of the matrix is 164 × 148. The matrix is obtained by spatial gridding of road network of 10 m × 10 m, and the gridding process is shown in Fig. 5.11.

DataInput_Lables dataset is a space-time matrix of speed data, with the size of 481 × 278. The horizontal axis represents 278 road sections, and the vertical axis represents the time axis of a day, with an interval of 2 min from 6:00:00 to 22:00:00.

BP neural network and support vector machine are respectively used to carry out speed regression prediction for a given road network, and 20 historical data are used to predict the speed in the next 2 min. The prediction results are shown in Figs. 5.12 and 5.13.

The mean absolute percentage error (MAPE) and root mean square error (RMSE) are used to evaluate the prediction effect. The calculation results of evaluation indexes are shown in Table 5.2.

It can be seen from the prediction results that both BP neural network and support vector machine can be used to predict the speed, and the prediction effect of the two

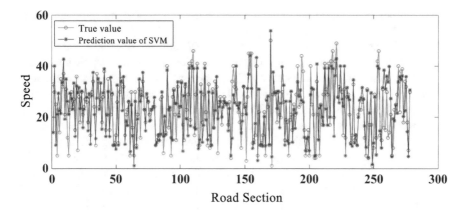

Fig. 5.13 Comparison of prediction results of support vector machine and actual values at 08: 00 am in the morning peak

Table 5.2 Evaluation index value of prediction effect

Prediction method	MAPE	RMSE
BP neural network	0.30325	8.9583
Support vector machine	0.31431	10.188

methods is relatively close. Compared with the prediction results of off-peak hours, it can be seen that the prediction accuracy of peak hours is higher. The reason is that the fluctuation range of road network speed in peak hours is smaller, which is conducive to the prediction of traffic flow parameters.

5.3.2.4   Road Network Traffic State Prediction Based on Deep Learning

Under the background of traffic big data, it is difficult for traditional machine learning to obtain valuable information from large-scale, complex events, heterogeneous and diverse big data with high and low value density. Compared with traditional machine learning model, deep learning model is more complex, has stronger network expression ability, learning ability, and generalization ability. It can mine the temporal and spatial evolution law of road network state from the complex traffic big data, so as to make an accurate prediction of the future road network state. Therefore, it is an important research direction to study road network state prediction based on deep learning theory. It has important theoretical research value and practical significance to grasp the evolution law of complex road network state through deep learning. At present, state prediction algorithms based on deep learning mainly focus on recurrent neural network, convolutional neural network, and their combination. The following will focus on several mainstream algorithms.

1. The prediction algorithm based on long-term and short-term memory neural network.

In the artificial neural network, the former and the latter two samples are relatively independent and do not interfere with each other, which makes the traditional neural network (such as BP neural network) have many shortcomings in processing sequence information. Therefore, Jürgen Schmiduber proposed recurrent neural network in 1992. The main contribution of recurrent neural network is that it allows to connect with neurons in the same hidden layer. Because the training method of recurrent neural network is gradient descent, the chain derivation rule is needed in gradient descent. The essence of chain derivation is back-propagation error. When the length of the sequence is long, gradient explosion or gradient disappearance is easy to occur. Many RNN-based topologies have been used for traffic prediction, such as Time-Delay Neural Network (TDNN) and State-Space Neural Network (SSNN). However, the traditional RNN model has two problems in traffic prediction: the traditional RNN cannot train the time series with long time delay, which is common in traffic prediction tasks; the traditional RNN relies on the preset time delay learning time series processing, but it is difficult to automatically find the best time delay window. In order to solve these shortcomings, Xiaolei Ma and others proposed a speed prediction algorithm based on long short-term memory neural network (LSTM NN). The RNN structure was originally proposed by Hochreiter and Schmidhuber. Its main goal is to model long-term correlation and determine the optimal time delay of time series problems. These characteristics are particularly important for traffic forecasting.

LSTM NN consists of an input layer, a cyclic hidden layer, and an output layer. Different from traditional neural networks, the basic unit of hidden layer is memory block. The memory block contains a self-connected memory unit to store the time state and a pair of adaptive multiplication gating units to control the information flow in the memory block, and its structure is shown in Fig. 5.14.

Each memory unit has three inputs and two outputs, including x_t, h_{t-1}, c_{t-1}, output h_t, c_t. They are controlled by three gates, namely input gate, forgetting gate, and output gate, and the output value of the three gates is controlled by activation function. At t time:

$$i_t = \text{sigmoid}(W_{ix}x_t + W_{im}m_{t-1} + W_{ic}c_{t-1} + b_i) \tag{5.20}$$

$$f_t = \text{sigmoid}(W_{fx}x_t + W_{fm}m_{t-1} + W_{fc}c_{t-1} + b_f) \tag{5.21}$$

$$c_t = f_t \odot c_{t-1} + i_t \odot g(W_{cx}x_t + W_{cm}m_{t-1} + b_c) \tag{5.22}$$

$$o_t = \text{sigmoid}(W_{ox}x_t + W_{om}m_{t-1} + W_{oc}c_{t-1} + b_o) \tag{5.23}$$

$$m_t = o_t \odot h(c_t) \tag{5.24}$$

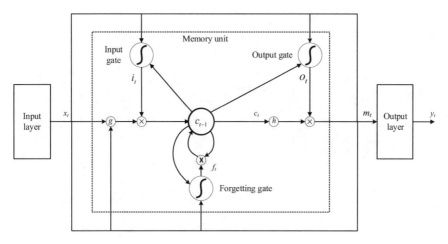

Fig. 5.14 LSTM neural network structure

$$y_t = W_{ym}m_t + b_y \qquad (5.25)$$

Among them, i_t, f_t, o_t represent the output of input gate, forgetting gate, and output gate respectively. W and b are coefficient matrix and deviation vector respectively, $\odot$ are point multiplication operations, sigmoid is the standard logical sigmoid function, g $(\cdot)$ is a center sigmoid function with a range of $[-2,2]$, h $(\cdot)$ is a center sigmoid function with a range of $[-1,1]$. The training process of long and short-term memory neural network is mainly based on truncated back propagation time and real-time recursive learning using the modified version of gradient descent optimization method.

2. Traffic state prediction of road network based on convolution neural network.

Most of the above deep learning models only consider the temporal correlation of traffic evolution of single or multiple road sections, and do not consider the spatial correlation between road sections from the perspective of the whole road network. In order to make up for this defect, some scholars introduced a method based on image processing, which represented the road network traffic state as a picture, and used the deep learning architecture of convolution neural network (CNN) to extract the spatiotemporal traffic features contained in the image. CNN is an efficient and effective image processing algorithm, which has been widely used in the field of computer vision and image recognition, and has achieved remarkable results.

Figure 5.15 shows the CNN structure applied to traffic prediction, which mainly includes four parts, namely model input, traffic feature extraction, prediction, and model output. The explanation of each part is as follows.

Input   Traffic feature extraction   Prediction   Output

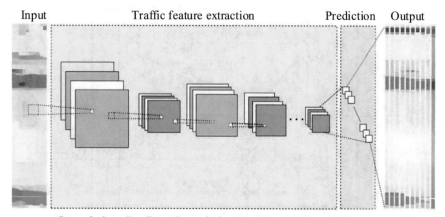

Convolution Pooling Convolution Pooling Pooling Fully-connected

Fig. 5.15 CNN structure

First of all, the input of the model is the image transformed from the traffic network with spatiotemporal characteristics. The time interval of input and output is F and P respectively, and the input of model can be written as

$$x^i = [m_i, m_{i+1}, \cdots, m_{i+P-1}], \quad i \in [1, N - P - F + 1] \tag{5.26}$$

where i is the sample marker, N is the length of the time interval, and m_i is the column vector of the traffic speed of all links in the traffic network in a time unit.

Then, traffic features are extracted through convolution layer and pooling layer, which is also the core part of CNN model. Pool process is represented by pool, and L is the depth of CNN. The input, output, and parameters of layer l are represented by x^j_l, o^j_l, and $(W)^j_l$, b^j_l, respectively, where j is the number of channels considering multiple convolution filters in the convolution layer. The number of convolution filters in layer l is represented by c_l, and the outputs of layer l and pooling layer can be written as

$$o^j_l = \text{pool}\left(\sigma\left(\sum_{k=1}^{cl-1} \left(W^j_l x^k_l + b^j_l\right)\right)\right), \quad j \in [1, c_l] \tag{5.27}$$

Among them, σ is the activation function. In the model prediction, through traffic feature extraction and learning, the final output road network features will be expressed as a dense vector in series, and flatten is used to represent the process of series connection

$$o^{\text{flatten}}_L = \text{flatten}\left(\left[o^1_L, o^2_L, \cdots, o^J_L\right]\right), \quad j \in c_L \tag{5.28}$$

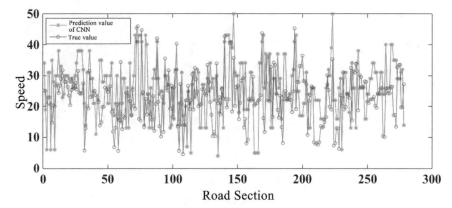

Fig. 5.16 Comparison of prediction results by CNN with actual values at 18:00 in the evening peak

Finally, the vector is converted into model output through the full connection layer, and the final model output can be written as

$$\widehat{y} = W_f o_L^{\text{flatten}} + b_f$$

$$= W_f \left(\text{flatten} \left(\text{pool} \left(\sigma \left(\sum_{k=1}^{cl-1} (W)_l^j x_l^k + b_l^j \right) \right) \right) \right) + b_f \qquad (5.29)$$

Among them, W_f is the parameter of the whole link layer, and $\widehat{y}$ is the predicted traffic state of the road network.

The weight-sharing mechanism of convolutional neural network can reduce some parameters of the model, but there are multiple convolution cores. Because the convolution core is for local feature extraction, the output of convolution can be pooled to reduce the size of parameters. In addition, the training mode and loss function of convolution neural network are the same as those of BP neural network. In the road network, the spatial association between adjacent road sections often has similarity, which is consistent with the translation invariance of convolution neural network. Therefore, convolution neural network can be used to extract spatial features of different distance levels.

Example 5.3 Take the prediction speed as an example to analyze the above algorithms (convolution neural network and recurrent neural network), select the same research object as the previous example, and give the comparison results and conclusions.

CNN neural network and LSTM neural network are respectively used to predict the speed of a given road network, and 20 min historical data is used to predict the speed of the next 2 min. the prediction results are shown in Figs. 5.16 and 5.17.

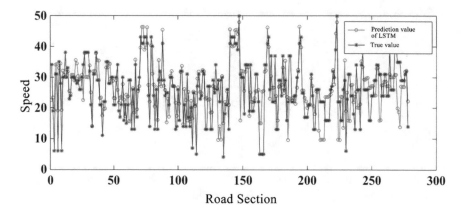

Fig. 5.17 Comparison of prediction results by LSTM with actual values at 18:00 in the evening peak

Prediction method	MAPE	RMSE
CNN neural network	0.21855	8.5523
LSTM neural network	0.20281	6.1653

Table 5.3 Evaluation index value of prediction effect

The average absolute percentage error and root mean square error are used to evaluate the prediction effect. The calculation results of evaluation indexes are shown in Table 5.3.

It can be seen from the prediction results that both CNN neural network and LSTM neural network can be used to predict the road network speed, and the prediction effect of the two methods is also very close. Compared with BP neural network and support vector machine, CNN and LSTM neural network have better prediction effect. It should be noted that when the number of data samples is small, the prediction effect of deep learning algorithm is inferior to machine learning, because deep learning algorithm needs to rely on more training samples for learning. Therefore, when the amount of data is small, the rule prediction effect made by traditional machine learning algorithm will be better.

5.4 Congestion Prediction Method and Application

5.4.1 Congestion Discrimination

5.4.1.1 Definition of Traffic Congestion

Traffic congestion is human subjective perception, which has certain subjectivity. At present, there is no unified definition standard for the specific concept of traffic congestion. The specific definition of traffic congestion given by Chicago traffic management department is: on the road, the road traffic state in which the lane occupancy exceeds 30% in 5 min. In the local standard of "Urban Road Traffic Congestion Evaluation Index System" of Beijing, traffic congestion is defined as the traffic phenomenon that the travel time of motor vehicles in the road network is prolonged and the travel delay is increased due to the contradiction between traffic supply and demand, or the influence of external complex factors such as bad weather, construction conditions, traffic events, and traffic control. To sum up, the essence of traffic congestion can be summarized as a traffic phenomenon that the demand of road traffic exceeds the traffic supply due to some factors, resulting in the imbalance of traffic supply and demand and traffic bottleneck.

5.4.1.2 Classification of Traffic Congestion

According to the severity of traffic congestion, traffic congestion can be divided into mild, moderate, and severe traffic congestion. Slight congestion refers to the phenomenon that the traffic speed of the road slows down for some reason, but there is no queue; moderate congestion refers to the phenomenon that the traffic can return to normal in a short time due to the low traffic capacity and the low speed of road vehicles. Severe congestion is due to some reasons, resulting in extremely low road capacity and speed, vehicles on the road queue for a long time, traffic is paralyzed, short time cannot return to normal road phenomenon.

According to the causes of congestion, it can be divided into frequent traffic congestion and occasional traffic congestion. Occasional traffic congestion is a temporary or short-term occurrence, and the specific time and location of the congestion events are difficult to predict. After the incident occurs, only through the coordination and cooperation of relevant departments and timely remedial measures can the harm caused by the traffic congestion incident be minimized. For frequent traffic congestion events, through observation and analysis, we can accurately grasp the specific location of traffic congestion, and even estimate the possible time range of congestion occurrence and diffusion according to the traffic law of bottleneck location, so as to provide reliable basis for traffic management. The main research object of this section is frequent traffic congestion, which is used to identify and predict the state of frequent traffic congestion.

5.4.1.3 Quantitative Standard of Traffic Congestion

Traffic congestion is related to road demand and supply capacity. To a certain extent, it is also related to the subjective feelings of road users. Therefore, it has quantitative and qualitative characteristics, and can be analyzed from both qualitative and quantitative aspects. From a qualitative point of view, due to some reasons, the operation of the road network will fail, resulting in a decline in service levels and failure of normal operations. Therefore, the managers need to intervene to resume the operation. However, in the actual traffic situation, everyone has different feelings about the degree of traffic congestion, therefore, the quantitative analysis of traffic status plays an important role in the identification of traffic jams. Since changes in traffic parameters are the most direct manifestation of changes in traffic conditions, quantitative analysis of traffic conditions can be carried out through changes in traffic parameters. So far, the standards of traffic state quantification are uneven, and the academic circles at home and abroad have their own quantitative standards.

Taking speed, queue length, and queue time as the comprehensive quantitative criteria of expressway, the Japanese highway administration puts forward the speed of 40 km/h as the driving standard, and calls the traffic condition that the queue length is more than 1 km and the queue duration is more than 15 min, which fluctuates repeatedly on the road, as traffic congestion.

Chicago Department of transportation believes that the traffic state corresponding to the 5 min occupancy rate of 30% or more lanes belongs to traffic congestion.

Texas Department of transportation believes that severe traffic congestion will occur when the actual travel time of road users is greater than the normal travel time, and the time delay exceeds the time delay threshold acceptable to the public.

In 2002, the Ministry of public security of China quantified the degree of traffic congestion on urban trunk roads by using the average travel speed of vehicles driving on urban trunk roads in the evaluation index system of urban traffic management, and divided it into four reference standards.

If $\bar{v} \geq 30$ km/h of the motor vehicle, it is unblocked.
If $\bar{v} < 30$ km/h, but ≥ 20 km/h, it belongs to slight congestion.
If $\bar{v} < 20$ km/h, but ≥ 10 km/h, it belongs to congestion.
If $\bar{v} < 10$ km/h, it belongs to severe congestion.

5.4.1.4 Discrimination Method of Traffic Congestion

The first is congestion identification method based on single traffic parameters, and the specific methods are as follows:

1. Congestion identification method based on vehicle speed. The algorithm is to measure the travel time data of vehicles between adjacent detectors by tracking detector, calculate the travel speed of vehicles by the ratio of road length and travel time, and compare the obtained speed value with the corresponding speed threshold, so as to determine the road traffic congestion. Obviously, the key of this algorithm is vehicle speed detection and speed threshold setting. In the speed based congestion identification, there are many aspects about the speed threshold research: in 1983, the California Department of transportation used the speed of 56 km/h as the speed threshold of the highway; in 1990, the Washington Department of transportation used the speed of 64 km/h as the speed threshold; in 1994, in Japan, if the speed on the road is lower than 40 km/h, it is regarded as the congestion state; in 2004, the federal highway of the United States The bureau uses the average travel speed of 72 km/h and 48 km/h as the standard to divide the congestion and severe congestion. Since 2000, Chinese scholars have also studied the speed threshold. Jifu Guo and others have investigated, taking 25, 20,

15 and 10 km/h as the speed threshold of serious congestion on urban express-way, urban trunk road, secondary trunk road, and branch road; Juan Liu and others think that 35 km/h is the speed threshold of urban expressway. As for the speed detection methods, the main applications are infrared detection method, ultrasonic detection method, induction coil detection method, laser detection method, radar speed measurement method, video speed measurement method, GPS satellite speed measurement method, etc.

2. Congestion identification method based on delay index. The algorithm is: according to the road traffic conditions, by measuring the travel delay of vehicles, the travel delay is digitized and compared with the corresponding threshold, so as to determine the road traffic congestion. Obviously, the key to the application of this algorithm is the detection of vehicle delay index and the setting of threshold. For example, Levinson and Lomax take the value of 0–10 as the value range of delay level in their literature.

3. Congestion identification method based on saturation. The algorithm refers to: according to the road traffic conditions, by calculating the saturation of the road section, the saturation value of the road section is compared with the corresponding threshold value, so as to determine the road traffic congestion; obviously, the key of the application of this algorithm is the detection of the saturation of the road section and the setting of the threshold value. Different parameters, such as congestion degree coefficient CSI or urban trunk road congestion time mileage LMDI, can be used to judge the degree of road congestion.

4. Congestion identification method based on occupancy. The algorithm refers to: using induction coil technology, through the detection equipment set on the road, collecting road occupancy data, and comparing the obtained occupancy data with the corresponding threshold value. If the road occupancy value exceeds the corresponding threshold value in several consecutive moments, it can be deter-mined that the road section is congested. Obviously, the key to the application of this algorithm is the detection of road occupancy and the setting of threshold. Collinsetal developed the occupancy algorithm for the first time in 1979 by using the induction coil technology. Through the loop induction coil detector set on the road section for real-time monitoring, the real-time occupancy data is obtained, and compared with the predetermined occupancy threshold, so as to realize the judgment of traffic events.

The second type is congestion identification method based on multi-parameter and modern advanced calculation method. The representative methods are as follows:

1. Typical traffic state automatic identification methods, such as California algo-rithm and exponential smoothing method mentioned in Sect. 5.2.1.

2. Congestion identification method based on artificial neural network. The recog-nition principle of artificial neural network is: when congestion occurs, the traffic parameters change significantly. From this, we know that there is correlation between traffic parameters and traffic state, which can be described by functional relationship. However, because the influence of traffic parameters on traffic state

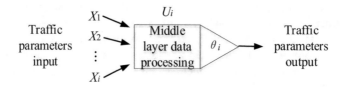

Fig. 5.18 Artificial neural network congestion identification process

is inconsistent, it is difficult to describe the functional relationship between them by simple linear function. In order to solve this key problem effectively, an ANN model is established. In 1995, Cheu developed a neural network algorithm based on multi-layer feedforward. The algorithm takes the traffic flow, occupancy, and speed of the upstream and downstream sections as the input items. The middle layer is the processing process for data processing, and the output layer outputs the final state of traffic—congestion and unblocked. Its structure is shown in Fig. 5.18.

3. Congestion identification method based on fluctuation analysis. In 2000, Adeh and Karim of the University of California, Berkeley developed a wave analysis algorithm based on wave theory to identify traffic congestion. The algorithm takes the difference of the cumulative occupancy rate of the detectors set up in the upstream and downstream of the road as the basic criterion for congestion identification. Under normal traffic conditions, if the difference of the cumulative occupancy rate detected by the detector has a continuous deviation, then the traffic state has congestion.

4. Congestion identification method based on queuing theory. In 2008, M. Farazy Fahmy [5] used queuing model to distinguish traffic state. Firstly, the collected traffic flow data is divided into arrival and departure traffic flow. Then the relationship between arrival flow and departure flow is compared, and the relationship between them is used to identify whether the traffic is congested or not. By comparison, if we find that: when the arrival flow is less than the departure flow, the traffic is smooth; when the arrival flow is greater than the departure flow, the congestion phenomenon is gradually emerging; when the departure flow is equal to zero, the traffic condition is in the congestion stage.

5. Congestion identification method based on catastrophe theory. The principle of this algorithm is: the traffic data obtained from the same detection station can judge the traffic state by comparing the relationship among speed, flow, and occupancy. The algorithm stipulates that the traffic data in three consecutive sampling periods must be compared. When the speed, occupancy, and flow of the sampling data detected by the detector reach the threshold in the sampling period, or two of the parameters reach the threshold, congestion can also be considered.

5.4.2   Basic Framework of Congestion State Prediction Method and Application

The related algorithms of road congestion state prediction can be extended by state prediction algorithm. One way is to preprocess traffic flow parameters based on the discriminant theory of congestion state, and then express it as the expression of congestion state, and then predict; the other is to judge the congestion by the discriminant method after the traditional traffic flow parameters are predicted. The congestion prediction results are got in comparison with the reasonable threshold.

5.4.2.1   Congestion Prediction Method

The research of traffic congestion prediction model belongs to the pattern recognition problem of information science. At present, many research results of traffic congestion prediction at home and abroad are mainly concentrated in this field, including prediction analysis based on time series correlation, neural network prediction, Bayesian network prediction, and multi classifier combination prediction. The research of this kind of method is mostly seen in the theoretical research, and because the basic theory based on it lacks robustness in dealing with big data, the model generally lacks long-term effect and scalability. As the congestion state prediction belongs to the discrete traffic state prediction category in traffic prediction, its corresponding output is only a few discrete categories, which is highly consistent with the prediction label in machine learning. Therefore, for congestion prediction, the effect of machine learning method will be very significant. At present, most of the more popular and accurate algorithms are based on artificial intelligence-related methods such as neural network, whose accuracy is much higher than the traditional basic theoretical model. Please refer to Sect. 5.3 analysis of traffic state prediction method for specific method contents.

Next, we will focus on a frontier congestion prediction algorithm based on restricted Boltzmann machine and recurrent neural network.

In this algorithm, restricted Boltzmann machine (RBM) and recurrent neural network (RNN) are combined to build a deep learning framework for high-dimensional time series prediction.

The restricted Boltzmann machine (RBM) usually consists of a visible layer and an implicit layer. The elements of each layer are linked with each other. The probability distribution function of RBM can be modified to conditional probability distribution function according to its previous state, so a variety of different RBM models can be established for complex time series modeling. Recurrent neural network (RNN) is a special form of neural network series, which contains at least one feedback connection as the internal state from the output to the input of the neuron. The loop structure also gives the network the ability of time processing and sequence learning. Because of the short-term memory of RNN, it is widely used in nonlinear time series data. Training RNN is similar to traditional multilayer feed-

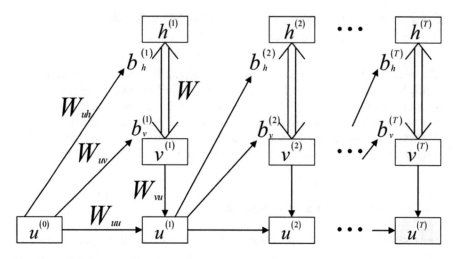

Fig. 5.19 RBM-RNN structure

forward neural network (FFNN). As time went by each feedback loop will be extended to single-layer feed-forward neural network at each time stamp with the development of RNN structure. In this case, the reverse propagation algorithm can be used to train RNN with repeated and effective.

In view of the time series prediction ability of RBM and RNN models. To get the most out of their advantages, the deep architecture of RNN-RBM model is built to describe the time dependence in high-dimensional series. The structure of the model is shown in Fig. 5.19.

The model is constructed by stacking conditional RBM and RNN. Conditional RBM is an extension of traditional RBM. It processes time series by providing feedback loops between the visible layer and the hidden layer. The deviation values of the visible layer and the hidden layer are updated according to the previous visible units. It is also applied to RBM-RNN model.

b^t_v and b^t_h represent the deviation vectors of the visible layer and the hidden layer in the RBM model at t time, respectively, and are updated by the hidden unit in the RNN model at t-1 time. RNN model and RBM model are connected by weight matrix w_{uv} and w_{uh}. The above process is as follows:

$$b^t_v = b_v + w_{uv}u^{t-1} \tag{5.30}$$

$$b^t_h = b_h + w_{uh}u^{t-1} \tag{5.31}$$

Among them, b_v and b_h are the initial deviation values of visible and hidden layers in RBM model. RNN model is developed over time, and the former implicit state in RBM model is generated based on input layer v^t and hidden layer u^t in the model. The activation calculation of hidden elements in the hidden layer is as follows:

$$u^t = \text{sigmoid}\left(b_u + w_{uv}u^{t-1} + w_{vu}v^t\right) \tag{5.32}$$

The implementation process of the algorithm is summarized as follows:

Step 1: Generate the value of hidden unit in RNN model.
Step 2: Update the bias in the RBM model based on the estimated value u^{t-1} in step 1, and calculate the RBM parameters.
Step 3: Calculate the log likelihood gradient in RBM model.
Step 4: Propagate the estimation gradient to the RNN model, and update the weights w_{uv} and w_{vu} over time to train the RNN model for prediction.

5.4.2.2   Case Analysis

Example 5.4 This part analyzes the above congestion prediction method based on RBM-RNN.

1. *Data input*: The selected object is a certain area of Ningbo City, as shown in Fig. 5.20, including 515 road sections. The main traffic flow parameters are GPS speed data with 2 min update frequency.

   Input section. Firstly, the congestion state is identified according to GPS data, and the speed threshold is divided by 20 km/h. Then, the congestion state of each road section is transformed into binary, which is defined as congestion below 20 km/h and represented by 1, and unblocked above 20 km/h and represented by 0. As the traffic state of a road section changes over time, its traffic state becomes a binary sequence with increasing length, and the congestion of the whole road network can be expressed as a high-dimensional matrix arranged in time and space. The key of the whole congestion evolution prediction model is to mine the evolution law of each matrix element in space and time.

Fig. 5.20 Study road network

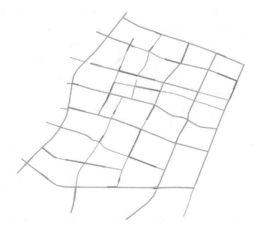

Algorithms	Runtime (seconds)	Prediction Accuracy (%)	Sensitivity (%)	Specificity (%)
RNN-RBM	354	88.2%	64.1%	91.1%
BPNN	13498	69.7%	38.2%	77.5%
SVM	14979	71.0%	36.6%	80.3%

Fig. 5.21 Comparison of traffic congestion prediction results of different algorithms

For each road segment n, the congestion state in time t is expressed as c^t_n. According to this, the traffic congestion state of the road network with N road segments in time T can be expressed as a binary high-order matrix.

$$\begin{bmatrix} c_1^1 & \cdots & c_1^T \\ \vdots & c_n^t & \vdots \\ c_N^1 & \cdots & c_N^T \end{bmatrix} \tag{5.33}$$

2. *Model building*: In view of the above-mentioned expression matrix, RNN-RBM model is used to predict the temporal and spatial congestion evolution of road network. A small batch gradient descent optimization method is used for realizing RNN-RBM model effectively: the training set is grouped into several small samples, and each sample set can be processed in parallel by standard gradient descent optimization method. GPU is designed on parallel architecture, so it can perform multiple tasks at the same time. This property is especially suitable for accelerating the calculation of RNN-RBM model. CUDA library, a parallel computing platform which was developed by NVIDIA, is used to execute the computing units in GPU. The parameters of RNN-RBM model are setup as follows: the amount of hidden units in RNN model is 100, the amount of hidden units in RBM model is 150, the learning rate of gradient descent optimization method is 0.05, the weight matrix W of RNN-RBM model is initialized by normal distribution with average value of 0 and variance of 0.01, and other weight matrices are initialized by normal distribution with variance of 0.0001. All deviation vectors are set to zero at the beginning of the model.

3. *Error calculation and training*: In the training process of RNN-RBM model, the objective of optimization is to minimize the cross entropy error. Because the cross entropy represents the distance between the probability distribution of the calculated output and the target output, it is appropriate to use the mean square error as the neural network classifier with binary value. The definition of cross entropy error (CEE) is as follows:

$$\text{CEE} = -\frac{1}{T} \sum_{n=1}^{N} \sum_{t=1}^{T} c_n^t \ln \mid \left(\widehat{c}_n^t\right) + \left(1 - c_n^t\right) \ln \left(1 - \widehat{c}_n^t\right) \tag{5.34}$$

The model is in comparison with the traditional prediction models BPNN and SVM, and the result is shown in Fig. 5.21. In comparison with the traditional NN and SVM algorithm, the accuracy of the deep learning prediction algorithm is improved by 10%, which fully proves the excellent performance of the deep learning model in congestion state prediction.

5.4.2.3 Basic Framework of Congestion Prediction Algorithm Application

In the background of big data, the "platform + application" structure, which is mostly presented by modern software system, can be used for the application of traffic congestion prediction algorithm. Application is a computer program that directly serves users to achieve specific purposes and fulfill specific needs. Due to the wide variety of user requirements, the differences between applications can not be described only by different configurations of parameters, but also involve the differences of business logic. However, different applications may present some uniform characteristics, such as the applications on Windows system contain at least one view window.

Therefore, the establishment of secondary development platform can effectively improve the application differences and meet the needs of diversified users, while reducing the development difficulty and cost of common features of various applications. The shortening of the secondary development time can also promote the rapid satisfaction of users' new needs, which is also the fundamental reason for the vigorous development of software platform technology.

The design of the system is very important for its future scalability and application performance. Figure 5.22 is a congestion prediction platform design based on multi-source heterogeneous traffic data, reflecting the current understanding and future expectation of the platform. It adopts an open structure, which is an open source project, so that the design of the system can be continuously improved with the expansion of functions. The system architecture is mainly made up of three parts: heterogeneous data sources from different institutions, data warehouse of intelligent transportation application and research laboratory, and network server running on the system server.

Among them, data warehouse is in charge of data archiving, with multiple data retrieval function provided by the system. The database schema is designed in advance to ensure the efficiency of data management and query. All kinds of traffic data can be systematically stored in the database management system, and the attribute relationship between the data follows the pre-designed pattern, which can also be easily maintained. Web server can present and disseminate data, and perform analysis algorithm according to user's task. It mainly serves traffic engineers, researchers, and travelers. For example, some download functions are limited to specific user groups. At the same time, the platform can connect multiple data servers by using different data communication technologies. If necessary, another server can be added to the system.

In the aspect of system design, the multi-layer architecture commonly used in software engineering is adopted. The main advantage of multi-tier architecture is that

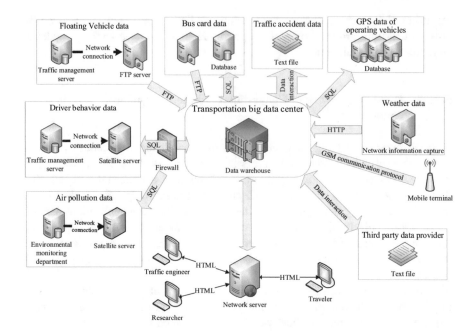

Fig. 5.22 System architecture (*FTP* file transfer protocol, *SQL* structured query language, *HTTP* hypertext transfer protocol, *GSM* global system for mobile communications, *HTML* hypertext markup language)

developers can modify or add a specific tier without rewriting the entire application. The model includes client presentation layer (client Web browser), server data layer (data warehouse), and two server logic layers (middleware and computing module). In comparison with the traditional three-tier master-slave model, an additional logic layer can deal with the quality of data. The computing layer can be used to control data sharing and execute algorithms. Middleware layer can alleviate the burden of computing layer, such as excessive database access, calculation of analysis algorithm, and data quality control (DQC). The client presentation layer (Web browser) is in charge of displaying the interface, visualizing the output, and receiving the user's input. The whole system flow is shown in Fig. 5.23.

The following is a detailed introduction of its main structure:

1. *Data quality control (DQC)*: The problem of data quality is widely concerned by transportation researchers and organizations. An automatic and stable DQC process facilitates traffic-related research. For ensuring the data quality, a two-step DQC data cleaning mechanism is designed to detect and delete errors and inconsistent data. The first step of data cleaning occurs in the process of data retrieval from different data sources. The wrong data will be marked or deleted. For example, the zero occupancy and negative flow in the ring detector data, and the GPS data offset in the freight database. Another data cleaning process takes place in the DQC module of the middleware layer.

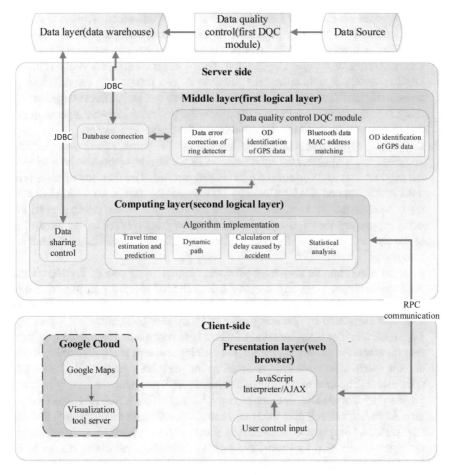

Fig. 5.23 Design system flow chart (*OD* start and end, *AJAX* asynchronous JavaScript and XML)

Apart from checking for errors, DQC will also perform preliminary data analysis and processing in the middleware layer to reduce the computing burden of the computing layer. For example, in some intersections, the early ring detectors are connected together, which will lead to the under counting problem, while the nonlinear probability model incorporating the second DQC module will correct the number of vehicles under counting. In addition, a software-based error detection and correction algorithm is also embedded in the middleware layer.

Another example of DQC is OD identification algorithm, which can integrate and extract OD information of private trucks and evaluate freight performance. Similarly, the original Bluetooth media access control address collected by the Bluetooth detector will be returned to the platform. Redundant data will be filtered in the first DQC module, and travel time will be calculated in the second DQC module of middleware layer.

2. *Intermediate device layer*: Middleware is a kind of computer program that runs independently in the server. As mentioned earlier, the purpose of building middleware layer is to utilize computing power to manage resources between server (data and two logical layers) and client (presentation layer). Aside from the DQC module, the data connection module is also installed in the middleware layer. In fact, this module is a program interface to connect multiple databases by using Java database connection API, which allows the middleware layer to query and receive results from the data warehouse for further processing.

3. *Computing layer*: The computing layer of the network server will perform complex algorithms after the completion of DQC. This layer can also help archive the original data and control the data sharing service. The function of asynchronous JavaScript and XML technology is to reduce the data transmission between the server and the browser, and minimize the conflict between the display on the existing page and the ongoing activities. This design can reduce the response time of the server and improve the performance of displaying dynamic and interactive web pages.

   Many algorithms in the design platform use this asynchronous JavaScript and XML technology, including iterative calculation of shortest path (travel time), statistical measurement of transport performance index, and delay related to accidents calculated by queuing theory and time series algorithm.

4. *Display layer*: The main function of client is to provide an interactive graphical user interface. As shown in Fig. 5.2, the user input is sent to the computing layer, and the calculation results are returned to the Web browser through the remote call process. The final result is visualization through two main third-party controls provided by Google Cloud, Google map API, and visualization API. The Google map API allows developers to visualize results on Google maps through Google map servers. Google visualization API allows users to visualize statistical charts through visualization tool services, such as histogram, pie chart, etc. so as to implement the platform, the open source integrated development environment, which combines Google network development kit and Eclipse, can provide a powerful development environment for the platform. Google network tools package includes Java API library, which permits developers to code web applications in Java language, and after that compile source code in JavaScript. In this case, development costs and time-consuming will be significantly lower than traditional web development methods, such as JavaScript or JavaScript and PHP. In addition, debugging of Google Web toolkit makes the traditional JavaScript web page development more convenient. Developers can access the existing widget template library in Google network toolkit to design web interface, or use Java-to-JavaScript compiler to transform and optimize java code in JavaScript.

   Through the theoretical framework of the above-mentioned traffic congestion prediction platform, the real-time regional traffic information can be accessed, data sharing and visualization can be promoted, and real-time and intuitive traffic congestion side information can be provided for travelers and traffic managers. Compared with traditional platform, it can not only realize the sharing, visualization, and analysis of traffic data, but also provide an online database network, which can greatly promote the scientific discovery and education achievements in the field of traffic engineering.

References

1. Payne HJ, Tignor SC (1978) Freeway incident-detection algorithms based on decision trees with states. In: Transportation research record, vol 682. TRB, National Research Council, Washington, DC, pp 30–37
2. Box GE, Jenkins GM (1976) Time series analysis: forecasting and control. J Time 31(4):238–242
3. Rumelhart DE, Hinton GE, Williams RJ (1988) Learning representations by back-propagating errors. Nature 533–536
4. Lecun Y, Boser B, Denker JS, Henderson D, Howard RE, Hubbard W, Jackel LD (1990) Handwritten digit recognition with a back-propagation network. In: Touretzky D (ed) Advances in neural information processing systems (NIPS 1989), Denver, CO, vol 2. Morgan Kaufmann
5. Fahmy MF, Ranasinghe DN (2008) Discovering dynamic vehicular congestion using VANETs. In: Proceedings of the 4th International conference on information & automation for sustainability, Colombo, Sri Lanka, December

Chapter 6
Intelligent Transportation Information Service Technology

ZHANG Hui, SUN Yifan, ZHANG Qi, LI Shaopeng, HOU Ninghao, and ZHANG Yijun

With the continuous progress of information technology and the rising status of transportation in the social economy, the transportation information service system, which aims to realize the informatization and intellectualization of travel and traffic smanagement, has become an indispensable part of modern transportation system and Intelligent Transportation Information Service Systems (ITS) research field.

At present, the transportation information service system has become a hot research topic of ITS in the world. And Intelligent Transportation Information Service Systems is also a widely used field in ITS. Its development and application are rapidly developing in Europe, America, and Japan. In the countries and regions that have carried out research and development in the world, its demonstration projects and research contents almost all involve various targeted traffic information service systems.

ZHANG Hui (✉) · SUN Yifan · ZHANG Qi · LI Shaopeng · HOU Ninghao · ZHANG Yijun
Intelligent Transportation Systems Research Center, Wuhan University of Technology, Wuhan, China
e-mail: zhanghuiits@whut.edu.cn

© Tsinghua University Press 2022
W. Yunpeng et al. (eds.), *Intelligent Road Transport Systems*,
https://doi.org/10.1007/978-981-16-5776-4_6

6.1 The Connotation of Intelligent Transportation Information Service

6.1.1 Concept of Intelligent Transportation Information Service Systems

6.1.1.1 Concept of Intelligent Transportation Information Service Systems

The advanced traffic information service system should cover a variety of transportation modes, comprehensively use a variety of advanced technologies, meet the needs of drivers, passengers, the public and traffic management departments for traffic information, make the traffic behavior of traffic participants more scientific, planned and rational, ensure the mobility, convenience and safety of travel, and ultimately improve social and economic benefits of the whole traffic system.

Advanced traveler information system (ATIS) is defined in the ITS system framework of developed countries such as America, Japan, and Europe. Many countries, including China, used this name in the first edition of ITS system framework. This part is defined as advanced transportation information system in some material and places. It refers to providing real-time and comprehensive traffic information, such as road traffic information, public traffic information, transfer information, traffic weather information, parking information, and travel-related information, to the traffic information center through sensors and transmission equipment equipped on roads, motor vehicles, transfer stations, parking lots, and meteorological center, based on a perfect information network. Users can determine their travel mode and route according to the information provided by the system [1].

Many experts and scholars at home and abroad believe that the traveler information system can't effectively cover the content of the traffic information service, and the extension of the traffic information system is too wide, which is too repetitive with other parts of the system framework. At present, a more consistent idea is to define this part as an advanced traffic information service system.

Traffic information service system can be simply defined as "an information service system composed of information terminal, traffic information center, wide area communication network, etc., which takes individual travelers as the main service object, provides travel information according to their needs, and optimizes the route to shorten travel time or reduce costs". It can also be comprehensively defined as an important part of intelligent transportation system. The system comprehensively uses a variety of advanced technologies to provide real-time and dynamic travel-related traffic information in the form of text, voice, graphics, video, and other multimedia for users through wireless and wired communication, so that travelers (including drivers and passengers) can get the information about the road traffic situation, the time required, the best transfer mode, the cost required, and the destination can be obtained at any time when traveling. Therefore, the travelers can be guided to choose the appropriate traffic mode (private car, train, bus, etc.),

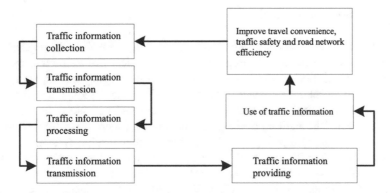

Fig. 6.1 The framework of traffic information service system

travel route and departure time, and complete the travel process with the highest efficiency and the best way.

According to different attributes, the ways and means of information collection, transmission, processing, and various traffic information release involved in the traffic information service system are different, such as:

1. Bus timetable and operation status information can be obtained from the bus management system.
2. Most of the information related to the road is collected by the monitoring system (vehicle detector, camera, vehicle automatic positioning system, etc.), after being processed by the traffic information processing center, the travelers can obtain it through the information display device (such as variable information board) on the roadside, or directly from various on-board devices.
3. Other static information, such as map database, emergency service information, driver service information, tourist attractions, and service information, can be inquired, received, and exchanged by travelers or the public at home, office, travel vehicle, commercial vehicle, bus, bus station, or personal communication facilities carried with them.

In these methods, due to the different information media and information providing methods, the amount and the form of information expression provided for travelers, the public and transportation departments are also very different. So the guiding effect on travelers is also different. From the respect of information flow, the information processing process is basically consistent. Namely, the collected raw data, such as road conditions, traffic flow status, etc., are comprehensively analyzed and processed. Finally, appropriate traffic information is provided to travelers and the public to affect the choice of travel routes and modes of traffic participants. Therefore, it can dredge the traffic flow and maintain the best traffic and improve traffic safety, so as to ultimately improve social and economic benefits. The structural framework of the traffic information service system is shown in Fig. 6.1 [2].

6.1.1.2 Development Course

With the rapid development of information technology, as well as the requirements of scientific and technological progress and high efficiency of traffic management, the traffic information service system is constantly developing and improving. For example, the visual development from the road signs and markings which provide static information to the variable information board and variable signs are widely used at present. The auditory development is from the general radio traffic information broadcast program to the roadside real-time communication system.

The traditional traffic information system serves for the whole traffic flow and can be used as the basis of traffic management system. The amount of information it can transmit and provide is limited. Since the 1970s, in the research of seeking to alleviate traffic congestion in Europe, America, Japan, and other developed countries, a comprehensive traffic information service system with individual travelers as the service object has emerged. Travelers can always drive on the shortest path (distance or time) with the help of their portable information system communicating with traffic information center, and avoid the blocked road section, accident section or bad environment section, so as to reduce the delay and ease the traffic congestion.

On the whole, the development of the integrated traffic information service system for individual travelers (Traveler information system) can be divided into two stages.

The First Generation of System Traveler Information System (TIS) is developed on the basis of computer technology and traffic monitoring system in the 1970s, which reflects people's initial desire to release information by using communication technology. These systems are mainly used to improve the local capacity of the road network, such as seriously congested intersections, or some intersections and sections blocked by special events and traffic accidents.

Release Methods Variable Message Sign (VMS) and Highway Advisory Radio (HAR).

Features VMS and HAR are one-way communication systems, which are used to transmit general travel information to vehicles. The information is filtered by individual travelers to select useful information (if any). At present, VMS is still an important method to release traffic information.

The Second-Generation System It is called Advanced Traveler Information System (ATIS). It adopts the latest technological achievements in information collection, transmission, processing, and release, and can provide a variety of real-time traffic information and dynamic route guidance functions for a wider range of traffic participants.

Release Means Vehicle terminal, cellular phone, cable phone, cable TV, large screen display, Internet, etc.

Features The high development of communication electronic map, computer, and multimedia technology makes it possible for ATIS to provide personalized travel assistance for travelers; ATIS focuses on providing travelers with the information they want, so it can greatly reduce the workload of information screening for travelers (for example, vehicles equipped with on-board computers can input travel start and end points, then query and select the route. The route is the shortest path calculated by the traveler information system according to the current real-time road information.

With the development of information technology and computer network, the construction of ATIS based on mobile communication technology has been studied, and various high-tech achievements such as multimedia and communication have been adopted. The automobile will gradually develop into a mobile information center and office. This will greatly strengthen the service function and service field of ATIS, extend the service object from travelers to all traffic participants, the public and traffic management departments.

6.1.2   Function of Intelligent Transportation Information Service System

The advanced traffic information service system can provide travel planning and route guidance for a variety of traffic modes, provide advisory services for various types of drivers and other travelers, allow travelers to confirm and pay for the services they enjoy, and have personal alarm function. Traffic information service can be provided before or during travel. The service before travel can provide travelers with traffic information for choosing travel mode, route and time, including road conditions, traffic status, public transport information, etc. Travelers can make such service requests and get help at home, workplace, parking lot, transfer station, and other places. Traffic information service in travel can provide travelers with travel information during the travel, such as traffic status, road conditions, bus information, route guidance information, adverse travel conditions, special events, parking lot location, and so on. The functions of advanced traffic information service system are mainly reflected in the following aspects.

1. Travel plans for multiple modes of transportation. It can provide regional infor-mation to help travelers choose and make travel plans including walking, private cars, buses, and other travel modes, even including railway transportation, water transportation, and air transportation.
2. User consulting services. It can provide a wide range of consulting services, including accident warning, delay notice, the estimated arriving time or transfer station under the current traffic condition, adverse travel conditions, the connec-tion between traffic modes and their schedules, the restrictions of Commercial Vehicle Operations (CVO) (height limit, weight limit, etc.), parking lot

information, bus station location information, and the information of upcoming toll stations.

3. Route guidance service. The guidance service based on real-time dynamic traffic information can provide dynamic route navigation and road travel time information, which can help drivers choose the best route to avoid serious congestion or other adverse traffic conditions.

4. Interface with related systems. The interface with the regional traffic management system can obtain the traffic information, accident information, and road information of Expressway and urban trunk road, and the interface with the regional bus management system can obtain the bus information, including bus timetable and bus operation status information. This information can be fused with monitoring information and real-time information from other sources to work together.

Through the investigation of travelers, the public, traffic management departments, as well as the design and application staff of traffic information service system, we know that the characteristics of traffic information service system are mainly reflected in the following aspects:

1. The information provided should be timely, accurate, and reliable, and the correlation of travel decision-making should be great.

2. It can provide relevant traffic information for the whole region, which requires the participation of public institutions across administrative regions.

3. It is easy to combine with other ITS systems, such as emergency management system, expressway management system, traffic signal control system, public transport management system, etc., in order to obtain a lot of traffic information.

4. The operators must be specially trained.

5. It is easy to be accepted and used by traffic participants and the public.

6. It is easy to maintain and does not need high operating cost and long operation time.

7. The end user can afford the cost of the services provided.

Traffic information service system should meet the development goals of specific countries and regions, under normal circumstances, the main objectives of the traffic information service system are shown in the following six aspects:

1. Promote travel mode selection based on real-time and accurate traffic state.

2. Reduce the pressure of travelers in unfamiliar areas.

3. Reduce the travel time and delay of individual travelers in multi-mode travel.

4. Reduce the travel time and delay of the whole transportation system.

5. Improve the overall efficiency of the transportation system, reduce the overall cost of the transportation system in many ways.

6. Reduce the risk of collision and the degree of casualties (e.g., reduce the energy dispersion of travelers in strange areas).

Table 6.1 Implementation effect of Traveler Information System (U.S. Department of Transportation)

Index	Effect
Risk of collision	Expected to reduce driver stress (4% ~ 10%)
Casualties	Combined with an emergency management system with GPS positioning and route guidance functions, it can reduce the degree of casualties
Travel time	Reduced by 4% ~ 20%, it will be more obvious in severe crowding
Capacity	The simulation shows that when 30% of the vehicles receive real-time traffic information, the capacity can increase by 10%
Delay	Save 1900 vehicle · hour during peak hours and 300,000 vehicle · hour per year
Emission estimates	HC emissions reduced by 16% ~ 25% Reduce CO emissions by 7%–35%
Traveler satisfaction	Reduce conscious pressure; wireless communication with rescue centers can increase safety by 70% to 95%

Practice shows that the traffic information service system has significant benefits in travel time, traveler satisfaction, road network capacity, and environmental impact, and can also reduce road congestion and traffic accidents.

The implementation effect of the US traveler information system reported by the US Department of transportation is shown in Table 6.1 [3, 4].

6.1.3   Classification of Intelligent Transportation Information Service System

The service content of the traffic information service system (TISS) is diverse, and the service methods are also various. According to different classification standards, the TISS can be divided into different types and each has its characteristics.

1. According to the providing information services time to traffic participants, the classification and characteristics are shown in Table 6.2.

2. According to the different content of the information provided, the classification and characteristics are shown in Table 6.3.

3. According to the integration degree of information flow three elements—information collection, processing and transmission, and the distribution of system functions, they are classified. The classification and characteristics are shown in Table 6.4.

Table 6.2 Classification of TISS (1)

Name	Features
Pre-trip information system	Pre-trip information services enable travelers to obtain information about travel routes, modes, time, road traffic, and public transportation before traveling and provide auxiliary decision-making services for travelers planning the best travel mode
In-trip driver information system	Provide drivers with information about travel choices and vehicle operating status, road condition information and warning information through various means, and provide route guidance functions for drivers and passengers in need
In-trip traveler transfer information system	Use multiple forms to provide travelers with transfer information services at stations, bus transfer points, and other places, such as departure time, travel cost, and travel time, to optimize the travel routes of travelers

6.2 Key Technologies of Intelligent Transportation Information Service

6.2.1 The Composition and Working Principle of the Traffic Information Service System

The advanced traffic information service system is mainly composed of three functional units: traffic information center, communication network, and user information terminal. The system composition is shown in Fig. 6.2.

Among them, the traffic information center refers to providing data processing, display, and interface functions for the entire system control, including the collection, classification, processing, analysis, provision of road traffic transportation data and social public information, as well as the implementation of optimal path search and other algorithms.

The communication network refers to the wired and wireless two-way data transmission provided between the user information terminal and the traffic information center, and the optical fiber data transmission between the information flow and the information center.

User information terminals refer to on-board information and navigation terminals, public display terminals for various road traffic information, personal information terminals including PC terminals and handhelds, and public information kiosks.

6.2.1.1 Traffic Information Center, TIC

TIC is the core of the advanced traffic information service system. It provides a central communication interface for vehicles and related traffic information resources and builds a comprehensive traffic information database on this basis to provide the various traffic information service functions described in the previous section.

Table 6.3   Classification of TISS (2)

Name	Features
Route guidance system	The route guidance system uses advanced information and communication technologies to provide drivers with rich driving information and guide them to drive on an optimized path, thereby reducing the vehicle's residence time in the road network, alleviating traffic pressure, and reducing traffic congestion and delays. This service is mainly for individual vehicles on the urban road network
Traffic flow guidance system	The traffic flow guidance system collects and sends traffic information in real-time, guides the reasonable distribution of traffic flow on time, to achieve an active traffic control method that uses the road network with high efficiency. The core and foundation of the traffic flow guidance system is the accuracy and timeliness of traffic information. This information includes road condition information, real-time dynamic traffic information (including traffic time and traffic flow, lane occupancy, vehicle speed, travel time and other traffic characteristics, traffic incidents, and congestion information), weather, and other environmental information. Traffic flow guidance is based on traffic flow prediction and real-time dynamic traffic allocation and applies modern communication technology, electronic technology, computer technology, etc., to provide necessary traffic information for travelers on the road network and provide information for their current travel decision-making and route selection. To avoid traffic congestion caused by blind travel and achieve the purpose of smooth and efficient operation of the road network, it faces all vehicles on the road network.
Parking lot information guidance system	The parking lot information guidance system provides parkers with the location information of all parking lots in a certain area and their parking space utilization information, to help drivers make reasonable parking choices and reduce roundabout driving and the resulting unnecessary traffic and environmental pollution.
Personalized information service system	Information about comprehensive social services and facilities related to travel can be obtained, such as addresses, business or office hours of catering services, parking lots, auto repair shops, hospitals, police stations, etc. after the travelers know this information, they can make a suitable travel plan and choose a suitable route.

The implementation of TIC generally adopts a distributed B/S or C/S architecture to establish an open system and provide a good operating platform, software, simple network, and standard interfaces to effectively enhance the traffic information service system to meet future needs and flexibility for expansion.

The traffic information center has the function of analyzing and processing the traffic and transportation data, can generate the traffic and transportation information database within the relevant area, and complete the work of searching for the optimal route. The basic data processing functions of the traffic information center include the following items.

Table 6.4 Classification of TISS (3)

Name	Features
Autonomous navigation system	The autonomous navigation system integrates the three elements of information flow and can conduct real-time navigation to the moving vehicle. It is a static system, equipped with positioning equipment and a historical map database on an independent vehicle. The vehicle does not communicate with the information center but uses a separate database that records past traffic conditions and road network information. The information collection, analysis, processing, and transmission of the system are all done independently on the car, without any contact with the traffic information center, so it can neither provide automatic navigation based on real-time dynamic traffic conditions nor report the passage time of the road section to the traffic control center.
Centralized navigation system (one-way communication system)	This kind of system indicates that the traffic information center unilaterally provides real-time dynamic traffic information to traffic participants. The traffic information center will collect road traffic information through various channels and send it to the vehicles on the road network through the signal transmission system after processing. Drivers and passengers use the on-board information receiving device to obtain current information about the traffic jam area and general road conditions to determine the driving route in real-time. The fatal weakness of the one-way communication system is that the traffic information center only provides information unilaterally, but cannot receive feedback from travelers to use the results and on-site traffic information obtained by traffic participants. But the system is improved compared with the autonomous navigation system
Centralized navigation system (two-way communication system)	The two-way communication system realizes two-way information exchange between the information center and travelers. Based on this, dynamic traffic flow distribution, dynamic traffic monitoring, and real-time traffic prediction can be carried out. An important feature of this system is that vehicles and traffic participants are no longer just passive information receivers, but can also become active traffic information collectors. It can not only give the travel time and speed of the vehicle through a certain road section but also obtain traffic information such as lane occupancy rate and traffic flow through a certain scale of detection vehicles. Countries are committed to this type of system research. Because the research and implementation of the system involve many units, institutions, industries, disciplines and specialties, and a large professional joint span, it has a high degree of difficulty, but it can be predicted that once the implementation of the system will greatly improve the traffic conditions and the resulting environmental and energy conditions

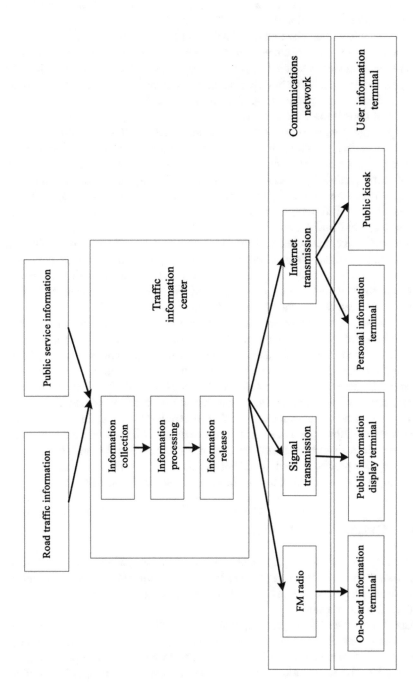

Fig. 6.2 Composition of the traffic information service system

1. According to the real-time dynamic road traffic conditions, update the traffic and transportation information database.
2. Generate and periodically update a database of estimated travel time of each road section in history. For example, at any time in the day, in the absence of current real-time traffic information, this is the best estimate of the travel time of the road segment and uses this as a basis to send the current road segment travel time to the TIC and other transportation information sources. The report comprehensively judges abnormal traffic conditions.
3. Compare and combine current road traffic information and historical road section transit time data, integrate all aspects of information to establish the best current traffic condition prediction model, which is used to provide vehicles with the best road section transit time estimation. Current road traffic information includes the actual travel time reported by the vehicle, data from the traffic signal system, and other dynamic information such as data collected by road patrol cars, vehicle detectors, and floating cars.
4. Using the above-mentioned real-time traffic condition prediction model, TIC can calculate the estimated transit time and select the optimal route from the departure to the destination, send the estimated result to the vehicles in the road network through the communication network, the drivers and passengers can use it. This information can choose the best route to the destination.

    At present, a variety of optimal path search algorithms have been researched and developed at home and abroad. Typical ones include the Dijkstra algorithm, heuristic search algorithm, neural network method, and ant colony algorithm. During vehicle driving, the route search algorithm should be able to continuously correct the selected optimal route and the transit time of the road section in a relatively short period according to the current road traffic conditions.
5. Carry out accident investigation work to determine the cause of the abnormal passage time. Abnormal means that the travel time on a certain day, a certain time, or a certain road section greatly exceeds the historical travel time. The accident investigation process is used to find out the cause of the sudden change in the road section's travel time and notify the traveler of relevant information.

6.2.1.2 Communication Network

Communication network, COM refers to the wired and wireless two-way data transmission provided between the user information terminal and the traffic information center, and the optical fiber data transmission between the information source and the information center.

Data is transmitted through the optical fiber digital network, between the information center and the fixed information source. The original traffic information is collected and transmitted to the information center for analysis and processing, is transmitted to the information sending device through the corresponding optical fiber network.

The two-way dynamic wireless data transmission system is responsible for completing the data exchange between the information center and the vehicle. On the one hand, the vehicle uses the receiving device to obtain real-time traffic information sent from TIC, such as travel time estimation, congestion or accident location in the current road section, etc.; on the other hand, the vehicle is a mobile traffic information detector, which transmits on-site traffic information (actual road section transit time) back to the information center through the on-board transmission device. At present, the widely used wireless communication methods include radio frequency communication (radio broadcast communication), microwave communication, infrared communication, mobile phone communication, and so on.

The information center sends real-time road traffic information to all vehicles equipped with a vehicle navigation assistance system, and the information is received by the vehicle through the on-board equipment, and furthermore is displayed after being processed (visual or auditory). At the same time, the vehicle automatically sends travel time reports and other related information to the traffic information center after completing a section of driving, which enables the information center to obtain real-time dynamic information from vehicles.

6.2.1.3   User Information Terminal

There are many types of user information terminals; on-board information and navigation terminals are one of the most widely used user terminals, the structure and functions of which are shown in Fig. 6.3 [2]. The on-board terminal includes a navigation assistance system and a radio data communication transceiver, while the

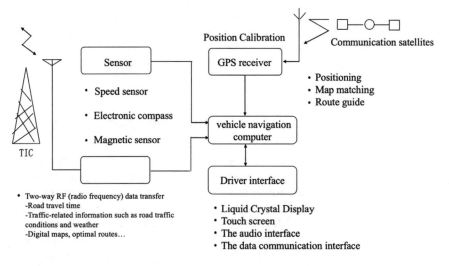

Fig. 6.3 Construction of typical on-board information and navigation terminals

navigation assistance system includes a vehicle navigation and positioning module, a vehicle computer and a display screen.

The vehicle navigation and positioning module includes various vehicle sensors such as speedometers, direction sensors, and GPS (Global Positioning System) receivers, which are used to provide accurate geographic location and time data for navigation. The position of the vehicle could be calculated with calibration and tracked in the road network through dead-reckoning, map-matching, and absolute position calibration.

The sensors commonly used in dead-reckoning include speed sensors installed on non-driven wheels, electronic compasses for determining direction, etc. Those sensors can provide distance, speed, and direction data to calculate vehicle position. Due to wheel tilt, tire pressure changes, etc., dead-reckoning is prone to have errors and has a cumulative effect. To correct the errors generated, the map matching technology and global positioning system are necessary.

Map matching refers to the use of various data (sensor data, GPS positioning data, digital map database, known best route, etc.) to determine the relative position of the vehicle on the map. This process compares the route obtained by dead-reckoning with the road alignment and optimal route in the map database to determine the position of the vehicle, feeds back the result to the dead-reckoning to correct the error of the dead-reckoning process.

The function of measuring, calculating, and reporting traffic information such as travel time, speed, and flow of road sections makes a vehicle equipped with on-board information and navigation terminals an active traffic detector, which is complete and important in providing real-time traffic information to TIC, this function is automatic for drivers and passengers without any additional burden.

On the entire analysis graph of traffic congestion, a single vehicle travel time report is almost meaningless, but reports sent by more than hundreds of vehicles equipped with on-board information and navigation terminals will generate an accurate and real-time graph of congestion and travel time of road sections all over a certain area. This information is transmitted to all the vehicles operating within the system, and the optimal route guidance instructions corresponding to the current location of each vehicle and the certain destination can be generated.

6.2.2 Theoretical Basis of Traffic Information Service System

The main function of traffic information service system is to generate information conducive to the traveler's travel based on traffic flow information collected in real time and transmits this information to the traveler in time. The research of traffic information service system mainly includes two aspects: implementation technology research and basic theoretical research. Implementation technology mainly refers to the integrated application of electronic control technology, communication technology, computer processing technology, GPS positioning and navigation system, etc., in the transportation system. They are essential hardware means and technical conditions for realizing urban traffic flow guidance. The key theory and model of urban traffic prediction and guidance is the core of the traffic information service system and the basic theory of ITS. Therefore, dynamic traffic distribution and driving route optimization design related to route guidance are the key technologies of the system.

At present, China has made rapid progress in implementing technology, but basic theory and model research is relatively lagging. For this reason, some research projects have not been applied in practice, and this is the main reason why countries all over the world attach importance to theoretical model research.

6.2.2.1 Dynamic Traffic Assignment

The characteristic that traffic demand changes with time makes the traffic flow on the traffic network dynamic. Therefore, in order to accurately describe various traffic phenomena on the transportation network, it is necessary to adopt a dynamic traffic model. One of the most important technologies is the use of dynamic traffic assignment. Generally, the static traffic assignment used in traffic planning assumes that the traffic demand during the model time period is constant, that is, the shape of the traffic flow distribution is fixed, and the maximum traffic demand in a certain time period would be calculated, so as to make a plan that meets the demand to achieve the purpose of planning. However, when it is necessary to describe the congestion characteristics of the urban transportation network, formulate urban traffic management measures, or release information on urban traffic conditions to travelers, it is necessary to study the dynamic distribution of urban traffic flow, because it determines the location and degree of urban traffic congestion. Dynamic traffic assignment considers the characteristics that traffic demand changes over time, and figures the instantaneous traffic flow distribution state, so it can be used to analyze traffic congestion characteristics, implement optimal control of traffic flow, and perform traffic information prediction and route guidance. From this aspect, dynamic traffic assignment is also an important technical foundation of ITS.

The early dynamic random assignment model can only handle the two-dimensional selection problem of travel time and travel route selection by a single OD. After development, it is now able to deal with the two-dimensional selection problem of the road network with the assumption that the traffic is evenly distributed and the speed remains same in the road section, which is obviously not consistent with the actual situation.

In recent years, the research on real-time dynamic traffic assignment theory has made great progress, the research methods include computer simulation, optimization theory, optimal control theory, inequality variational principle, etc., and the problems that can be dealt with have changed from single travel time or travel path selection to comprehensively travel time and travel path selection. There are two types of models: continuous-time model and discrete-time model. The optimal control dynamic traffic assignment theory and model based on genetic algorithm are introduced in the following part.

Although experts and scholars in the field of transportation have developed a variety of dynamic traffic assignment models, they generally assume that the demand is fixed rather than changing with time, the network has only a single end point, and the selected path is fixed. The key variables of the dynamic traffic network model established by Papageorgiou M. are the distribution proportion and composition proportion of traffic sub flow for specific terminal nodes, which is suitable for dynamic traffic assignment of multi-terminal traffic network which traffic demand changes with time. The model system has made great progress on the possibility of solving and similarity with the actual road network, but it takes a long time to use the direct optimization algorithm, so it is only suitable for small-scale traffic network.

In the mid-1970s, scientists in the United States, Germany, and other countries studied the global optimization method of simulating the biological evolution process to solve complex optimization problems, collectively referred to as simulated evolutionary optimization algorithm, also known as genetic algorithm (GA), which has been effectively applied in many fields. Because the genetic algorithm only needs the objective value of each feasible solution and does not need to assume that the objective function is continuous or differentiable, it adopts the parallel search mode of multiple clues to optimize and with no special requirements for the search space, and the optimization time is saved, so it is easy to use, and has strong adaptability.

Combining the model system established by Papageorgiou M. with the fast global optimization algorithm—genetic algorithm can greatly improve the practical value of the model. However, the model system established by Papageorgiou M. still needs dynamic OD information, which is difficult to achieve in the actual road network and traffic flow guidance. In addition, the model system uses a recursive method to calculate various traffic parameters, when the collected dynamic traffic information is subject to multiple interference, it will cause serious error accumulation effects and reduce the reliability of the assignment results. Therefore, for the practical application of traffic information service systems and urban traffic flow guidance systems, the development of new and more practical dynamic traffic assignment model algorithms is the focus of future research.

In addition, the real-time adaptive traffic control system and the route guidance system themselves will affect the route choice behavior of travelers, and even affect the distribution of traffic flow, so how to integrate the influence of traffic control and route guidance systems into dynamic traffic assignment has also become a new problem that needs to be studied and solved in dynamic traffic assignment when dynamic traffic assignment simultaneously provides a technical basis for the traffic control system and the route guidance system.

6.2.2.2   Route Selection and Optimization

Dynamic route guidance is to provide the most convenient and fast route from the current location to the destination for vehicles driving in the road network, which is called shortest route. The shortest route here has two meanings: one is the shortest distance based on the existing road. This shortest route can be found based on the knowledge of the existing road network structure and graph theory, and it is static. The other is the path with the shortest travel time or the least road resistance considering the real-time road traffic flow conditions. The calculation of this shortest path has the following two prerequisites: the first is to obtain the real-time traffic flow distribution from the aforementioned dynamic traffic assignment; and the second is to establish a certain travel time function or road resistance function, and plus the distribution of traffic flow or other factors to the calculation of travel time or road resistance, so as to select the shortest route according to the calculation result and provide it as an induction route to the driver. In theory, this shortest route changes dynamically and continuously with traffic flow, but the actual operation can only be discrete processing by dividing the system work into several time periods, finding multiple dynamic shortest routes in one time period for the driver to choose.

In the process of route selection and optimization, the following issues must be paid attention to.

1. Dealing with delays at traffic intersections. For the delay of the traffic intersection, the intersection can be divided into multiple virtual points, and the corresponding virtual road section can be established, and the control center sets the generalized road resistance of the virtual road section according to different timing schemes. In addition, by adding virtual points to split the intersection, the abstraction of no turning left and waking up the vehicle at the intersection can be realized.
2. Determination of evaluation indicators. The evaluation index is mainly reflected in the road resistance function. In order to meet the needs of multi-objective route optimization, generalized road resistance should be used in the dynamic route guidance system. The generalized road resistance (travel cost) refers to the quantified value of the price paid by the traveler in order to complete the trip and the negative impact on the society. According to the actual situation, there are generally five types of road resistance: travel time, travel distance, congestion, road quality, and comprehensive cost. The five types of road resistance are not contradictory, but according to different goals, the best route derived from this type of road resistance is generally different.
3. The user's travel characteristics have many influences on the dynamic route guidance system. The user's route selection based on the road level, pavement quality, and comfort level will affect the optimal route.

At present, the most popular shortest route algorithms are Dijkstra, Bellman-Ford-Moore, Floyd, Heuristic Search—A* algorithm, SPFA, ant colony algorithm, and so on.

The Dijkstra algorithm is a shortest algorithm proposed by E.W. Dijkstra that applies to all arcs with non-negative weights. It is also one of the classic algorithms for solving the shortest route problem. It can give the shortest route from a specified node to all other nodes in the graph. Its time complexity is $o(n^2)$, and n is the number of nodes.

The Bellman-Ford-Moore algorithm was proposed by Bellman, Ford, and Moore in the 1950s and 1960s, and its time complexity is o(nm), and m is the number of edges/arcs. At present, this time complexity is the best among all the shortest route algorithms with negative weight arcs, but its actual calculation effect is often not as good as Dijkstra's algorithm.

The Floyd algorithm is an algorithm for finding the shortest route between all pairs of nodes in the graph. It was proposed by Floyd in 1962. Its time complexity is $o(n^2)$. Although it is the same as the time complexity of doing Dijkstra's algorithm for each node, but the actual calculation effect is better than the Dijkstra's algorithm.

The more popular heuristic search algorithm is the A* algorithm which is first proposed by Hart, Nilsson, Raphael, and others. The innovation of this algorithm is that when the next node to be checked is selected, known global information has been introduced to estimate the distance of the current node, which is treated as a measure of the possibility of evaluating the possibility that the node is on the optimal route, so that the nodes with a higher possibility could be searched for, which improves the search efficiency.

The full name of the SPFA algorithm for finding the single-source shortest route is Shortest Path Faster Algorithm, which was proposed by Duan Fanding in 1994. It also has an important function of judging negative loops (which will be reflected in the differential constraint system). On the basis of the Bellman-Ford algorithm, a queue optimization is added to reduce redundant slack operations, which is an efficient shortest path algorithm (SPFA will be blocked by malicious data, if it is not necessary to judge the negative loop, Dijkstra is recommended).

Ant colony algorithm is a probabilistic algorithm used to find optimal routes. It was proposed by Marco Dorigo in 1992, whose inspiration comes from the behavior of ants finding routes in the process of searching for food.

In addition, for automatic navigation systems in practical applications, the storage and calculations capacity of the on-board computer are limited. In the face of huge road networks and information, it is very necessary to seek algorithms with small storage capacity; for real-time navigation systems, timeliness is highly expected. Therefore, in many cases, the precision is exchanged for time to realize the application of the algorithm in the actual situation. For the characteristics of automatic vehicle navigation, some progress has been made in the shortest route in recent years, of which mainly including improvements in data structure, two-way search, hierarchical search, K-shortest route algorithm, neural network-based algorithm, genetic algorithm. and TC-B Method algorithms based on travel characteristics, in addition. Interested readers please refer to related contributions.

6.2.2.3 Traffic Information Release

After the formation of traffic information that is conducive to travel, how to provide it to the traveler in an effective and intuitive way, so that the traveler can easily receive this information, without bringing more driving load, this is the traffic information display to solve problem. This requires that, on the one hand, a good human-machine interface design based on ergonomics should be implemented to make human-computer interaction functions easy to implement. On the other hand, the information itself should be displayed in a clear and intuitive form, such as expressing roads in different colors. The crowded condition of the traveler does not have to spend too much energy to receive this information, which will affect driving. In terms of dynamic route guidance, how to combine the guidance information with the existing electronic map to give a clear and intuitive route guidance is also a problem to be solved in information release.

6.2.3   Technical Basis of Traffic Information Service System

6.2.3.1   Global Positioning System

The global positioning system (GPS) was originally called the NAVSTAR system (Fig. 6.4). GPS is an omnidirectional, all-weather, all-time, high-precision satellite navigation system developed and established by the U.S. Department of Defense. Global users provide low-cost, high-precision navigation information such as three-dimensional position, speed, and precise timing. The entire system requires 24 satellites to provide high-precision positioning and continuous global coverage. It is a model application of satellite communication technology in the navigation field.

The basic working principle of GPS positioning is that satellites continue to send information such as time and ephemeris parameters to the earth. After the receiver receives the information, it processes and calculates the three-dimensional position, three-dimensional speed, and time of the receiver to provide vehicle position information. GPS uses CDMA code division multiple access technology, and different satellites have different modulation codes, but have same carrier frequencies. The positioning accuracy of civil GPS is generally about 10 m, while the positioning accuracy of military is relatively high, which can reach about 1 m.

GPS is mainly composed of GPS space constellation part, ground monitoring part, and GPS signal receiver part. The space part is the GPS satellite system. The

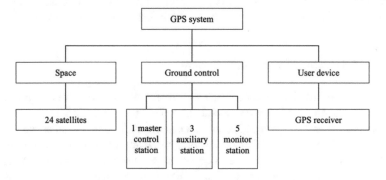

Fig. 6.4 Composition of GPS

GPS satellite constellation is also called the 24GPS constellation. It consists of 21 working satellites and 3 orbital backup satellites, with an orbital altitude of 20,183 km. When using GPS signals for navigation and positioning, only four GPS satellites can be observed to calculate the three-dimensional space coordinates of the observation station, which is called a positioning constellation. The ground control part is the ground monitoring system. The position of GPS satellites in positioning and navigation is calculated through the ephemeris launched by the satellites. Ephemeris is a parameter describing the movement and orbit of a satellite, which is provided by the ground monitoring system. The GPS satellite ground control station system mainly includes the main control station located in Colorado, USA, and 3 auxiliary stations and 5 monitoring stations distributed around the world, through which the GPS satellites are monitored. The user equipment part is the GPS signal receiver. The main task of GPS signal receivers is to capture and track satellite signals, and process the signals in order to calculate the observed three-dimensional space position, speed, and time in real time. The process of GPS receivers capturing and tracking GPS satellites in static positioning is fixed, while dynamic positioning uses GPS receivers to determine the trajectory of a moving object.

Currently, GPS positioning technology has widely penetrated into many fields such as economy, science, and technology. With the development of intelligent transportation systems, GPS technology is also widely used in the positioning and navigation of moving targets such as vehicles. In the development history of vehicle navigation and positioning technology, the GPS system is currently the most widely used and most mature satellite positioning system, followed by my country's "Beidou" navigation and positioning system, which is expected to be completed in 2020 and consists of more than 30 satellites. It will provide a full range of global positioning, navigation, and positioning services covering the world by sea, land, and air. The "Beidou" navigation and positioning system has three main functions: rapid positioning, providing users in the service area with real-time, all-weather positioning services, with positioning accuracy equivalent to that of GPS civilian positioning; short message communication, which can transmit up to 120 Chinese characters at a time Information; precision timing, accuracy up to 20 ns. With the success of the "Beidou" network, it has received more and more attention in the on-board positioning technology of the traffic information service system.

6.2.3.2   Geographic Information System

Geographic information system (GIS) is also known as geo-science information system or resource and environmental information system. GIS is an important information system for geospatial data management. It is a computer-based management platform for spatial database of geoscience information. It has unique functions that other database systems do not have, except for database general data input, storage, query, display, etc. In addition to the general functions, it also integrates the traditional and unique geoscience information record carrier—map

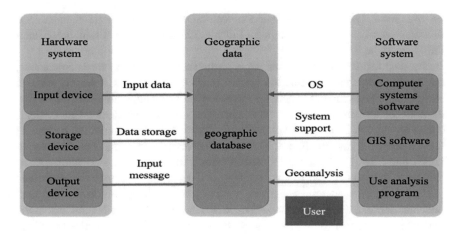

Fig. 6.5 Composition of GIS

and geographic analysis functions with the general database operation functions to realize the storage, query, and statistical analysis of spatial geographic information. Geographical information systems usually consist of hardware systems, geographic data, software systems, and users, as shown in Fig. 6.5.

At present, people use GIS technology to manage and analyze multi-source composite information such as the human economy, natural resources, and environment on the earth's surface, and use GIS technology to discover the spatial distribution, spatial structure, spatial connection, and spatial process of the natural environment, economic, and geographic elements of the region. The law of evolution of the region serves regional development such as regional macro decision-making and multi-objective development. The combination of GIS technology and traditional traffic information analysis and processing technology extends the geographic information system for transportation (GIS-T). GIS-T is the application of geographic information system in the fields of traffic survey and design, traffic management, and traffic planning, and it is the basis for the operation of the traffic information service system.

The most basic function of GIS-T is to edit and measure layers, and to support the display of layers. The most important thing is to edit its attributes and spatial data, and it also has input and storage functions. The function is to analyze and map the time and space of traffic geographic information. In the editing function, it includes the modification of attributes and the deletion or addition of points, lines, and areas required by the user; and the mapping function, whether it is production or for the display of the map, is flexible and diverse. In the display, different belonging traffic information objects can be input and output (display); for the measurement function, the area of the belonging area can be measured or the line segment on the map that needs to be measured can be measured. Through this function, GIS-T constructs a spatial digital model of the traffic road network, which provides a basis for the generation of digital traffic maps and the loading of traffic attribute data. In addition, on this basis, GIS-T also has superimposition function, dynamic segmentation function, terrain analysis function, grid display function, shortest time route function, and so on.

Traffic information has obvious regional characteristics. In practical applications, the traffic geographic information system (GIS-T) data model is mainly used to model and express the traffic network, so as to realize the integration and interaction of traffic information and road network spatial information. The GIS-T data model is a data model that is further developed and expanded on the basis of the traditional GIS arc-node data model. Since the arc-node data model adopts plane reinforcement, nodes must be generated where arcs intersect, so that it needs to interrupt the geometric data of the road when processing and describing dynamic multi-attribute traffic information, thereby generating a road network that is not in the real road network. The existence of road nodes increases the number of nodes and arcs in the database, resulting in data redundancy. In order to avoid the deficiencies of the arc-node data model, the GIS-T data model based on linear referencing method (LRM) and dynamic segmentation (DC) technology is the mainstream basic technology applied in today's traffic information system. In a traffic information system based on dynamic traffic information, massive amounts of traffic flow information, road condition information, and traffic control information are all efficiently and accurately integrated with road network spatial information on the GIS-T data model.

6.2.3.3 Map Matching Technology

The purpose of data collection in the transportation system is to fuse the data obtained by the sensors in the transportation environment and collect the dynamic characteristic data information of the transportation system state. The most basic map matching technology is the analysis and processing of the dynamic position data of vehicles in GIS-T, and the technology to obtain the position data of moving vehicles is called online map matching. Online map matching (Map Matching) algorithm uses a small amount of simple information, such as GPS data latitude and longitude of satellite positioning, etc., to calculate the position of the moving vehicle on the road network segment. Calculating the position of the moving vehicle on the road network segment is a dynamic fusion process of the GPS data and map network data model such as GIS-T. The map matching algorithm of the traffic information service system requires high 660 accuracy and efficiency. According to the technical methods used in the algorithm, the map matching algorithm is divided into three categories: geometric algorithm, topological algorithm, and advanced algorithm. The geometric algorithm matches the GPS point to the nearest road endpoint or shape point, that is, point-to-point matching; or the GPS point is matched to the nearest road, that is, point-to-line matching; or the GPS point list is matched to the nearest road, that is line-to-line matching. The topology algorithm uses the topological relationship of the road network: connectivity, adjacency, and association, combined with road attributes (such as steering restriction and road direction) to infer the section of the vehicle traveling, and it can be used when the sampling rate is low and the sampling interval is not high. The advanced matching algorithm uses Kalman filtering, Bayesian inference, and fuzzy logic technology to infer the position of GPS points in the road network model. The three algorithms have their own advantages and disadvantages. In actual research, the design of map matching algorithms usually uses these three technologies to complement each other and improve the accuracy of the algorithm to obtain accurate vehicle location.

Online map matching technology provides continuous vehicle location data for the traffic information system, which corrects the display deviation of the vehicle on the road network model (road network digital map data model) caused by GPS positioning error, so that the traffic information system can learn accurate road location information of driving provides data of the vehicle in real time and support for traffic information analysis such as road traffic flow operation status and vehicle driving route optimization.

6.2.3.4 Advanced Vehicle Networking Traffic Information Transmission Technology

The Internet of Things (IoT) technology is showing off in daily life, such as smart wearable systems, smart homes, smart industries, and smart transportation. Its main goal is to enable interoperability between heterogeneous devices and objects. The goal of the Vehicular Adhoc Network (VANET) is to achieve real-time communication in order to enhance traffic safety and management effectiveness. The Internet of Vehicles (IoV) is an important branch of the Internet of Things research, which developed from VANET. And LoV as the basis for the future development of transportation information system transmission has become a hotspot of transportation information system research in recent years.

The on-board ad hoc network VANET aims to improve transportation efficiency and improve people's travel comfort. VANET serves the mutual communication between vehicles. It has the characteristics of node, mobility, and data flow, turning the vehicles used in the role of transportation into intelligent terminals.

VANET refers to the open mobile Adhoc network formed by the communication between vehicles, between vehicles and fixed access points, and between vehicles and pedestrians in the traffic environment. It aims to build a self-organizing, open, structured network on the road. A vehicle communication network that is easy to deploy and low in cost. Ad hoc network is a wireless distributed structure, which has the ability of self-organizing, centerless, and multi-hop data transmission, so as to realize applications such as assisting driving, accident warning, workshop communication, and road traffic information query. Compared with other networks, VANET has the advantages of high dynamic topology, high node moving speed, predictable trajectory, accurate positioning, etc. It has extraordinary versatility and practicality in the field of transportation.

VANET communication architecture can generally be divided into three categories: Wi-Fi based on wireless access in the vehicle environment (WAVE), self-

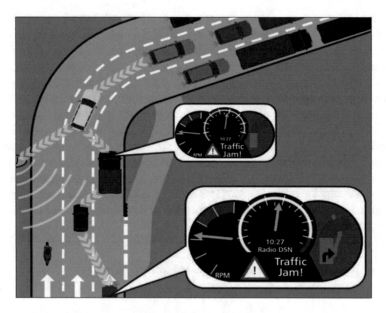

Fig. 6.6 Avoid traffic jams and accidents through connected cars

organizing network (Adhoc), and a hybrid architecture that combines the two. Under the Wi-Fi-driven structure, the roadside units RSUs set on the road network are used as wireless access points to provide communication functions for vehicles within the coverage area; the second type is formed by a group of vehicles driving on the road, the self-organizing network uses WAVE for communication. These networks are independent of each other and do not need any infrastructure assistance; and under the hybrid structure, cellular and self-organizing networks use WAVE for communication at the same time. As one of the standards of V2V communication in the Internet of Vehicles, the IEEE802.11p protocol has been supported by the US government and business circles. At the same time, based on the basis of cellular communication, Chinese enterprises have led the formulation of LTE-V2X related standards and the research of subsequent evolution technologies. In addition, in order to strengthen the cooperation between the automobile and communication industries, the global communication industry and some automobile companies jointly established 5GAA (5G Automotive Association). In order to actively promote the industrialization of LTE-V2X, China has established a cellular car networking C-V2X project team to cooperate with the development of DSRC/5G work. Fig. 6.6 shows the application scenarios of the Internet of Vehicles transmitting road congestion information, assisting safety warnings and vehicle route re-planning guidance.

6.2.3.5   V2X Communication Technology Based on Cellular Network C-V2X

Vehicle to Everything (V2X) is a new generation of traffic information and communication technology that connects vehicles with everything. V stands for traffic unit---vehicle, and X stands for any object that interacts with the vehicle. The current X mainly includes vehicles, people, traffic roadside infrastructure, and network nodes. V2X information interaction unit combination modes include: Vehicle to Vehicle(V2V), Vehicle to Infrastructure(V2I), Vehicle to Pedestrian(V2P), Vehicle to Network(V2N), as shown in Fig. 6.7.

The C in C-V2X refers to Cellular, which is a technology for wireless communication between vehicles based on the evolution of cellular network communication technologies such as 3G/4G/5G. It includes two communication interface standards, one of the standards is the short-distance direct communication interface standard (PC5) between cars, pedestrians, and roads, and the other one is the communication interface standard (Uu) between the terminal and the base station, which can realize long-distance and larger-range reliable communication between vehicles. C-V2X is a vehicle networking communication technology based on the 3GPP global unified standard, including two technological development processes of LTE-V2X and 5G-V2X. From the perspective of technological evolution, LTE-V2X supports smooth evolution to 5G-V2X. At present, LTE-V2X technology is a mature vehicle networking communication technology.

V2V refers to the communication between the vehicle and the vehicle through the on-board terminal. On-board terminals can obtain information about the velocity, location, and driving conditions of surrounding vehicles in real time. Vehicles can also form an information exchange platform to interactively transmit texts, pictures, and videos and other digital information in real time. V2V communication is mainly used to avoid or reduce traffic accidents, vehicle supervision and management, etc. V2I refers to the communication between on-board equipment and roadside infrastructure (such as traffic lights, traffic cameras, roadside units, etc.).

The roadside infrastructure can also obtain information about vehicles in nearby areas and release various real-time information. V2I communication is mainly used in real-time traffic information services, vehicle monitoring and management, and

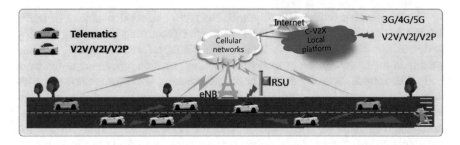

Fig. 6.7 Wireless communication technology for vehicles

non-stop toll collection. V2P means that vulnerable traffic groups (including pedestrians, cyclists, etc.) use equipments (such as mobile phones, laptops, etc.) to communicate with on-board devices. V2P communication is mainly used to avoid or reduce traffic accidents between people and vehicles, non-motor vehicles and motor vehicles, and traffic information services. V2N refers to the connection of on-board equipment to the cloud platform through the access network/core network, the data information interaction between the cloud platform and the vehicle, the storage and processing of the acquired data information, and the provision of various information application services required for vehicle travel. V2N communication is mainly used in vehicle navigation, remote vehicle monitoring, emergency rescue, and entertainment information services. Therefore, V2X organically connects people, vehicles, roads, cloud platforms, and other transportation participation elements. It not only supports the Internet of Vehicles to form an interactive perception body, and obtains more traffic perception information than single vehicle, but also helps build an intelligent transportation system. Promoting the innovation and development of automobile and traffic information service models is of great significance to improve traffic efficiency, energy saving, and consumption reduction, reduce pollution, reduce accident rates, and improve traffic management.

With the help of efficient and all-round information interconnection between people, vehicles, roads, and cloud platforms, C-V2X is currently developing from information service applications to traffic safety and efficiency applications, will gradually support the realization of autonomous driving information collaboration applications evolution. C-V2X is the foundation and important part of traffic information service application scenarios. Typical traffic information service application scenarios include emergency call services and so on. The emergency call service means that when the vehicle is in an emergency (such as a traffic accident, mechanical failure, etc.), the vehicle can automatically or manually initiate an emergency call for help through V2X communication network, and send out basic information data for help, including the type of vehicle and the traffic accident (failure) location and time, accident type, casualties, and help content, so that service providers such as medical assistance, accident handling groups, insurance services, operator rescue or third-party emergency rescue centers can receive help information and take emergency assistance measures. This scenario requires the vehicle to have the ability of V2X communication in the traffic information service system to establish a communication link with the network.

The introduction of C-V2X will enrich the types and service content of traffic information services, especially in local information processing and sharing with fog nodes as the main body of calculation, and global information processing and sharing in the transportation system with cloud information processing as the main body. The introduction of local fog nodes will improve the effectiveness and real-time performance of local information, and strongly support the evolution of autonomous driving services. With the in-depth integration of information sources, information services will be further refined and personalized to fully support the individual needs of users. Its evolution process is closely related to platform integration, AI optimization, and computing optimization, which will fully affect the information service of the traffic information service system and provide methods and development directions.

6.2.3.6   Variable Information Sign Information System

The vehicle self-organizing network is composed of a vehicle and a roadside unit equipped with an on-board unit, uses Wi-Fi and dedicated short-distance communication technologies to achieve vehicle-to-vehicle communication and communication between vehicle and roadside unit. Among them, dedicated short-range communication (DSRC) is a technology developed by the automotive industry. It was originally designed to achieve frequent data exchange between vehicles and between vehicles and roadside units. DSRC is released in the form of protocol stacks. It is a collection of a set of communication protocols, and these collections are a set of standards dedicated to short-range communications in the V2V communication including the IEEE 802.11p protocol, the IEEE 1609 protocol series, and the SAE series. In addition, DSRC also added a special processing mechanism for security messages to meet the needs of security application services. At present, the DSRC protocol stack has not formed a unified international standard. The existing DSRC standards are mainly proposed by Europe, the United States, and Japan, which are ENV series, 900 MHz, and ARIBSTD-T75 standards. Among them, the DSRC standard developed and promulgated by the ASTM Research Institute of the United States has the greatest impact. The E2213–03 standard promulgated in 2002 sets the communication frequency band to 5.9GHz, which makes the data transmission speed up to 27Mbit/s, and the communication distance is close to kilometers. The United States has allocated a 75 MHz bandwidth range for DSRC in the 5.9GHz spectrum range. This spectrum range is divided into seven 10 MHz size channels. Except for a dedicated control channel for ordinary security communications, the remaining six channels are used as Service Channel (SCH) for non-safety applications. According to estimates by the US Department of Transportation, DSRC-based workshop communication has effectively avoided 82% of the country's traffic incidents such as collisions and saved tens of thousands of lives. In the United States, in addition to effectively preventing collisions between vehicles, DSRC is also widely used in traffic information services. The most of these applications involve communication between vehicles and roadside units, such as assisting navigation, electronic payment, increasing fuel power, and collection of traffic probes, and distribute traffic updates, and can even be used in entertainment and business areas.

A variety of wireless communication technologies can be used in the V2X communication mode. The existing communication technologies include DSRC, Wi-Fi, and Cellular Network. And only DSRC is a set of corresponding protocols and standards dedicated to one-way or two-way short-range to medium-range wireless communication between vehicles. Table 6.5 is the parameter performance comparison between DSRC technology and other wireless communication technologies. Through comparison, it is found that DSRC technology is more suitable for

Table 6.5 Comparison between DSRC and other wireless communication technologies

Wireless communication technology	DSRC	Wi-Fi	Cellular	WiMAX
IEEE standard	802.11p	802.11a	N/A	802.11e
Time delay/ms	<50	>100	>100	/
Mobility/(km/h)	>60	<5	>60	>60
Communication distance/km	<1	<0.1	<10	<15
Data transmission rate/(Mb/s)	3–27	6–54	<2	1–32
Communication bandwidth/MHz	10	20	<3	<10
Communication frequency band/GHz	5.86–5.925	2.4/5.2	0.8/19	2.5

short-range communication between vehicles in terms of time delay, mobility, communication distance, data transmission rate, communication bandwidth, and communication frequency band.

6.2.3.7 The New Generation of High-Speed Internet of Vehicles Communication System 5G

In intelligent transportation systems, in order to effectively avoid traffic accidents and improve traffic efficiency, millisecond-level delays and nearly zero transmission errors are essential. Information sharing in the Internet of Vehicles system is time-sensitive and requires a stable and quick-response network connection. The highly dynamic connection, sensitive information sharing, and time sensitivity of the Internet of Vehicles communication system require higher requirements on the reliability, low latency, and availability of the network. Therefore, the low-latency, high-reliability communication technology application scenario of the Internet of Vehicles has become an important scenario among the four technical scenarios of 5G.

International Telecommunications Union (ITU) proposed a time schedule for the 5G plan at the International Mobile Telecommunications Conference in October 2014, and promoted the implementation of the 5G plan in three phases. Phase 1 is to be completed by the end of 2015. ITU defines the macro requirements and key capabilities of 5G through the vision proposal. Phase 2 is from 2016 to mid-2017, completes the establishment of requirements and evaluation standards for key technologies in 5G, and collects new 5G technologies currently researched by worldwide institutions. Phase 3 is from 2017 to the end of 2020, soliciting new 5G technologies from all over the world, and completing the evaluation in accordance with the evaluation standards, and finally formulating 5G standards.

For the Internet of Vehicles, at least one party in the communication process is a moving vehicle, which makes the network topology change frequently and the calculation is complicated, and the data transmission communication link maintains a short time from establishment to disconnection, so it is time-aligned. The requirements for extension and reliability are very high, and the core business of the Internet of Vehicles is mostly related to traffic safety and information services. In the application of 5G mobile communication systems related to traffic safety and information services, the reliability is required to reach 99.999%, and the time delay from end to end is less than 5 ms.

5G needs to support both high-frequency (above 6GHz) and low-frequency (below 6GHz) transmission methods to meet the technical requirements and performance indicators of various scenarios. The use of high-frequency bands is mainly used to meet the needs of ultra-high data rates, because the high-frequency band has the advantages of continuous large bandwidth, and is a newly developed frequency band that does not require backward compatibility. Therefore, the high-frequency band corresponds to the newly designed high-frequency air interface. The low-frequency band is mainly to provide advantages such as ultra-low latency (air interface 1 ms), high-frequency efficiency, etc., but also to provide large connections and other capabilities to meet the needs of equipment for large-scale coverage and mobility. The low-frequency band is divided into 4G evolution air interface to provide low latency, large connection function and low-frequency new air interface to provide low latency, high-frequency efficiency function.

6.2.3.8   Cloud Computing-Based Traffic Big Data Mining Technology

Traffic data is the foundation of the traffic information service system. It has the geographic characteristics of the data source, the spatial range of data acquisition is wide, the frequency of generation is high, and it has the characteristics of continuous. TISS has the characteristics of diverse data sources, complex data structure, and large amount of data. The Transportation Information Center (TIC) conducts in-depth analysis and mining of the massive amount of information to achieve an accurate understanding of the physical phenomena of traffic, which is to accurately grasp the essence and the nature of traffic phenomena and the basis of the law of development and change. For the massive traffic information data owned by TISS, it is necessary to build a computing layer with huge data analysis, processing, computing, and storage capabilities, as shown in Fig. 6.8.

The cloud computing system has ultra-high computing performance, and a single device can process up to 20 million pictures per day. Cloud computing analysis has the structured intelligent analysis function of vehicle image information taken by cameras, electronic police, and some monitoring equipment, mainly including identifying the brand, model, producing year, body color, category, and abnormal characteristics (such as blocking face, blocking license plate), unique local features (such as annual inspection signs, car interior decorations), and so on. The vehicle license plate color and license plate number in the submitted images can be identified twice, and big data mining can be performed on multi-source massive data such as images through big data technology, and the time, geography, and trajectory of the vehicle in the toll-gate data can be compared and identified to obtain the analysis results of the operation status of the transportation system network.

There are two advantages of cloud computing technology in the traffic information system: one is to improve the resource utilization level of traffic information

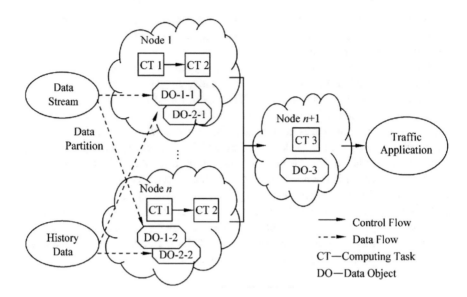

Fig. 6.8 Cloud computing processing model for traffic perception spatio-temporal data

equipment, and the application of cloud computing technology, and it can effectively improve the utilization rate of various equipment resources and reduce the cost of information platform construction; The second is to improve the processing level of traffic information data, cloud computing integrates distributed computing, through distributed computing, distributed processing of data can be realized, and the storage, mining, analysis, and processing of massive data can be realized in a very short time, which provides data support for scientific decision-making for the traffic information service system to make reasonable traffic information release.

Cloud computing technology has supercomputing power, dynamic resource scheduling, on-demand service provision, and massive information integrated management mechanism, which can effectively carry out massive data storage, calculation, and analysis, thereby enriching the content of traffic information services and improving the reachability and accuracy of information transmission, improve the quality of service from the core of the traffic information service system. The computing layer of the Transportation Information Center (TIC) uses cloud computing technology efficiently to analyze and mine the required information and laws from massive data, combine existing experience and mathematical models to generate higher-level decision-support information, and obtain various types of analysis and evaluation data, which data can provide decision-making support for traffic guidance, traffic control, traffic demand management, emergency management, etc., and bring more effective data support for traffic management, planning, operation, service, and active safety prevention, as well as for public safety and society management provides a new concept, model and means. For example, in response to the problem of periodic traffic congestion on urban roads, TISS uploads a large amount of data to the cloud, uses cloud computing to perform comprehensive analysis of massive data to eliminate various random interferences. At the same time, cloud computing describes the law of traffic operation based on traffic flow theory, and discovers the traffic jam mechanism, so as to provide information services for regional traffic guidance and control.

6.2.3.9   Variable Message Signs Information System

Variable Message Signs (VMS) is an important part of the intelligent transportation system and an important device for the release of traffic conditions and traffic guidance information. The main function of VMS is to provide travelers with richer and more timely information services, thereby helping the entire road network operate more efficiently and safely. The Variable Message Signs is the main realization tool of the traveler information system in the traffic guidance system. It provides traffic-related information and guidance to the driver through the electronic information display board installed on the side of the road or above the highway.

In the group vehicle guidance information system, the variable message signs information that needs to be collected and processed includes basic road network information, traffic control information, traffic condition information, and traffic condition prediction information.

The basic information of the road network includes road network structure information (such as road sections, nodes, number of lanes, road network topology parameters for vehicle positioning), road network attribute parameters (such as road section name, one-way traffic section, no turning section, loading quality, and clearance limitation) and other useful service information. In order to meet the needs of guidance, all these information need to be computerized, that is, the road network information is transformed into a road network digital map.

Traffic control information includes traffic control information, information on traffic emergencies, road construction and maintenance, road prohibition, etc. during major incidents. This information can be provided by the traffic control center and road management departments.

Traffic conditions and prediction information. The vehicle guidance system must have the function of providing drivers with real-time traffic conditions and can make necessary predictions on the traffic conditions to guide the vehicles. The main parameters of traffic conditions include traffic flow, traffic density, intersection saturation, delay, driving speed, road travel time, etc. Traffic flow, traffic density, and intersection saturation can be provided by the coil detection system and the video detection system, and can be obtained after processing by the traffic control center. The driving speed and road travel time are obtained by the detection vehicle and the beacon. The United States uses detection vehicles to first obtain the traffic conditions on the road network, equip a certain number of vehicles with a navigation system, and each vehicle sends back traffic information to the information center at any time. Europe and Japan send information to the receiver on the vehicle through a beacon on the roadside, and the receiver returns a signal to the beacon. The driving

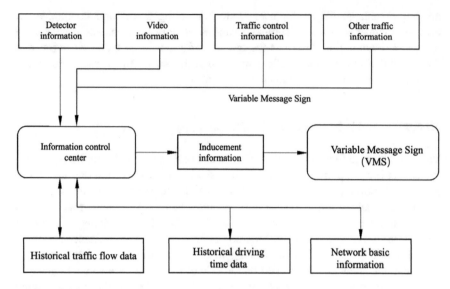

Fig. 6.9 VMS information display system information flow diagram

speed and road travel time are calculated based on the time difference between the two beacons.

When real-time traffic conditions are obtained, the Variable Message Signs Information System will make further predictions on the traffic conditions. Based on the prediction, the optimal driving route is calculated for the vehicle depended on the transportation analysis model, and through the VSM and other inducement equipment to provide users with audio and video prompts or induce instructions.

The composition of the Variable Message Signs Information System is similar to that of the traffic information service system, and both have an information control center and a communication system. The Variable Message Signs Information System is shown in Fig. 6.9.

1. Information Control Center. The information control center is the main control center for vehicle guidance. It integrates all traffic flow information data collection, processing, and transmission. Its main functions are database establishment and update, communication with other information sources, communication with VSM, and data analysis and processing for traffic information, calculation of the optimal route. Its hardware system is composed of computers and various communication equipment.

2. Communication System. The key to the good functioning of an information system is the communication system. The group vehicle guidance system uses a one-way communication method, and the one-way communication is only one-way transmission of information from the information center to the VSM.

3. VSM. The VSM is the terminal equipment of the group vehicle guidance information system, which provides drivers with good guidance information, and its display mode is generally words and graphics.

6.3   Advanced Traffic Information Service System

Advanced Traffic Information Service System is a very important field in ITS research. It has always been the focus and hot spot in ITS research, which attracted the attention of traffic engineering workers. Many countries and regions such as Europe, America, Japan, and others have invested a lot of manpower, material, and financial resources to conduct research, development, experimentation, and actively put into operation, accumulated rich experience, and achieved considerable results.

6.3.1   Typical Traffic Information Service System and Its Key Technologies

At present, the typical traffic information service systems that have been constructed, applied, are TravTek, TravLink, 511 system, IntelliDrive/VII, SafeTrip21, and [5–11] in the United States; VICS and SmartWay systems [12–14] in Japan; and RDS-TMC and Traffic-master systems [15–18] in Europe. There are also some systems that focus on the realization of the route navigation function, but the traffic information provision function is not perfect, such as the EURO-SCOUT system in Europe, the ADVANCE system in the United States, and the DRGS in Japan.

6.3.1.1   The U.S. Traffic Information Service System

The U.S. traffic information service system is shown in Table 6.6.

1. 511 travel information service system in the United States. The US 511 Travel Information Service System is a traffic information service hotline led by the US Department of Transportation. Users can dial the phone or log in to the website to obtain the required traffic travel information. The initial idea of establishing the 511 Travel Information Service System began with a petition from the U.-S. Department of Transportation, the American Highway and Transportation Public Servants Association, and the State Department of Transportation to the U.S. Communications Commission in March 1999. The petition pointed out the creation of a 511 traffic information system can effectively ensure that organizations release relevant traffic information to the public.

   The 511 system is a non-federal government-funded project. The federal government has no mandatory regulations and the specific rules are completely

Table 6.6 Traffic Information Service System

Country	System Name	Time	Main function	Key technology	Application status
The U.S.	TiavTek	1991– 1994	A demonstration project of a driver information system on a scale of 100 vehicles, providing drivers with navigation, route selection, real-time traffic information, local information, and mobile phone services	Customized mobile wireless communication system mobile phone, vehicle terminal, GPS, and speech synthesis navigation	Small-scale test
	TravLink	1994– 1996	Realize the computer-assisted dispatching and automatic vehicle positioning of the bus on the I-396 road in Duluth, and provide travelers with real-time bus information	GPS, variable information board, and query kiosk	Trial on specific road segment
	511 system	2000- now	Provide information on the construction, accidents, special events, and congestion of national arterial highways, bus timetable, cost information, and meteorological conditions affecting traffic, etc.	Camera photos of call centers, information portals, and highway traffic flow	Large-scale domestic application in the U.S.
	IntelliDrive/ VII	2004– 2009	Realize the integration of automobiles and road facilities through information and communication technology, use test vehicles to obtain real-time traffic data, support dynamic route planning and guidance, and improve driving safety and efficiency	DSRC, WAVE/ IEEE 1609, vehicle-to-vehicle communication and vehicle-to-road communication	Small-scale test
	SafeTrip21	2008- now	Provide drivers with soft safety warnings to enable them to	GPS, CAN/OBD II radar, acceleration sensor, camera, and	Small-scale test

(continued)

Table 6.6   (continued)

Country	System Name	Time	Main function	Key technology	Application status
			adjust the speed of vehicles in a timely manner and reduce the probability of accidents on highways	3G mobile communication	

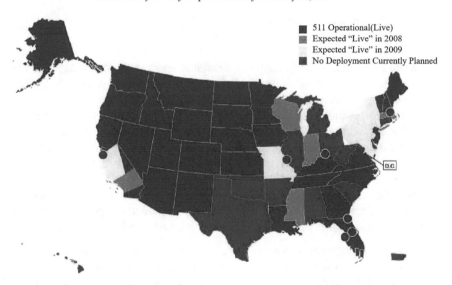

511 Deployment status
Accessible by 70% Population in 2009
Accessible by 47% of Population as of February 21,2008

Fig. 6.10   Application of 511 system

formulated by the state and local agencies. In 2001, the U.S. Department of Transportation began planning to fund state transportation departments, and a total of 46 states have received funding so far. Each state determines a reasonable business model for the operation of the system and provides users with valuable information services. In 2010, all states in the United States opened the 511 system, more than 90% of the population in the country understands and knows the 511 system, and the annual call volume will reach 40 million, as shown in Figs. 6.10 and 6.11 [5]. The cost of the 511 system mainly includes labor, equipment, data update, communication, and marketing expenses, and each state uses various financing channels to independently construct it. Encourage private organizations to provide value-added services, they can accept advertisements and sponsorships, but cannot reduce the quality of user services; they can

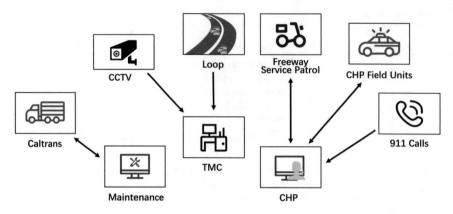

Fig. 6.11 511 System data source

ensure users' local calls through telecommunication control technology, and can exchange data and share applications with neighboring states. After data processing and format conversion, it is used by the local system. If you need to provide travel time information, you need to rely on automatic road detection equipment. At the same time, road users are encouraged to report road accident information through 511. Some 511 system also support tourist information service functions. The 511 system can also provide information services for hearing impaired users and non-English users; the peak call volume is mainly concentrated in severe weather conditions, serious traffic accidents, and holidays.

511 system services include road information, public transportation information, weather information, warning information, and travel information.

(a) Road information. From unexpected traffic conditions (such as reports of traffic incidents with a large impact) to traffic within a certain road section. Congestion status and driving time data usually provide information about road conditions and road construction in advance, and provide incident report and the latest development status report.

(b) Public transportation information. The user can obtain the public transportation information of any city from the city list according to the voice prompts, and finally obtain the information and contact number of the public transportation operator. The San Francisco Bay Area 511 system provides the most extensive public transportation information, including fares, service announcements, bicycle information, travel service information for the disabled, event hotlines, lost and found, and ferry service schedules.

(c) Weather information. Information can be provided to users 72 hours in advance including current and forecast weather conditions, temperature, road temperature, rain, snow, and icing.

(d) Early warning information. The 511 system is an effective way to release orange warning information. All those who dial the 511 system. Users can hear the orange warning message before entering other information service items.

(e) Travel information. The 511 system can transfer user calls to transport operators to provide information about tourism information. Some can be linked with travel service websites. There are 26,000 trips on the Virginia Travel Information Service website. Tourist information, some states also provide national park tours, concerts, theaters, local celebrations and festivals, conferences, and other information about the activity. At the same time, the 511 system is also connected to the service telephones of airlines and railway passenger transport companies.

(f) The future development of the 511 system will be developed from the following aspects: improve data sources; emergency alert/broadcast messages (evacuations, major accidents, homeland security, yellow alerts); time stamps; telephone information content filtering; consider regional differences.

2. TRAVTEK system. Among the ITS field experiments conducted in the United States, the TRAVTEK study conducted in Orlando, Florida is representative. This experiment aims at the practical application of real-time road guidance and information service systems, using the navigation system of the two-way communication function of the car phone.

As shown in Fig. 6.12 [7], TRAVTEK is composed of a traffic management center, an information service center, and a vehicle equipped with a navigation device. The traffic management center is responsible for collecting, managing, and providing road traffic information, as well as providing information management and services necessary for the operation of the system. The information service center collects various service information (yellow pages information) for tourist facilities, hotels, restaurants, etc. The car navigation device provides three functions such as vehicle positioning, route selection, and interface. The device can display a map of the Orlando area that contains information such as jam locations, accidents and constructions, route guidance that meets driver's requirements, and text messages with the possibility of use.

The user interface of the TRAVTEK system can provide navigation, yellow page information, road traffic accident information, and traffic status information. The interface design of the navigation system mainly achieves the following goals: Provide users with more effective navigation to save time and cost; easier to obtain valuable regional traffic information, reduce the pressure on drivers, and increase driving fun; maintain the safe driving of drivers through the forecast and information release of emergency events; improve the traffic capacity of the road network, improve traffic congestion.

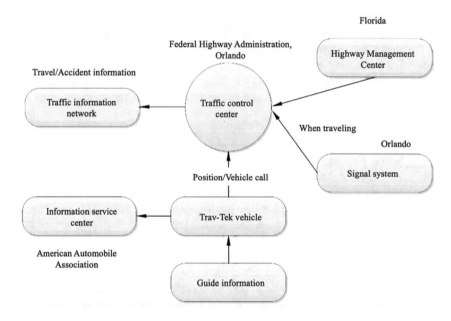

Fig. 6.12 TRAVTEK concept map

Users who are equipped with TRAVTEK on-board units in Orlando, Florida can access a rich information environment through the user interface. The functions of the on-board equipment include navigation information, route guidance, real-time traffic information, and regional service information and point of interest information. In the main menu of the system, the user can select the unmovable function items: enter the destination for route navigation, browse local service facility information and point of interest information, browse regional electronic maps, request emergency services, enter the user guide and modify the map display Information about the area where the vehicle is located.

In the TRAVTEK system, all pre-trip route guidance information and most functional information are provided to users through visual information. Detailed examples of related functions are as follows:

1. Some information functions before traveling. Before traveling, the user can input the destination query-related information through the visual interface and touch screen or other keys. After entering the new interface, the user can further make detailed settings.
2. Interested site query function. The TRAVTEK system can provide travelers with information on local service facilities and interested sites. The menu allows users to choose different interested sites, including stay site, hotels and camping sites, places to visit, meals, and other service sites. After selecting the above options, entering the next level menu can also provide more complete service information.
3. Information electronic map function. If the traveler does not enter the destination for guidance, the map screen can be used to browse the electronic map, and the

Table 6.7 Comparison of TRAVTEK system functions and ATIS/CVO expected functions

ATIS/CVO	Function	TravTek Realization
	Travel planning	*
	Multi-modal travel coordination	*
	Route and destination selection before travel	*
	Dynamic route selection	*
On-board route navigation system	Route navigation	*
	Route guidance	*
	Automatic charging	*
	Route itinerary (CVO unique)	
	Computer-aided route allocation (CVO unique)	
	Broadcasting service/point of interest	*
In-car personnel service information system	Service information/points of interest directory	*
	Destination coordination	*
	Message delivering	*
	Signal guidance along the way	*
Vehicle signal information system	Signal announcements along the way	*
	Signal adjustment along the way	*
	Danger warning	*
	Information along the way	*
On-board safety advisory and warning system	Rescue automation	*
	Manual rescue call	*
	Vehicle status monitoring	
	Vehicle and vessel monitoring (CVO unique)	
CVO unique functions	Vehicle allocation	
	Adjustment management	
	Adjust execution	

area can be selected by pressing the button. The electronic map also displays the location and heading of the vehicle.

4. Traffic flow information service. In the TRAVTEK system, real-time traffic information is transmitted wirelessly from the traffic control center to users. Colored signs have been added to the map on the information screen to display information on moderate road congestion, severe road congestion, lane closures, and traffic accidents or other accidents.

5. Route guidance function. When the traveler enters the destination and other relevant information, the system can provide users with visual information for route guidance.

The functions provided by the TRAVTEK system basically cover some aspects of the four subsystems pre-planned by ATIS in the United States. Table 6.7 shows the functions of the TRAVTEK system, which can be compared with the functions proposed by ATIS/CVO. It can be seen from the table that the TRAVTEK system satisfies the 18 functions in ATIS, which is quite good under the circumstances at the time.

6.3.1.2 European Traffic Information Service System

The introduction to the European traffic information service system is shown in Table 6.8.

The representative traffic information systems in Europe are SOCRATES, TrafficMaster, EUROSCOUT, and RDS/TMC. Among them, RDS/TMC is the most successful and widely used large-scale traffic information solution.

1. RDS/TMC. RDS is a European specification for data broadcasting systems formulated by the European Broadcasting Union (EBU) in 1984. In 1986, the International Radio Consultative Committee (CCIR) adopted recommendation No. 643 on RDS. Guidelines (EN50067) were formally proved and published in 1990. Since then, European countries have offered RDS broadcasting services [15].

   RDS-TMC (Radio Data Service/Traffic Message Channel) is one of the applications that uses RDS technology to achieve information release. Traffic information is encoded according to standards before broadcast and released using RDS technology. The on-board terminal equipment can receive the code information, and can choose the realization of the information, such as text, simple graphics, and language. Receiving RDS-TMC requires a special radio receiver, the most important part of which is the TMC card, which contains specific route information and so on. Assuming a traveler from the United Kingdom to Rome via Belgium, Germany, Switzerland, France, and Italy, the user only needs to buy or rent a TMC containing the route from the United Kingdom to Rome, and the receiver will automatically switch to another radio station that provides the TMC information stream so that users can receive the latest road information.

   The user can select the language type of the TMC receiving terminal to display traffic information. Users can also choose to filter information so that only information related to the route being driven is selected. There are many new transmission methods that can carry TMC services, including digital broadcasting, mobile Internet, paging, and GSM/GPRS mobile communication networks. RDS-TMC has the following characteristics:

   • The latest traffic information, real-time transmission.
   • Real-time getting hang of accidents, road works, and traffic jams.
   • The information is automatically filtered, and only the current route information is displayed.
   • The user chooses the language in which the information is expressed.
   • High-quality digital transmission.

Table 6.8 Introduction to European Traffic Information Service System

Nation	Name	Time	The main function	Key technology	Application status
Europe	SOCRATES	1989–1991	The traffic command center makes full use of the infrastructure of traditional cellular wireless phones to conduct two-way communication with vehicles in motion to realize the collection and release of traffic information	Mobile communication technology and GPS	Small-scale test
	RDS-TMC	1994 to now	The system encodes and publishes the traffic information according to the standard through the radio data system (RDS), and the on-board terminal receives the traffic information corresponding to the code through the broadcast	RDS technology, TMC encoding and transmission	Large-scale application in Europe
	CVIS	2006–2010	Create a comprehensive information platform that integrates hardware and software to achieve real-time traffic information acquisition and sharing, and improve traffic management efficiency, involving private cars, buses, and commercial transportation	Galileo satellite navigation, GPS, WLAN, and vehicle sensing	Small-scale test
	DRIVE C2X	2012–2013	A vehicle-road collaborative system test project covering seven European countries, through the tests of four typical applications of safety, traffic management, environmental protection, and business, to verify the traffic information service system based on vehicle/vehicle communication	ITS G5 access technology, system integration technology, privacy, and information security technology	Small-scale test

- Compatibility of receivers in Europe.
- European right to free or low-cost services.

For users, first RDS-TMC can receive silent FM data channels, which means users can listen to music or broadcast news, meanwhile, receiving TMC data transmission without interference. In addition, information is available for immediate display, and it is not necessary to wait for a fixed traffic news announcement or listening Scheme. TMC information services are continuous and will be submitted directly to drivers, unlike the roadside variable information signs just intermittently provide information services.

TMC traffic information service providers receive traffic flow, accidents, weather, and other information from the traffic monitoring system and emergency services from the traffic information center, traveler information requests, and other information. After processing, TMC traffic information is generated according to ALERT-C coding protocol. The basic broadcast traffic information that TMC users can accept is mainly as follows:

- Event description, detailed information on weather conditions or traffic problems and their severity.
- Location, region, road section, point location that be affected.
- Direction and range, indicating the location of the affected adjacent sections or points, and the affected traffic direction.
- Duration, expected duration of the problem.
- The divergence suggestion, whether the driver is advised to find alternative routes or not.

TMC traffic information services are most widely used in Europe and North America, with some European countries free of charge some traffic information services, including traffic accidents, traffic congestion, and bad weather. TMC free traffic information services almost availably covers the whole Europe well. There are also commercial TMC services, such as those increasingly used TMC - Built-in map navigation systems, as shown in Figs. 6.13, 6.14, and 6.15.

Most TMC products are based on vehicle GPS navigation system, RDS receivers receive TMC information for users' information and navigation services. Today, some small intersection-based information receivers can also provide intersection information services, as shown in Fig. 6.14.

With the advent of integrated or additional FM/RDS-TMC tuners, more and more manufacturers are starting to provide PDA-based navigation systems without permanent installation on vehicles Display device.

RDS-TMC technology originated in Europe and is also the most widely used in Europe. Since 1994, Sweden has implemented a nationwide coverage of RDS-TMC business. In 1995, Germany and the Netherlands have implemented it; in 1996 it developed to Paris. In 1997, France, Switzerland, Austria, Italy, and other countries have also implemented RDS-TMS. Currently, there are 18 countries that have implemented RDS-TMS project in Europe. After about 20 years of development, now RDS-TMC technology is mature and related products are available worldwide.

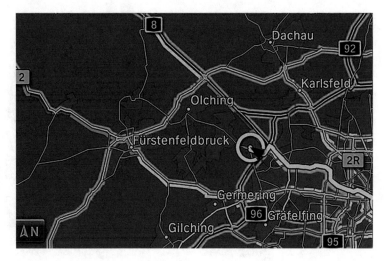

Fig. 6.13 Traffic condition display on TMC

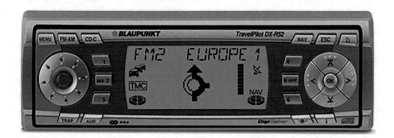

Fig. 6.14 Navigation service on TMC

The industrial scale of annual sales of tens of billions of euros has been formed. RDS-TMC service has been determined to be in Europe in the short term. Area's only large-scale transport information solution, the number of fully operational services increased from 10 to 15 in 2003, another 5 have entered the final stage of the experiment. In our country, the technology started lately, only individual regions and units established some special purpose RDS radio, but as a wireless broadcast data system which is low-cost, has mature technology and coverage widely, its breeding market opportunities are immeasurable.

2. Traffic-Master System. Traffic-master is a system which is centered in London, UK; and it has been widely practical on highway. The system effectively uses the existing paging network to provide traffic information. The system is composed of the Sensors for collecting highway traffic data, Control center for processing and sending information, the vehicle terminal that receives information and represents it on a display. And the system structure is shown in Fig. 6.16.

Fig. 6.15 TMC Device

The sensor emits two infrared rays to the front end and back end of the vehicle body respectively, and the speed of the vehicle is measured according to the time difference of the reverse wave reflected by the two rays through the vehicle. The computer of the sensor controller calculates the average speed of every 3 mins. If the speed is less than 30 mile/h, the information will be sent to the control center. The control center is composed of multiple computers, some are used to collect data, some are used to send processed information, and some are used to input text. The vehicle terminal includes a receiving device and a display device, which can display the low-speed range in all regions or locally enlarged regions. If it is switched to text mode, detailed information about accidents and construction can be known. If the vehicle device logins in advance to obtain the identification number, it can receive information for specific individuals and can be displayed on the screen, as shown in Fig. 6.17.

3. SOCRATE System. SOCRATES is the largest project in the European Community Research Program DRIVE. The purpose of SOCRATES is to study the use and feasibility of cellular radio-based road transport information system (RTI), and to propose the use of European cellular radio systems (such as GSM) as the basis of DRIVE.

   The overall concept is based on the collection, storage, processing of road traffic information in traffic control centers, and information flow between vehicles and roadside infrastructures: First send information to the driver and his in-car unit, and then collect real-time traffic data and other information from the vehicle.

   The concept of SOCRATES and its results and conclusions so far, including the main experimental sites in Gothenburg. Finally, the pilot projects in several

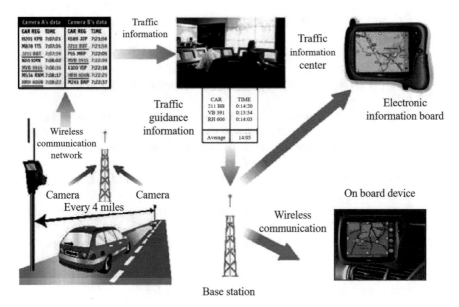

Fig. 6.16 Traffic-master system framework

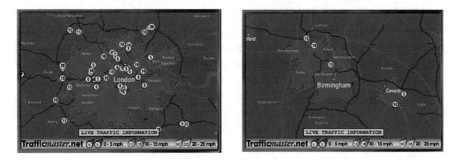

Fig. 6.17 Traffic-master on-board device display

European countries are described as the next step in commercialization of SOC-RATES in Europe in the 1990s.

SOCRATES is a transnational research project, which applies cellular radio to complete RTI system, including dynamic route guidance and many other applications. The project is the largest project (V1007) – "Special road infrastructure for vehicle safety in Europe" in the European Community Research Program DRIVE.

The main topics of this project are:

- Using high capacity duplex link to transmit the required information in cellular radio system.

- Dynamic update vehicle navigation equipment to provide interactive route guidance.
- Other business-related applications of duplex radio link.
- The project includes:
- It proves the feasibility of the dynamic route guidance system based on cellular radio.
- The displaying proposed communication link can also support other applications, such as danger warning, emergency call, automatic vehicle determination Location, road charges, hotel parking status, and so on.
- Change system capacity by number of users.
- It shows how the technology developed by cellular radio is used for the simplified equipment of DRIVE users.
- Providing computer simulation of data processing, vehicle information flow, and vehicle navigation system.
- Equipped with test sites and used laboratory models and prototype tests to verify theoretical predictions.
- Preliminary suggestions are proposed for the coherent systems of other applications with routing guidance and cellular radio support.

A basic assumption of IRTE is that the general communication infrastructure should provide all RTI applications services. The main applications considered in DRIVE and the main applications listed in SOCRATES are as follows.

Dynamic route orientation. This is the main RTI application, which is not only based on the detailed knowledge of the road network, but also provide in-car route advice to drivers based on current and projected traffic conditions. The estimation of the Laboratory for Transport and Road Research benefits of route guidance shows that there will be huge time saving and vehicle operating costs saving.

Connecting parking information will improve parking management, allowing drivers to find the most convenient parking space at the end of the journey, finally, parking spaces are reserved and even parking fees are paid.

Fleet management. IRTE will provide vehicle monitoring facilities for fleet operators, provide effective scheduling and control.

Public traffic management and information systems. Dynamic vehicle scheduling, passenger information service, and public transport fleet management will be included in IRTE.

Danger warning. Vehicle communication will provide warnings for drivers, including accidents, fog, ice, and other dangers.

Emergency call. The SOCRATES system will provide emergency call facilities for automatic collision sensing in an emergency activation.

Paging urgently. Allow individuals to pray through communications infrastructures for emergencies.

Automatic deduction. Road charges or capacity to provide services will be an important part of IRTE.

Driver information. Vehicle links will provide general information about the availability of specific services such as traffic conditions, specific problem points, hotels, and gas stations.

Tourism information. Information of particular interest to visitors will be included, including selection of scenic routes and the details of special events, etc.

Traffic management and traffic planning data. Improved IRTE traffic monitoring will provide important traffic data sources, it greatly saves the time of traditional traffic investigation.

Travel plan. Links between dynamic route guidance and public transport information will allow for pre-trip planning services so that travelers can not only plan time, but also plan travel mode.

The key to achieving IRTE is the extensive availability of two-way communication between vehicles and traffic information and control center networks. When this communication infrastructure can be used, most of the applications listed above will bring greater benefits. However, for most applications, the communication link does not need to be a one-to-one contact familiar to ordinary telephone calls.

The concept of SOCRATES will support all the applications listed above. In the current stage of development, dynamic route guidance is the focus and the most important application.

6.3.2   Traffic Information Service System of China

With the continuous development of China economy and the continuous improvement of people's living standards, people's demand for high-quality transportation services is becoming more and more urgent, and the expectation of service standard is getting higher and higher. In the reality, on the one hand, although road transportation infrastructure construction developing rapidly, the growth rate of transportation supply is still difficult to cope with the rapid growth of transportation demand brought about by the accelerated growth of motor vehicle ownership, and the contradiction between transportation supply and demand. On the other hand, compared with the development of road infrastructure, the development of our traffic information services is in a weaker position. Dynamic traffic and transportation information cannot be fully grasped, and there is a lack of effective release methods, which can improve the road foundation. Because of the failure to fully develop and effectively apply the transportation informatization method of facility operation efficiency, it has become the "bottleneck" that affects the improvement of transportation service level. Therefore, the development of public travel traffic information services has a very urgent need, has a good development prospect, is an important means to effectively improve the quality of public travel, and is an important area of traffic informatization that needs to be developed urgently. In 2005, the "Public Travel Traffic Information Service System" organized and implemented by the Ministry of Communications has become one of the three major informatization demonstration projects. The public travel traffic information service system is based on the information resources of the highway information resource integration system and the passenger station management information system, through the Internet, call centers, mobile phones, PDAs and other mobile terminals, traffic broadcasts, roadside broadcasts, graphic TV, and on-board terminals, variable information boards, warning signs, on-board rolling display screens, large screens, touch screens, and other display devices distributed in public places, to provide travelers with relatively complete travel information services; provide information on road conditions, emergencies, construction, along the way, weather, environment, etc., for driving travelers; provide ticketing, operation, station services, transfers, and other information for travelers who use public transportation; accordingly, travelers can make arrangements in advance travel plan, change travel route, make travel safer, more convenient and more reliable. At the same time, it integrates various information related to railways, civil aviation, tourism, meteorology, etc., and combines with radio and television to provide more comprehensive and more types of services, so that the public can feel the convenience of traffic information services. For details of Chinese Traffic Information Service System, see Table 6.9.

6.3.2.1   Traffic Public Travel Service Management System in Hubei Province

In recent years, the rapid development of the national economy, the rapid increase in the number of motor vehicles, and the frequent and diversified ways of public travel, and multimedia and multi-channel access to information have become the daily habit of the public. Therefore, the traveler information provided by the traffic management department service also is highly expected. How to strengthen the collection of traffic information, enrich the channels of information acquisition, improve the information release to meet the actual needs of the public, and facilitate travel are a new topics facing the transportation department. Practice has proved that the construction of the public travel service management system can better meet the needs of the public for travel information. Taking Hubei Province as an example, the basic situation of the public travel service management system [19] will be introduced.

1. System construction. The Hubei Provincial Traffic Public Travel Service Management System belongs to the second phase of the highway traffic information resource integration and service engineering application software development project carried out by the Hubei Provincial Department of Transportation. On the basis of the first phase of the construction, the system takes the integration of traffic information resources as the construction concept, and aims to further enhance service awareness, improve service capabilities, and improve service levels. It will change the convenience of managers to the convenience of the people and serve the people for safe and convenient travel.

   The public travel service management system of Hubei Province integrates existing highway digital information resources, such as highway road monitoring

Table 6.9 Introduction of domestic traffic information service system

Area	Information Service System (Platform)	Main highway traffic information service functions and content	Operation management entity
Nationwide	China highway information service network	Highway blockage, weather warning, highway travel planning, basic highway information, highway traffic information, highway map	Network monitoring and emergency response Center of the Ministry of transport
Beijing	Public travel network of Beijing	–	–
Tianjin	Expressway travel service network of Tianjin	Traffic conditions, travel planning, travel reference, travel guide	Expressway network management command Center of Tianjin
Hebei	Expressway travel information service system in Hebei Province	Real-time road conditions, route planning, attraction and hotel query, high-speed weather, high-speed service, ETC service	Highway administration command and dispatch Center of Hebei Province
Shanxi	Public travel traffic information service system of Shanxi Province	Traffic map, dynamic road conditions, unlimited traveling, fare inquiry, weather inquiry	Shanxi provincial department of transportation
Inner Mongolia	Public travel information service system of Inner Mongolia	Traffic travel planning, Inner Mongolia map, travel scenic spot query, passenger station query, gas station query, service area query, maintenance station query, road condition query, weather forecast	Region transportation and communication information center Inner Mongolia autonomous
Liaoning	Public travel information service system for expressways in Liaoning Province	Destination query, service area query, weather query, road condition query	Department of Transportation in Liaoning provincial
Jilin	Transportation public travel service network	Dynamic road conditions, travel planning, road passenger transportation High-speed toll, traffic tour sub-map, travel assistance (high-speed weather, traffic information), message board	Department of Transportation in Jilin provincial
Heilongjiang	No related system	–	–
Shanghai	Shanghai transportation network	Route planning, road query, facility query, real-time traffic emergencies, road construction, client download	Shanghai urban and rural construction and transportation development research institute

(continued)

Table 6.9 (continued)

Area	Information Service System (Platform)	Main highway traffic information service functions and content	Operation management entity
Jiangsu	Jiangsu high-speed public travel service network	Map line, real-time road conditions, high-speed weather, real-time traffic, route query, service facilities, high-speed snapshot, rate inquiry, card processing, online Q&A	Jiangsu expressway network operation management center
Zhejiang	Zhejiang transportation	Real-time road conditions, road construction, route query, and planning Inquiry about transportation expenses (tolls), traffic broadcast, service area location, travel Weibo	Zhejiang provincial department of transportation information center
Anhui	No related system	–	–
Fujian	Fujian transportation information and communication center	Traffic meteorology, travel dynamics, real-time road conditions Hotspot navigation, road facilities, travel common sense, travel inquiry, illegal inquiry	Fujian provincial transportation and communication information center
Jiangxi	Public travel service network of Jiangxi Province	Travel guide, map of Jiangxi, red tourism, traffic status, travel reference	Information Center of Jiangxi Provincial Department of transportation
Shandong	Shandong traffic travel network	Electronic map, road network indication, traffic information, high-speed infrastructure, and fee query Travel, inquiry center, Sina Weibo, customer phone	Shandong provincial department of transportation
Henan	Henan expressway travel service network	Real-time traffic, destination inquiry, route inquiry, service, parking area, Toll Station, civilized demonstration road, high speed common sense, customer phone	Highway Administration Bureau of Henan Provincial Department of transportation
Hubei	Hubei provincial transportation public travel service network	Traffic map, dynamic road conditions, highway information, travel planning, fee inquiry, Traffic & Tourism, traffic yellow pages, personalized service, call center, online broadcasting	Hubei provincial department of transportation

(continued)

Table 6.9 (continued)

Area	Information Service System (Platform)	Main highway traffic information service functions and content	Operation management entity
Hunan	Hunan expressway information service network	Weather forecast, road condition information, driving guide Electronic map, introduction to attractions, illegal inquiries, tolls, travel common sense, ETC application, report and complain	Hunan provincial expressway administration
Guangdong	Public travel information service system of Guangdong Province	Electronic map, travel guide, traffic query, traffic news, toll inquiry, suggestion feedback, road construction	Guangdong transportation archives information management center
Guangxi	Guangxi expressway travel information service network	Road network diagram, high-speed map, road condition information, rate query, service facilities, travel guide, travel guide	Guangxi Zhuang autonomous region expressway administration
Hainan	Hainan provincial transportation public travel information service system	Traffic road conditions, travel meteorology, expressway service area, tourist attractions	Hainan provincial department of transportation
Chongqing	Chongqing transportation public travel service network	Real-time road conditions, road maps, dynamic data, travel routes	Chongqing municipal traffic commission
Sichuan	Sichuan traffic public travel (website) service system	Expressway introduction, travel planning, toll query, road conditions Electronic map, my trip (line collection), scenic spot query, and route recommendation	Information Center of Sichuan Provincial Department of transportation
Guizhou	Guizhou transportation public travel (website) service network	Traffic information, travel planning Electronic map, travel weather, high-speed service query, general knowledge of travel, ETC introduction and use	Guizhou transportation information center
Yunnan	No related system	–	–
Xizang	No related system	–	–
Shaanxi	Shaanxi transportation public travel service network	Sanqin Tong card, road condition information, traffic announcement Traffic map, fare inquiry, miles inquiry, transportation & tourism, integrated services, travel common sense	Shaanxi expressway toll management center

(continued)

Table 6.9 (continued)

Area	Information Service System (Platform)	Main highway traffic information service functions and content	Operation management entity
Gansu	No related system	–	–
Qinghai	Qinghai traffic travel information service network	Maps Qinghai (road conditions, incidents, flow, traffic) self-driving travel, real-time traffic, toll guide and query, travel weather, SMS platform, tourist attractions	Qinghai provincial department of transportation
Ningxia	No related system	–	–
Xinjiang	Xinjiang Uygur autonomous region traffic public travel service system	Dynamic road conditions, travel planning, traffic information Transportation & Tourism, illegal inquiries, compensation standards Call center, SMS platform, online broadcasting	Xinjiang department of transportation

video, road condition and weather monitoring information, passenger station management information, and refines and organizes it according to public travel habits. System functions close to actual needs provide information services for travelers through the Internet, customer service hotlines, broadcasting, and roadside variable information boards. As a convenient and efficient service method for public travel, the public travel traffic information service system provides call center services. Through the above-mentioned resource integration, the public travel traffic information service system call center service platform will use rich travel information to provide real-time services to travelers, such as consulting travel weather conditions, travel routes, road conditions, vehicles, and ticket sales for civil aviation. Addresses of outlets, asking for directions, surrounding environmental facilities (hotels, hotels, attractions, shopping places, etc.), through the answers to these questions, truly meet the public's needs for travel services.

2. System functions. The public travel service management system has enriched public travel service information, enhanced the depth and breadth of traffic information services, and provided the public with travel information services such as traffic maps, weather, road conditions, tourist attractions, route planning, and schedules, which greatly facilitated the public for travel, the functional module structure diagram, and related function descriptions are shown in Fig. 6.18.

Compared with the traditional public travel information service system, the traffic public travel service management system of Hubei Province adopts advanced network technology, computer technology, and streaming technology in data collection, database construction and management, data exchange and sharing, and

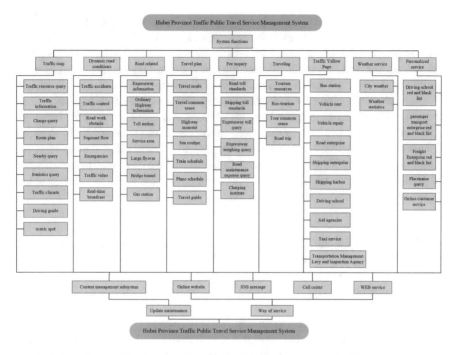

Fig. 6.18 Function module structure of public travel service management system

information release. Media technology and WebGIS technology, and breakthroughs in key technologies such as linear referencing, dynamic segmentation, and station number positioning, have ensured the usability, safety, stability, and scalability of system construction. Its specific characteristics are reflected in the following aspects.

The system integrates multi-service application systems such as expressway traffic adjustment system, expressway toll collection system, operation management system, and video monitoring system, which has realized the integration of transportation management, expressway, port and navigation, railway, aviation, and other information resources with meteorological and other related industry information resources in the transportation industry, providing travelers with all-round, multi-angle, and more accurate travel information and breaking the previous industry information-sharing barriers.

In terms of data presentation, GIS technology is more used in the system. Nearly all queries can be completed through GIS system, and all GIS-related information can be visually displayed on the GIS system, thus improving the friendliness and operability of the system and allowing users to have a more intuitive feeling about the information they query.

In order to facilitate the public to quickly obtain travel information from multiple channels, the system provides a variety of information release methods, such as mobile phone message, call hotline, broadcast, etc., and combines with the public before and during travel traffic information needs to meet the needs of various travel people to obtain travel information anytime and anywhere.

Since the completion of the public transportation travel service management system in Hubei province, it has been widely used and become one of the important ways of traveler for traffic information, its rich and real-time traffic information, friendly access interface system won the praise of the society. The system has 554,759 visitors in 2013, it provides convenience for the traveler. The construction of public travel information service system has greatly enhanced the service ability of the transportation industry to the public.

6.3.2.2 The Representative Enterprise-Level Road Traffic Information Service System in China

The representative enterprise-level road traffic information service system in China is shown in Table 6.10.

Take AutoNavi navigation as an example to introduce [21].

AutoNavi navigation has several plates, including self-driving navigation, travel sharing, public travel and information services.

- ·Self-driving navigation: Real-time dynamic road conditions, optimal route for the whole journey, 1.8 million electronic eyes and voice navigation.
- ·Travel sharing: ride-hailing services, hitch rides.
- Public transportation: Real-time bus, shared bikes, electric bikes, hybrid planning.
- Information Services: Discover surrounding areas, mass accurate dynamic data, search for different places, and integrate data sources.

1. AutoNavi has carried out extensive cooperation with local transport departments. Actively with the traffic administrative department of the map data sharing and integration, in the field of traffic data, Scott map has introduced a traffic information public service platform, traffic police platform, relying on traffic big data cloud for relevant institutions provide cities plugging point, business circle road, authority traffic events, plugging point anomaly monitoring traffic information analysis, not only improve the efficiency of the public travel, but also assisted the policies of the government management, to develop a more reasonable improvement measures, reduce urban traffic congestion.

   As of April 2015, Scott map has joint Beijing, Guangzhou, Shenzhen, Tianjin, Shenyang, Qingdao, Dalian, Wuxi, etc. More than 20 traffic administrative departments of the local government and Beijing communication radio and other authoritative media organization jointly launched the AutoNavi traffic information public service platform, providing transportation organizations with road traffic information analysis and avoiding congestion based on big data. On

Table 6.10 China's representative enterprise-level road traffic information service system [20]

Classes	Outstanding companies	Outstanding systems	Main service functions	Service range
Traditional companies	NAVINFO	SiWei map	Electronic map, navigation service	Urban roads and freeway
	Cennavi	Road condition traffic eye	Real-time traffic flow, simple graphical traffic information, traffic incident information, historical data, dynamic route planning, weather, map, traffic index	Urban roads and freeway
	ZhangCheng tech	ZhangCheng LuKuangTong	Traffic information, traffic events, dynamic route planning, violation inquiry, weather forecast	Urban roads
Navigation companies	AutoNavi	AMAP	AR real-time navigation, 3D real-time navigation, online and offline map, real-time dynamic road conditions display, cloud data synchronization, voice broadcast, monitoring reminding	Urban roads and freeway
Internet companies	Baidu	Baidu map	Map display, information search, location, navigation	Urban roads and freeway
	Tencent	WeChat	Three functional modules: High-speed road conditions, high-speed services, and more	Freeway
Freeway companies	Shandong freeway group	Yi freeway	Road information, monitoring overview, highway facilities/highway maps	Freeway
	Chongqing freeway group	Chongqing freeway	Monitor quick view, real-time road conditions, nearby navigation, route tolls, service area, and Chongqing specialty recommendation	Freeway

May 29, 2015, AutoNavi released a new version and launched the "Traffic Police Platform" project. Traffic departments can immediately release official information such as road control, construction, and traffic restriction at the end of the traffic number through the Amap traffic police platform. Shenzhen took the lead in entering the Amap traffic police platform, and Guangzhou, Dalian, Shenyang, and other cities also became the second batch of occupants, jointly exploring the construction of "Internet + Traffic" with Amap.

As of September 2016, AutoNavi has reached strategic cooperation with more than 40 local traffic management departments in Beijing, Guangzhou, Shenzhen, Wuhan, Nanjing, Hangzhou, Tianjin, Shenyang, Dalian, Wuxi, Qingdao, and Chongqing, and has reached business cooperation with more than 70 local traffic management departments.

Fig. 6.19 Three-dimensional real-world navigation map

2. Local cooperation. Water map. In many cities with frequent rainfall, the *Water Map* was first developed by AMap in China. In conjunction with many local traffic police departments, the city water map is released to ensure traffic safety during the flood season. The water map, which can be checked on normal days, also in case of rainstorm. Amap will timely push information to remind users to bypass the water point. The data source of the water map is the authoritative data provided by the traffic police. At the same time, AutoNavi map users can also upload their own water points around the information, through the official review can be online. In addition, AutoNavi map also jointly launched the traffic police around the "find the water point" award activities, to ensure that the data release is comprehensive, to provide reference for citizens to travel. In June 2016, the waterlogging maps pioneered by AutoNavi were launched in Beijing, Shanghai, Guangzhou, Shenzhen, Wuhan, Ningbo, Zhengzhou, Nanjing, Hangzhou, and many other places.

3D real navigation. In December 2015, AutoNavi officially launched the 3D real-time navigation function, which simulates real road scenes and driving routes in navigation products through the establishment of 3D real-time data model, so that drivers can get more immersive navigation guidance. Amap successively realized online 3D imaging navigation in Beijing, Shanghai, Guangzhou, Shenzhen, Xiamen, Hangzhou, compared to the last plane map navigation, 3D navigation can give drivers very clear 3D scene reduction and more accurately depict direct driving route in complex intersection and overpass area, as shown in Fig. 6.19.

Real road map. AutoNavi map and the traffic police departments of Guangzhou, Jiangmen, and other cities have jointly launched the function of real traffic picture, so that users can view the real traffic pictures provided by the traffic police, so as to more clearly understand the real-time traffic information, such as

Fig. 6.20 Real road map

road congestion, traffic accidents, road construction, and other road surface conditions, as shown in Fig. 6.20.

3. AutoNavi is in Chengdu. Based on the support of their own data, the functions of AutoNavi includes road traffic, electronic police, traffic control, traffic incidents, to analyze, high-speed traffic, congestion abnormal congestion, traffic information analysis and remind roads, traffic safety tips, real-time dynamic traffic event publishing, congestion ranking, real-time bus, etc.

6.3.2.3 Traffic Information Service System Based on Mobile Internet

1. System introduction. Intelligent Traffic Information Service System (hereinafter referred to as Intelligent Traffic Information Service System, as shown in Fig. 6.21), based on wireless video surveillance system, combines road video resources with public traffic travel demands to provide real-time, accurate, and intuitive road traffic information service for mobile phone users. When the user is driving or walking in the city, he or she can inquire the best route, the surrounding traffic information, the bus and subway information and the real-time arrival

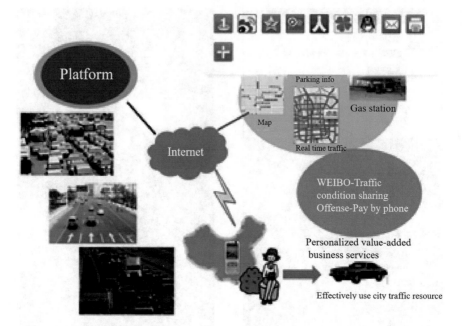

Fig. 6.21 Intelligent traffic information service system

information of the location, etc., through the input and selection of the mobile phone interface.

2. System architecture. Intelligent traffic information service system is divided into intelligent traffic information service system mobile client software and background business system two parts. According to the way of information acquisition, transmission, and use, the background business system can be divided into three levels: the basic layer, the shared information layer, and the service layer. Build a comprehensive information service platform in the command center to integrate all subsystems, as shown in Fig. 6.22.

The basic layer (as shown in Fig. 6.23) mainly includes the acquisition and transmission of various traffic information, signal control, operational vehicle management, electronic toll collection, emergency handling, traffic information management and release system, on-board navigation and positioning system, etc., to provide effective help for travelers to choose travel plans. Traffic information management and release system collect, transmit, and process dynamic traffic information through the geographic information system, provide real-time and predictive traffic information services for commercial transportation companies, government agencies, and the general public.

The shared information layer (as shown in Fig. 6.24) refers to a comprehensive urban intelligent transportation information service platform formed by the comprehensive integration of various components of the functional layer. It

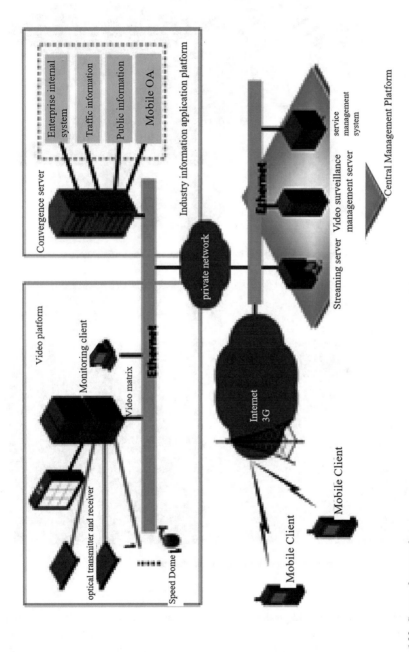

Fig. 6.22 System framework

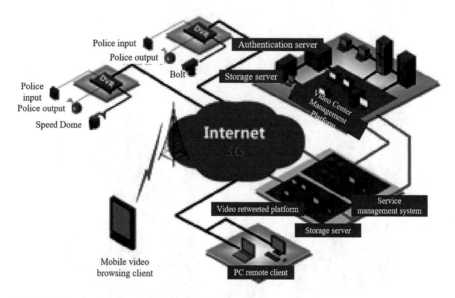

Fig. 6.23 The basic layer of the back-end business system

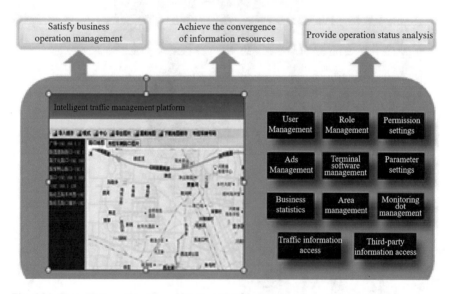

Fig. 6.24 Shared information layer of back-end business system

integrates, analyzes, and processes various traffic information collected from the basic layer, and is shared by various upper-level services. It also provides a basis for the cross-system linkage of traffic police, traffic, public security, and other systems. The shared information layer is mainly supported by the geographic

information system platform, which provides powerful assistance for traffic control, design of transportation schemes, road plan and design, etc.

The service layer is the highest layer of the entire system, and is the interface for the system to interact with travelers and traffic managers. The system provides a control plan for road control equipment through the service layer, provides road condition information for travelers, and assigns management tasks to traffic managers. At the same time, the service layer is also responsible for receiving information from travelers and managers, such as traffic accident alarms and road condition information provided by traffic managers. In addition, through the service layer, the best travel plan can be provided according to the requirements of travelers, so as to ensure the smooth flow of roads as much as possible and to improve the efficiency of the entire transportation system (Fig. 6.25).

3. System function. Mobile phone users can access the client software to view the road conditions pictures of the entire city, query the real-time road conditions of the city's main roads and bridges, highway event information, the optimal driving route between designated starting points, and the predicted driving time, etc., providing more enrichment and comprehensive traffic road services for the many users, and fully meet the traffic travel needs of customers (Figs. 6.26 and 6.27).

View road information. The user can view the road traffic video of the driving route in advance or in real time through the client, and learn the road traffic information at any time.

Dynamic traffic report. User positioning is carried out through GPS + base station + WIFI, proactively reminding the congestion of the line ahead according to the driving route, and providing congestion information broadcast in the form of voice, text, and images.

Reminder of parking lot vacancy. Obtain the location and dynamic vacancy information of the city's main parking lots, and actively remind the target parking

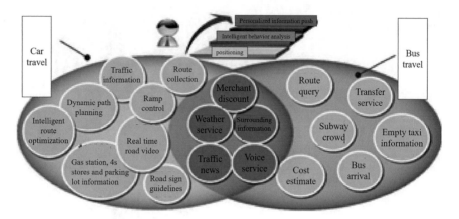

Fig. 6.26 System function

lot vacancy information with voice according to the user's destination and driving route.

Line reminder customization. Users can customize the road condition reminder service of the route. The system actively broadcasts the road condition information of the selected route every day according to the user's customization.

Inquiry of traffic service information. Provide traffic service information such as gas stations, local weather, and violations.

Bus station service system includes traffic geographic information query system, electronic stop sign system and waiting infrastructure, etc. The electronic stop sign includes a communication receiving module and a data processing module. It is connected to the monitoring and dispatching center through a wireless or wired system. Its basic function is to provide passengers with the operating conditions of public buses on the bus line. The transportation geographic information query system uses transportation GIS as the basic platform to provide travelers with various public transportation information and service information, so that passengers can get the information they need and feel the humanized information service in the whole process from waiting for the bus to arriving at the destination by bus.

Public transport information service system mainly includes three types of traffic information services: information on the driving status of public buses in the system (time, location, and speed); bus operating information (different departure intervals, the number of passengers in buses and other bus stops along the way, and emergencies); information on traffic conditions related to road system and transfer system. Bus information service system can provide the information of the required content at the time and place where the traffic users need the information, so that the public transportation users have sufficient basis for decision-making and judgment.

Fig. 6.27 System function

Parking guidance system. In order to reduce unnecessary detours and invalid driving distances of vehicles, and avoid slow driving and distraction of drivers caused by looking for parking lots, it is necessary to develop and apply parking guidance systems. There are usually two levels of parking guidance. The first level is a parking guidance and information service in a large area. It provides traffic travelers with information on the distribution of parking facilities in the destination area, distances, and the current utilization status of parking facilities, so that travelers can choose the mode of transportation. And the decision-making and judgment of the parking area; the secondary guidance is to induce the route of the specific parking facility and provide information on the current use of the parking facility, which is convenient for users to choose and reach the parking lot smoothly.

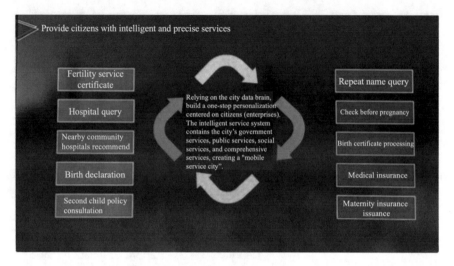

Fig. 6.28 Hangzhou City Data Brain

6.3.2.4   Hangzhou City Brain Information System

1. Definition of Hangzhou City Data Brain. The Urban Data Brain is a platform-based artificial intelligence center constructed based on the urban life form theory of urban science and the "Internet + modern governance" thinking, innovatively applying big data, cloud computing, artificial intelligence, and other cutting-edge technologies. It integrates and collects government, enterprise, and social data, integrating calculations in the field of urban governance to realize the functions of vital signs perception, public resource allocation, macro decision-making and command, accidents prediction and warning, and "urban disease" management [22], as Fig. 6.28 shows.

   This means that in the near future, the Hangzhou City Data Brain will bring convenience to everyone living in Hangzhou. For example, road dynamics, real-time parking space data, medical outpatient traffic, etc., can all be queried with one-click in the handheld "urban data brain" in the future.

2. The direction of urban data brain construction. In the past 2 years, the transportation system of Hangzhou City Data Brain has been piloted in Xiaoshan District and some roads in downtown Hangzhou. The average delay of the pilot Zhonghe-Shangtang viaduct was reduced by 15.3%; there were 208 intersections in the coverage area of Xiaoshan District, and the overall unblocked ratio increased by 5%. The city data brain realizes the batch exchange and sharing of 1.125 billion pieces of data.

   As a new infrastructure that supports the sustainable development of the city, the city data brain is not only to alleviate congestion, but also to comprehensively consider various fields related to city operation and management, and to fully open up data to improve the ability of various business systems.

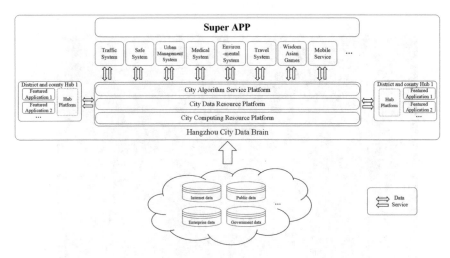

Fig. 6.29 The framework of Hangzhou City Data Brain

(a) Platform-based artificial intelligence hub. The overall architecture of Hangzhou City Data Brain includes brain platforms (including computing resource platforms, data resource platforms, and algorithm service platforms), industry systems, super applications, district and county centers, etc., as shown in Fig. 6.29.

   The brain platform is currently under construction and is expected to be online and trial run before the end of 2020.

(b) Urban brain planning in recent years. At present, the data of 16 city-level departments are incorporated into the brain platform; and a database of topics such as urban management and transportation has been built.

   In 2018, based on the stable operation of the transportation system V1.0, expand the scope of the pilot and further optimize the algorithm; improve the urban data brain data platform, build the urban data brain transportation system V2.0, and expand the coverage (Fig. 6.30).

   From 2019 to 2021, comprehensively promote the construction of the urban data brain transportation system in the main urban area of Hangzhou, build a safety system, and promote the construction of smart Asian Games and urban management, medical care, tourism, environmental protection, and other fields.

   In 2022, realize the full coverage of the "urban data brain" traffic governance, basically complete the construction of the "urban data brain" in various industry systems and put them into actual operation; carry out cross-industry and cross-domain data resource development, awaken more "sleeping" data, and serve people's production and life, so that the value of data can be fully reflected, basically complete the construction of the main scenes of Hangzhou City Data Brain, and provide data support and service guarantee for the successful hosting of the 2022 Asian Games.

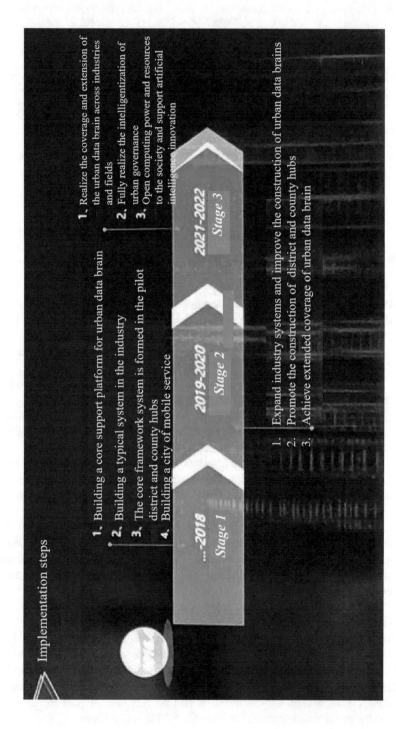

Fig. 6.30 City brain planning

(c) Establish multiple barriers to ensure data security. During the construction and operation of the urban data brain, the cloud service provider and the Alibaba security team are jointly responsible for the underlying data security, with a comprehensive data security strategy configured, and key sensitive data "available and invisible"; domestic leading information security agency is introduced to conduct third-party data security management control, audits, and realizes full life cycle monitoring from data generation, collection, storage, exchange, utilization to destruction, to ensure that the ins and outs of each piece of data are clear, the flow is compliant, to block and warn for abnormal situations and ensure the security of data resources.

3. The application of urban brain in Hangzhou.

(a) Four steps of urban traffic development in Hangzhou. Hangzhou, like the major cities in the world, has gone through the following steps of development in the whole process of urban development.

Step 1: Standardization of transportation facilities

With the intelligent distribution of urban road traffic signal lights and the standardization of traffic signs and lines, the whole urban area of traffic signs and lines, traffic signals have been combed and upgraded, promoting the standardization of traffic facilities, standardization.

Step 2: The channelization of traffic organization.

After the standardization of traffic facilities, some unreasonable lane distribution, improper signal timing, and interference from the entrances and exits of some road sections have been collected. A series of traffic organization optimization works, such as variable lanes at intersections, left-turn straight waiting areas, tidal lanes, and micro-circulation traffic, have been carried out.

Step 3: The orderly stage of traffic order regulation.

After the standardization of traffic facilities and the optimization of traffic organization, the regulation of traffic order was carried out. Transport facilities have been optimized, and more than 14 million traffic violations were punished in 2017, ranking first among 36 major cities in the country for six consecutive months. By improving the road traffic order, the traffic efficiency has been significantly improved.

Step 4: The intelligent phase of big data, cloud computing, and artificial intelligence.

With the expansion and rapid changes of the city, the contradiction of inadequate police numerical strength is becoming more and more prominent. In this context, Hangzhou uses artificial intelligence and machine intelligence to more accurately watch every detail of urban traffic, and more accurately regulate each stage of urban traffic nodes. It is estimated that 464 km of expressways will be built in Hangzhou by 2022, and 281 km of new expressways will be built in 4 years. The total length of the rail network is 446 km, and 329 km will be built in 4 years.

(b) The city data brain has been piloted in two places in Hangzhou.

Pilot 1: Zhonghe-Shangtang Viaduct and Moganshan Road.

Before 2015, the Zhonghe-Shangtang Viaduct was rated by the AutoNavi Map App as one of the most congested expressways in China with the lowest speed during rush hours. After half a year's practice of urban data brain, the average delay of Zhonghe-Shangtang Expressway was reduced by 15.3%, and the travel time was saved by 4.6 mins. Average delays on Moganshan Road were reduced by 8.5% and travel time was saved by 1 min. Now during the daytime, the speed of the entire expressway is basically maintained at 40 ~ 45 km/h.

Pilot 2: 5 square kilometers area of Xiaoshan District.

In the pilot area of 5 square kilometers in Xiaoshan District, the average traffic speed increased by more than 15% and the average travel time was saved by 3 min. The speed of special vehicles such as 120, 119, and 110 increased by more than 50% and the rescue time was reduced by more than 7 min.

According to data from the AutoNavi Map App, Hangzhou's congestion ranking dropped from eighth place in 2016 to 48th in 2017, ranking the first in the country in terms of congestion alleviation, and the trend of traffic congestion decline has been formed.

(c) In the pilot process of Hangzhou City Data Brain, the following four tasks are mainly carried out.

• Plan four routes.

Set up four realization routes: comprehensive perception, strategy leading, intelligent imitation, feedback system.

On the strategic level, we should pay attention to the strategic and macroscopic management of urban traffic, realize the transformation of the characteristics of traffic points, lines, and planes to the macro indicators, and solve the problems of repetition and misalignment in traffic management and safety prevention and control.

On the tactical level, the data technology is used as the support to find out the influence factor and the correlation factor of the non-direct relationship of traffic management, and the traffic characteristic analysis module is established based on the detection technology such as video, signal, ramp, and section. Using the regularity, repetition, and periodicity of statistical science to predict traffic flow, congestion, security risks, and other trends, then achieve urban traffic management early warning, pre-judgment, pre-decision.

• Data fusion to find traffic jams and disorderly spots.

The first is to carry on the analysis to the blocking point. Based on the analysis of traffic flow theory and traffic characteristics, external data fusion and traffic police after the internal resources to build traffic blocking spot algorithm, by counting imbalance degree and delay ratio, etc., detect the intersection, expressway, ramp, and road section every 2 min just like CT slices, find urban traffic blocking spot compared with historical data. For

Fig. 6.31   63 alarms a day at the entrance of Xinhua Hospital

example, in the viaduct pilot, all the ramps were screened in slices and compared with historical data to determine which ramps should be sealed off and which should be signaled.

The second is video spotting. When 144 monitors are included in the data analysis, the monitoring points will automatically alarm. In the original traffic organization, the treatment of random points and blocking points and the optimization of traffic required manual judgment on the spot, which led to the failure of real-time discovery and real-time governance. Automatic alarm through video detection can be more targeted and efficient to control the blocking and random points, as shown in Figs. 6.31 and 6.32.

- Speed dome monitoring instead of traffic police road inspection to detect all kinds of traffic incidents.

In the construction of urban data brain, extract empirical features to form algorithms and scenes to realize machine intelligence, such as traffic accidents, vehicle broken down, illegal parking, etc., as shown in Fig. 6.33.

Hangzhou's road mileage is more than 1900 km, with more than 1700 traffic light junctions and less than 1000 traffic police on the road. According to the current schedule of three shifts, only about 260 intersections can be inspected at most during peak hours, and more than 1500 intersections are out of control. Therefore, the different attributes of each event can be accurately judged by monitoring the automatic patrolling of substitute personnel on the road surface, so as to automatically alarm traffic events such as traffic jam points, chaotic points and traffic

Fig. 6.32 156 alarms a day at the entrance of the People's Hospital

Fig. 6.33 Video surveillance automatic inspection

accidents, traffic jams, and so on. At present, Hangzhou urban area has more than 3400 monitoring, if all automatic patrol, then the whole video monitoring will produce more than 50,000 alarm data, an average of more than 200 every day.

Innovate the application of bayonet equipment. Originally, the bayonet device was mainly used for vehicle track and face search in public security accidents. Now, the bayonet application is upgraded to off-site enforcement through the urban data brain, including drivers talking on mobile phones while driving, not using seat belts, key vehicle alarm, etc., all of which can be extracted from the bayonet system. This off-site enforcement method through the extraction of bayonet data can liberate the limited police force to a large extent, so that the traffic police can quickly deal with accidents and broken down vehicles, and quickly manage pedestrians and motor vehicles.

Use city big data to innovate police affairs. The original traditional traffic police service mode based on fixed point is innovated and upgraded to the current mode of intelligent supervision of the intersection and quick processing of road section rolling. For this reason, the traffic police mobile team is set up to quickly deal with all kinds of incidents discovered by the urban data brain alarm. Calculating according to the 260 police, for example, in the rush hour, all as a traffic police team, each policeman could be in charge of 7 km area in Hangzhou, that is to say, every police can reach as far as 3.5 km place, according to the motorcycle, they can arrive at the scene as long as 3–4 mins, then dispose all kinds of events.

6.4 Summary

This chapter first introduces the concept and meaning of the traffic information service system, briefly describes the development background and course, as well as its function, characteristics, effects, and classification. The service content, system composition, and working principle of the traffic information service system are further elaborated, and have discussed the key technology and development of traffic information service system. On the basis of expounding the basic concept, composition, and key technology, taking Europe, America, and Japan as examples, this paper introduces the construction, operation, and development of traffic information system in the world at present, and introduces several typical traffic information service systems in detail. Finally, the application of traffic information service system in China is introduced, and several typical examples are given.

Initially, the development of traffic information service system is mostly a small-scale system with simple information service content and single service mode within the scope of smaller areas. With the continuous development of ITS, the concept of intelligent transportation system has been expanded from the traditional road traffic scope to railway, aviation, water transport, and other fields, and the concept and content of traffic information service has also been expanded to various kinds of information service content of various transportation modes. And, with the development of the other subsystems in the field of ITS and the demands of ITS integrated, systematic and the development of intelligent, transportation information service system is no longer an independent of ITS subsystems, instead, it participates in the entire ITS system on the basis of a comprehensive traffic information platform, and improves the intelligent and systematic level of the system through integration with other ITS subsystems, thereby improving the function and efficiency of ITS to a greater extent, and improving integration the safety and efficiency of the transportation system to promote the improvement of the intelligence level of the transportation system.

References

1. Lu XP, Li RM, Zhu Y (2004) Introduction to Intelligent Transportation System[M]. CHINA RAILWAY PUBLISHING HOUSE, Beijing. (Published in Chinese)
2. Huang W, Chen LD (2003) Introduction to Intelligent Transportation System[M]. China Communications Press, Beijing. (Published in Chinese)
3. Yang ZS (2003) Generality of Intelligent Transportation System[M]. China Communications Press, Beijing. (Published in Chinese)
4. Yang ZS (2004) Urban Traffic Flow Guidance System[M]. CHINA RAILWAY PUBLISHING HOUSE, Beijing. (Published in Chinese)
5. ITS AMERICA: http://www.itsa.org/
6. Taylor KB (2006) TravTek – Information and services center[C]. In: Vehicle Navigation and Information Systems Conference. IEEE, Piscataway, pp 763–774
7. TRAVELINK. https://www.travelink.com/
8. INTELLIDRIVE. http://www.intellidrive.co.za/
9. Grau G, Heyn T, Vaccaro A et al (2012) Developing ITS Services for the Open SafeTRIP Platform[J]. Procedia – Social and Behavioral Sciences 48:2728–2737
10. Celidonio M, Zenobio DD, Fionda E et al (2012) SafeTRIP: A Bi-directional Communication System Operating in S-band for Road Safety and Incident Prevention[C]. In: Vtc. DLR, pp 1–6
11. Bell E, Dinning M, Kay M et al (2008) Gearing up for SafeTrip-21[J]. Public Roads 72(2)
12. Vehicle Information and Communication System Center: http://www.vics.or.jp/
13. HuaQiang Electronics Net. Intelligent Transportation System (ITS) development status in Japan [EB/OL]. http://tech.hqew.com/fangan_1201373 (Published in Chinese)
14. Introduction to Japan's advanced roads: http://news.makepolo.com/5967431.html (Published in Chinese)
15. Traffic Message Channel(TMC) Introduction. https://wenku.baidu.com/view/ce6bc8a7f524ccbff1218470.html (Published in Chinese)
16. ERTICO. DRIVE C2X. http://ertico.com/projects/drive-c2x/#
17. Flament M (2011) Using EuroFOT Operational Experience in Drive C2X[C]//. World Congress on Intelligent Transport Systems.
18. European Commission. News from CVIS. http://www.cvisproject.org/
19. Hubei Provincial Transportation Public Travel Service Network: http://gzcx.hbjt.gov.cn:808/ (Published in Chinese)
20. China Intelligent Transportation Systems Association. China's Intelligent Transportation Industry Development Yearbook [M]. IChina, 2016(23):206-221. (Published in Chinese)
21. AMAP. http://a.autonavi.com/outer/index.jsp (Published in Chinese)
22. ZHEJIANG NEWS. Hangzhou "City Brain": https://zj.zjol.com.cn/news/771594.html. (Published in Chinese)

Chapter 7
Intelligent Traffic Management and Control Technology

ZHANG Cunbao, CHEN Feng, QIN Ruiyang, LI Xuemei, and WANG Houyi

Transportation system is a complex and huge system which is composed of people, vehicles, roads, and their environment. Therefore, it is limited to solving traffic problems unilaterally only from people, vehicles, and roads. It is necessary to start from the point of view of the system, comprehensively consider factors such as people, vehicles, and roads, and apply various advanced technologies and scientific methods to traffic management and control. On the one hand, the "node" in traffic systems, such as people and vehicles, has a full understanding of the system. On the other hand, the traffic managers can understand and monitor the state of the traffic system in real time, and make scientific management and optimal regulation, so that the system can work at maximum efficiency and realize the intelligence of the whole system.

7.1 Overview of Intelligent Traffic Management System (ITMS)

7.1.1 Concept and Development Status of ITMS

Intelligent Traffic Management System (ITMS) is an important part of intelligent transportation systems. Intelligent traffic management is defined as follows: According to the urban road traffic information collection, processing, release, and decision-making process, various advanced technologies and scientific methods are used to achieve the automation, modernization, and intelligence of traffic management. The system constructed to achieve the goal of intelligent traffic management is

ZHANG Cunbao (✉) · CHEN Feng · QIN Ruiyang · LI Xuemei · WANG Houyi
Intelligent Transportation System Research Center, Wuhan University of Technology, Wuhan, China

© Tsinghua University Press 2022
W. Yunpeng et al. (eds.), *Intelligent Road Transport Systems*,
https://doi.org/10.1007/978-981-16-5776-4_7

ITMS. The related technologies used in the realization of intelligent traffic management are called intelligent traffic management technology, which includes not only various IT technologies of electronic and information processing, but also related theories and methods of traffic science, system engineering, and management science.

The development of ITMS not only promotes the improvement of urban traffic management, but also strengthens the effective coordination between various traffic modes through information technology, starting from improving the transport capacity and transport efficiency of existing traffic facilities to enhance the control capabilities of road traffic and rapid response and disposal capabilities for emergencies. By improving the traffic efficiency and load-carrying capacity of the road network to alleviate traffic congestion, reduce traffic accidents, reduce time delays, energy consumption, and exhaust emissions caused by traffic congestion, conditions are created for the sustainable development of cities. At the same time, it also promotes the development of transportation and related industries, which is of great significance to the construction of urban modernization.

The ITMS system in the United States integrates highway and urban traffic management to reduce travel time, improve efficiency, better automatically detect accidents and establish accident response system. Metropolitan Area Guidance Information and Control (MAGIC) is a typical ITMS system. MAGIC is an important project of ITMS in New Jersey. Its goal is to reduce road traffic congestion, thereby reducing vehicle exhaust emissions. MAGIC includes content such as road monitoring, traffic management, traffic information services. Various sensors installed in the MAGIC system receive information, which is processed by the main computer to control road traffic conditions and display relevant data and deal with accidents. The expert system provides alternative paths to vehicles and recommendations to mitigate traffic congestion. The MAGIC system also uses highway information broadcast to provide service information for vehicle drivers, inform traffic, road and weather conditions, and recommend path selection. The accident detection system uses the data from the road monitoring system to automatically alarm for possible accidents after calculation by the computer software system. In addition, MAGIC can use the ramp monitoring system to detect the occupancy rate of the ramp and the vehicle speed on the ramp of the loop coil, and cooperate CCTV cameras monitor and control vehicles entering the ramp area.

Japan's Universal Traffic Management System (UTMS) was built in 1993 to adapt to the increasing demand for traffic in Japan, and is committed to achieving a safe, comfortable, and environmentally friendly traffic society. The key of UTMS is to realize the interactive two-way communication between the vehicle and the control center. The communication system uses infrared beacon to accurately transmit the control information of the management center on traffic demand and traffic flow to the driver, and comprehensively manage the traffic flow to avoid traffic congestion and ensure the safety of the vehicle. UTMS is mainly composed of six systems:

- The integrated traffic control system is set in the traffic control center, which is the information center for two-way communication with the vehicle device, and the traffic signal is controlled through this system.
- Advanced traffic information system, which can use the existing mobile communication network and vehicle communication media in UTMS to realize bidirectional information communication in traffic management.
- Dynamic path guidance system, which can guide vehicles to reach the destination along the optimized path, so as to achieve the purpose of dispersing vehicle driving routes and reducing traffic congestion.
- Vehicle driving management system, by accurately providing the traffic police with the driving positions and road conditions of buses, taxis, and trucks, etc., and through the signal management at the intersection, it helps vehicles to drive effectively and promotes smooth traffic.
- Public transit priority system, which can ensure the priority of bus using the road by controlling the priority signal and setting the priority route.
- Environmental protection management system, through which the traffic control center can provide traffic information such as weather and air quality to vehicle drivers and travelers through this system, and can implement vehicle operation management by controlling traffic signals to reduce exhaust, reduce traffic noise pollution and protect the environment.

Although China's intelligent traffic management system started late, after nearly two decades of unremitting efforts, considerable progress has been made. At present, domestic first-tier cities such as Beijing, Shanghai, Shenzhen, etc. have built ITMS with intelligent control center, control system, and control platform as the core, which integrates Advanced Traveler Information System (ATIS), Advanced Traffic Management System (ATMS), Advanced Public Transport System (APTS), Advanced Vehicle Control System (AVCS), Freight Management System, Electronic Toll Collection System (ETC), Emergency Medical Service (EMS), and other application subsystems. It has a high level of application in traffic information collection, traffic control, grid vehicle recognition, traffic guidance, traffic incident processing, intelligent traffic violation management, video surveillance, and other terms.

7.1.2 Framework and Function of ITMS

As an application system for traffic management department, ITMS takes ATMS and ATIS applied in urban road traffic management as the main body, and is interrelated with other ITS subsystems. The overall framework of ITMS is shown in Fig. 7.1.

ITMS mainly includes the following components and functions.

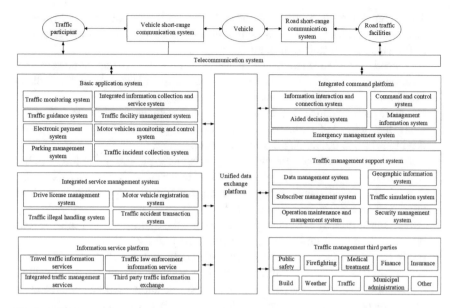

Fig. 7.1 ITMS overall framework

7.1.2.1 Basic Application System

1. Basic traffic data processing: Through data and external equipment, the collection of traffic static data is realized, and the information including urban map, population, economy, and road resources. The digital processing of text data is realized through the central server, and the basic database is obtained, and the data storage and reading can be realized.
2. Traffic status monitoring: The real-time traffic situation on the road is mastered by the on-road equipment, and information including passing traffic data, traffic flow data, meteorological information, and traffic events is collected to realize the comprehensive monitoring of the road in the center.
3. Traffic organization and management control: Realize the collection of traffic travel demand data, complete the analysis of historical traffic flow data, and optimize the traffic organization scheme according to the traffic situation. Through the analysis and judgment of the existing traffic data, the traffic situation can be predicted, and the priority control of public transport and other special vehicles can be realized.
4. Motor vehicle investigation and control: Carry out the investigation and control of illegal vehicles, and can give feedback within the specified time. It can realize early warning management and analyze vehicles trajectory.
5. Electronic toll collection and electronic payment: The system can provide an interface to realize the functions of online payment of certificates, vehicle inspection, traffic illegal fines, and other related business expenses, to realize the automation and convenience of payment.

7.1.2.2 Integrated Command Platform

1. Emergency command: In a state of emergency, effective control of the situation can be achieved through the ITMS, including traffic dispersion for large-scale activities, traffic congestion command and dispatching, traffic incident disposal, plan management and evaluation, video connection with the scene, control and dispatching of special service tasks, etc.
2. Traffic safety situation assessment: It can evaluate and warn the traffic safety status of the road network, road sections, and important nodes. It can judge the operation status of the road network, road section, and important nodes, and analyze the influence of the weather situation on traffic. Road safety audits can be conducted to comprehensively assess the safety situation.
3. Service management: Through the ITMS, the internal service status of the traffic police can be managed, and can realize the positioning of police officers and police vehicles, as well as the manual and automatic arrangement and assessment of service, and the emergency management and special service task management can be completed.

7.1.2.3 Integrated Business Management System

1. Basic information business management: The management of motor vehicle information, driver information, and road traffic illegal accident processing information can be realized, and the query of traffic management information such as public security network and police traffic can be realized.
2. Comprehensive business analysis: The driver information business, vehicle information business, traffic accident information business, and traffic illegal handling information business are analyzed. It can complete the research and judgment of the traffic accident situation and traffic illegal situation, and prevent road traffic accidents through research and judgment.

7.1.2.4 Information Service Platform

1. Road traffic information release for travelers: The real-time information of road traffic can be released to the traveler through the terminal, which can respond to the query request of the terminal user, release the real-time progress of road traffic control at any time, and provide parking guidance and reservation service.
2. Comprehensive traffic safety services for travelers: Drivers can query basic information of drivers and motor vehicles, illegal information of drivers, and motor vehicles through the ITMS. System can be accessed through the terminal to achieve the relevant business processes and laws and regulations query.
3. Information exchanges for third parties: Road maintenance departments and construction departments can collect and transmit information through the system. It can realize the collection and transmission of meteorological and

environmental information. First aid resources and fire resources can be collected and updated. It can also provide traffic information services to transportation, fire protection, construction, financial insurance, and other related business units.

7.2 Typical ITMS and Application

ITMS involves a wide range of areas, and there are many related application systems. In the actual traffic management, common typical application systems mainly include intelligent traffic monitoring system, electronic police system, intelligent public transport management system, parking guidance system, and emergency management system. This section will describe the typical application systems in detail.

7.2.1 Intelligent Traffic Monitoring System

7.2.1.1 Overview of Intelligent Traffic Monitoring System

Intelligent traffic monitoring system is currently one of the most widely used systems in the field of traffic management. Intelligent traffic monitoring system can be divided into three parts according to the information flow, namely information collection, information processing and information release. Monitoring information collection can be regarded as an information management system, including information collection, transmission, and classification storage. Information processing is the key part of the whole intelligent traffic monitoring system. According to various data and information collected and detected, information processing provides control strategies through processing, analysis, and judgment, and regulates the relevant traffic operation through corresponding equipment. As the terminal of information flow output of intelligent traffic monitoring system, information release shoulders the task of dialogue with the driver. For the control strategy and warning information of intelligent traffic monitoring system can be transmitted to the user in time, it is necessary that information release can affect the user in a timely manner through various ways. Intelligent traffic monitoring system is suitable for highway, important bridge tunnels, urban road traffic management, and other occasions.

7.2.1.2 Composition and Function of Intelligent Traffic Monitoring System

Intelligent traffic monitoring system has practical application in both urban roads and highways. The urban traffic monitoring system in real time detects the traffic flow and vehicle traffic conditions of the main traffic arteries and intersections. According to the feedback information of each monitoring point, it predicts the possible blockages in some traffic arteries and intersections, so as to adjust and induce the traffic flow in time, reduce the traffic congestion, maximize the utilization rate of the road system, and create a safe and comfortable road traffic environment. The highway monitoring system implements the traffic control scheme through the detection of traffic flow, the monitoring of traffic conditions, the environmental meteorological detection, and the monitoring of operation conditions of the whole highway, so as to achieve the purpose of controlling traffic flow, improving traffic environment, and reducing accidents, so that the highway operation can achieve higher service levels.

1. Urban Traffic Monitoring System. The development direction of modern urban traffic monitoring is mainly reflected in network and intelligence. In urban traffic monitoring system, modern computer technology, electronic technology and information technology are comprehensively used to realize intelligent management. A complete urban road traffic monitoring system is actually an integration of multiple subsystems. Based on the characteristics of urban road traffic, the composition of modern urban traffic monitoring system is shown in Figure 7.2.

    Traffic Monitoring Subsystem: Traffic monitoring subsystem consists of a front-end image collection unit, a signal transmission unit, and a central control unit. Its function is to transmit the situation of each intersection to the central control unit in real time and quickly, so that the traffic control management personnel can observe the real situation of each intersection in the city at any time, so that the managers can make correct judgments and provide information support for road traffic guidance.

    122 Alarm Accepting and Processing System: 122 alarm accepting and processing system consists of telephone communication dispatching subsystem, telephone digital recording subsystem, data management subsystem, head terminal subsystem, and electronic map subsystem (optional). Among them, the communication dispatching subsystem and the digital recording subsystem can run on the same computer, and the alarm subsystem and the electronic map subsystem can also run on the same computer.

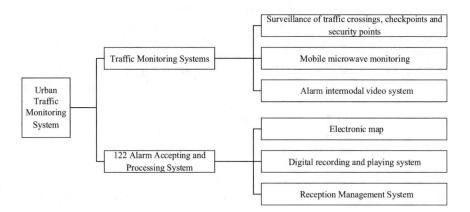

Fig. 7.2 Function Composition of Urban Traffic Monitoring System

Command Center Computer Management System: The main task of the command center computer management system is to connect the above independent application systems together to achieve orderly management, information dissemination and information communication for higher authorities and society. The traffic command center is the center of traffic monitoring system, which needs to be integrated through traffic integrated information management system.

2. Freeway Traffic Monitoring System. The main purpose of freeway traffic monitoring system is to provide effective means of traffic management, control, monitoring and inducing traffic operation for freeway. It collects various traffic data through the use of advanced electronic technology, helps the freeway management department to grasp the various data and information of the traffic flow in time, processes and analyzes the information through computer software, forms control strategies, and issues various control commands in a timely manner. Traffic information, deal with various traffic accidents and congestion phenomena, divert traffic flow, formulate effective measures to reduce traffic accidents and traffic congestion, and provide various services to relevant departments in a timely manner (including rescue of injured persons, dragging off faulty vehicles, repairing damaged roads and damaged equipment, etc.) to ensure the safety and smoothness of freeways.

The working principle of freeway traffic monitoring system is to collect freeway road, traffic, accidents, and other information through vehicle detectors, emergency telephones, closed-circuit television, and patrol vehicles along the road, and transmit them to the monitoring center for calculation, processing, and recording. At the same time, the road utilization, vehicle operation, and equipment working state are displayed by using the map plate, graphic display, Chinese character display, television, and other facilities of the monitoring center. Based on the above information, the on-duty personnel put forward the control scheme, issued the necessary commands or instructions, and transmit warnings or suggestions to road users through the variable intelligence board and variable speed limit signs set along the line.

The operation of traffic monitoring system especially emphasizes the role of monitoring center. Under normal traffic conditions, the information processing computer of the monitoring center conducts analysis according to traffic flow and meteorological conditions to form a decision-making control scheme, and uses variable information boards and variable speed limit signs to guide and control traffic. In the case of abnormal traffic conditions, the information processing computer of the monitoring center displays the corresponding control scheme to the operator according to the data processing results. After the operator confirms the event according to the field image, emergency telephone, patrol car, etc., monitoring center issues control instructions and induces control scheme.

The freeway monitoring system is mainly composed of a traffic information collection system, a central control system, and an information release system.

Traffic Information Collection System: The ways of information collection in traffic information collection system are manual and automatic. The main information collection methods are meteorological detection device, vehicle detection device, closed-circuit television, emergency telephone, radio equipment, etc.

Central Control System: The central control system is an intermediate link between the information collection system and the information release system, which is the core part of the monitoring system. Its main functions is to perform real-time operation, processing, and analysis on the data from the information collection system. According to the analysis results, the control scheme is decided, the corresponding control command is issued, and the event handling is commanded. Traffic conditions of each main road section are monitored through the closed-circuit television system. Responsible for communication within the jurisdiction area. Real-time monitoring of the working status of the entire system equipment. The central control system is usually composed of a computer system, an indoor display equipment and monitoring system console.

Information Release System: The information release system is a facility installed on freeways to provide road traffic information and guidance control instructions to road users, as well as to provide help orders or road traffic information to management, rescue departments and the society. Its main equipment includes variable information and variable speed limit signs, lane control signs, command telephones and traffic broadcasting systems, etc. The system mainly includes the following aspects, providing information to road users, such as traffic congestion, accident alarm, weather, road construction, etc. Providing advice or control commands to road users, such as the best driving route, the best speed limit, lane control signal, ramp control signal, etc. Providing information to the management and relief departments and providing information to society.

7.2.2 Electronic Police System

7.2.2.1 Overview of Electronic Police System

As a modern high-tech traffic management means, electronic police can not only effectively solve the contradiction between the police force and the management task, monitor the road 24 h a day, and improve the awareness of traffic participants to abide by the law consciously, but also effectively regulate law enforcement behavior and promote law enforcement justice. It also provides a powerful reconnaissance means for combating illegal and criminal behaviors such as avoiding annual inspection, causing accidents and escaping, and stealing and robbing vehicles. Because the electronic police system can work all the time, the blind spots of road traffic management in time and space are eliminated to a certain extent, the monitoring period and scope of traffic management are expanded, and the illegal behavior of motor vehicle drivers is effectively suppressed. It has been recognized by road traffic management departments at all levels and is widely used in cities at all levels in China.

7.2.2.2 Composition and Function of Electronic Police System

1. System Composition: The electronic police system consists of three parts: the front-end subsystem, the network transmission subsystem, and the back-end management subsystem. It can automatically capture, record, transmit, and process traffic violations such as running red lights, pressing lines, retrograde, not following the guiding lanes, illegal parking, not wearing seat belts, and talking on mobile phones while driving. At the same time, it also has the function of card port and can record the traffic information in real time.

   The front-end subsystem is responsible for completing the collection, analysis, processing, storage, and upload of the front-end data, which is mainly composed of electronic police capture unit, light compensation unit, video analysis and recording unit and other related components. Intersection traffic illegal information and bayonet information are transmitted by IP. The front-end subsystem includes the following parts.

   (a) Image detection plays the role of vehicle induction in the system, mainly including loop coil detector, video detector, ultrasonic or microwave (radar wave) detector, infrared detector, etc.
   (b) Image capture plays the role of image capture in the system, mainly including camera and video cameras.
   (c) Image collection is to digitize analog video images. Utilizing multichannel video image collection card makes the multichannel analog video image convert into digital video information after a multiplexer, A/D converter, cropping, and compression coding. Video compression coding methods commonly used internationally include MJPEG, Wavelet (wavelet transform), MPEG-1 (such as VCD), MPEG-2 (such as DVD), and MPEG-4, etc.
   (d) Image processing includes two parts, control host and system application software, which play the role of control, image recognition, storage, and management in the system.

   The network transmission subsystem is responsible for the transmission and exchange of data, pictures, and videos, and the construction of video private network. The intersection LAN is mainly composed of field industrial switches, point-to-point bare optical fiber and optical fiber transceiver. The central network is mainly composed of access layer switch and core switch.

   The back-end management subsystem is responsible for the aggregation, processing, storage, application, management, and sharing of relevant data within the jurisdiction, which is composed of central management platform and storage system. The central management platform is composed of servers carried by platform software modules, including management servers, application servers, web servers, picture servers, video servers, and database servers, etc.

2. System Functions. Red Light Running Monitoring and Recording Function. The electronic police system automatically perceives the traffic light signal through video triggering. At the red light signal, when the vehicle passes, the system host will quickly detect these changes, and judge whether there is a vehicle passing by

Fig. 7.3 Snapshot of vehicles running red light

analyzing and processing this change. When it is detected that there is a vehicle passing under the red light state, it will automatically capture a picture of the illegal vehicle, take three pictures of the illegal process of the vehicle and a close-up high-definition picture (Figure 7.3). As shown in Fig. 7.3, the image can clearly distinguish the red light state, red light time, parking line, illegal time, illegal location, illegal type, vehicle type, license plate color, license plate number, body color, etc. Three consecutive images can accurately and clearly reflect the process of vehicles breaking the red light illegally. Four images will be combined into an evidence image. At the same time, a set of illegal images has a high-definition video corresponding to it.

Monitoring and Recording Function for Driving in Noncompliant Lanes: The electronic police system uses direct video analysis to monitor the driving state of the vehicle, which can accurately judge the illegal behavior of the straight lane left/right turn, right-turn lane straight going or left-turn, left-turn lane straight going, or right-turn (as showed in Fig. 7.4). At the same time, a group of illegal pictures corresponds to a high-definition video.

Vehicle Retrograde Automatic Monitoring and Recording Function: The system uses video tracking technology to track each motor vehicle in the screen, which can directly detect vehicles driving in the reverse direction in the captured lane and directly identify the vehicles license plate information (as showed in Fig. 7.5). At the same time, a group of illegal pictures corresponds to a high-definition video.

Interval Speed Measurement and Recording Function: The system sets two adjacent monitoring points on the same road section, calculates the average driving

Fig. 7.4 Snapshot of vehicles driving in noncompliant lanes

speed of the vehicle on the road section based on the time of the vehicle passing through the two monitoring points before and after, and determines whether the vehicle is speeding or not according to the speed limit standard on the road section. At the same time, it publishes the information of traffic illegal vehicles in real time on the LED screen to inform and warn more vehicles (as showed in Fig. 7.6).

Monitoring and Recording Functions for Vehicles Occupying Lanes Illegally: Based on vehicle trajectory tracking and judgment, the system can detect and record the illegal driving and parking behaviors of vehicles in the monitoring area in bus lanes, nonmotorized lanes, emergency lanes, etc. The capture pictures include forbidden signs, lane lines, vehicle location, license plate, etc.

License Plate Number Automatic Identification Function: The system can locate and identify the license plate number automatically. Possessing license plate calculation automatic recognition ability of the civil vehicle, police vehicle, military vehicle, armed police vehicle license plate calculation automatic recognition ability.

Bayonet Recording Function: The system adopts video detection technology to detect, capture, record, save, and identify all vehicles passing through each lane at

Fig. 7.5 Snapshot of retrograde vehicle

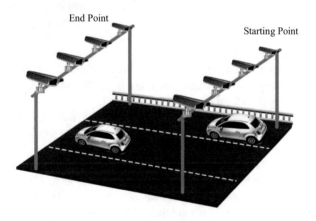

Fig. 7.6 Schematic diagram of interval speed measurement

red, green, and yellow lights, and identify license plate information. The bayonet image is a composite, the left half is a panoramic image of the vehicle, and the right half is a close-up image of the vehicle. All recorded vehicle passing information includes time, location, direction, vehicle type, vehicle license plate number, license plate color, etc.

7.2.2.3 Key Technologies for Electronic Police Systems

Cameras are installed at traffic intersections to take videos of the traffic intersections, and the task of automatic illegal recognition and automatic license plate recognition is completed by computer threshold comparison and analysis of digital image signal.

Automatic violation identification is to identify dynamic objects (such as vehicles), track their trajectory, and then automatically determine vehicle violation by relying on road signs and intersection signal lights in the monitoring area. The key links involved are as follows.

1. Detection of Moving Target: The main technical means used in this link include background image difference method, inter-frame different method, and optical flow method. These three methods each has their own characteristics, combined with different actual situation can achieve better application effect.

   (a) Background Image Difference Method: The basic principle is to first store the static background in the monitoring field of vision, and then subtract the real-time image from the static background in the actual operation. The value of each pixel of the difference is compared with the predetermined threshold. If the pixel value is greater than the threshold, it is considered to be the foreground point, otherwise it is the background point. When using this technical solution, it is greatly affected by external conditions such as light and weather, because in different weather and different light conditions, the background image is not static, so it is difficult to construct an ideal static background image as the basis. Another problem is to determine the predetermined threshold. Only the appropriate threshold can correctly segment the target area, which also needs to be adjusted and determined according to the actual situation.

   (b) Inter-frame Difference Method: This solution is to make a difference between a series of adjacent two frames in the recorded video image. The target contour is obtained by dividing the operation. In the recognition of multiple moving targets, this scheme can achieve good results. Since the interval time of the two images in the video is so short, so the change of the background is also little, and the influence on the difference is little. The disadvantage is that the differenced image is not composed of an ideally closed contour area, and the obtained target is often local and discontinuous, which is detrimental to the recognition of moving targets.

   (c) Optical Flow Method: The basic principle is to set a velocity vector for each pixel in the video to form a whole image motion field. At the fixed time of motion, the points on the image correspond to the points on the three-dimensional object one by one, and then the image is dynamically analyzed according to the velocity vector characteristics of each pixel. According to the optical flow vector of the image, continuous change of this area is analyzed to determine whether there is a moving target. Optical flow method avoids the selection of basic static background and threshold in differential time, but the calculation of vector analysis for each pixel is large and vulnerable to video noise. The real-time processing of images cannot be achieved according to the existing hardware processing ability, and the application range is limited.

2. Multi-target Tracking: Segment the object in the video stream, then segment the object with the object of the previous image the tag is matched to achieve the purpose of tracking. The basic principle is to match the region in the current image frame and the target region of the known image. If the known target region is represented as a target list, all the regions in the current image frame are represented as a measurement list. For each element in the measurement list, the most similar element is found in the target list. This method is suitable for the case of small interaction between targets, and has a large relationship with the selection of target features.

3. Discrimiation of Vehicle Violations: After the vehicle trajectory is extracted, the vehicle violations are automatically identified according to the traffic light signals and road traffic sign lines. When dealing with traffic violations, three important evidence must be recorded: the violation screen, the panoramic position of the illegal vehicle, and clear license plate number. Among them, panoramic position of the illegal vehicle indicates that the vehicle was indeed in the illegal position, and a clear license plate number, which clearly identifies the violation vehicle. Therefore, only the video surveillance system based on computer technology can complete automatic monitoring. As an important part of the current intelligent transportation system, automatic vehicle recognition has gradually formed several mature and effective recognition technologies, such as radio frequency identification, license plate recognition, and bar code recognition. Barcode recognition and radio frequency recognition belong to indirect recognition, it is difficult to effectively identify whether the vehicle is true and consistent with its license plate information. License plate recognition is a direct recognition no corresponding barcodes or other radio frequency identification signs need to be installed in the vehicle, and it is easy to maintain and use. Automatic license plate recognition uses the vehicle picture taken by the camera as the input image, and uses computer image processing and pattern recognition technology to recognize license plate characters.

License plate recognition system has two triggering methods: one is peripheral triggering and the other is video triggering. The working mode of peripheral trigger means that the coil, infrared, or other detectors are used to detect the vehicle after receiving the vehicle trigger signal through the signal and license plate recognition system, and then the vehicle image is collected, the license plate is automatically identified and the subsequent processing is carried out. Video trigger means that the license plate recognition system uses dynamic moving target sequence image analysis and processing technology to detect the movement of vehicles on the lane in real time, capture the vehicle image when the vehicle passes, identify the license plate, and follow-up processing. Video trigger mode does not require coil, infrared, or other hardware vehicle detectors.

7.2.3   Intelligent Public Transport Management System

7.2.3.1   Overview of Intelligent Public Transport Management System

Intelligent public transport management system integrates modern communication, information, electronics, control, computer, network, GPS, GIS, and other technologies into the public transport system. Under the premise of key basic theoretical research such as public transport network distribution and public transportation dispatching, through the establishment of public transport intelligent transport dispatching system, geographic information system, public transport information service system, and public transportation electronic toll collection system, etc., make it more convenient for managers, operators and individual travelers to coordinate with each other and make clearer decisions. Through the construction and implementation of intelligent public transport system, the purpose of alleviating the pressure of public transport passenger flow, balancing the full load of public transport vehicles, reducing the cost of public transport operation, and improving the efficiency of travel, so as to establish a convenient, efficient, comfortable, environmentally friendly, and safe public transport operation system. The informatization, modernization, and intelligence of public transport dispatching, operation, and management will attract public transport, alleviate urban traffic congestion, and effectively solve urban traffic problems.

Intelligent public transport management system mainly takes travelers and public transport companies as the service objects. For travelers, the intelligent public transport management system collects and processes dynamic data (such as passenger flow, traffic flow, bus location, bus station waiting status, etc.) and static traffic information (such as traffic laws, moral management measures, location of large bus travel generation sites, etc.), and provides dynamic and static public traffic information (such as departure schedule, transfer route, optimal route guidance, etc.) for travelers through various media, so as to realize the purpose of planning travel, optimal route selection, avoiding traffic congestion, and saving travel time. For public transport enterprises, the intelligent public transport management system mainly realizes the dynamic monitoring, real-time dispatching, scientific management and other functions of public transport vehicles, and realizes the modernization and information management of public transport enterprises, so as to improve the service level of public transportation and the operating efficiency of public transport enterprises.

7.2.3.2   Composition and Function of Intelligent Public Transport Management System

For public transportation enterprises in domestic cities, the organic combination of the intelligent operation system of public transportation and intelligent management system of public transportation enterprises can fully realize the sharing and application of public transportation information resources. The aforementioned two systems share data and other related business operations through the bus communication subsystem and data center. The system architecture of urban intelligent public transport management system is shown in Fig. 7.7.

1. Intelligent Operation System of Public Transport

   The intelligent operation system of public transport is mainly composed of the following eight parts:

   (a) Data Center: The data center is the hub of information interaction between each subsystem of the urban intelligent traffic management system, which mainly completes the collection, storage, and processing functions of public transport–related information. The data center logically consists of a central database system, data processing system, and database management system. The central database system is responsible for the collection and storage of data, and the data processing system is responsible for the fusion and preliminary data processing of the data collected by the bus information collection subsystem.

   (b) Public Transport GIS Platform: The public transport GIS platform system uses information processing technology to establish the integration of vector electronic map database, relational database, and other multi-platform and multi-database based on public transport information. It has strong spatial analysis and spatial data processing ability of traffic network system, realizing the best route selection mode under the condition of the least number of transfer and shortest distance, etc., and has the query function of common passenger traffic information such as aviation, railway, waterway, and highway.

   (c) Public Transport Communication Subsystem: According to the data transmission requirements between subsystems, the bus communication subsystem can choose to use various wired communication modes such as optical fiber network, Ethernet, ADSL, MODEM, and wireless communication modes such as wireless conventional communication, wireless cluster communication, GSM, CDMA, and satellite communication.

   (d) Public Transport Information Collection Subsystem: The public transport information collection subsystem is a prerequisite for the normal operation of public transport dispatching subsystem, public transport information service subsystem, public transport evaluation subsystem, and other modules. The real-time dynamic information collected by the bus information collection subsystem provides the basis for bus enterprises to achieve real-time vehicle dispatching and provides basic data sources for bus information services. At the same time, the historical information of the central database of public transport management can also provide data support for long-term bus dispatching and bus system evaluation.

   (e) Public Transport Information Service Subsystem: Public transport information service subsystem is an effective way to improve the reliability of public transport travel modes, and it is also an effective means to guide the balanced distribution of traffic demand. A complete public transport information

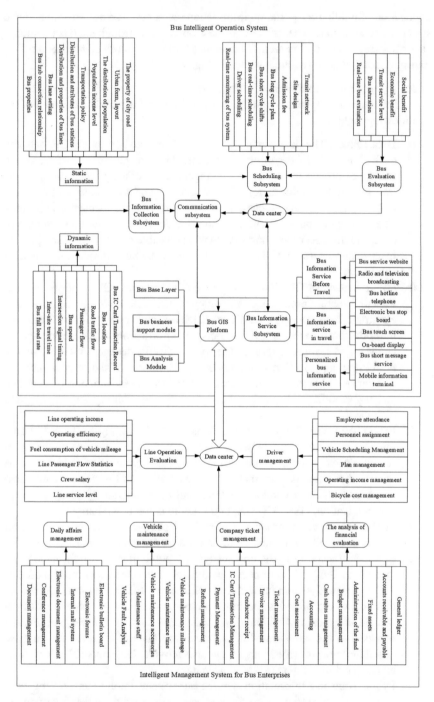

Fig. 7.7 Frame diagram of urban intelligent public transport management system [1]

service system can provide multifaceted information services for passengers, including pretravel information services, travel information services, personalized information services. At the same time, the public transport information service subsystem will provide public transport information services for other transport enterprises and relevant management departments and government departments through data interfaces.

(f) Public Transport Dispatching Subsystem: Public transport dispatching subsystem is the core subsystem of the whole intelligent urban public transport dispatching system. In terms of classification, it divides into line dispatching and regional dispatching. The bus dispatching subsystem includes four modules: intelligent public transport dispatching module, real-time public transport dispatching module, public transport dispatching optimization module, and intelligent public transport regional dispatching module. Among them, the intelligent public transport dispatching module will determine the dispatching of each route and the time when the public transport arrives at each station according to the passenger flow prediction of the line and the public transport real-time operation dispatching module in the process of bus operation according to the change of passenger carrying rate and public transport travel time prediction of each public transport real-time dispatching. Public transport is dispatching optimization module independently optimizes the real-time public transport dispatching module according to the historical data of public transport enterprise operation, urban planning, traffic management, and other aspects of information. The regional dispatching module integrates the management and dispatching of public transport lines in the region on the basis of intelligent dispatching of public transport lines, makes full use of the resources of public transport enterprises, and also improves the service level of public transport.

(g) Public Transport Evaluation Subsystem: The public transport evaluation subsystem mainly evaluates the operation of public transport enterprises according to the satisfaction index of the social system to the operational performance of public transport enterprises.

(h) Public Transport Tolling Subsystem: Public transport tolling subsystem introduces the IC card as a bill medium into the operation of public transport enterprises, which avoid the process of cash transactions and change for flight attendants.

2. Intelligent Management of Public Transport Enterprise

Intelligent management system of public transport enterprise is the management system of public transport enterprise itself. It is an intelligent information management system covering all relevant departments of the public transport enterprises, and provides information exchange with the central database system, including the functions of the MIS system and OA system. It mainly realizes the automatic management and paperless office of public transport enterprises, and its main functional modules include line operation evaluation, company ticket management, daily affairs management, vehicle maintenance management, fleet and bus station management, crew management, and financial evaluation and analysis.

7.2.3.3 Key Technologies of Intelligent Public Transport Management System

For the dispatching problem of public transportation, the commonly used methods include static dispatching and dynamic dispatching. Among them, static dispatching mainly refers to the departure time according to the prearranged timings. Dynamic dispatching includes vehicle dispatching, driver dispatching, and vehicle control and adjustment. Normalization of public transport operation can be realized by dynamic dispatching. It is difficult to respond to the actual situation in the operation of public transport vehicles by static dispatching. Because in the general public transport dispatching system, the travel and time of the vehicle are based on experience, without considering the actual operation of the unexpected situation. Therefore, during the process of vehicle design and operation, it is difficult to deal with various emergencies and ensure optimal dispatching only by static dispatching. When the actual situation changes, the flexibility of the static method is not good. Dynamic dispatching can reduce the waiting time of passengers and improve the level of public transport services through real-time dispatching of public transport.

This section lists a simple dynamic dispatching scheme. Considering that the actual situation of the vehicle in the operation process is very complex, in order to effectively simplify the research process, the following assumptions are made.

1. Human factors, road traffic, and other factors during the operation of vehicles are not considered, so that the operation time of different vehicles between the same adjacent stations is the same.
2. The parking time of a bus at each station is fixed, and is not affected by factors such as the number of passengers.

In the dynamic dispatching model, the various parameters are defined as following. Station set is $K \in \{K | K = 1, 2, \ldots, N\}$. The vehicle set is $I_m = \{i, i + 1, \ldots, i + m - 1\}$, and $i \leq m \leq M$. M represents the maximum number of vehicles that the bus system can provide, d_{ik} represents the time when the vehicle comes out, δ represents the time when the vehicle enters and leaves the station in minutes. Therefore, if the vehicle can save the inbound and outbound time, you can reduce the operation time, $\delta = 0$, c_0 is used to indicate the parking time of the bus at the station, and the unit is also in minutes, the time that can be saved without parking is expressed as $\Delta = c_0 + 2\delta$.

To achieve dynamic dispatching, the fundamental problem is to determine the objective function, that is, to ensure that the time and minimum required for all passengers in the station while waiting for the vehicle. The model is defined as min (W). Assuming that the number of all passengers in a time period h_{ik} is $h_{ik}r_k$, the average waiting time for all passengers in that time period can be expressed as $h_{ik}/2$.

After derivation, the time cost required for passengers waiting for the ith car at station k can be set as:

$$w_{ik} = r_k h_{ik}^2/2 + h_{ik}P_{i-1k} \tag{7.1}$$

In the above formula, w_{ik} represents the time cost of passengers waiting for the ith car at k station. r_k represents the passenger arrival rate of a bus station. P_{i-1k} represents the number of remaining passengers of ith vehicle at station k. In the absence of dynamic dispatching, $P_{i-1k} = 0$. h_{ik} represents the departure time interval between vehicle i and vehicle $i-1$ at station k. h_i represents the time interval between vehicle i and vehicle $i-1$ at any station. Therefore, according to this definition, $h_{ik} = d_{ik} - d_{i-1k}$. Therefore, the waiting time cost of passengers at each site of the line can be converted to:

$$W = \sum_{i=1}^{m} \sum_{k=1}^{K_C} \left(r_k h_{ik}^2/2 + P_{i-1k}h_{ik} \right) \tag{7.2}$$

From Eq. (7.2), the waiting time cost of passengers can be calculated, and the dynamic dispatching model can be further constructed. That is, the dynamic dispatching is carried out between the vehicle i and the vehicle $i-1$. If the number of stations that the two vehicles do not stop during the operation is assumed to be n_1 and n_2, the time saved by the two vehicles through non-stopping can be expressed as $n_1\Delta$ and $n_2\Delta$. In general, in order to reduce the waiting time of passengers at non-stop stations and promote the uniformity of the vehicle interval on the entire line, $n_1 \geq n_2$ should be set. As shown (Fig. 7.8), the dotted line represents the departure time without dispatching, while the solid line represents the departure time after dynamic dispatching. The mainly vehicles that can affect passengers waiting time cost are $i, i+1, i+2$. If w_1, w_2, w_3 are used to represent the waiting time cost of stations and $0\sim n_2$, $n_2\sim n_1$ and $n_1\sim N$ refer to the difference of waiting time cost

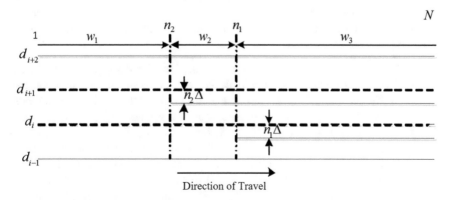

Fig. 7.8 Dynamic Dispatching Model Diagram of Bus

without parking, respectively. Thus, the entire waiting cost can be expressed as $w = w_1 + w_2 + w_3$.

7.2.4   Parking Guidance Information System

7.2.4.1   Overview of Parking Guidance Information System

Parking Guidance Information System (PGIS) includes large parking intelligent guidance system and urban parking guidance system. It is a system that uses multilevel information release as the carrier to provide real-time information such as the location of the parking lot (garage) location, the number of parking spaces, and the state of empty space, so as to guide the driver to park. It plays an important role in regulating the uneven distribution of parking demand in time and space, improving the utilization rate of parking facilities, reducing the road traffic caused by looking for the parking lot, reducing the waiting time caused by parking, improving the efficiency of the entire transportation system, improving the operating conditions of the parking lot, and increasing the economic vitality of the commercial area.

7.2.4.2   Composition and Functions of Parking Guidance Information System

The general PGIS consists of four subsystems, namely the information collection subsystem, the information release subsystem, the information processing subsystem, and the information transmission subsystem. The overall function of PGIS in big cities is to release parking information and provide basic data for urban intelligent transportation systems. The direct function of PGIS is to provide parking information to traffic managers who have parking needs.

The overall structure of PGIS adopts centralized-distributed system architecture. The collection and processing of data information and the layout of the database are distributed, and the sharing and fusion of data and the maintenance and management of consistency are centralized.

1. Information Collection. The information collection of PGIS is an important part. It is also the basic work of parking lot design, road infrastructure construction and even traffic planning. Through the collection of information such as the number, location and utilization status of parking spaces, it can not only provide information guarantee for the release of PGIS, but also grasp the status and laws of parking, clarify the nature of parking problems, and propose targeted solutions.

   Information Collection Classification: The information collection of PGIS can be divided into static data collection and dynamic data collection. Static data collection is a stable and constant information in the PGIS for a period of time. It mainly completes the statistics and input of the location, type, and rate of each parking lot or roadside parking space, as well as the information of related stations with parking-transfer functions. Dynamic data collection is information about the relative changes in time, such as the utilization of parking spaces and the opening and closing of parking lots. According to data sources, parking information collection can be divided into direct collection and indirect collection. Direct collection through the parking management host to obtain parking information, and indirect collection through other intelligent transportation system each collection data node integration of traffic industry information.

Principle of Data Collection. Data collection can be divided into the following three categories:

(a) Manual Collection: Manual collection is a relatively traditional nonautomatic collection, which does not require complex equipment, but the accuracy and timeliness of information is difficult to control.

(b) Collect According to Vehicle Characteristics: Traffic information collection is mainly to detect vehicles, and convert the existence and movement of vehicles into electrical signals output. The vehicle is a component that contains a lot of iron. An entity with quality and geometric shapes, and has certain characteristics of light, heat, and electricity. According to these characteristics, information collection includes methods such as magnetic detection, ultrasonic detection, electromagnetic wave detection, thermal detection, quality detection, and video image detection.

(c) Collect with the Help of External Objects: With the development of transportation, the PGIS has higher requirements for information, not only the number of vehicles, but also the distinction of different collection objects, and then different strategies are adopted. The more common ones are number identification and IC card.

2. Information Processing: The information processing of parking guidance system not only provides parking information, but also undertakes the tasks of storing parking information or roadside parking information and processing the change mode of parking usage. These functions will lay the foundation for future parking demand forecasting and parking reservation. The parking information processing of parking guidance system is realized by management software in two steps, namely front-end processing system and management center system.

The front-end processing system generally refers to the parking management system, which mainly has the following functions.

(a) Import and export data of vehicles are collected, such as vehicle properties, vehicle number, import and export time.

(b) Utilization of parking space.

(c) Other functions required by parking management, such as charging statistics.

The functional composition of the management center system is shown in Fig. 7.9.

Change of Parking Space during Travel Time: From the current collection technology, the parking space collector for the current parking space collection is relatively clear, but it cannot make an accurate prediction for the future parking

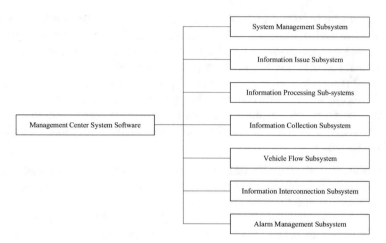

Fig. 7.9 Function composition of management center system

space changes. In order to prevent the driver from seeing the information in the parking lot with a certain distance between the screen and the parking lot, it is necessary to process the information in the parking lot with a reasonable distance.

System Optimization: In the regional PGIS, the system always considers how to provide parking information to drivers accurately and timely. Although this model is simple and practical, there are also some shortcomings, such as the limited information provided, unable to consider the different requirements of different drivers, that is, unable to achieve the interaction between the system and users. With the establishment of urban intelligent transportation information platform, traffic flow guidance system, GPS, and multi-level PGIS, the information processing of PGIS has higher requirements. The system optimization of PGIS is to draw up a reasonable traffic control strategy, that is, the optimal model of the system, based on the parking management and smooth traffic flow of the whole city, and then take corresponding countermeasures according to different system objectives.

Information Release: Parking information release is the main part of PGIS, which can be divided into fixed guidance information and variable guidance information according to whether the guidance information is variable. The fixed guidance information is mainly based on parking signs. Because of its low cost, it can be used as a useful supplement for the guidance information release of parking lots. The variable guidance information release plate can provide the changing parking space or parking lot information, and some fixed guidance information is attached to the variable information release plate, which can save the cost or improve the stability of the release system.

Information Forms: Common parking guidance screen information are shown in Table 7.1.

Guidance Information Classification Requirements: The induction system should be designed into a three-level or four-level induction system according to the

Table 7.1 Common information forms of parking guidance screen

Level	Types of signs	Static content	Dynamic content
A	Text + arrow	P vacancy + induced region name + arrow	Total regional remaining berths
B	Map	P vacancy + road network	Total number of remaining berths
C	Text + arrow (combined type)	P vacancy + parking name + arrow	Total number of remaining parking lots
D	Text	P vacancy + parking name	Total number of remaining parking lots

Table 7.2 Classification system of parking guidance system

Property	Level	Effect	Proposed location
Regional level predictability induction markers	Level 1(A)	Display the location and control information of the induced area	Regional peripheral main road
	Level 1(B)	Zone induction, guiding the location and vacancy letter of adjacent zones	Main roads outside the subregion
Street level induction signs	Level 2(C)	A sign indicating the parking lot along the road information	Intraregional roads
Parking level indicator sign	Level 3 (D)	Guide the parking location and total blank information	Parking entrance

characteristics of the induction region. Generally, the use of level 3 inducers are as shown in Table 7.2.

The district that publishes the regional parking space information. In the multilevel PGIS, information release often involves statistics and processing of regional space data. In view of the cross-regional nature of information release, in order to improve the guidance effect, in the sub-management center, the distribution of regional information is distributed and the physical area is processed. The division of information collection should be different. The partitions for releasing regional parking space information mainly go along the following principles: The range of each partition should not be too large, it should be limited to 6–8 blocks, and it is best to control within a rectangular area with a side length of about 500 m; the name can be distinguished and easy to identify; the capacity of the parking lot in each zone is roughly equal to the parking demand; the road with heavy pedestrian traffic should be avoided to cross the zone; the induction route to the parking lot should be avoided as far as possible to make a left turn; each zone is best It can be separated by arterial road.

3. Information Transmission

Classification of Communication: Transportation communications technologies include radio broadcasting, cable communications, microwave communications, mobile communications, optical fibers communications, digital baseband communications, digital carrier communications, infrared and ultrasonic communications, and satellite communications. Different communications technologies are applied in different scopes. The above communication technologies in ITS do not all exist independently, and most of them are mutual penetration and cross each other.

According to the management level of sender and receiver, the information transmission of parking guidance system can be divided into the following:

(a) Parking lot or roadside parking space—center of control.
(b) Management center—LED publishing screen.
(c) Management center—other information release methods.

4. Communications structure.

According to different transmission tasks, the communication of the parking guidance system can adopt different methods. At present, the more common wired optical cables, optical fibers, and wireless methods. Cables include solid wires, DPN data networks, etc., and wireless include GPS, GPRS, CDMA, etc.

7.2.5 Emergency Management System

7.2.5.1 Overview of Emergency Management System

Emergency management system for emergencies is a human-oriented; guided by scientific management theory; and on the basis of a scientific management system, it uses computer hardware, software, and network communication equipment to monitor and control the operating conditions of the freeway around the clock. Quickly detect and judge emergencies, and quickly take appropriate incident response measures to avoid traffic accidents (or secondary accidents) and ensure timely rescue and accident elimination after the accident, and support the drivers and passengers, integrated human–machine system for traffic vehicles and grassroots operations. The general goal of traffic emergency management is to take preventive measures before traffic accidents to reduce and avoid the occurrence of abnormal traffic accidents. When traffic incidents occur, timely discover and take appropriate emergency rescue measures to minimize casualties and property losses. And to minimize the impact of traffic delays caused by the incident, and return to normal traffic conditions in the shortest time.

7.2.5.2 Composition and Function of Emergency Management System

Emergency management is an important part of traffic safety management, which can significantly improve traffic safety and operational efficiency. Traffic safety emergency management is a systematic work, which includes emergency warning, emergency detection, emergency analysis, emergency decision-making, rescue execution system, emergency assessment, plan management, etc. Each link is related to

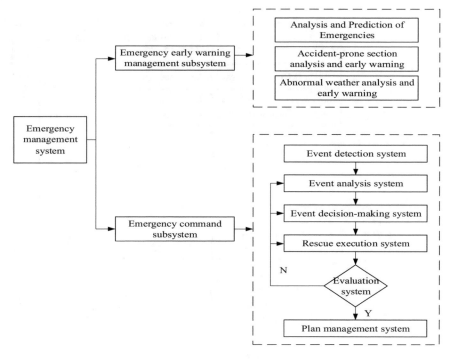

Fig. 7.10 Overall framework of emergency management

the efficiency and effectiveness of emergency management. The overall framework of emergency management is shown in Fig. 7.10.

1. Emergency Early Warning Management Subsystem. The main functions of the emergency early warning management subsystem include data mining, information collection, statistics and analysis of real-time traffic conditions and real environment, and find out the distribution characteristics of traffic accidents and their causes and influencing factors, such as traffic safety control in severe weather and risk factors in traffic accident-prone areas. The early warning system should immediately monitor these risk factors, issue early warning to the traffic safety management department, and take early warning programs in time to prevent accidents to the greatest extent. Emergency early warning management subsystem has three functional modules: traffic information processing and prediction module, accident-prone section analysis and early warning module, abnormal weather analysis and early warning module. According to the objective process of road operation activities, the working mode of emergency warning management subsystem is shown in Figure 7.11.

The development of the road early warning emergency management subsystem is based on the operating company's monitoring of the traffic operation indicators of the roads under its jurisdiction. According to the obtained

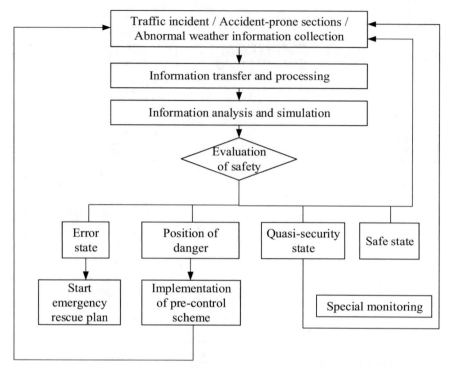

Fig. 7.11 Emergency warning management subsystem working mode diagram

information, the management department confirms that the monitoring indicators are in abnormal, dangerous, quasi-safe, or safe state by identifying, diagnosing, and evaluating the traffic phenomena and implements corresponding control countermeasures. When the indicator is in a safe state, continue daily monitoring. When the monitoring indicators are in a quasi-safe or dangerous state, the early warning department enters the special monitoring stage according to the specific situation or instructions, or proposes pre-control measures and is executed by the decision-making or the dispatching department until it is restored to a safe state, while the dispatching department enters the response plan into the response library for future reference. When the monitoring index enters the abnormal state, the entire road management organization enters the emergency rescue procedure, and starts the emergency plan until the road operation enters a safe state again.

2. Emergency Command Subsystem

Incident Detection System: Whether urban traffic emergencies can be detected in time is the first problem to be solved in traffic emergency management. The incident detection subsystem is used to detect and determine the nature of the incident. It integrates various detection and monitoring equipment (such as induction coil, video monitoring equipment, microwave detection equipment, floating car information collection equipment), and coordinates with relevant information departments to collect relevant meteorological, road environment, traffic flow conditions (traffic flow, velocity, interval travel time) and other information, timely predict, discover and analyze the location, scale and development trend of emergencies, and provide reliable basis for emergency response decision-making and command. Incident detection technology is the theoretical basis of event detection subsystem, which is not only related to whether the role of monitoring system (hardware part) can be fully played, but also has great significance for the treatment of accidents. For example, the rapid detection after the accident, the rapid response to the accident, taking appropriate traffic control measures to prevent the occurrence of secondary accidents, rapid processing of accidents, reducing accident losses, etc., all have a direct relationship with the incident detection algorithm.

Incident Analysis Subsystem: After the incident detection subsystem detects the occurrence and upcoming traffic emergencies, the original incident information, environmental information, and traffic information of the incident location are summarized into the incident analysis subsystem for further filtering and analysis. Within the incident analysis subsystem, it is necessary to complete the identification of the type, severity, cause and other factors of the event, and to determine the degree of decline in the capacity of the bottleneck caused by the incident, the possible blockage, and the degree of spread of the blockage. Perform analysis and prediction. Provide a basic basis for the next incident decision. The process of incident analysis is the process of incident confirmation first, because there will be a certain false alarm rate during the incident detection process, and the attribute data of the incident provided by the incident detection may be incomplete. Therefore, the incident analysis first needs to determine whether the incident is present or not. Make judgments, and then need to use various predesigned models and plans and special mathematical algorithms to classify and analyze the incident, and finally get the incident's characteristic information, severity, impact index, and other important parameters. When the data provided by the detection subsystem is not complete or incomplete, it is necessary to use data mining technology and data fusion technology to further process the attribute data of the incident.

Incident Decision-Making Subsystem: Decision analysis is the difficulty of traffic emergency management. It is responsible for generating rescue plans and notifying relevant departments to dispatch rescue resources. The module uses the information collected by the event detection subsystem and the preliminary results of the analysis subsystem to generate rescue strategies, including lane control strategy and ramp control strategy. The generation of these strategies should consider the traffic capacity of the relevant road network, the matching between the traffic capacity of each section, and the predicted driving time of each section. At the same time, the relevant rescue departments should be informed to implement the accident rescue process. The plan content generated by the incident decision-making subsystem should include the composition and responsibilities of emergency agencies, emergency communications support, organization of rescue personnel and preparation of funds and materials, preparation of emergency and rescue equipment, disaster assessment preparation, several parts of the emergency action plan. On the basis of the plan, the departments needed to be called for incident rescue should be designated, the responsibilities and permissions of each department should be clarified, the equipment needed for emergency rescue (medical equipment, fire equipment, vehicle traction hoisting equipment, etc.) should be determined, the specific steps for rescue implementation should be formulated, the emergency traffic control scheme should be put forward, the green channel for rescue vehicles should be formed, and the strategy of traffic guidance should be put forward. After receiving the emergency information report of the information system, the types and severity of emergencies are analyzed through the resource database and expert assistant decision-making system, and then the corresponding emergency treatment plan is launched for rescue, which provides the best route to the location of the incident for the relevant departments, and the emergency rescue instructions are issued through the communication system. In the emergency scene, according to remote-sensing images, video images and feedback information, emergency rescue measures are responded and adjusted at any time. According to different rescue needs and the division of labor of various functional departments, the relevant departments are informed of accidents and rescue needs information, and the rescue work is coordinated.

Rescue Execution Subsystem: The rescue execution subsystem is to coordinate various rescue departments (traffic police, hospitals, fire, road maintenance departments, armed police, military, etc.) after receiving emergency notifications, according to the plan division of the decision-making system, and perform their own duties to deal with emergencies and organize rescue, the traffic command department implements necessary traffic control on the scene, according to the rescue green channel and traffic guidance provided by the decision-making system, assist other departments to implement emergency rescue at the fastest speed, and eliminate emergencies in the shortest time. It also feeds back the situation of the incident site and the changes in the needs of various rescue departments to the incident decision-making subsystem in time, so that the emergency system can modify the emergency rescue plan in time, better implement the emergency rescue work, and make the road network traffic return to normal faster.

Effect Evaluation Subsystem: The main function of the evaluation subsystem is to evaluate the effect of solving traffic emergencies. Through the implementation of the rescue execution process, the traffic emergencies are either solved in time, or have not been effectively solved, or have not played the effect of reducing the loss of events. These need to be evaluated by the evaluation subsystem. The content of the evaluation includes the evaluation of the event itself and the evaluation of the traffic condition of the incident section. If the effect does not meet the requirements, the specific information that does not meet the requirements is fed back to the corresponding subsystem (including the incident analysis subsystem, the incident decision subsystem and the rescue execution subsystem) for reanalysis, decision-making, and rescue. If the evaluation results meet the requirements, the case of this traffic emergency is closed, and the relevant information of the whole event (including the time, location, type, severity, emergency plan, duration, emergency resources consumed, and emergency effect) is input into the plan management subsystem of urban traffic emergency to form a historical case database, which provides a useful reference for future traffic emergency management.

Plan Management Subsystem: The plan management subsystem is the intelligent warehouse of urban traffic emergency management system, which includes the model base, knowledge base, and historical database of traffic emergency management. These databases have practical and effective reference value for the management of emergencies. Establishing and improving the plan management subsystem of traffic emergencies is of great significance for traffic emergency management. Of course, the ideal database is not built at the beginning, needs to be gradually improved, and perfected in the daily incident management process. The plan management subsystem is mainly responsible for the collation and archiving of the basic information of emergencies and the whole rescue process information. It analyzes the causes of emergencies, evaluates the rescue effect, generates rescue reports, and provides historical basis for future event management.

7.2.6   Typical Urban Intelligent Traffic Management System and Its Application

7.2.6.1   Intelligent Traffic Management System of Beijing

In recent years, Beijing has constructed an ITSM framework with "one center, three platforms and eight systems" as the core, which is highly integrates nearly 100 application subsystems such as video surveillance, individual positioning, 122 alarm and command, GPS police vehicle positioning, signal control, cluster communication, and strengthens the practical ability of intelligent traffic management. The structure of ITMS of Beijing is shown in Fig. 7.12.

1. Upgrade and Transformation System of Traffic Operation Monitoring and Intelligent Analysis Platform: The upgrade and transformation system of the transportation operation monitoring and intelligent analysis platform is considering the needs of transportation decision-making, and fully integrates real-time dynamic traffic flow data, public transportation operation data, industry statistics data, travel characteristic data, infrastructure data, transportation geographic information data, and urban background data to build a traffic decision database. The intelligent analysis platform is mainly based on the private cloud platform of the core calculation engine of Hadoop road network operation monitoring traffic operation evaluation and decision analysis system. It realizes special analysis of urban road network operation monitoring, public transport operation monitoring,

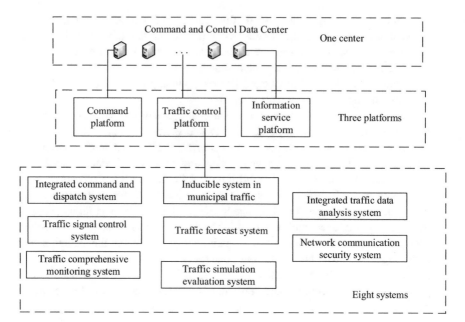

Fig. 7.12 ITMS of Beijing

and highway operation monitoring analysis, and establishes the traffic operation monitoring and intelligent analysis platform upgrading and transformation system that meets the stability requirements of TB data level and industrial application system.

2. Wireless Geomagnetic Vehicle Detection System: The wireless geomagnetic vehicle detection system (MVDS) uses fully automatic adaptive geomagnetic detection technology. By laying wireless magnetic induction intensity sensors on the road, it can perceive the change of spatial magnetic field when the vehicle passes through. It has the advantages of standard adaptive tracking and high recognition rate. In 2016, led by the Beijing Transportation Commission, the thirty-eighth Institute of China Electronic Technology Group and the Highway Research Institute of the Ministry of Transportation carried out the technical verification test of parking monitoring equipment. The wireless geomagnetic monitoring system has outstanding performance in accuracy, stability, and durability. It has passed the test contents such as emission spectrum, magnetic field sensitivity, waterproof and dustproof, high and low temperature and humidity, anti-interference, etc. Finally, the comprehensive accuracy of 91.4% is obtained and passed the test smoothly.

3. Video Surveillance System for Public Transport Vehicles: The video surveillance system for public buses has achieved full coverage of 1087 public buses. It is planned to complete the 3G/4G wireless dedicated line network access of 1087 buses on the existing basis, and complete the construction of supporting projects such as the company's image management center computer room and other

facilities and network expansion in 2018. Based on the original image information management system center and seven subcenters, a public transport vehicle video monitoring and management platform is constructed, which includes the upgrading of the image management center platform, and the increase of video control, management, and alarm functions. On the basis of integrating image resources, improve the image information sharing system with the Municipal Transportation Committee, the Municipal Public Security Bureau's Bus Corps, and the Municipal Public Security Transportation Administration to realize the effective sharing and utilization of image information resources.

4. Intelligent Operation Dispatching System for Public Transport Vehicles: In 2016, on the basis of the original function of intelligent dispatching system, the bus group company upgraded and improved the dispatching system, and completed the docking of important basic data such as people, vehicles, and lines of the dispatching system with the group information resource management platform. Besides, it finished the localization of dispatching system, line access of multiple operation modes and optimization of real-time computing module. The optimization of the report system realizes the unification and standardization of the basic data sources, the support of the dispatching system for the state without network, and the compatibility of the departure mode of all the registered lines in the group. It strengthens the usability and adaptability of the dispatching system, and the statistical data of the system is timelier and more reliable.

5. Image Management Platform of Public Transport Group: The image management platform of public transport group has built a security monitoring system of 1 center, 7 sub-centers, 16,000 vehicles, 215 midway stations, more than 100 bus stations, and 390 bus stations and 757 midway stations are under construction. It strives to realize the full coverage of bus and bus station monitoring system and the full coverage of bus midway station monitoring system in key areas, to realize the all-round, all-weather, and three-dimensional control of the ground bus system, and to realize the image and video docking with the transport administration, traffic administration and public security, so as to improve the overall level of public security order in the field of public transport. Since the establishment of public transport image system, more than 1000 videos have been collected for public security organs every year, which plays an important role in building a "Safe Beijing."

6. Customized Public Transport E-Commerce Platform: Beijing Public Transport Group launched a customized bus platform. Passengers in Beijing can participate in travel demand survey through the customized public transport platform, and book business public transport seats and online payment through the customized public transport platform. Completed custom public transport IC card identification system development and more than 130 business bus on-board equipment installation work, through the card to complete authentication. Complete the development of large customer group purchase, spare seat reservation, passenger automatic distribution, and other functions, which laid the foundation for business expansion. The development and online of commercial vehicle charter channel are completed. Customized bus mobile app officially launched,

passengers can be communicated and collected through the phone from line browsing, order submission to mobile payment process operation, etc. Customized public transport APP and WeChat also have the functions of passenger self-checking tickets, free ticket issuance and transfer, arrival forecast of fast direct line module and passenger circle.

7. Public Transport Route Inquiry Service App: "Public transport online" is a free real-time public transport information query app based on smart phones released by Beijing Public Transport Group. The public transport online APP realizes the real-time arrival broadcast of all public transport routes. At present, the download has reached 410,000 times. The arrival forecast function, after selecting a certain station on a certain line, you can query the distance from the nearest three vehicles to the station and the expected arrival time.

   Line customization function: Common lines can be customized, and the forecast information of the station on the line after customization will be displayed on the front page of APP, which is convenient and quick.

   The arrival reminder function: Set the line arrival alarm clock, when the vehicle arrives at a station a few minutes before reminding passengers to go out in advance.

   Transfer query module can query white/night public transport transfer information, and passengers can directly view real-time bus information in the transfer interface, more convenient to use. The user center module includes feedback, public transport news query, offline map download, version query, and other functions. In addition, the APP can be shared to more people through WeChat, circle of friends, QQ Zone, microblog QQ, and other ways.

8. Beijing Traffic Police Mobile App: With the help of Internet + technology, the mobile phone app of Beijing traffic police has built a comprehensive platform for traffic management information service, business office, and police–civilian interaction. It has opened 15 major functions, such as accident handling, entering Beijing for licenses, and illegal collection. By the end of the year, 3.8 million registered users, 230 million visits, and more than 300,000 daily business transactions have been registered. It ranks first in the same kind of mobile phone government software in the country, realizing more data and less running leg for the masses. It not only facilitates the handling of the masses, alleviates the window pressure and road congestion, but also provides data support for further improving traffic management services.

9. Online Passenger Flow Analysis and Decision Support Platform for Urban Rail Transit. Beijing Transportation Industry Data Center and Beijing Transportation Information Center have developed an online passenger flow analysis and decision support platform for urban rail transit based on the application demonstration project of the Internet of Things for rail transit safety and prevention of Beijing Transportation Committee. The platform is the first domestic rail transit integrated big data processing and application platform that integrates government industry supervision, enterprise operation and auxiliary decision-making and public travel information service. It covers eight functions of rail transit basic information management, online passenger flow prediction, four-layer

congestion information release based on station, interval, line and road network, new line access passenger flow prediction, emergency passenger flow prediction, train operation plan evaluation, passenger flow statistics and analysis and emergency disposal. The platform has the models and algorithms of historical passenger flow sorting, real-time passenger flow prediction, new line access passenger flow prediction, and rail transit passenger flow simulation with independent intellectual property rights, and is the first, leading and suitable rail transit passenger flow analysis and decision-making platform in China. Based on the B/S architecture, this platform uses ROLAP, MOLAP, HOLAP and other branch processing technologies to optimize the database management and application, and forms an efficient, fast, and low resource consumption rail transit passenger flow big data solution. The platform provides strong decision support for rail transit operation monitoring, operation management, emergency management and information services, and is an important means to realize the leap of rail transit from network construction to network operation. The comprehensive promotion of the platform in China has important guiding significance for the realization of rail transit network operation in various cities.

7.2.6.2 Intelligent Traffic Management System of Shenzhen

Shenzhen has built an ITMS that including "one platform, eight systems." One platform is traffic information sharing platform, eight systems include traffic information collection system, traffic control system, grid vehicle identification integrated application system, arterial traffic guidance system, parking guidance system, traffic incident system, intelligent traffic violation management system, closed-circuit television monitoring system, etc. Among them, traffic information collection includes traffic control system, grid vehicle identification integrated application system, intelligent traffic violation management system, CCTV monitoring system, traffic incident system, etc. Traffic information integrated platform includes common information platform. Applications and services include arterial traffic guidance system, parking guidance system, intelligent traffic control system, etc.

1. Intelligent Traffic Control System: Intelligent traffic control system is the core system of ITS, which plays an important role in ITS system and is the main data source of platform and other systems. At present, there are 1308 signal intersections in the city, including 828 intelligent traffic signal control network intersections and 707 coil detectors. 46 adaptive control intersections can be realized. Shenzhen intelligent traffic signal control system is a distributed adaptive control system. Signal phase timing can be automatically carried out according to traffic flow. The classification of system function modules is clear, and the classification of system function modules is clear. Shenzhen intelligent traffic signal control system is in a leading position in China, and it can realize multi-period coordination control and dynamic optimization coordination control. The traffic management department of Shenzhen takes the lead in breaking through the technical threshold of countdown time. According to the change of traffic signal control

each cycle, the correct cycle and green light release time are calculated in advance, which solves the related technical problems of traditional countdown time. This technology is the first in the country, and its scientific and technological content is in the forefront of the country. It has high real-time and integrity, and it is shown in different directions according to the new national standard. It combines the intelligent pedestrian countdown with the motor vehicle countdown, which fully reflects the humanized management.

2. Grid Vehicle Identification Integrated Application System: Grid vehicle recognition integrated application project is a comprehensive information application system based on real-time monitoring and data collection of vehicle license recognition information. It provides vehicle information related application services for multiple units or departments, including public security traffic police, public security command, public security criminal investigation, public security technical investigation, municipal grid office, environmental protection bureau, land planning bureau, transportation bureau, etc. It realizes the monitoring, analysis, and prediction of road traffic state, the monitoring, management, and early warning of specific vehicles, the real-time calculation of various traffic violations, and the comprehensive analysis of vehicle information. Up to now, a total of 118 monitoring sections of the construction monitoring points have been completed, and the grid has been initially formed. Its scope basically covers the main roads in the city, as well as the second and the third line gateways in the city. It has realized the traffic illegal monitoring functions such as overspeed, off-road driving, fake license plate vehicles, yellow mark vehicles, single and double license plate limit. As well as the blacklist warning control, vehicle monitoring, and other social security functions, has achieved great economic and social benefits.

3. Traffic Incident Detection System: Traffic incident detection system is a part of traffic information collection and a high-level data detection system. The system will obtain the processed basic data and combine with manual detection to achieve timely warning of traffic incidents. The event detection system transmits the relevant information to the comprehensive control system, and then the comprehensive control system is released to the guidance system, command system and other related systems, which provides information guarantee for the rapid positioning, rapid alarm, and rapid evacuation of traffic events.

4. Arterial Traffic Guidance System: As a part of Shenzhen's traffic guidance system, the arterial traffic guidance system generates guidance information according to the guidance strategy and publishes it in time by using the collected real-time traffic information, so as to effectively guide the travel vehicles, improve the utilization rate of existing roads, realize the balanced distribution of traffic flow in the road network, and provide good services for the safe and rapid driving of drivers. Combined with the actual situation of Shenzhen City, the system takes the shared information platform as the data support, integrates the advanced traffic concept and mathematical model concept, establishes the arterial traffic guidance model, optimizes and calculates all kinds of information such as traffic flow, delay time and travel time, automatically generates guidance

information, and automatically publishes large screens to travelers in real time through VMS, which lays the foundation for the construction of vehicle guidance, and can implement the dynamic information publishing function.

5. Parking Guidance System: Parking guidance system is a part of the whole urban traffic guidance system. Through the comprehensive analysis of the road network, arterial traffic flow, parking distribution, and parking spaces in the area, the driver is guided to choose parking spaces and driving lines from blind driving, which effectively reduces the invalid driving caused by parking, so as to comprehensively alleviate local traffic congestion and balance road network traffic flow. At present, the construction of Prince Shucheng subdistrict and People's Nanzi subdistrict has been completed, including 9 parking lots and 18 parking guidance cards. South of the people's access to 39 parking lots, set up 58 parking guidance signs.

6. Common Information Platform: Common information platform is the basic subsystem of ITS, an important basis for decision-making of other subsystems, and a key technology to realize the transformation of traffic management from simple static management to intelligent dynamic management. The shared information platform is a communication network platform established to meet various traffic information needs by formulating the interface and functional connection requirements between subsystems of intelligent traffic management system. The shared information platform strengthens the comprehensive processing capacity of traffic information and facilitates the decision-making of traffic managers and travelers. The common information platform takes the traffic model library as the main body, uses the data warehouse technology, carries on the quantitative analysis to the traffic data, through the quantitative analysis, enhances the decision-making science, provides the plan or the data support for the traffic control and the management decision-making, provides the traffic condition and so on various kinds of basic data for ITS each system, provides the traffic data report for the traffic management department such as the monitoring center, the command center, and provides the basic statistical analysis report for the traffic decision-makers.

7.2.6.3 Intelligent Traffic Management System of Wuhan

Wuhan city actively promotes the construction of intelligent traffic management system, establishes the "five centers" of urban traffic fine management, and realizes the transformation from administrative order-driven work to information instruction-driven work. The five centers are mainly composed of the following five parts:

1. Big Data Center. The main duties of big data center include:
    Collecting data: Implement intelligent perception plan, vigorously promote the construction of vehicle networking and electronic identification of transportation facilities, and improve the front-end traffic data collection capacity; integrating resources, integrating data of functional departments such as transportation, planning, urban management, water affairs, education, and meteorology with

the Municipal Network Information Office. They signed strategic cooperation agreements with Didi Travel and other Internet companies to collect and summarize Internet data. At present, a total of more than 200 types of traffic data are collected to form a "big data pool." Nearly 700 million new data are added daily, which lays a solid foundation for the three-dimensional description of the relationship between people, vehicles, and roads in Wuhan.

Data Analysis: The construction of Wuhan Traffic Management Cloud Platform and the industry optimal algorithm is used to provide accurate analysis of big data for traffic travel, government affairs, and management. At present, 22 functions such as traffic information release, government service, and traffic management have been realized.

Application Data: Through data analysis, the data are transformed into intelligence and instructions, which directly serve accident prevention, blocking removal and order management, and truly make decisions with data and speak with data.

Leading Innovation: In the big data center, two joint venture studios, Jiangbei and Jiangnan, are set up to invite enterprises and universities to enter, deeply integrate practical needs with frontier technologies, and jointly innovate high-end and deep applications of big data in the field of transportation and management.

2. Information Command Center: The command room, research and judgment room, video studio, traffic order optimization room "four rooms in one," so that the intelligence information network collection, full source combing, full judgment, on the basis of the original command center function, the "intelligence, command, service" integration, the formation of intelligence information research and judgment, command dispatching, service organization, rapid response, evaluation feedback closed loop mechanism. The city's traffic police, auxiliary police, and social rescue forces do "one map display, one key dispatching, one call."

3. Safety Risk Control Center: Establishing risk control center, carrying out traffic safety prediction, monitoring and evaluation in advance, responsible for risk prediction and early warning, supervision of source management, guidance of precision strike, accident depth investigation, promoting coordinated governance, security theory research, and so on.

4. Law Enforcement Supervision Center: The "data police gauge" platform is built to correlate and compare the data of mobile law enforcement terminals such as digital radio, PDA, alcohol tester, and law enforcement recorder, so as to realize the real-time upload of law enforcement video, law enforcement data, and real-time alarm of law enforcement anomalies. The whole process of police law enforcement on duty is traceable and can be closed-loop managed. Next, the functions of law enforcement supervision center will gradually expand to team management, discipline inspection and supervision, performance management, and become the performance evaluation center of the traffic police force.

5. Public Relations Linkage Center: The traffic management department of Wuhan has set up a public relations linkage center, set up a public opinion cloud platform, adhere to the "people have to call, I have to respond," from the masses of the most

resentment, the most hope, the most urgent things. Through the public opinion cloud platform to gather 14 different channels of public opinion, so that the full collection of public opinion, full processing, full feedback. The establishment of news public opinion studio, on the one hand, to strengthen the propaganda report. The organization carried out the "three concessions" concentrated publicity in the city, advocating motor vehicle courtesy zebra line, courtesy emergency channel, courtesy special channel around the campus (convenient for students to take the vehicle to go the special channel, allowing stop and go). Through the "Wuhan good driver" selection activities, traffic illegal rectification "micro broadcast" and other activities, we can improve the citizens' awareness of civilized traffic, establish the authority of the rule of law, and advocate civilized travel. On the other hand, strengthen public opinion guidance. Through authoritative publication, speech guidance, and hot spot transformation, the emotions of citizens and netizens are guided to the fact.

Based on the "five centers" of urban traffic fine management, the main systems established in Wuhan are as follows:

1. Wuhan Public Security Bureau Public Security Traffic Intelligent Integrated Control System: Through the unified interface specification and adaptive access service, the system collects individual vehicle (identity/behavior) traffic information of each card port and electric police manufacturer located in the provincial/city/county (each core area) entrances and exits, high-speed national and provincial highway entrances, urban sections or intersections, and realizes the aggregation, analysis, storage and application of vehicle information in large-scale, high-density, continuous, and long-term road network. For the business application of public security traffic management, the system provides quasi-real-time inspection and control alarm (interval speeding, annual inspection, to be scrapped) and group vehicle traffic characteristics analysis (traffic flow, average travel time, OD) applications for individual traffic illegal suspected vehicles. For public security, the system provides early warning and screening applications for suspected vehicles (high-risk suspected vehicle warning, series and parallel suspected vehicle analysis, accompanying vehicle analysis, and suspected vehicles at the scene of the incident). The system uses cloud computing and big data technology, which is suitable for cities with more installations and large traffic volumes of trucks and electric police equipment. The system can deeply mine massive data, and has strong parallel expansion ability. The calculation ability of the system does not decrease with the increase of time and data volume.

2. Wuhan Static Traffic Management Comprehensive Analysis System: The static traffic management comprehensive analysis platform of Wuhan Traffic Management Bureau is an important business application system of Wuhan Traffic Management Bureau. Under the constraints of the unified standard and standard system of Wuhan Traffic Management Bureau, it integrates the unified system management and safety management system of Wuhan Traffic Management Bureau, and shares the existing public support environment and resources of Wuhan Traffic Management Bureau. The system releases parking information

such as the number of available parking spaces and parking lot information through a variety of channels to provide parking guidance services for the public, including roadside parking guidance screens, and Internet applications such as auto-navigation. Full information sharing is achieved with traffic control command and dispatch system, inspection and distribution control system, public security and other application systems, public security bureau application system, and the police magic cube to be built. The static traffic management comprehensive analysis platform is responsible for the collection of static traffic information within the scope of traffic management, and provides data support for other systems to maximize the benefits of data applications. Form a close business linkage with the inspection and control system, strengthen vehicle-related security management capabilities, and make full use of the powerful capabilities of the inspection and control system to strengthen static traffic management capabilities.

3. App System of Wuhan Yixing River City: The app system of Wuhan Yixing River City in cloud is an intelligent traffic comprehensive service platform developed and constructed by using advanced computer, GIS, mobile Internet, and cloud computing technology, which realizes information sharing and exchange between traffic-related business departments. The project platform is based on the urgent problems of urban traffic (such as congestion, public travel information service, self-service business processing, etc.), with the basic purpose of improving government service ability and management level, ensuring smooth road and traffic safety, serving public travel, and facilitating public life. Provide comprehensive travel information services for the public to promote the harmonious development of urban people, vehicles, and roads. The system mainly provides comprehensive information services for public security traffic control users, social travelers, and third-party company users. With the thinking of "Internet + transportation," the system focuses on public travel demand, integrates transportation travel service information, provides comprehensive transportation services for the public, expands the coverage of information services, realizes intelligent transportation guidance, and makes public travel more convenient and efficient.

4. Wuhan Traffic Operation Coordination Command Center: Wuhan Traffic Operations Coordination Command Center (TOCC) specifically includes one large-screen display system, two infrastructure projects, and three business application platforms. The comprehensive traffic operation monitoring platform realizes the comprehensive monitoring of the city's traffic operation status and the thematic monitoring of administrative behavior supervision; Integrated traffic vehicle positioning information analysis platform to achieve the city 's ground bus, taxi, "two passengers and one management," integrated traffic emergency daily management, emergency disposal and mobile applications, and display TOCC project results on large screen and computer desktop.

7.3 Overview of Intelligent Traffic Signal Control System

7.3.1 Brief Introduction of Traffic Signal Control System

7.3.1.1 Development Process

Traffic signal control is a traffic management measure that assigns traffic rights to conflicting traffic flows in time by means of various control hardware and software devices (such as artificial, traffic signal lights, electronic computers, etc.) in places where traffic flow space separation cannot be achieved (mainly intersections). Reasonable traffic signal control can reduce traffic congestion, ensure smooth urban road and avoid traffic accidents.

As early as the nineteenth century, people began to use signal lights to direct vehicles on the road and control the order of vehicles passing through intersections. In 1868, the British pioneered signal control on the road. In London, a gas lamp that alternately sheltered red and green glass was used as a signal lamp for traffic control. In 1914, Cleveland in the United States began using an electric light source timing signal machine. In 1917, Salt Lake City introduced an interconnected signal system, and then New York, Chicago, and other cities began to appear manually controlled red, yellow, and green three-color signal lights. In 1926, the British set up the first automatic traffic signal machine in Wolverhampton, which laid the foundation of urban traffic signal automatic control. In 1928, the United States developed the world's first inductive signal machine, and the flexible coordinated timing control system was born. For the first time, it realized the adjustment of traffic signal time according to traffic flow. In 1952, Denver developed a signal network timing scheme selective signal control system by using analog computers and traffic detectors. In 1959, Toronto, Canada, carried out experiments and established a set of traffic signal control system (UTC) controlled by IBM650 computer in 1963. For the first time, computer technology was applied to traffic control, greatly improving the performance and level of the control system, marking the development of urban traffic signal control has entered a new stage.

In the 1960s, all countries in the world successively applied computer technology to traffic control, studied the signal linkage coordinated control system with large control range, established the mathematical model to simulate the traffic flow situation, so as to effectively alleviate the increasingly tense urban traffic problems, and developed many signal control systems. Typically, the Traffic Network Study Tool (TRANSYT) system developed by the British Institute of Transport and Road in 1966, the Sydney Coordinated Adaptive Traffic System (SCATS) system developed by Australia since the 1970s, and the Split Cycle Optimization Technique (SCOOT) system developed by the British Institute of Transport and Road in conjunction with three companies in the early 1970s. After the 1980s, with the development of information technology, urban traffic control began to develop toward informatization and intelligence. In the 1990s, developed countries have begun to appear intelligent traffic control system, and urban traffic control system into the intelligent transportation system, become an important subsystem of advanced traffic management system. As of 2000, more than 480 major cities in the world have adopted advanced traffic signal control systems.

Since the end of the twentieth century, with the continuous development of information technology and control technology, a variety of new control systems have emerged in response to various traffic conditions, such as Real-Time Adaptive Control System (RT-TRACS) system, Strategic Real Time Control for Megalopolis Traffic (STREAM) system, Method for the Optimization of Traffic Signal in Online Controlled Network (MOTION) system, Signal Management in Real Time for Urban Traffic Networks/Traffic-responsive Urban Control (SMART NETS/TUC), etc. China has basically established regional traffic control systems in cities at the provincial capital level through introduction or independent research and development.

7.3.1.2   Genealogical Classification

The development of traffic control is a process of continuous practice. In the process of practice, people put forward and developed many different types of traffic control methods and control systems, and classify traffic signal control systems from different angles.

1. Classification by control method

Traffic control can be divided into timing control, induction control, and adaptive control according to the methods method.

Timing Control: Timing control refers to the operation of the traffic signal controller according to the preset timing scheme. The scheme divides the day into several periods. According to the historical average traffic flow data of different periods in the day, the signal control parameters such as long cycle and green signal ratio corresponding to different periods are calculated offline. Single-stage timing control with only one timing scheme a day. One day according to the traffic volume of different periods of several timing scheme is called multistage timing control.

Actuated Control: Actuated control refers to a traffic control method that sets vehicle detectors on the intersection inlet, and the traffic signal controller can adopt appropriate signal display time to meet the traffic demand according to the real-time traffic flow at the intersection detected by the detector. The timing scheme of the signal lamp in this way is calculated by a computer or an intelligent signal controller, and can be changed at any time with the traffic information detected by the detector. Depending on the location of the detector, it can be divided into semi-actuated control and fully-actuated control.

(a) Semi-actuated Control: Semi-actuated control refers to the actuated control that only detectors are set on some inlet roads (generally secondary roads) intersections.

(b) Fully-actuated Control: Fully-actuated control refers to the actuated control that detectors are set on all inlets of the intersection.

Adaptive Control: Adaptive control refers to a control method that regards the traffic system as an uncertain system, and the system can continuously measure its state (such as traffic flow, parking times, delay time, queue length, etc.), track and predict the change trend of traffic state, change the adjustable parameters of the system or generate a control scheme for a certain control goal, so as to achieve the optimal or suboptimal control effect. According to the different control methods, it can be divided into decision-selection and decision-generation.

(a) Decision-selection: The decision-selection refers to that the corresponding control strategies and schemes are stored in the computer in advance corresponding to different traffic flows. When the system is running, the most suitable control strategies and schemes are selected according to the real-time collected traffic flow data to implement traffic signal control.
(b) Decision-generation: The decision-generation refers to calculating the optimal traffic control parameters in real time according to the real-time collected traffic flow data, forming the signal control timing scheme, and immediately manipulating the signal controller to operate the traffic signal lamp according to this Scheme.

2. Classified by control range: Traffic control can be divided into single point control, arterial coordination control, and regional coordination control according to the spatial range involved in the control.

Single Point Control: Single point control is referred to as point control, which means that the traffic signal control at each intersection only operates independently according to the traffic situation of the intersection, and does not exchange any information with its adjacent intersections. It is the most basic form of traffic signal control at the intersection. The point control is applicable to the situation where the distance between adjacent intersections is far, the line control effect is not large, or because the traffic demand of each phase changes significantly, the long cycle of the intersection and the independent control of the green signal ratio are more effective than the line control.

Arterial Coordination Control: The arterial coordination control is referred to as line control, also known as green wave control. It refers to connecting the traffic signals of several continuous intersections on the arterial line in a certain way, coordinating the green light start time and signal timing scheme of traffic signal lights at each intersection, so that the vehicles do not often encounter red lights when passing through these intersections. The basic idea of line control is to set the speed of the section on the specified traffic line, and hope that the vehicle will meet the green light every time it reaches the first intersection. However, in fact, it is difficult to meet the requirements that all roads have green lights because the speed of each vehicle is different and varies at any time, and there are left and right such as left and right turns and vehicle access at the intersection. However, it is possible to make the vehicles along the road encounter few red lights, reduce the parking and queuing delay of most vehicles, and maintain the continuous traffic on the arterial line. The key to line control is that the signal cycle of each intersection must be the same.

Regional Coordinated Control, referred to as "face control," refers to the multiple signal intersections in a region as a whole for mutual coordination, control area of traffic signals are controlled by the traffic control center centralized management control. For regions with small scope, the whole region can be centralized controlled. For a wide range of areas, it can be partitioned hierarchical control. According to the different control strategy and control structure, the regional coordinated control system can be classified as follows:

1. Classified by Control Strategy.

   (a). Timing Off-line Control System: The timing off-line control system uses the historical and current statistical data of traffic flow to optimize off-line processing. The optimal signal timing scheme of multi-period is obtained and stored in the controller or control computer to implement multi-period timing control for the whole regional traffic.

   Timing off-line control system uses historical and current traffic flow statistics to optimize off-line processing, and obtains the optimal signal timing scheme of multi-period, which is stored in the controller or control computer to implement multi-period timing control for the whole regional traffic. The timing offline control system is simple, reliable, and efficient investment ratio is high, but it cannot adapt to the random changes of traffic flow, especially when the traffic data is outdated, the control effect is significantly reduced. When the optimal timing scheme is re-established, traffic investigation will consume a lot of labor. At present, the mature timing offline control systems include TRANSYT, Corridor Simulation (CORSIM), Progression Analysis and Signal System Evaluation Routine (PASSER), etc.

   (b). Actuated Online Control System: Actuated online control system sets up a vehicle detector in the control area traffic network, collects traffic data in real time, identifies the traffic model, and then obtains the optimization problem related to the timing parameters, solves the timing scheme of the problem online, and implements on-line optimal control.

   The actuated online control system can respond to the random changes of traffic flow in a timely manner, and the control effect is good, but the control structure is complex, the investment is high, and the reliability of equipment is high. Currently more mature online control systems are SCATS system, SCOOT system.

2. Classification by Control Structure.

   (a) Centralized Control Structure: Centralized traffic signal control at all inter-sections in the region is achieved in a control center by connecting all the signal machines in the region into a network, using one small and medium-sized computer or multiple computers online, as shown in Fig. 7.13.

   (b) Hierarchical Control Structure: The entire traffic signal control system is divided into upper and lower subsystems. The upper subsystem receives the

Fig. 7.13 Centralized
traffic control system
structure diagram [2]

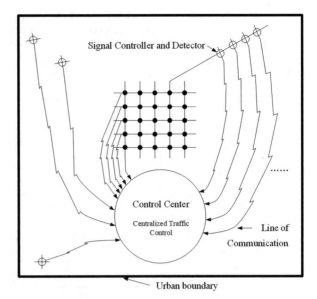

Signal Controller and Detector

Control Center

Centralized Traffic Control

← Line of Communication

Urban boundary

intersection timing scheme from the lower subsystem. These timing schemes are coordinated and analyzed from the overall perspective, so that the timing scheme of the lower subsystem is corrected. The lower subsystem makes necessary adjustments according to the revised scheme. The upper subsystem mainly completes the coordination and optimization task of the whole system, and the lower subsystem mainly completes the execution task of intersection timing adjustment in the region. The hierarchical control structure is generally divided into three levels, as shown in Fig. 7.14.

The first level is located at the intersection and is controlled by a signal controller. Its functions should include: (a.) monitoring equipment failures (detectors, signal lights, and other local control facilities); (b.) collection and aggregation of test data; (c.) transmission of data on traffic flows and equipment performance to second level of control; (d.) receive and operate instructions issued by superiors.

The second level is located in a relatively central position in the controlled area, and its functions should include: (a.) monitoring the data of traffic flow and equipment performance from the first-level control and transmitting them to the third-level control center; (b.) manipulate the first-level control, determine the type of control to be implemented (single-point control or regional control), select control methods, and coordinate the first level control.

The third level is located in a reasonable center of the city, and should act as a command and control center, responsible for the coordinated control of the whole system. The control center can monitor the data of any signalized intersection in the control area, receive and process the data of traffic flow conditions, determine the control strategy of the second-level control, and provide monitoring and display

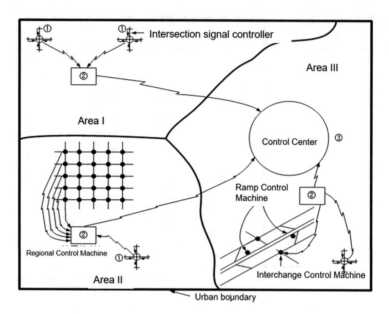

Fig. 7.14 Layered traffic control system structure diagram [2]

equipment. In addition, the control center can receive information on equipment failures in order to take appropriate measures.

7.3.2 Typical Intelligent Traffic Control System

In order to adapt to the rapid increase in traffic volume and alleviate road traffic congestion, many cities in China and abroad have invested a lot of labor and material resources in the research and development of traffic signal control system, and have achieved a series of results. Among them, the more successful systems are SCATS (Australia), SCOOT (UK), Real-time Hierarchical Optimized Distributed and Effective System (RHODES, USA), Signal Progression Optimization Technology (SPOT)/Urban Traffic Optimization by Integrated Automation (UTOPIA, Italy).

7.3.2.1 SCATS System

SCATS is a real-time scheme selection adaptive control system developed by the Roads and Transport Agency (RTA) of New South Wales, Australia. It has been studied since the 1970 s and has been put into use in Sydney and other cities since 1980.

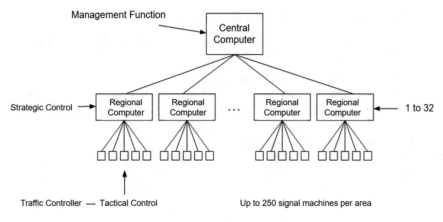

Fig. 7.15 SCATS system structure diagram [3]

1. Structure of SCATS System

The control structure of SCATS is hierarchical three-level control. The highest level is the central monitoring center, the middle level is the regional control center, and the lowest level is the signal control machine. When a signal controller is controlled by a regional control center, usually every 1–10 signal controllers are combined into a "subsystem," and several subsystems are combined as a relatively independent system. The systems are basically irrelevant of each other, and there is a certain coordination relationship between the various subsystems within the system. With the real-time changes in traffic conditions, the subsystems can be combined or re-separated. The choice of the three basic timing parameters is based on the subsystem as the accounting unit. The structure of SCATS system is shown in Fig. 7.15.

The central monitoring center, in addition to the centralized monitoring of the operating status of the entire control system and the working status of various equipment, also has a computer dedicated to the management of the system database to monitor the various data of all regional control centers and the control of each signal control machine. Operating parameters are dynamically stored (in the form of a constantly updated dynamic database). Traffic engineers can not only use these data for system development, but also all the development and design work can be completed on the machine (off-line work mode).

While SCATS implements the overall coordinated control of several subsystems, it also allows each intersection to "work independently" and independently implement vehicle induction control. The former is called "strategic control," and the latter is called "tactical control." The organic combination of strategic control and tactical control greatly improves the control efficiency of the system. SCATS uses the vehicle detection device set near the parking line to control the traffic effectively and flexibly in this way. Therefore, SCATS is actually a scheme selection control system that uses induction control to make partial adjustments to the timing Scheme.

2. Signal Timing Parameter Optimization of SCATS System

The following part briefly introduces the main links of the signal timing parameter optimization of the SCATS system.

(a) Division and Combination of Subsystems

The division of subsystems by SCATS is determined by traffic engineers according to the historical and current data of traffic flow, and the environment and geometric conditions of the transportation network, and the established subsystems are the basic units of the control system. In the process of optimizing timing parameters, SCATS uses the "merging index" to determine whether adjacent subsystems need to be merged. In each signal cycle, the "merging index" needs to be calculated once. When the difference of signal cycle time required by the two adjacent subsystems is not more than 9 s, the cumulative value of "merging index" is "+1," otherwise is "−1." If the cumulative value of the "merge index" reaches "4," it is considered that the two subsystems have reached the merged standard. The merged subsystem can be automatically redivided into the original two subsystems when necessary, as long as the cumulative value of the "merging index" drops to "0."

After the subsystems are merged, the signal cycle duration of the new subsystem will adopt the longer of the signal cycle durations performed by the original two subsystems, and the other of the original two subsystems will then slow down or accelerate the growth of its signal cycle. Speed until the "external" phase difference scheme of these two subsystems is realized.

(b) SCATS Timing Parameter Optimization Algorithm

SCATS takes the subsystem composed of 1–10 intersections as the basic control unit, and these intersections have a common cycle length. On each entrance lane of all intersections, a vehicle detection device is installed, and the detection device (such as an inductance coil) is separately installed behind the stop line of each lane. According to the real-time traffic volume data provided by the vehicle detection device and the actual traffic volume of the parking line section during the green light period, the algorithm system selects the common cycle time of each intersection in the subsystem, the green signal ratio and phase difference of each intersection. Considering the possibility of merging adjacent subsystems, it is also necessary to select an appropriate phase difference for them (the phase difference outside the subsystem).

As a real-time scheme selection control system, SCATS requires that the control area be divided into several subregions in advance according to traffic conditions by using offline calculations in advance. According to the measured traffic data inside the subregions, four green signal ratio schemes, five internal phase difference schemes (referring to the relative phase difference between the intersections in the subsystem), and five external phase difference schemes (referring to the phase difference between adjacent subsystems) are developed for each intersection. The real-time selection of signal cycle and green signal ratio is based on the overall needs of the subsystem, that is, the length of the shared cycle is determined according to the needs of key intersections in the subsystem. The corresponding green time of the intersection determines the percentage of each phase of green light in the signal cycle according to the principle of equal or close saturation of each phase. Obviously, with the adjustment of signal cycle, the green time of each phase also changes, and under the condition that the scheme remains unchanged, according to the actual measurement situation, the green signal ratio and phase can be slightly adjusted to adapt to changes in traffic flow.

SCATS selects the signal period, green signal ratio, and phase difference as independent parameters, and the algorithm used in the optimization process is based on "class saturation" and "comprehensive flow."

- Degree of Saturation

The Degree of Saturation (DS) used in SCATS refers to the ratio of green light time which is effectively used by traffic flow g′ to green light display time g. The calculation formulas of DS and g′ is as follows:

$$DS = \frac{g'}{g} \tag{7.3}$$

$$g' = g - (T - th) \tag{7.4}$$

where DS refers to the Degree of Saturation, g means the total time of display green light for vehicle traffic (s), g' means green light time effectively used by vehicles (s), T means the time when there is no car on the parking line during the green light (s), t refers to a necessary idle time between the front and rear vehicles when the traffic normally passes the section of the parking line (s), h means the essential null number, and parameters g, T, and h can be provided directly by the system.

- Comprehensive Flow

Since the vehicle detection coil used in SCATS system is as long as 4.5 m, the detection accuracy of the vehicle throughput is affected, especially when traffic congestion occurs at the intersection. In order to avoid using parameters directly related to vehicle type (body length) to represent traffic flow, SCATS introduces a virtual parameter called "comprehensive flow" to reflect the number of mixed traffic flows through the parking line. The integrated flow rate q′ refers to the converted equivalent of vehicles passing through the stop line during a green light period. It is determined by the directly measured class saturation DS and the maximum flow rate S that actually occurred during the green light period. The calculation formula is as follows.

$$q' = \frac{DS \times g \times S}{3600} \tag{7.5}$$

where q' means comprehensive flow (veh), and S refers to the maximum demand recorder flow rate (veh/h).

(c) Selection of Signal Cycle Time

The selection of the signal cycle time is based on the subsystem, that is, within a subsystem, the cycle time that the entire subsystem should use is determined according to the intersection with the highest class saturation. SCATS is equipped with vehicle detectors on each inlet lane of each intersection. The largest DS value measured directly by each detector in the previous cycle is taken out, and the cycle length that should be used in the next cycle is determined accordingly.

In order to maintain the continuity of signal control at intersections, the signal cycle is adjusted in continuous small steps, that is, the length of a new signal cycle is limited to ± 6 s compared with the previous cycle.

For each subsystem range, SCATS requires four limits of signal cycle change, namely, the minimum value of signal cycle (C_{min}) and the maximum value of signal cycle (C_{max}), which can obtain the medium signal cycle duration (C_s) with good continuity of two-way traffic in the subsystem range and the signal cycle (C_x) slightly longer than C_s. In general, the selection range of signal period is only between C_{max} and Cs. Only when the traffic flow arrival detected by the vehicle detector at the key position is lower than the predetermined limit, the signal period value less than Cs and even C_{min} is adopted. The signal period higher than C_x is determined by the "critical" import lane detection data (DS value). These "critical" lanes are significantly higher than other lanes, requiring more green light release time, and thus need to get "preferential" lanes from the increase of the signal cycle.

(d) Choice of Green Ratio Scheme.

In SCATS, the selection of green ratio schemes is also based on subsystems. Four green ratio schemes were prepared for each intersection for real-time selection. These four schemes aim at the ratio of each phase green light time to the signal cycle length (usually expressed as percentage) under four possible traffic loads. In each green signal ratio scheme, not only the time of green lights in each phase is stipulated, but also the order of green lights in each phase is stipulated. In different green signal ratio schemes, the order of signal phase may be different, that is, in SCATS, the order of intersection signal phase is variable.

In the green signal ratio scheme of the SCATS system, the flexibility of multiple choices is also provided for local tactical control (vehicle induction control mode at unit intersection). Affected by fluctuations in traffic arrival rates, some phases may have surplus green time under the established green signal ratio scheme, while others may have insufficient green time allocated. Therefore, without lengthening and reducing the signal cycle time, it is possible and necessary to make a reasonable residual and deficiency adjustment for the green light time of each phase changing with the real-time traffic load. This requires specific provisions on possible adjustment methods in the green credit ratio scheme. At some intersections, the green light time of some phases may not be suitable for shortening due to the requirements of vehicle induction control. Then, in the plan, pay special attention to the green light time of these phases, which can only be lengthened but not shortened.

The selection of green signal ratio scheme should be carried out once in each signal cycle, and the general process is as follows. In each signal cycle, four green signal ratio schemes are compared and their "selection" is "voted." If a program is "elected" twice in three consecutive cycles, it is selected as the implementation program for the next cycle. In an inlet lane, only the lane with the highest DS value is considered as the green signal ratio.

The selection of the green signal ratio scheme is interleaved with the adjustment of the signal period. Combining the two, the result of continuous adjustment of the green light time of each phase, so that the saturation DS of each phase is maintained at approximately the same level, which is the "equal saturation" principle.

(e) Selection of Phase Difference Scheme.

In SCATS, the internal and external phase difference schemes must be determined in advance and stored in the central control computer. Each category contains five different programs. During the operation of the system, the phase difference must be selected in real time for each signal cycle. The specific steps are as follows.

In the subarea, the first scheme of the five phase difference schemes is only used in the case that the signal cycle time is exactly equal to C_{min}. The second scheme is only used when the signal period satisfies $C_s < C < C_s + 10$. The rest three schemes are selected according to the real-time detected comprehensive flow value. For each relevant entrance road, the traffic flow and saturation that the inlet can release should be calculated when three phase difference schemes (the third, fourth, and fifth schemes) are implemented respectively, which is essentially similar to the widest passband method. SCATS is to compare the green wave bandwidth provided to each entrance road by the above three schemes. The larger the passband width can be provided, the more obvious the superiority of this scheme is. After comprehensive comparison, the most superior phase difference scheme is selected. Four selected options in five consecutive cycles were selected as new ones for implementation.

The external phase difference scheme is also selected in the same method as the internal Scheme.

3. Characteristics of SCATS System.
   SCATS systems have the following characteristics.

   (a) The detector is installed in the stop line and there is need to establish a traffic model, so its control scheme is not model based.
   (b) The optimization of periodic length, green ratio, and phase difference is based on the measured class saturation value in a number of predetermined schemes.
   (c) The system can change the phase sequence according to the traffic demand or skip the next phase (if there is no traffic request for this phase), so it can respond to the traffic demand of each cycle in time.
   (d) It can automatically divide the control sub-zone and has the function of local vehicle induction control.
   (e) Each cycle can change the cycle time.

The disadvantages of SCATS system include the following.

(a) The traffic model is not used, which is essentially a real-time plan selection system, which limits the degree of optimization of the timing plan and is not flexible enough.
(b) The detector is installed near the stop line, and it is difficult to monitor the movement of the vehicle fleet. Therefore, the optimal reliability of green time difference is poor.

7.3.2.2   SCOOT System

SCOOT stands for "green signal ratio-periodic-phase difference optimization technique." And it is a real-time scheme generative adaptive control system. It was jointly developed by the British TRL (Transport Research Laboratory, TRL) and three companies in 1973, and was officially put into use in 1979. After years of development, the SCOOT system has been upgraded many times, and more than 200 cities around the world are currently using the system.

Principle and Structure of SCOOT System

SCOOT processes the vehicle arrival information collected by the vehicle detector installed at the exit of the upstream intersection online, forms the control scheme, and adjusts the parameters such as green signal ratio, cycle time, and phase difference in real time to adapt to the changing traffic flow. The system is mainly composed of four parts: the collection and analysis of traffic data, traffic model, optimization and adjustment of traffic signal timing parameters, control of the signal system.

All calculation and analysis in SCOOT system are based on periodic flow diagram, which can be calculated by SCOOT program. The diagram of periodic flow change is a histogram of the change of traffic volume with time in a cycle, which is represented by the ordinate and the abscissa (limited by the length of a cycle).

In order to facilitate calculation, a signal cycle is usually divided into several periods, each period is about 1 s to 3 s. In the SCOOT traffic model, the basic data of all the calculation process are the average traffic volume, turn traffic volume and turn length of each period. In order to describe the whole process of traffic flow running on a line, SCOOT uses the following three types of cycle flow diagrams.

1. Arrival flow diagram (referred to as the arrival diagram). This diagram shows the change in the arrival rate of the traffic to the downstream stop line without being blocked.
2. Exit flow diagram (referred to as exit diagram). This graphic depicts the changes in actual traffic flow as it leaves the upstream intersection.

3. Saturated exit diagram (referred to as full flow diagram). The "full flow" diagram is actually a flow diagram that leaves the stop line at a saturated flow rate. This icon appears only when the traffic flow passing through during the green light is saturated.

SCOOT system is based on real-time measured traffic data and uses traffic models to optimize timing. The system processes the traffic volume information collected by the vehicle detector to form the Cyclic Flow Profiles (CFP), which is then combined with the static parameters pre-stored in the computer, such as the running time of the fleet on the connection, the signal phase sequence and the phase time wait for calculation in the traffic model together. The SCOOT optimization program thus calculates the best combination of signal timing, and the obtained optimal timing plan is immediately sent to the signal machine for execution.

SCOOT optimization uses a small step size asymptotic optimization method, which does not require too much calculation, so that it can follow the instantaneous change of CFP, that is, the signal timing can be changed slightly with the change of CFP, which can ensure the timing of the timing plan. The adjustment will not cause major disturbance to the operation of traffic flow, but it can add up to produce a new coordinated control mode in time.

In addition, SCOOT has special monitoring and response measures for the possible traffic congestion and congestion on the road network. It can not only monitor the working state of each component of the system at any time, and give an automatic alarm for the fault, but also provide operators with details of the signal timing scheme being executed at each intersection, the queuing situation of vehicles in each cycle (including the actual position of the queue tail) and the flow arrival schema and other information at any time. It can also automatically display these information on the terminal equipment.

SCOOT system is a two-level structure. The upper level is the central computer, and the lower level is the intersection signal machine. Traffic volume prediction and timing optimization are completed on the central computer, signal control, data collection, processing, and communication are completed on the signal machine. The structure of the system is shown in Fig. 7.16.

Signal Timing Parameters Optimization of SCOOT System

The following is a brief introduction to the four main links of the SCOOT system's optimized timing plan.

1. Detection

• Detector

SCOOT uses ring coil detector to detect traffic data in real time. In order to avoid leakage and retest, the coil adopts 2 m × 2 m square ring. When parking is not allowed at the roadside, it can be buried in the middle of the lane. Sensors should be

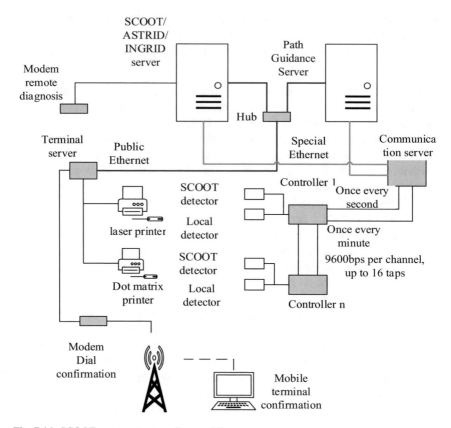

Fig. 7.16 SCOOT system structure diagram [4]

buried in all lanes. One sensor detects one or two lanes. When two lanes share one sensor, the sensor can be located across the middle of the lane.

- Appropriate Location of the Detector

Through real-time detection, SCOOT can achieve the purpose of real-time prediction of traffic flow pattern and system performance index (PI value) on the parking line. Therefore, the appropriate location of the detector is the place as far as possible from the downstream parking line, which is usually set at the exit of the upstream intersection. When choosing the location of the detector, the following factors should be considered.

(a) When there are branch lines or intermediate entrances and exits between the two intersections, and the traffic volume is greater than 10% of the traffic flow of the arterial line, the detector should be installed as far as possible at the downstream of the branch or intermediate entry and exit, otherwise supplementary detectors should be set at the branch or intermediate entrances and exits.

(b) The detector should be located downstream of the bus stop to avoid other vehicles missing detection due to detours.
(c) The detector should be located downstream of the sidewalk. Considering that the speed requirement of the vehicle passing through the detector is basically equal to the average speed of the section, the sensor should be at least 30 m away from the pedestrian crossing.
(d) The detector is set at a distance from the downstream parking line at least equivalent to 8 ~ 12 s of driving time or more than the maximum queue length of vehicles in a cycle.

The advantages of setting detectors in this way include:

(a) It can detect the current cycle flow in real time, and predict the cycle flow chart to the parking line in real time.
(b) Real-time detection of the current cycle queue length to avoid aggravating traffic jams caused by vehicles passing the upstream intersection.
(c) It can detect the degree of vehicle congestion in real time.
      The disadvantage of setting up the detector in this way is that it is not as good as setting it close to the parking line to detect the saturated flow and perform the function of actuated control in real time.

• Collection of Traffic Data

The traffic data that SCOOT detector can collect include the following:

(a) Traffic volume.
(b) Occupy time and occupancy rate. Occupied time is the time when the detector senses the vehicle passing, and the occupancy rate is the ratio of occupancy time to the entire cycle time.
(c) The degree of congestion. Measured by the occupancy rate of the hindered vehicle fleet, SCOOT divides congestion degree into eight grades (0–7) according to occupancy rate, which is called congestion coefficient. Congestion coefficient is sometimes one of the goals of SCOOT timing optimization.

In order to accurately collect the time when the sensor passes through the vehicle or not, the sampling period should be short enough. The SCOOT detector automatically collects the induction signals of each sensor every 0.25 s, and analyze and processes it.

2. Division of Subregions

The division of SCOOT system subregions is predetermined by the traffic engineers, and the system operation is based on the delimited subregions. It cannot be combined or divided during operation, but SCOOT can have dual-cycle intersections in the subregions.

3. Model

• Cycle Flow Chart—Vehicle-Group Forecast.

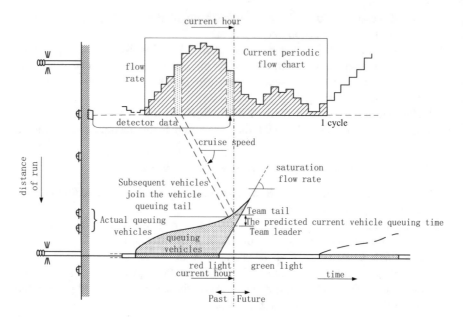

Fig. 7.17 Vehicle queuing forecast schematic [2]

The SCOOT system is processed according to the traffic information (traffic volume and occupancy time) detected by the detector, and then drawn in real time into a vehicle arrival cycle flow diagram on the cross-section of the detector. Then, on the periodic flow chart of the sensor section, through the discrete model of traffic flow, the periodic flow chart that arrives at the stop line is predicted, that is, the "arrival" pattern. The unit of the ordinate of the SCOOT cycle flow chart is lpu (connected traffic chart unit), which is a mixed measurement unit of traffic volume and occupancy time, and its function is equivalent to the conversion function of pcu. Corresponding to the unit of measurement of traffic volume, lpu, the unit of saturated traffic on the parking line is also changed to lpu.

- Prediction of Queue

Figure 7.17 is the principle diagram of vehicle queue length prediction. The upper right side is the "arrival" pattern on the detector section measured by the detector, which is updated every cycle; the lower right side is the predicted queuing diagram on the stop line section. SCOOT controls the time of the signal light by the computer, so the computer understands the current state of the signal light and adds the vehicles during the red light period to the queue. After the green light is on, the vehicle will drive out of the parking line at a certain "saturated flow rate" (prestored in the computer database) until all the vehicles in the queue are dissipated. Since vehicle speed and fleet dispersion are difficult to accurately estimate, the predicted queuing must be inspected and corrected on the spot. The inspection usually compares the actual observed vehicle queuing length with the displayed predicted queuing length, for example, the predicted queuing length has not reached the detection The cross-section of the detector, but the detector is actually occupied by vehicles, indicating that the SCOOT model underestimates the queue length.

- Congestion Prediction

In order to control the queue extending to the upstream intersection, the blocked queue length must be controlled. The traffic model calculates the congestion coefficient according to the occupancy rate of the detection, which can reflect the degree of vehicle obstruction. At the same time, since the SCOOT detector is located on the exit road near the upstream intersection, when the detector detects that there is a car parked on the detector, it indicates that the queue is about to extend to the upstream intersection.

- Effectiveness Prediction

SCOOT uses the weighted sum of delay and parking times, or fuel consumption as a comprehensive efficiency index (PI), but SCOOT sometimes uses congestion coefficient as one of the efficiency indexes.

From the above queuing prediction, SCOOT can predict the delay and parking times under each timing scheme.

The influence of congestion on signal timing optimization increases with the increase of congestion. Consider reducing the degree of congestion in the timing optimization, and the congestion coefficient can also be listed as one of the comprehensive performance indicators. The index used in the comprehensive efficiency index should be determined according to the control strategy. For example, during peak hours, reducing vehicle delay is the main control objective; at short-distance intersections, considering the need to avoid queuing up and blocking upstream intersections, the congestion coefficient can be one of the control objectives.

In addition, SCOOT regards saturation as the basis for the optimal period length, because the saturation decreases (increases) as the period length increases (shortens). When the saturation reaches 100%, serious traffic congestion is bound to occur, so SCOOT controls the saturation not to exceed 90%.

4. Optimization

The signal parameter optimization strategy of SCOOT system is to make frequent and appropriate adjustments to the optimal timing parameters as the traffic arrival volume changes. The appropriate amount of adjustment is small, but due to the frequent adjustments, the continuous accumulation of these frequent adjustments can be used to adapt to the traffic trend in a period of time. This optimization strategy is one of the main reasons for the success of SCOOT, with the following four major benefits.

- Appropriate adjustment of the timing parameters will prevent excessive ups and downs, which can avoid the instability of traffic flow caused by sudden changes in timing.

- Since the timing parameters only need to be adjusted quantitatively, the optimization algorithm is greatly simplified, and the adaptive control of real-time operation can be realized.
- Frequent adjustments avoid the problem of long-term prediction of traffic flow.
- The amount of adjustment of timing parameters each time is not large, but because of frequent adjustments, it can always track the trend of adapting to traffic changes.

The optimization methods of signal parameters (green ratio, phase difference, signal cycle, etc.) are introduced below.

1. Optimization of Green Ratio
   The key points of the optimization of the green ratio are as follows.

   - SCOOT optimizes the green ratio of each intersection individually.
   - A few seconds before the start of each phase, recalculate whether the current green time needs to be adjusted.
   - The adjustment of green light duration takes ±4 s as the step length to optimize the green light duration. In other words, the PI value after adjusting ±4 s is compared with the PI value that maintains the original state. The scheme with the smallest PI value is selected and sent to the signal control machine, which is a temporary change.
   - With each temporary change, the system controller makes a permanent change of ±1 s of the green light duration. After storage, it is used as the starting point of the next change. This trend adjustment is conducive to tracking the traffic change trend in a period of time.
   - In addition, SCOOT also needs to consider the minimum total saturation at the intersection, the length of vehicle queue, the degree of congestion and the limitation of the shortest green light duration when optimizing the green ratio.

2. Optimization of Phase Difference
   The main points of phase difference optimization are as follows:

   - When SCOOT optimizes the phase difference, the unit is the subregion.
   - SCOOT performs a phase difference optimization calculation for each intersection (whether the phase start time changes or not) before each cycle.
   - The adjustment of phase difference is also ±4 s.
   - The method of optimizing the phase difference is the same as that of optimizing the green signal ratio, but the minimum sum of PI values on all adjacent roads is taken as the optimization objective.
   - When optimizing the phase difference, the queue between the short-distance intersections must be considered to avoid the queue tail of the downstream intersection blocking the traffic of the upstream intersection. SCOOT first considers the continuity of traffic between these intersections, and if necessary, it can sacrifice the coordinated control between signals on the long line (which can accommodate large queue vehicles) to ensure that there is no queuing and blocking the upstream intersection on the short line.

3. Length Optimization

   The key points of cycle length optimization are as follows.

   - When SCOOT optimizes the cycle length, the unit is the subregion.
   - SCOOT calculates the cycle length of each intersection in the subregion every 2.5 to 5 minutes. The cycle length of the key intersection is taken as the shared period length in the subregion.
   - The goal of cycle length optimization is to limit the saturation of key intersections in the sub-region to 90%. When the saturation is small, the cycle length decreases, and the capacity decreases, which can increase the saturation. Saturation close to 90%, stop reducing cycle length. When the saturation is large, the cycle length increases, and the capacity increases, which can reduce the saturation.
   - The adjustment of period length is ± 4 s–± 8 s.
   - SCOOT considers the selection of double periodic signals when adjusting the cycle length. If the overall PI value is optimal due to the double periodic signals, the selected period length can be adjusted separately.
   - SCOOT also needs to consider the upper and lower limits of the signal cycle. The lower limit is mainly for the consideration of traffic safety, taking into account the minimum time required for pedestrians to cross the street safely and the minimum green light time for vehicles to be released at one time. Generally, 30–40 s is the lower limit of the signal cycle. The upper limit is to minimize vehicle delay on the premise of meeting the maximum traffic requirements. The upper limit is determined by the local actual traffic conditions (vehicle type composition, crossroad plane size, and traffic load size, etc.). Overall 90–120 s is commonly used as the upper limit in foreign countries. Due to the large number of bicycles and low-speed large vehicles in China, it can be appropriately relaxed (120–200 s).
   - The influence of traffic congestion coefficient is not considered in the cycle length optimization, so the SCOOT system only considers the congestion coefficient in the green signal ratio and phase difference optimization.

4. Improvement

   In order to deal with saturated or supersaturated states, SCOOT version 2.4 and above versions have corresponding improvements, mainly including the following points.

   - Gate Control

     The main purpose of gate control is to limit the flow of traffic to sensitive areas and redistribute the convoys to roads that can accommodate longer convoys, so as to prevent the formation of long convoys or congestion in the region. In order to realize the sluice control, SCOOT must be able to modify the signal timing of the intersection, which may be far from the relevant area or even in another subarea. Gate logic allows one or more lines to be defined as critical or bottleneck lines. The gate connection is designated as the connection of the storage fleet. Without these connections, the bottleneck connection will be blocked. When the bottleneck connection reaches a predetermined saturation, the green light of the gate connection will decrease.

All control logic is contained in the green ratio optimizer. For a bottleneck connection, traffic engineers need to determine its critical saturation, which can be problematic if exceeded. This critical saturation is used to trigger the gate algorithm. The possible role of the latter is that if the saturation is less than or equal to the critical saturation, and both judgments are so, the gate does not work. If the saturation of the gate is greater than the critical saturation, and the two judgments are the same, the gate will work, usually to reduce the green light time of the gate connection. However, the gate logic may also cause the increase of the green light time of the gate connection downstream of the bottleneck, so as to release the fleet of the gate connection as soon as possible. All changes are governed by the normal green ratio optimizer.

- Saturated Phase Difference

  Under saturation conditions, the phase difference requirement for a connection is different from the requirement under normal conditions (to minimize the delay of the connection). At this time, the setting requirement of phase difference is to maximize the traffic capacity. When the upstream intersection shows the green light to the critical entrance, this connection will not be saturated. When a line is measured saturation, the saturation phase difference will be forced to take, and the phase difference optimizer will abandon its optimization results.

- Use the Information of Adjacent Connections to Deal with the Saturation Problem.

  To solve the saturation problem, one connection can use its own information and the saturation information from another connection together, or only use the latter. If the convoy of a link is too long and reaches the detector of the upstream link, the saturation of the upstream link can be regarded as the result of the saturation of the link. At this time, the upstream link should be regarded as the saturation information source of the link, and the timing plan of the downstream intersection of the link should be adjusted.

Characteristics of SCOOT System

SCOOT system has the following characteristics.

- A flexible and accurate real-time traffic model can be used not only to determine the signal timing scheme, but also to provide various information such as delay, number of stops, and congestion data for traffic management and traffic planning.
- Adopting short-term forecasting form, only forecasting the traffic conditions in the next cycle, and controlling according to the forecasting results, greatly improving the accuracy of forecasting and the effectiveness of control.
- The frequent small increment form is used in the optimization and adjustment of signal parameters, which not only avoids the disturbance caused by the mutation of signal parameters to the vehicle operation in the controlled road network, but

also adapts to the large changes of traffic conditions through frequent cumulative changes.

- The vehicle detector of the system is buried at the exit of the upstream intersection, which provides sufficient time for the optimization and adjustment of the signal timing at the downstream intersection. At the same time, it can prevent the convoy from blocking to the upstream intersection (measures are taken to avoid it before this happens).
- It has the ability to identify the state of the detector. Once the detector fails, it can take corresponding measures in time to reduce the impact of detector failure on the system.

The shortcomings of SCOOT system include:

- The establishment of traffic model requires a large number of road network geometry and traffic flow data, so it is time-consuming and laborious.
- Signal phase cannot be automatically increased or decreased, and phase sequence cannot be automatically changed.
- Saturated flow rate check is not automated, so site installation and debugging are more cumbersome.
- The problem of automatic partition of control subregions has not been solved.

7.3.2.3 RHODES System

Real-time Hierarchical Optimized Distributed and Effective System (RHODES) system was successfully developed by Arizona University in 1996, and field tests were successively carried out in Tucson and Tempe, Arizona. The results show that the system has good control effect on semi-crowded traffic network. The core technology of RHODES system mainly includes three parts, controlled optimization of phase (COP), effective green band (Realband), and prediction algorithms (road network load prediction, road network traffic prediction, and intersection traffic prediction). In the version after 2000, the bus priority function was added by introducing the concept of "Busband."

Principle and Structure of RHODES System

The Main Principle of the RHODES System: Due to technological advances in the fields of communication, control, and computers, it is possible to quickly transmit and process information and implement multiple control strategies flexibly. The traffic flow changes randomly, and the number of vehicles arriving in the same time interval is also random. This change provides an opportunity to improve the performance of the traffic control system at the intersection control layer; similarly, the number of vehicles in the fleet also changes. This small change also provides an opportunity to improve the performance of the traffic control system. The effect of these opportunities alone to improve the performance of the control system may be

Fig. 7.18 RHODES system
principle [5]

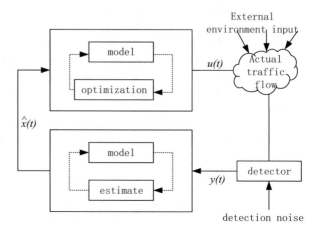

insignificant, but if each opportunity is fully utilized, the overall improvement effect will be quite objective. The RHODES system makes full use of this point, obtains the necessary information of the traffic flow in advance through the prediction model, and makes timely and effective responses to it in advance. The system principle is shown in Fig. 7.18.

The Structure of the RHODES System

The RHODES system is a two-level structure in terms of physical structure, namely the central computer level and the signal controller level. From this point of view, it is similar to the SCOOT system. But it decomposes the system control problem into a three-layer hierarchical structure, namely, the intersection control layer, network control layer, and network load distribution layer, as shown in Fig. 7.19.

At the intersection control layer, the traffic flow prediction, phase sequence, and green light duration control are mainly based on the detected traffic flow and various constraints. This control must be performed every second. At the network control layer, it mainly predicts the driving situation of the fleet, so as to establish coordination constraints for each intersection in the network. This prediction is carried out every 200–300 s. At the network load distribution layer, the total traffic demand in a long period (usually 1 h) is mainly predicted, so as to determine the upper bound of the queue length in the future. Many technologies in Advanced Traveler Information System (ATIS) and dynamic traffic flow allocation can be implemented at this layer. Although this distributed structure increases the communication tasks between the intersection controllers, it reduces the calculation tasks of the central controller and the communication tasks with the intersection controllers, and it is easier to implement the communication between the intersection controllers, which makes RHODES's real-time adaptive optimal control of traffic flow becomes possible.

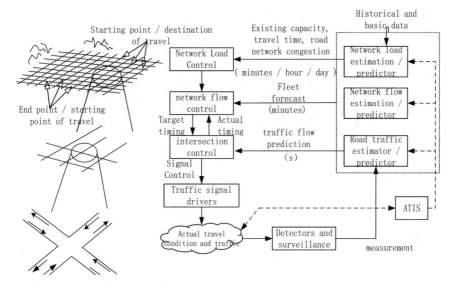

Fig. 7.19 Hierarchical structure of RHODES system [4]

Control Method of RHODES System

The RHODES system tracks the fluctuation of traffic flow in a short period of time and the change trend of traffic flow in a long period of time by predicting the arrival of a single vehicle and the movement of the vehicle fleet. According to the principle of optimizing the given performance index (minimum average delay, minimum average queue length at the intersection or minimum number of stops), the phase is divided, and a non-parametric control model is proposed. In other words, the signal timing scheme is not formulated by using parameters such as period, phase difference and green signal ratio, but by using phase sequence and phase length.

Intersection Control

The intersection control layer proposes a concept of phase controllable optimization (COP). According to the predicted value of arriving vehicles, the dynamic programming (DP) method is used to find the optimal phase sequence and phase length, so that the given performance index can reach the optimal, and the phase sequence can also be a pre-given fixed sequence. In order to apply dynamic programming in real time, the RHODES system uses a sliding time window to reduce the amount of calculation, and uses a dynamic programming algorithm to optimize the phase sequence and phase length. Before optimization, the phase sequence should be given (which can be given by the network flow control of the upper layer), so that each phase corresponds to a phase, and the number of phases is generally greater than the number of phases, which may be the same phase corresponds to different stages. If the traffic engineer has no restriction, it is allowed to skip the phase by setting a certain phase length to 0, so as to achieve the purpose of optimizing the phase sequence. The COP is carried out in a sliding time window of 45 ~ 60 s. Each possible decision in each stage is evaluated in a forward recursive manner, and then the backward recursive decision is made to minimize the phase sequence and phase length of the system performance index in the optimal sliding time window. The decision in the first stage is implemented. Then, before the current phase is about to end, the next optimization will begin with the current phase based on recent observations and projections. COP can use different performance indicators (including delays, queue length, and number of stops) to make the control algorithm more flexible.

Network Flow Control

The network flow control layer proposes an algorithm called real-time green band (Realband). Its principle is based on the current fleet forecast and comprehensively consider the possible conflicts of the fleet in all directions of the network, and use the decision tree method to control the network. The traffic signal is coordinated and optimized, and a traveling green wave band is generated. Its width and speed value can optimize the network objective function, that is, the number of delays and stops is the least. When the RHODES system predicts that there is a conflict between the green signal demands of two or more fleets, it first generates a decision tree, and each branch of the decision tree represents a conflict resolution strategy. Use the corresponding model to evaluate each strategy, and select the strategy with the best performance as the control strategy of the network control layer. The control decision of the network layer is used to establish coordination constraints for the intersection control layer.

The real-time green band algorithm combines the advantages of the coordinated control system based on the delay model (such as SCOOT) and the coordinated control system based on the bandwidth model (such as SCATS), takes full account of the continuity of the vehicle fleet in all directions, and generates the green band online according to the predicted value of the arrival time, size and speed of the vehicle fleet, as well as the degree of dispersion or compression and the interference of the steering vehicle flow, so as to ensure the continuity of the vehicle fleet as much as possible and optimize the system performance index.

Realband identifies the fleet and predicts its movement in the network (including the time to reach the intersection, the size and speed of the fleet) by filtering and fusing the traffic flow data obtained from different information sources in recent minutes, and then uses the APPES-NET model to evolve the predicted movement of the fleet through the network within a given time window. Traffic signals provide a suitable green time for the predicted fleet to optimize the given performance indicators. If there is a conflict between the two teams arriving at the intersection at the same time on the demand for green signals, one of the teams is given priority to the right of passage or one of the teams is divided into two parts, so that the given performance index is optimal. The main goal of Realband is to resolve the conflict between the vehicle groups' demands for green signals.

Characteristics of RHODES System

The RHODES system has the following characteristics.

1. RHODES is a two-level structure in hardware, namely the central computer level and signal controller level. But the control structure adopts a three-tier hierarchical structure, namely the intersection control layer, network control layer, and network load distribution layer. The intersection control layer tracks the fluctuation of traffic flow in a short period of time (45–60 s) for intersection traffic control. The network control layer tracks the change trend of traffic flow in a long time (200–300 s) to establish coordination constraints between intersections. The network load distribution layer tracks the long-term traffic flow variation and provides interfaces with other modules of the intelligent transportation system.
2. Proposed a new coordinated real-time adaptive control strategy between intersections: "Real-time green band." According to the current fleet forecast value, comprehensively consider the possible conflicts of the fleet in various directions on the network, use the decision tree method to optimize the network, and generate the traveling green wave belt in real time. Its width and speed value can optimize the regional objective function, that is, delay and the least number of stops.
3. Install the vehicle detector at the vehicle entrance of the intersection to detect the traffic flow in the three directions (turn left, turn right, and go straight) at the upstream intersection. The detection results can be used for downstream predictions as well as the original predictions at the entrance. Verification and correction of results.
4. A concept of controllable phase optimization is proposed. According to the predicted value of arriving vehicle, the optimal phase sequence and phase length are found by dynamic programming method. In order to apply dynamic programming in real time, the system adopts sliding time window to improve the accuracy of control.
5. The system optimization objective can be to minimize the average vehicle delay, the average queue length at the intersection or the number of parking. Flexible and diverse system objectives make the control algorithm more flexible.
6. The interface with traffic analysis software is provided, which can offline evaluate the advantages and disadvantages of timing schemes or be used as a research tool.

7.3.2.4 SPOT/UTOPIA System

SPOT/UTOPIA system is a distributed real-time traffic control system developed by Mizar Automazione in Italy. The earliest version was installed in Turin, Italy, in 1985 and achieved satisfactory results. The SPOT/UTOPIA system consists of two parts, SPOT (local) and UTOPIA (region), and its structure is shown in Fig. 7.20. UTOPIA is a relatively advanced area control, which optimizes the use of a macro-traffic model based on historical data. SPOT completes the local optimization work,

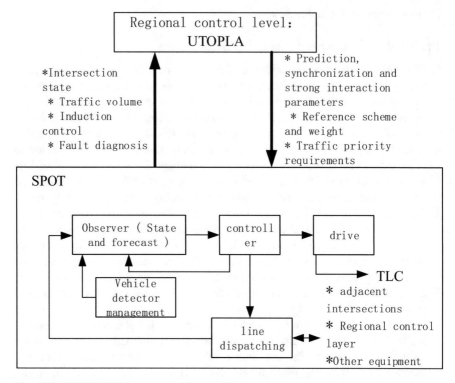

Fig. 7.20 SPOT/UTOPIA system architecture [6]

uses the micro model on each traffic control machine, and uses the data from the local control machine and the area model to optimize a single intersection. The system adopts the concept of "strong interaction" to ensure the optimality and robustness of the area control.

SPOT System

When SPOT works independently, it is a small distributed traffic control system, and there are generally no more than six intersections managed by SPOT system. There is no need to install the central computer when it is used below six intersections. When the system is large, the UTOPIA central computer control system needs to be added. Each intersection machine must install a SPOT unit, which can communicate with the traffic light controller and other intersection machines, so the communication mode of each intersection is equivalent.

At each intersection, SPOT seeks to use the minimum total cost function. This "cost" function mainly considers delays, number of stops, remaining capacity, area control strategies recommended by superior control, priority of public transportation and special vehicles, pedestrian crossing requests, and other special circumstances.

SPOT continuously repeats and adjusts to minimize the intersection cost function. In each calculation cycle, all SPOT units exchange information about traffic status and priority strategies with their neighboring SPOT units. The optimization goal is to ensure that the total travel time of the private car is the shortest under the premise that the bus does not encounter the red light, so the maximum weight is given to the lost time item of the bus at the intersection.

SPOT has two control principles when optimizing, prediction (each SPOT unit receives real-time arrival prediction from upstream intersections) and strong interaction (in local optimization, each control unit considers the negative impact it may bring to the downstream intersection). Adjacent intersections exchange data every 3 s, and at the same time, each intersection signal controller performs an optimization on the sliding time window. The length of the time window is 2 min, which means that the result of each optimization can only run for 3 s. When necessary, the regional control machine UTOPIA can participate in the adjustment of traffic demand forecasts and modify local parameters.

Both SPOT and UTOPIA have a complete set of operational rules for estimating traffic demand and traffic model parameters. In SPOT, the queue length is estimated every 3 s, and the turn ratio and road capacity are improved every cycle. There is also a congestion detection model, which greatly reduces the system adjustment and repeated time.

SPOT can provide good adaptive control for all traffic conditions, and it can play a greater role when traffic flow has large, sudden and unpredictable changes. Without sacrificing adaptive performance, SPOT can give absolute priority to bus or other special vehicles.

The SPOT unit does not directly control the signal lamp, but controls the traffic signal controller, and obtains sensor data from the traffic signal controller or directly from the vehicle detector. SPOT control unit can also be connected with beacon controller of variable information flag and navigation system.

The SPOT unit is connected to a LAN, and the communication structure can be radial, raster, or layered. Each SPOT unit is a node of the network and has complete message link capability. Therefore, any SPOT unit can transmit information to the appropriate terminal, which greatly reduces the installation and maintenance costs and enhances the robustness of the system.

UTOPIA System

UTOPIA uses a central computer in a network to run. UTOPIA calculates an optimized control strategy for each subarea in the network, and each subarea has a common cycle length. UTOPIA provides an event scheduler and a conversion model, which can provide TCP/IP protocol interfaces for other systems. During operation, the system maintains a historical database, including regular flow, turning ratio, saturated flow, and life cycle. UTOPIA also maintains a statistical database, mainly for specific statistical calculations on historical databases.

UTOPIA provides the following features:

1. Equipment diagnosis: Real-time monitoring of the status of intersection equipment (traffic light controller, roadside units, detectors, communication lines, other peripherals).
2. Traffic data detection: Continuous real-time monitoring of traffic data in the control area (cycle length, queue length, saturation flow rate, traffic flow, and turn ratio) and regular updates.
3. Performance monitoring: Monitor system performance in real time through validity index and failure indicator of intersection equipment.
4. Historical data analysis: The files about the measured and estimated traffic data and diagnostic information are automatically managed, including the analysis detection data, visualization of historical data, and comparison of historical data with current data.
5. Generation and transmission of regular reports: Through e-mail regular transmission of statistical reports on system and equipment functions, the form, and configuration of reports can be customized by users.
6. Automatic alarm generation: The alarm can be automatically generated by SMS (Short Message Service) and e-mail. Alarm generation can be customized in the following aspects—the selection of crossroads related to alarm, the definition of alarm threshold for selected equipment and parameters, the selection of alarm types for each recipient, the number of alarm receivers and e-mail addresses.
7. Service management: This function allows permission setting for other personnel.
8. Provide simulation environment for planning: Online network simulation tools can compare and analyze different scenarios. System control and UTOPIA can be simulated to evaluate the impact of UTOPIA installation or the impact on existing system improvements.
9. Flexible interaction with the system: While monitoring system and machine performance, authorized users can directly set functional parameters and send commands to UTOPIA system. According to the architecture of UTOPIA, the crossroad controller interacts with the adjacent crossroad controller and intersects with the regional control reference strategy calculated by the central system mutual, so as to control the signal intersection.

Characteristics of SPOT/UTOPIA System

SPOT/UTOPIA system has the following characteristics:

1. The function of bus priority is considered in the design and development of the system, so its control goal is to ensure that the bus does not encounter red light as far as possible, so that the total travel time of private cars is the shortest. In order to achieve this goal, the system introduces the concept of weight in minimizing the objective function, and the maximum weight gives the bus loss time term at the intersection.

2. In order to ensure the optimality and robustness of subarea control, the system adopts the concept of strong interaction, that is, the objective function of this intersection should consider the state of adjacent intersections and the constraints given by the regional control level.
3. UTOPIA is to make optimal control decisions for the whole network. It receives the intersection state information from each SPOT unit, determines the division of subregions, the optimal period of subregions (each subregion runs under the same period), and the optimal weight. In addition to communicating with SPOT, the system can also exchange information with a higher layer of system (such as traffic command center) through TCP/IP interface. UTOPIA also established a database of actual state information during operation, so that backup scheme selection can be carried out when real-time system faults occur.

7.4   Application and Development Trend of Intelligent Traffic Control System in China

The research on traffic signal control system in China started late. In the late 1970s, Beijing adopted DJS-130 computer to study the coordination control of arterial lines. Since the 1980s, on the one hand, the state has adopted the policy of combining introduction and development to establish some urban road traffic control systems. On the other hand, investment in urban traffic signal control technology research and development to adapt to China's mixed traffic as the main characteristics of intelligent traffic control system. Such as Beijing introduced SCOOT system, Shanghai introduced SCATS system. Shenyang Institute of Automation, Chinese Academy of Sciences, built and implemented the first urban traffic adaptive control system in Dalian. The Ministry of Communications, the Ministry of Public Security and Nanjing completed the "Seventh Five-Year" Nanjing urban traffic control system, the key project, has laid a good foundation for the development of traffic signal control system in China. At present, China's city annual increase and update of urban traffic signal control intersections are in ten thousand scale, the emergence of Wuxi Huatong, Nanjing Rice, Beijing Yihualu, Zhejiang University Central Control, Shanghai Baokang, Lianyungang Jerry, Shanghai Electric Jun Code, Nanjing Toronto, Nanjing Luopu, Zhejiang Dahua, and other mainstream signal control manufacturers. At the same time, China has gradually formed some representative traffic signal control systems, such as Nanjing Rice urban traffic control system, Qingdao HiCon traffic signal control system, Shenzhen Greenway SMOOTH traffic signal control system, etc. In China urban road traffic management and control plays an increasingly important role.

7.4.1   Nanjing Rice Urban Traffic Control System

Nanjing Rice Urban Traffic Control System (NUTCS) is the first real-time adaptive urban traffic signal control system developed by China. With the approval of the former National Planning Commission and the National Science and Technology Commission, NUTCS was jointly developed by the Institute of Traffic Management Science of the Ministry of Public Security, Tongji University, 28 Institutes of the Ministry of Electronics (now the 28th Institute of China Electronics Technology Group Co., Ltd.) and Nanjing Traffic Police. It is a key national science and technology research project (No. 2443) in the "Seventh Five-Year Plan" and has won many technical awards from the Ministry of Public Security and the state.

NUTCS combines the advantages of SCOOT and SCATS to meet and adapt to the domestic road network density with low density and wide intersection spacing and the outstanding traffic characteristics of mixed traffic. It can automatically coordinate and control the traffic signal timing plan in the area, and balance the traffic flow of the road network. Operation to minimize the number of stops, delays and environmental pollution, and give full play to the traffic benefits of the road system. Usually, the two-level control structure at the intersection level and the regional level can be expanded to the three-level distributed hierarchical control structure at the intersection level, the regional level and the central level if necessary. The system set up real-time adaptive, fixed timing and no cable linkage control three modes, with security, fire protection, ambulance, bus signal, and manual designated functions. When necessary, manual intervention can be used to directly control the intersection signal to execute the designated phase to ensure the smooth flow of urban road traffic and the priority passage of special vehicles. The working mode is flexible and the function is complete.

In order to further study the signal control system adapted to the characteristics of urban traffic flow in China, Nanjing Rice Large Electronic Systems Engineering Co. Ltd. was established in July 1988 by the 28th Institute of China Electronic Science and Technology Group Corporation. Nanjing Rice independently developed the first generation of signal machine technology through years of efforts. After the first set of networking signal machines suitable for China's national conditions was put into the market, it was first popularized and applied in Zhuzhou City, Hunan Province, effectively improving the local traffic congestion, reducing the intensity of traffic police work, and obtaining user recognition.

In 2014, Nanjing launched the construction of urban traffic signal machine networking and bus signal priority control system. During the Nanjing Youth Olympic Games, Nanjing Rice made important contributions to traffic security and special service tasks, and ensured the traffic safety of the National Public Sacrifice Day on December 13. After the bus signal priority system is completed, the average speed of the bus increases by 15%, and the bus is the number of parking is reduced by 30%, which creates great social and economic benefits. At the same time, Nanjing has strengthened the signal coordination and control, and built 154 green wave belts and 1063 intersections. Among them, 41 tidal green wave belts and 287 intersections have been built. The urban green wave signal control rate reaches 80%, which effectively improves the urban traffic efficiency.

In 2017, Nanjing Rice developed an intelligent control system of urban road network signal based on holographic detection. The "spatiotemporal three-section holographic collection" mode was applied to obtain the queuing information of vehicles on the import road in space, as well as the passing information of vehicles on the stop line, the middle part of the channel, and the end of the channel. The traffic demand of each direction of the green light at the beginning, the middle, and the end of the green light was analyzed in time. The multidimensional spatiotemporal information was deeply integrated to master the law of vehicle arrival and departure, and the formation and dissipation of queuing. At the same time, the advanced solutions of holographic control are proposed for single intersection and arterial road, which can minimize the delay of intersection control, eliminate the phenomenon of green light in flat peak, balance the queue of each flow in peak, and ensure the coordinated control of green wave at upstream and downstream intersections. The system has been applied in Taishan Road and Huangshan Road of Hexi Street in Nanjing. According to statistics, in the full intelligent control mode, the signal light in the flat peak period to reduce empty waiting, peak period to reduce queuing reflects a more obvious advantage.

After promoting the use of Rice road traffic signal control system and signal products, Nanjing has obtained the first-class management level city title of "unimpeded project" for nine consecutive times. At present, there are more than 1500 intersections in Nanjing using Rice traffic signal control machine to manage the intersection traffic signal. More than 150 roads in the urban area have realized green wave control, and the green wave signal control rate is 80%. At present, Nanjing Rice's traffic signal control system has developed to the third generation of products, mainly distributed in Nanjing, Zhuzhou, Changshu, Foshan, Qinhuang-dao, and other 120 domestic large and medium-sized cities, and has been extended to Kenya, Pakistan, and Cote d'Ivoire and other overseas countries.

7.4.2 Qingdao Haixin HiCon Traffic Signal Control System

HiCon adaptive traffic signal control system is an intelligent transportation solution developed by Qingdao Haixin Network Technology Co., Ltd., including HSC-100 series traffic signal machine, HiCon traffic signal control system software, and CMT traffic signal machine configuration and maintenance tool software. According to the current situation of mixed traffic, HiCon system establishes a machine-non-mixed control model to control mixed traffic flow. The multilevel distributed control structure is divided into control platform layer, control center layer, communication layer and intersection layer. The system has a complete algorithm system, including regional coordination control algorithm, inductive coordination control algorithm, pedestrian secondary crossing algorithm, coordinated control algorithm of urban rapid access and urban intersection, and emergency detection algorithm. It supports NTCIP open protocol and meets the latest national standards.

In 2002, about 50 development staff of Haixin Network Technology Co., Ltd. took more than 3 years. According to the characteristics of China's transportation, a centralized coordinated signal machine-HSC-100 signal machine was developed. At the end of 2003, the maritime communication network technology successfully won the bid of Qingdao Huangdao District and Longkou City traffic signal control system, which opened the prelude of the maritime traffic signal control system entering the market. In December 2005, HiCon adaptive traffic signal control system won the bid of Beijing intelligent traffic management investment construction project, completely breaking the monopoly situation of foreign companies in high-end signal controller, which is a milestone in the history of China's traffic signal control development. At the beginning of 2006, Haixin again won the bid for the construction project of Beijing expressway traffic signal control system, which played a good role in promoting the comprehensive improvement of Beijing traffic conditions. Its system characteristics are as follows, using NTCIP communication protocol, the system is complete, and has good versatility and compatibility. Efficient, reliable, and open communication subsystem ensures the reliability of internal real-time communication performance, efficiency, scalability. At the same time, the openness of the system is truly realized. The system interface is transparent, providing secondary development capabilities and facilitating the integration of multiple systems. The system has good fault diagnosis function, can display the fault condition of the intersection equipment in real time, and can realize the remote maintenance function of the signal machine through the network. The system adopts a real-time optimization algorithm combining scheme selection and scheme generation, and uses advanced prediction and degradation technology to greatly reduce the dependence of the system on the detector. Traffic signal machine CPU uses 32-bit chip, control function is strong.

The winning bid of Beijing Olympic Games and Beijing expressway traffic signal control system project makes the sea signal machine gradually move toward the national market. After a long period of accumulation, the signal machines of HITIC network technology have been distributed throughout the country, and are the third largest international signal control system widely used after SCATS and SCOOT.

In 2014, the HiCon traffic signal control system was applied to the Qingdao World Garden surrounding road intelligent traffic management service system project, Jinan City traffic police detachment traffic signal upgrade project, Foshan Chancheng upgrade traffic signal lamp project, Jiangmen City Pengjiang and Jianghai two district public security intelligent traffic management system phase II construction project. In 2015, the system was applied to the traffic signal control system construction project of Jinan traffic police detachment, the intelligent transportation phase III project of Shouguang City, the intelligent transportation system construction project of Heshan City, the intelligent transportation system construction project of Taishan City, the traffic signal control system upgrading project of Nanchang Traffic Administration Bureau, and the intelligent transportation control and guidance system project of Baoding central city. The intelligent traffic management system of Wuhan Donghu New Technology Development Zone—the upgrading and transformation project of regional traffic signal control system, and the intelligent application demonstration project of Xining urban public transport.

At present, Haixin has installed traffic signal control system for more than 80 cities in the domestic market, including Beijing, Qingdao, Jinan, Wuhan, Fuzhou, Xiamen, Guiyang, Lanzhou, Taiyuan, Yinchuan, Nanchang, Changsha, Urumqi, Xining, Changchun, Nanchang, Guilin, Foshan, Jiangmen, Zibo, Yantai, Zhenjiang, Suzhou and other large and medium-sized cities. The successful cases of HiCon adaptive traffic signal control system application are mainly in Taiyuan, Nanchang, Changchun, Fuzhou, Xiamen, Suzhou, Huzhou, Zibo, Heze and other cities.

7.4.3 Shenzhen SMOOTH Traffic Signal Control System

Shenzhen regional traffic signal control system was established in 1989. Around 1999, in view of Shenzhen high saturation, high complexity, high expectations of traffic status and regularity, variability, random combination of traffic characteristics, Shenzhen traffic police proposed the development of SMOOTH traffic signal control system needs. SMOOTH system was developed in early 1999. In the middle of 2001, the prototype of signal machine and coil vehicle detector was put into trial and operated induction control. At the end of 2002, the system platform on-line trial operation, to achieve green wave control and other functions, the first use of GPRS wireless networking. At the end of 2003, the signal machine and vehicle detector were upgraded to embedded platform to realize adaptive control and bus priority control, and the system function tends to be perfect.

Subsequently, Shenzhen began to promote the use of SMOOTH system in 2003. In 2007, the system realized the release overflow specific energy such as bottleneck control and close-distance intersection group control. In 2009, the application of full-range countdown in adaptive control was realized, with 105 adaptive control intersections and 38 inductive control intersections. In recent years, mainly through the green wave control benefits, green wave control over the intersection of more than 76%.

SMOOTH traffic signal control system adopts distributed control mode, three-tier architecture, large database, multi-server collaborative processing. According to the traffic demand and traffic characteristics of Shenzhen, a flexible and effective control strategy is adopted to maximize traffic capacity and minimize congestion during peak hours. In the development process of SMOOTH system, Japan's KATNET system, the UK's SCOOT system, Australia's SCATS system, and other systems that were most widely used in the world at that time were studied and analyzed, aiming at fully drawing on the advantages of famous systems, abandoning limitations and taking the road of technological innovation. The SMOOTH system inherits the method of KATNET system to identify traffic state, draws on the strategy of SCOOT system proximity prediction, and introduces the means of SCATS system tactical fine-tuning. In view of the current situation and development trend of traffic in China, a multi-objective decision-making control strategy based on traffic state recognition and a comprehensive solution combining single intersection adaptive control and road network regional coordination control are proposed.

The SMOOTH system has the characteristics of highly intelligent automatic control It can adjust the phase and timing of signal control at any time according to the change of traffic flow, so as to maximize the improvement of intersection capacity and reduce the queuing and delay of motor vehicles. In addition, the system also has the characteristics of wide control range, central and intersection wireless networking, good scalability and openness, high stability and strong anti-interference ability.

SMOOTH system is mainly used in the following aspects, variable lane and variable signal lamp control, countdown indication, pedestrian crossing demand sensing, signal control via left turn, decision support of tidal lane signal control system, compared with the Internet big data integration, and realize signal tuning and timing optimization.

The SMOOTH system is mainly used in Shenzhen and Kunming. At present, the SMOOTH system in Shenzhen has a total of about 2800 intersections. There are 235 single-machine operation intersections, and more than 2500 networked control intersections. Wireless networking is mainly used. There are more than 4000 registered devices for vehicle detectors, and more than 500 intersections in Shenzhen's history have achieved adaptive control of intersections during the peak period of application. The application results show that the system meets the design goals and application requirements, effectively reduces the traffic delay of the road network, improves the capacity, and significantly improves the traffic jam. In addition, the SMOOTH system integrates with the application brain of the Shenzhen traffic police, and realizes the self-selection of the whole process of traffic flow perception and control mode based on the data perception of Shenzhen traffic brain and the computing power of artificial intelligence.

References

1. Zhang J, Wang J, Li M (2008) City intelligent public traffic management system. China Construction Industry Press, Beijing. (Published in Chinese)
2. Wu B, Li Y (2005) Traffic management and control. China Communications Press, Beijing. (Published in Chinese)
3. Yan X, Wu C (2014) Intelligent transportation system - principle, method and application. Wuhan University of Technology Press, Wuhan. (Published in Chinese)
4. Liu Z (2003) Intelligent traffic control theory and application. Science Press, Beijing. (Published in Chinese)
5. Zhai R (2011) Principle and application of road traffic control. China People's public security. University Press, Beijing. (Published in Chinese)
6. H. Lu, R. Li and Y. Zhu. China Railway Press, Beijing, 2004. (Published in Chinese)

Chapter 8
Vehicle Intelligent Driving Technology

YU Guizhen, ZHOU Bin, WANG Zhangyu, LI Han, LIAO Yaping, and LIU Pengfei

The era of artificial intelligence is coming quietly, and it affects and changes every aspect of every industry. In particular, vehicles, which are closely related to people's life and travel, have been significantly affected. Artificial intelligence is helping people to move toward an intelligent unmanned vehicle revolution. Intelligent vehicles will greatly improve the way people travel. It uses the most cutting-edge technologies such as environmental perception, decision-making, and planning to interact and integrate the information of people, vehicles, and roads, and continuously collect and process various useful information to make vehicles more intelligent. It will reduce the burden of human driving, avoid more traffic accidents, and make vehicles truly serve people's lives.

This chapter will introduce intelligent vehicles through four aspects: intelligent vehicles themselves, environmental perception, decision-making planning and control, and the application of new technologies for future intelligent vehicles.

8.1 Intelligent Vehicles

8.1.1 Introduction of Intelligent Vehicles

The arrival of the automobile era has expanded the scope of people's choice of travel tools and changed people's concept of time and space. However, with the increase of vehicle ownership, the contradiction between vehicles and vehicles, between vehicles and people, and between vehicles and the environment is increasingly prominent. The rapid development of the automobile industry and the surge of social

YU Guizhen (✉) · ZHOU Bin · WANG Zhangyu · LI Han · LIAO Yaping · LIU Pengfei
School of Transportation Science and Engineering, Beihang University, Beijing, China
e-mail: yugz@buaa.edu.cn

© Tsinghua University Press 2022   399
W. Yunpeng et al. (eds.), *Intelligent Road Transport Systems*,
https://doi.org/10.1007/978-981-16-5776-4_8

vehicle ownership are increasingly contradictory with urbanization construction, traffic management, air pollution control, etc., and the development of society, economy, and environment are facing unprecedented challenges. The intelligent vehicle with electronic electrification technology and intelligent technology as its core provides an effective way to solve the four major public hazard problems of "safety, energy, pollution and congestion."

An intelligent vehicle [1] is an ordinary vehicle that realizes intelligent information exchange with people, vehicles, roads, etc. through the on-board sensor system and information terminals, so that the vehicle has the ability to think, including intelligent environmental perception and judgment capabilities—it can automatically analyze the vehicle driving safety and risk status, as well as the absolute capacity to act—the vehicles in accordance with the wishes of people reach their destinations, and ultimately replace human purpose driving operation. Intelligent vehicles use computers, modern sensing, information fusion, communications, artificial intelligence, and automatic control technologies. They are a typical high-tech complex. In recent years, intelligent vehicles have become a research hotspot in the field of vehicle engineering in the world and a new driving force for the growth of the automotive industry. Many developed countries have incorporated them into their respective intelligent transportation systems.

Different from autonomous driving in general, intelligent vehicle refers to the use of a variety of sensors and intelligent road technology to achieve autonomous driving. Intelligent vehicles first have a set of navigation information database, storing the information of national highways, ordinary highways, urban roads, and various service facilities (restaurants, hotels, gas stations, scenic spots, parking lots). Secondly, the GPS positioning system is used to accurately locate the position of the vehicle, and the data in the road database is compared to determine the future driving direction. Then through the road status information system, the traffic management center provides real-time road status information ahead, such as traffic jams, accidents, etc., and changes the driving route in time when necessary. Then, the vehicle collision prevention system, including detection radar, information processing system, driving control system, control the distance with other vehicles, in the detection of obstacles in time to slow down or brake, and transmit the information to the command center and other vehicles. And emergency alarm system and wireless communication system, after an accident, automatically report to the command center for rescue, or used for the contact between the vehicle and the command center. The automatic driving system is used to control the start of the vehicle, change the speed and steering, etc.

Through the research and development of intelligent vehicle technology, intelligent vehicle improves the level of vehicle control and driving, and ensures the safe, smooth, and efficient running of vehicles. The continuous research and improvement of the intelligent vehicle control system is equivalent to extending the driver's control, visual and sensory functions, which can greatly improve the safety of road traffic. The main feature of intelligent vehicle is to make up for the defects of human factors with technology, so that even in very complex road conditions, it can also automatically control and drive the vehicle to avoid obstacles and drive along the predetermined road trajectory, so as to achieve the purpose of serving people.

8.1.2 Classification of Intelligent Vehicles

From a development perspective, intelligent vehicles will go through two stages [2]: the first stage is the primary stage of intelligent vehicles, that is, assisted driving; the second stage is the ultimate stage of the development of intelligent vehicles, that is, fully replacing people's unmanned driving. The National Highway Traffic Safety Administration (NHTSA) defines intelligent vehicles as the following five levels.

1. No Automation (L0)

    At this level, the driver has complete control of the original underlying structure of the vehicle, including brakes, steering gear, accelerator pedal, and engine, all at all times.
2. Function-Specific Automation (L1)

    This level of vehicle has one or more special automatic control functions, through the warning to prevent the accident, can be called the assisted driving stage. Many of the technologies at this stage are already familiar, such as Lane Departure Warning System (LDW), Forward Collision Warning System (FCW), Blind Spot Information System (BLIS).
3. Combined Function Automation (L2)

    This level of vehicle has a system that integrates at least two original control functions together, and does not require the driver to control these functions at all. It can be called a semi-autonomous driving stage. At this stage, the vehicle will intelligently judge whether the driver has responded to the warning hazard. If not, it will take actions on behalf of the driver, through measures such as Autonomous Emergency Braking (AEB), Emergency Lane Assist System (ELA), etc.
4. Limited Self-Driving Automation (L3)

    This level of vehicle can make the driver completely do not need to control the vehicle in a certain driving traffic environment, and the vehicle can automatically detect the change of the environment to determine whether to return to the driver's driving mode, which can be called the highly automatic driving stage. For example, Google driverless vehicles are basically at this level.
5. Full Self-Driving Automation (L4)

    This level of vehicle can fully control the vehicle, detect the traffic environment throughout, and achieve all driving goals. The driver only needs to provide the destination or enter the navigation information, and does not need to control the vehicle at any time, which can be called complete autonomous driving stage or unmanned driving stage.

At present, in addition to the classification standards introduced by the National Highway Traffic Safety Administration (NHTSA), there are also classification standards designated by the Society of Automotive Engineers (SAE). According to

SAE standards, intelligent vehicles are classified into six levels according to the level of human intervention: no automation (L0), driving assistance (L1), partial automation (L2), conditional automation (L3), high automation (L4), and fully automation (L5). The main difference between the two different classification standards is the fully automated driving scenario, SAE further subdivides the scope of the automated driving system. The detailed differences and standards are shown in Table 8.1.

8.1.3   Composition of Intelligent Vehicles

Intelligent vehicle is an integrated system that integrates environmental perception, planning and decision-making, multilevel driving assistance, etc. From a specific and realistic perspective, the more mature and predictable functions and systems in this system mainly include intelligent driving system, life service system, safety protection system, location service system, and vehicle service system.

8.1.3.1   The System that Improves the Performance of the Vehicle Itself

1. Intelligent driving system [3]: The intelligent driving system is a comprehensive system that uses advanced information control technology and integrates functions such as environment perception and multilevel driving assistance. According to the hierarchical control structure theory and the hierarchical structure characteristics of the traffic system, the logic framework of the intelligent driving system based on the Internet thinking application can be divided from bottom to top into the perception layer, the network layer, the analysis layer, and the application layer.

Perception is to collect driving information for drivers in the process of driving. The perception layer, namely the data collection layer, is mainly composed of the information of various factors affecting driving, namely, the information collection of people, vehicles, and roads, as well as the mutual connection and cross-influence of the three information, which can be mainly divided into the following two points.

  (a) The collection of road condition information, such as road geometry, road surface conditions, road disasters, road network conditions, and traffic conditions, can generally be collected through high-precision navigation systems such as GPS or BeiDou system.
  (b) Vehicle information, which mainly includes original vehicle data, such as vehicle model, vehicle theoretical parameters, etc., as well as vehicle driving dynamic data, such as driving speed, driving time, driving trajectory, etc., generally can be collected through the CAN bus.

Table 8.1 Classification standards of intelligent vehicles

Automatic driving level					Main			
NHTSA	SAE	Name (SAE)	SAE definition		Driving operation	Around the monitoring	Support	System scope
0	0	No automation	The human driver has full authority to operate the vehicle, and can be assisted by warning and protection systems during driving		Human driver	Human driver	Human driver	n/a
1	1	Driving assistance	Provide driving support for one operation of the steering wheel and acceleration/deceleration through the driving environment, and other driving actions are operated by the human driver		Human driver and system	Human driver	Human driver	Limited
2	2	Partial automation	Provide driving support for multiple operations of the steering wheel and acceleration through the driving environment, and other driving actions are operated by the human driver		System	Human driver	Human driver	Limited
3	3	Conditional automation	All driving operations are completed by the unmanned driving system, and the human driver provides appropriate responses according to the system request		System	System	Human driver	Limited
4	4	High automation	All driving operations are completed by the unmanned driving system. According to system requests, human drivers do not necessarily need to respond to all system requests or limit road and environmental conditions, etc.		System	System	System	Limited
4	5	Fully automation	All driving operations are completed by the unmanned driving system, and the human driver takes over when possible, driving under all road and environmental conditions		System	System	System	Unlimited

The network layer is the data transmission and scheduling layer, which is specifically interpreted as the transmission, scheduling, and storage of driving information. The road condition information is transmitted through message communication after data collection by the navigation system. The vehicle information is collected by the CAN bus and then transmitted by the GPRS communication module. After the data is transmitted to this layer, it will be summarized and integrated by this layer and then transmitted to the analysis layer.

The analysis layer is the analysis and processing layer of big data, which is specifically explained as the background big data processing technology of driving information. Due to the disorder of big data collection and processing, the data affecting driving is calculated and processed under the defined function model. The processing results will be transmitted to the application layer, and at the same time will be returned to the network layer for storage and recall, and a driving database will be established in the network layer.

The application layer is the application service layer, which is specifically explained as the feedback control of the data analysis results and its application. It can carry out cross-application and cross-system information sharing and information coordination through data interface based on the results of data collection and processing. Under the big data application thinking of the Internet and the concept of interconnection, the application of intelligent driving system is mainly divided into three modules: user service system, traffic management system, and automobile marketing system.

(a) User service system: The intelligent driving system based on Internet thinking takes the driver's driving safety, comfort, etc. as constraints, and uses the Internet's cloud processing and computing platform to obtain recommended vehicle safety evaluation values, warning suggestions, appropriate speeds, and other driving control data. The vehicle body receives data through the CAN bus, automatically performs data signal conversion, carries out driving control and adjustment, and proposes a visual interface for auxiliary guidance of driving countermeasures, human–computer interaction coordinates the relationship between vehicles, guarantees driving safety, and improves people's driving pleasure.

(b) Traffic management system: Through the call to the driving database, the traffic management department can accurately and real-time grasp driving conditions; better organize, plan, coordinate, and direct transportation activities; improve road driving efficiency; and reduce the traffic loss rate.

(c) The driving database can provide enterprise data services for automobile enterprises, improve the quality of vehicle bodies, and promote the directional development of enterprises.

2. Life service system: Through some on-board hardware, programs, and software, the life service system satisfies some needs of drivers in the driving process under certain conditions, such as video and audio entertainment, information inquiry, service subscription, and all kinds of life services.

3. Safety protection system: The safety protection system includes functions such as vehicle anti-theft, vehicle tracking, etc., which can provide certain safety protection for the driver in terms of the performance of the vehicle itself, so as to reduce the loss caused by intrusion and theft during use.
4. Location service system: In addition to providing accurate vehicle positioning functions, the location service system can also allow the vehicle to realize automatic location intercommunication with another vehicle, so as to achieve the agreed goal of driving and improve the driver's driving comfort.
5. Use vehicle assist system: Vehicle assistance systems include maintenance reminders, abnormal warnings, remote guidance, etc.

The combination of these systems is the equivalent of TV cameras, computers and automatic control systems that equip the vehicle with "eyes," "brains," and "feet."

8.1.3.2 The System that Enhances Driver's Ability

In addition to systems that improve the performance of the vehicle itself, the current intelligent vehicles also have systems that enhance the ability of the driver, mainly as follows.

1. Route Guidance System.
   The Route Guidance System (RGS) has the following divisions.

   (a) Positioning system: With an antenna connected to a global positioning system, the vehicle's position and direction on the road can be shown on a small screen or on a windshield.
   (b) Communication systems: In addition to the general radio communication system for the vehicle to communicate with the outside world, a special antenna must be configured to connect to the Advanced Travelers Information Systems (ATIS) of the traffic management center and carry out two-way communication to obtain the road network traffic and climate conditions of the destination. At the same time, it can also obtain the future traffic and climate conditions predicted by the traffic center, as well as the best route to the destination, which can also be displayed on the electronic map.
   (c) Real-time navigation system: In addition to the drivers watching the live traffic and variable information signs along the way with their eyes, there is also a dedicated antenna associated with the cellular communication system along the highway to receive real-time traffic conditions and hazard warnings and driving instructions along the road ahead, and it is provided to the driver through the audiovisual terminal.

2. Automatic collision avoidance system: Automatic collision avoidance system is an on-board device to assist drivers in driving safety. It mainly monitors the collision avoidance distance between the front and rear sides of the vehicle and other vehicles or buildings in real time through the sensors of radar, infrared ray, and camera at the front and rear ends and left and right sides of the vehicle. Sound

warning will be issued when the safety range is exceeded, prompting the driver to pay attention to control the movement of the vehicle. In the case of unmanned driving or automatic driving on the automatic highway, the anti-collision information will be fed back to the computer, and the automatic driving system will automatically adjust the actions of the vehicle after processing, so as to avoid collision.

The composition of the system is as follows:

(a) Signal acquisition system: Using radar, laser, sonar, and other technologies to automatically detect the speed of the vehicle, the speed of the vehicle in front, and the distance between the two vehicles.
(b) Data processing system: After the computer chip processes the distance between the two vehicles, it judges the safe distance between the two vehicles. If the distance between the two vehicles is less than the safe distance, the data processing system will issue an instruction.
(c) Controller: It is responsible for analyzing and filtering the instructions sent by the data processing system and issuing instructions to the executing actuator.
(d) Actuator: It is responsible for issuing an alarm to remind the driver to brake. If the driver does not execute the instruction, the actuator will take measures, such as closing the windows, adjusting the seat position, automatic deceleration, automatic braking, etc.

The working principle of the system is as follows:

(a) Tracking identification: The radar recognition system is used to effectively monitor the dynamic and static targets in front of the vehicle in real time, and transmit the measurement data to the central processing system in time.
(b) Intelligent processing: The central processing system analyzes, calculates, and processes the information collected by the radar; effectively analyzes and judges the obstacles in front of the vehicle; quickly processes and issues instructions; converts the radar acquisition signals into executable signals; and transmits them to the alarm display system and brake execution system in time.
(c) Alarm reminder: The display and alarm system of the anti-collision device can promptly display the speed of the vehicle and the distance of the most threatening vehicle or object ahead. When receiving an instruction from the central processing system, it can remind the driver of dangerous information ahead of him by voice, and he needs to drive cautiously. The alarm function is displayed, which greatly improves the safety of people and vehicles.
(d) Slow down and brake: In case of danger, when the brake execution system receives the braking instruction from the central processing system, it can quickly take actions such as intelligent deceleration, emergency braking, and parking on the vehicle according to the danger, as shown in Fig. 8.1.

Fig. 8.1 Radar detection system

3. Autopilot system [4]: When a vehicle is driving unmanned or autonomously on an automatic highway, the first is the lane-following system, which relies on the magnetic sensors of the front and rear devices of the vehicle to monitor in real time the extent of the vehicle's deviation from the magnetic nail navigation line on the automatic road lane. When the value exceeds the specified value, the computer management system will issue an instruction to direct the automatic steering actuator to turn, and how much to turn left to right is determined by the computer management system after processing based on the information collected from the front and back ends. The second is automatic transmission, automatic throttle, automatic braking, and other systems. These systems receive the information provided by the computer management system from the collision avoidance system to determine the instructions to control the vehicle's actions, and the relevant actuators will accelerate, decelerate, stop, and reverse. Generally, the vehicle movements determined by the anti-collision system information are all fine-tuned. In addition, the information such as the speed change received by the automatic highway coaxial leaking cable is operated by the same command and related actuators to automatically change the vehicle's movements.

At present, the United States has three main controls on the dynamic performance of intelligent vehicles in autonomous driving on autonomous highways: One is the ability to follow the road, that is, the ability to automatically drive along the magnetic nail navigation line on the road surface; the other is the performance of the group of vehicles, that is, whether The ability to keep a certain safe distance from the front, rear, left, and right vehicles and automatically follow the vehicle in front. The third is the obstacle avoidance performance, that is, the ability to automatically avoid, continue forward, and reverse when encountering obstacles on the front, rear, left, and right.

4. Computer management system: The computer management system is the computer platform. All the electronic control systems on the vehicle, including those that improve the performance of the vehicle and enhance the ability of the driver, are managed by the computer management system, and all the electronic control systems are organically combined into one. As a whole, make full and correct use of technology.

With the rapid development of intelligent vehicles, system functions are becoming more and more complex, real-time requirements are getting higher, and security levels are getting higher and higher. The traditional automobile distributed control architecture based on CAN bus can no longer meet future needs. The new electronic and electrical architecture with a standardized backbone network and multi-domain control has become the best choice for the development of intelligent vehicles in the future, which is very necessary for

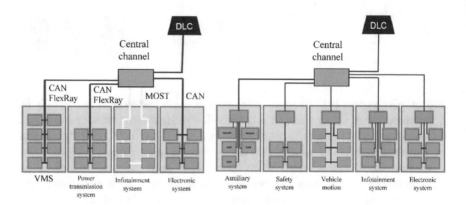

Fig. 8.2 Current automotive electronic and electrical architecture and future automotive electronic and electrical architecture

the efficient transmission and management of complex functions of intelligent vehicles and a large amount of interconnected information, as well as system security.

The intelligent driving computing platform takes environmental perception data, GPS information, vehicle real-time data and V2X interactive data as input, and outputs drive, transmission, steering, and braking based on core control algorithms such as environmental perception positioning, path decision planning, and vehicle motion control. The control command realizes the automatic control of the vehicle, and realizes the human–computer interaction of the automatic driving information through the human–computer interaction interface (such as the instrument). In order to achieve the high-performance and high-safety control requirements of the intelligent driving system, the intelligent vehicle computing platform brings together a number of key technologies, including basic hardware/software platform technology, system safety platform technology, vehicle communication platform technology, cloud computing platform technology, core control algorithm technology, etc., as shown in Fig. 8.2.

5. Other systems: Vehicles with different functions will also be selectively installed with different systems. For example, the electronic payment system is a necessary system for all intelligent vehicles; the operating vehicle dispatch system CVOS is only available for intelligent vehicles of transportation companies; the on-board passenger service system is only available for public passenger intelligent vehicles; as for the intelligent instrument display system and electronic anti-theft system, etc. It belongs to the connotation of intelligent vehicles and is indispensable.

8.1.4 Development at Home and Abroad

Judging from the overall development of autonomous driving at home and abroad, the United States and Germany led the development of the autonomous driving industry, Japan and South Korea quickly awakened, and my country is catching up with the following trends.

8.1.4.1 With the Goal of Commercialization as Soon as Possible, Accelerate the Introduction of Road Testing and Regulations

In terms of road testing, the United States, Germany, Japan, South Korea, and China have all actively promoted road testing as the basis for the application of autonomous vehicles. From an international point of view, all countries have regarded 2020 as an important time node, hoping to realize the full deployment of autonomous vehicles by then.

The United States is actively enacting autonomous driving legislation at the state level. To date, more than 20 states have passed relevant bills or administrative orders to clarify test conditions and requirements. In September 2017, the US House of Representatives approved the Self-Driving Act (SELF DRIVE Act). The draft bill aims to perform federal functions to ensure vehicle safety by encouraging the testing and research and development of self-driving vehicles.

In 2015, the German government allowed the autonomous vehicle test project to be carried out on the A9 highway connecting Munich and Berlin. In March 2019, the transportation department also issued subsidies to the Diginet-PS autonomous driving pilot project in Berlin for the development of processing systems and provide real-time traffic information for autonomous driving, and in June 2017 issued the world's first autonomous driving ethics code.

Japan's Nissan has already tested its autonomous vehicle LEAF in Tokyo, Silicon Valley, and London, hoping to accumulate safety test records as soon as possible. South Korea plans to commercialize Level 3 autonomous vehicles in 2020.

The Ministry of Industry and Information Technology of my country launched a pilot demonstration of Shanghai Smart Connected Vehicles in Shanghai in 2016, and launched a broadband mobile Internet-based intelligent vehicle and smart transportation application demonstration in Zhejiang, Beijing, Hebei, Chongqing, Jilin, Hubei, and other places to promote autonomous driving Test work. Beijing has issued a "five-year action plan" for the demonstration of intelligent vehicles and smart transportation applications, and will complete the smart road network transformation of all arterial roads within the Beijing Development Zone by the end of 2020, and deploy 1000 fully autonomous vehicles in phases.

In the summer of 2017, my country also held discussions on the technical route and standard framework of autonomous driving for the first time. The Ministry of Industry and Information Technology and the National Standardization Management Committee jointly issued the "Guidelines for the Construction of the National

Table 8.2 The exploration of autonomous driving legislation in various countries around the world

Country	Status
United Kingdom	Regarding the exploration of autonomous driving at the test, standard and legislative level, the UK Department of Transport has issued the "autonomous vehicle test operation rules" and "key principles for networked automated vehicle cyber safety"; it plans to start testing on highways in 2020
Australia	The National Transportation Commission takes the lead in discussing the supervision plan for the safety management of autonomous vehicles
Canada	Allow drive test and initiate demonstration of relevant legislation
France	Approve testing of autonomous vehicles on highways; set up an interdepartmental joint team to review existing laws and start to revise laws, safety standards, etc.
Finland	Revise current road traffic rules; allow autonomous vehicles to be tested in specific areas of public roads after approval; approval of driverless buses on the road test
Sweden	Current laws allow testing of highly automated driving vehicles; revise vehicle regulations, driving license rules and liability regulations, etc.; adjust current vehicle standards and performance testing regulations; set up new driving licenses to adapt to autonomous vehicles
Finland	Start to review current traffic laws; propose tests on highways; carry out research on responsibilities, driving skills requirements, data protection, and impact on infrastructure
Japan	Allow road testing; work with the European Union to formulate a globally unified technical standard for autonomous vehicles, and is committed to urging the United States to adopt the same standards and policies
Korea	Initiate the revision of current road traffic regulations; designate test operation areas for autonomous vehicles and open dedicated experimental roads
Singapore	Allow self-driving vehicles to be tested within a certain range; carry out pilots of driverless taxis

Internet of Vehicles Industry System (Intelligent Connected Vehicles) (2017)" (Draft for Solicitation of Comments), my country's autonomous driving technology has chosen the strategic development path of "intelligence + connectivity."

In addition to the United States and Germany, other countries listed in Table 8.2 also strongly support the development of autonomous driving and have initiated exploration at the level of testing, standards, and legislation.

8.1.4.2 With the Direction of China Unicom, Promote the Research and Development of the System and the Communication Standard System

Judging from the current industry trends, most companies have adopted the development path of Connected Cars, speeding up the development of chip processing capabilities and autonomous driving cognitive systems, and promoting the introduction of unified vehicle communication standards.

In terms of research and development, the German Bosch Group and NVIDIA are cooperating to develop artificial intelligence autonomous driving systems. NVIDIA provides deep learning software and hardware. Bosch AI will be based on NVIDIA Drive PX technology and the company's upcoming super chip Xavier, which will provide level 4 automatic driving. Driving technology IBM announced that its scientists have obtained a patent for a machine learning system that can dynamically change the autonomous vehicle control rights between the human driver and the vehicle control processor in a potential emergency situation, thereby preventing accidents from becoming harder.

In terms of vehicle communication standards, communication technologies such as LTE-V and 5G have become the key to communication standards for autonomous vehicles, and will provide high-speed and low-latency network support for autonomous driving.

On the one hand, the coordinated promotion of LTE-V2X at home and abroad has become an important development direction of 3GPP 4.5G. Datang, Huawei, China Mobile, China Academy of Information and Communications Technology, and other joint efforts have made positive progress in the standardization of V2V.V2I.

On the other hand, LTE-V2X technology is gradually evolving to 5G and V2X with the development of autonomous driving requirements. 5G- and V2X-dedicated communications can extend the sensing range beyond the working boundary of on-board sensors, realize safe and high-bandwidth business applications and autonomous driving, complete the transformation of vehicles from transportation tools to information platforms and entertainment platforms, and help further enrich business scenarios.

Currently, the 5G Automobile Association (5GAA) and the European Automobile and Telecommunications Association (EATA) have signed a memorandum of understanding to jointly promote the C-V2X industry, using cellular-based communication technology for standardization, spectrum, and pre-deployment projects. China Mobile has cooperated with BAIC, GM, Audi, and others to promote 5G joint innovation, while Huawei has cooperated with BMW and Audi to promote the development of 5G-based services.

In addition, the standard system plan for intelligent connected vehicles organized and drafted by the Ministry of Industry and Information Technology will be released soon. The standard system for connected vehicles is also gradually being improved, which is crucial to the development of intelligent connected vehicles.

8.1.4.3   Leading by Innovative Formats, Internet Companies have Become an Important Driving Force

Internet companies are born with genes for business innovation and development, and now they are also involved in the autonomous driving industry, becoming an important driving force in the industry.

In the United States, Google started the research and development of unmanned driving companies in 2009 [5]. From December 2015 to December 2016, it recorded

635,868 miles on California roads. It is not only the company with the most test mileage in California, but also the lowest system outage rate enterprise. Uber, the largest US ride-hailing service provider, has been approved to conduct driverless road tests in Pittsburgh, Tempe, San Francisco, and California. Lyft, the second-largest ride-hailing service provider, announced a three-phase development plan for autonomous vehicles in September 2018. Testing has also been carried out in Pittsburgh. Apple also obtained a California testing license in April 2019. South Korea approved the Korean Internet company Naver to test self-driving vehicles on the road, becoming the 13th self-driving vehicle research and development company to be licensed, and plans to commercialize level 3 self-driving vehicles in 2020.

In my country, Baidu obtained a test permit in California, USA, in September 2018, and started a trial operation of unmanned vehicles on ordinary open roads in Wuzhen, Zhejiang, in November. Baidu President and Chief Operating Officer Lu Qi released the Apollo plan on April 19, 2019. He plans to open the company's autonomous driving technology to the industry and open up functions such as environment perception, path planning, vehicle control, and in-vehicle operating systems. Code or ability, and provide complete development and testing tools, the purpose is to further reduce the threshold for the development of unmanned vehicles, and promote the rapid popularization of technology.

Tencent established an autonomous driving laboratory in the second half of 2016, relying on the accumulation of technologies in 360° surround view, high-precision maps, point cloud information processing, and fusion positioning to focus on the research and development of core technologies for autonomous driving. Alibaba, LeTV, etc. have also begun to cooperate with SAIC and other auto companies to develop Internet cars.

8.1.4.4 Taking Corporate Mergers and Acquisitions as a Breakthrough, Start-Ups, and Leading Companies Become Targets

The main targets of mergers and acquisitions by companies with faster development of autonomous driving are leading companies or start-ups that master the key technologies of autonomous driving.

In July 2016, General Motors acquired Cruise Automation, a Silicon Valley startup company, for more than US$1 billion. The RP-1 highway autonomous driving system developed by the latter has the potential for highly automated driving applications.

In March 2017, Intel acquired Israeli technology company Mobileye for USD 15.3 billion. The latter is dedicated to the development of software and hardware systems related to autonomous driving. It held a series of patent on image recognition and is the main camera supplier of driving assistance systems for companies such as Tesla and BMW.

Uber acquired deCarta, a startup company that provides location APIs, in 2015, and also acquired employees proficient in image and data collection from Microsoft's Bing department.

In April 2017, Baidu announced the wholly owned acquisition of xPerception, an American technology company focusing on machine vision software and hardware solutions. The company provides machines with stereo inertial cameras as the core to customers in robotics, AR/VR, and intelligent blind guidance industries. Vision software and hardware products can realize the positioning of the intelligent hardware in an unfamiliar environment, the calculation of the three-dimensional structure of the space, and the path planning. According to industry analysis, Baidu's move may be to strengthen software and hardware capabilities in the field of visual perception.

8.2   Perception

8.2.1   Environmental Perception

Perception refers to the ability of driverless system to collect information from the environment and extract relevant knowledge [6]. Among them, environment perception refers to the ability to understand the environment, such as the location of obstacles, traffic signs, marking, pedestrian and other data semantic classification. Generally speaking, positioning is also a part of perception. Positioning is the ability of self-driving vehicle to determine its position relative to the environment.

Environmental perception is the core technology of intelligent vehicles. It detects and identifies roads, vehicles, pedestrians, traffic signs, traffic signals, etc. by sensors or self-organizing networks installed on intelligent vehicles. It is mainly used in advanced driving assistant systems, such as adaptive cruise control system, lane departure alarm system, road keeping assistant system, vehicle parallel assistant system, automatic braking assistant system, etc., to ensure the intelligent vehicle to reach the destination safely and accurately.

The core of driverless system can be summarized as three parts: perception, planning, and control. The interaction of these parts and the interaction with vehicle hardware and other vehicles can be represented by Fig. 8.3.

In order to ensure the understanding of the environment of the self-driving cars, the environment perception part of the self-driving cars usually needs to obtain a large amount of information about the surrounding environment, including the location, speed, and possible behavior of the obstacles; drivable area; the traffic rules, etc. Self-driving cars usually obtain the information by fusing LiDAR, camera, millimeter wave radar, and other sensors. As shown in the Fig. 8.4.

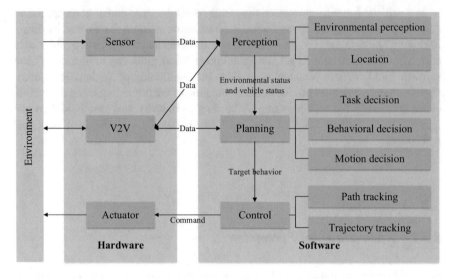

Fig. 8.3 Interaction between several key parts of automatic driving system

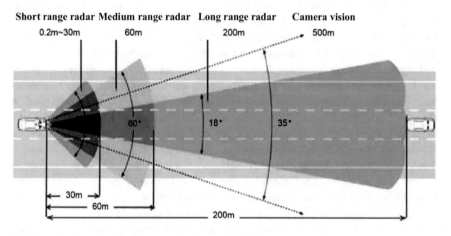

Fig. 8.4 Configuration requirements for automatic driving

8.2.2 Composition of Environmental Perception System

The intelligent vehicle environment sensing system consists of information acquisition unit, information processing unit, and information transmission unit, as shown in Fig. 8.5.

1. Information acquisition unit

   The perception of the environment is the premise and foundation of intelligent vehicle. The real-time and stability of the sensing system to obtain the

Information acquisition unit

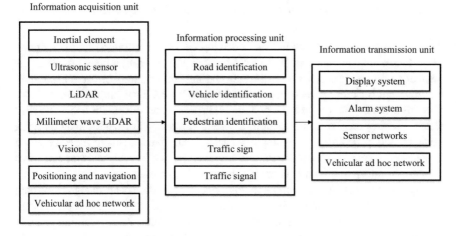

Fig. 8.5 Composition of intelligent vehicle environment perception system

surrounding environment and vehicle information is directly related to the accuracy and effectiveness of subsequent detection. Information acquisition technology mainly includes ultrasonic sensor, LiDAR, millimeter wave radar, vision sensor, positioning and navigation system, vehicle self-organizing network technology, and so on.

2. Information processing unit.

The information processing unit is mainly used to identify the road, vehicles, pedestrians, traffic signs, traffic lights, etc.

3. Information transmission unit.

After analyzing the environmental perception signal, the information processing unit sends the information to the transmission unit, and the transmission unit performs different operations according to the specific situation. If the analyzed information determines that there are obstacles in front and the distance between the vehicle and the obstacles is less than the safe distance, the information will be sent to the control execution module. The control execution module automatically adjusts the speed and direction of the intelligent vehicle according to the speed, acceleration, steering angle, and other factors of the vehicle, so as to achieve automatic obstacle avoidance or automatic braking. If the information transmission unit transmits the information to the sensor network, the vehicle internal resource sharing can be realized. The information can also be transmitted to other vehicles around the vehicle through the self-organizing network to realize the information sharing between vehicles.

8.2.3 Overall Function of Environmental Perception

The overall functional structure of self-driving cars can be divided into perception layer, function layer, and task layer. Among them, the perception layer processes the data from the vehicle sensors to provide the key information of the surrounding environment for other parts of the system. For example, local information includes state information such as vehicle posture and speed, road information such as road shape, parking area and intersection, dynamic obstacle information such as other vehicles and pedestrians around the vehicle, local static obstacle map, showing obstacle-free area, dangerous area, and impassable area in real environment, road congestion information, and estimating the area that cannot pass.

Algorithm set is included in each component of traffic scene recognition, path planning, and vehicle control. For example, traffic scene recognition requires localization, target detection, and target tracking algorithms. Path planning usually includes task and motion planning. Vehicle control corresponds to path tracking algorithm. The basic control and data flow of the algorithm are shown in Fig. 8.6.

8.2.4 LiDAR

8.2.4.1 Introduction of LiDAR

LiDAR (Fig. 8.7) is a kind of equipment that uses laser to detect and range. It can send millions of light pulses to the environment every second. Its internal structure is a rotating structure, which enables LiDAR to build a real-time three-dimensional map of the surrounding environment.

Fig. 8.6 Basic control and data flow of automatic driving algorithm

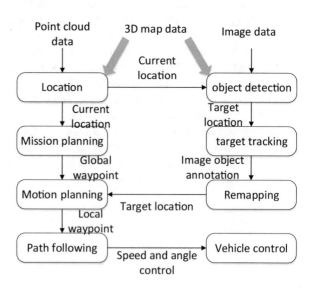

Fig. 8.7 LiDAR

 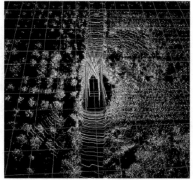

Fig. 8.8 Point cloud maps created by LiDAR

Generally speaking, LiDAR scans the surrounding environment at a frequency of 10 Hz. The result of one scan is a three-dimensional graph composed of dense points, each point has (x, y, z) information, which is called point cloud graph. Figure 8.8 shows a point cloud map created using Velodyne VLP-32c LiDAR.

Because of its reliability, LiDAR is still the most important sensor in automatic driving system. However, in practical use, LiDAR is not perfect. The point cloud is often too sparse, even missing some points. For irregular surface, it is difficult to distinguish its mode by using LiDAR. And Lidar is also not available in conditions such as heavy rain. The parameters of LiDAR are shown in Table 8.3.

8.2.4.2   Principle Analysis and Design of Ranging Module

1. The principle of trigonometric distance measurement. The trigonometric ranging model is shown in Fig. 8.9.

According to the principle of similar triangle, we can know

Table 8.3 Parameters of lidar

Parameter name	Unit	parameter range	note
Laser wavelength	Nm	775 ~ 795	Infrared band
Ranging range	M	0.15 ~ 15	Measurement based on white highly reflective object
Range resolution	Mm	1% ~ 2%	Higher accuracy of short distance ranging
Measuring angle	°	0 ~ 360	
Angular resolution	°	≤ 1	
Measuring frequency	Hz	≥ 3000	3000 by default, customizable
Scanning frequency	Hz	1 ~ 10	Customizable

Fig. 8.9 Triangular ranging model

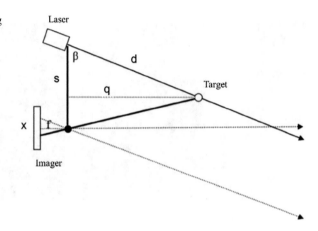

$$q = \frac{fs}{x} \tag{8.1}$$

where q is the measured distance, s is the distance between the laser head and the lens, f is the focal length of the lens, and x corresponds to s; x is the variable, assuming that the angle β is a constant.

The relationship between angle β and q is

$$d = q/\sin\beta \tag{8.2}$$

By deriving the function (8.1), we can get

$$\frac{dq}{dx} = \frac{fs}{x^2} \tag{8.3}$$

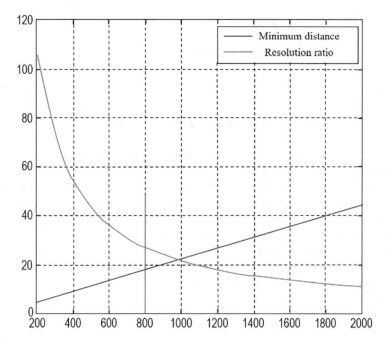

Fig. 8.10 System resolution and ranging function

2. Ranging module design: Many factors need to be considered in the design of ranging module. The evaluation function (standard) of ranging module design is system resolution, minimum and maximum of ranging.

The key variables corresponding to the evaluation function are lens focal length (f in Fig. 8.9), mechanical dimension of ranging module (s in Fig. 8.9), angle between laser head central axis and lens central axis (residual angle of angle β in Fig. 8.9). In addition to the key variables, there are lens field angle, transmittance, aperture, the size of the chip, resolution, and so on.

Firstly, a concept of system resolution is introduced: the measured distance change corresponding to the translation of a pixel.

From Eq. (8.3), when dx is one pixel, dq is the system resolution, that is, the system resolution $r = -kqq/fs$.

In Fig. 8.10, the light color line represents the system resolution, and the dark color line represents the minimum measurement distance, that is, the blind area.

The parameters of the photosensitive chip corresponding to the curve in Fig. 8.10 are 752 pixels, 6 μm for each pixel. Because the expected minimum distance is no more than 20 cm, fs should not be greater than 900; if the system resolution is no more than 30 cm at 6 m, fs should be greater than 700; therefore, $fs = 800$ is selected. Considering the standard condition of lens focal length and the size of LiDAR, $f = 16$ mm and $s = 50$ mm are selected.

As for the angle β, it depends on the size and resolution of the chip, for example

$$\beta = \arctan\left(f/(376 \times 6) \right) \approx 82° \qquad (8.4)$$

8.2.4.3   Engineering Realization of Ranging Module

1. Module calibration: The main error sources of ranging module are system resolution and calibration error.

   The purpose of calibration is to match the actual module with the ideal model as much as possible. As shown in Fig. 8.11, the pointing angle of the laser head, the pointing angle of the lens, and the lens distortion are the key parts to be adjusted. Among them, the laser head and lens can be adjusted by the mechanical device to achieve the ideal position. There are two steps to deal with the lens distortion: positioning the imaging pixel of laser point to sub-pixel level; compensating the corresponding distance for different measured distances, and fitting the curve $1/x$, as shown in Eq. (8.1).

   As can be seen from Fig. 8.12, the longer the measurement distance is, the greater the error will be.

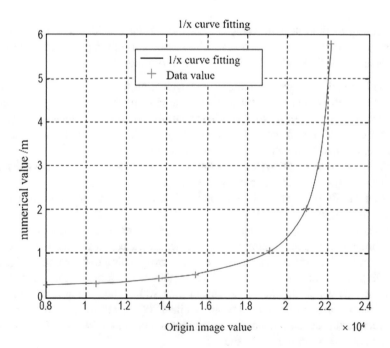

Fig. 8.11 Calibration curve

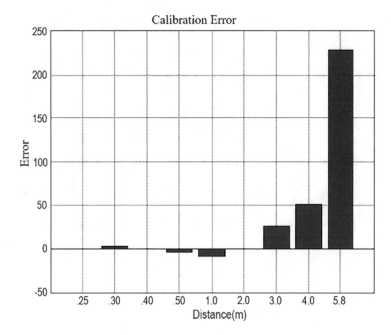

Fig. 8.12 Function of calibration error and measurement distance (the greater the sensitivity, the greater the error)

The calibration method mentioned above is based on a basic premise: The thermal stress and mechanical vibration are relatively small. In practice, thermal stress and mechanical vibration cannot be ignored. The material with better rigidity and less influence of thermal stress should be selected in design.

2. Ranging algorithm: The energy distribution of the spot can be approximated as a Gaussian model. Therefore, in order to reduce the error, the gray centroid method is selected to estimate the pixel position of the spot imaging point:

$$\sum_i I(i)/i/\sum_i I(i) \tag{8.5}$$

3. Ambient light interference: When 650 nm red laser is used in the experiment, the harm to human eyes, ranging signal-to-noise ratio and anti-interference ability to ambient light should be considered.

The laser safety level of IMLidar is class-I. The higher the laser power is, the higher the Signal-to-Noise Ratio (SNR) of ranging is. The results show that the IMLidar is invalid in the case of direct sunlight, but the ambient light has no effect on the measurement distance in the outdoor and indoor environment without direct sunlight.

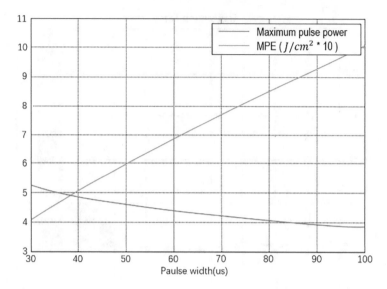

Fig. 8.13 Laser power

Fig. 8.14 Hardware logic
of LiDAR

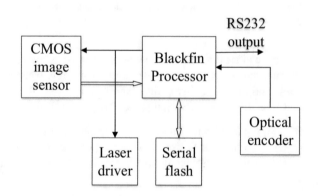

There are two ways to deal with the influence of ambient light: one is temporal filtering, the other is narrowband filtering.

Figure 8.13 shows the relationship between the laser pulse width and the maximum allowable light intensity. The choice of parameters is the result of a game, which needs to weigh the pros and cons in many aspects.

4. Hardware logic: The working process of LiDAR is emitting laser and exposing the photosensitive chip at the same time; reading pixel data; calculating the center (centroid) position of pixel; converting the calculation result (pixel position) in C into distance information.

The logic of the hardware is shown in Fig. 8.14.

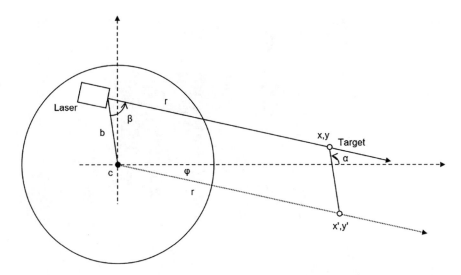

Fig. 8.15 LiDAR rotation (rotation geometry of LDS, coordinate system centered on rotation center *c*)

8.2.4.4   Engineering Realization of LiDAR Scanning

1. LiDAR rotation scanning: The reference point of LiDAR ranging is defined as the rotation center of LiDAR.
   It can be seen from Fig. 8.15.

$$x' = \gamma \cos \varphi, y' = \gamma \sin \varphi$$
$$\alpha = \pi - \beta + \varphi \qquad\qquad (8.6)$$
$$x = x' + b \cos \alpha, y = y' + b \sin \alpha$$

In other words, after the LiDAR rotates, the reference point of ranging is different from that of the original ranging module, so it is necessary to convert the ranging data into formula (8.6).

2. The angle is synchronized with the range. The data provided by lidar sensor is range and angle information, which is a polar coordinate information. Angle acquisition and ranging need to be synchronized to reflect the environmental information.

3. Life problem: The traditional low-cost LiDAR uses slip ring to supply power and communicate with the rotating body. The life of the slip ring is relatively short, only about 1000 h, so the life of the whole radar is affected. IMLidar uses wireless power supply, so its life will be longer, but the specific data is difficult to measure.

Fig. 8.16 Millimeter wave
radar schematic

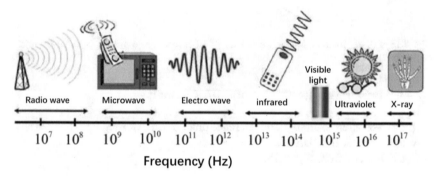

Frequency (Hz)

Fig. 8.17 Light wavelength indication

8.2.5   Millimeter Wave Radar

8.2.5.1   Introduction to Millimeter WaveRadar

As shown in Fig. 8.16, Millimeter-WaveRadar (mmWave Radar) is a detection radar
working in the millimeterwave band. Millimeter-Wave (MMW) refers to electromagnetic waves with a length of 1–10 mm, and corresponding frequency range is
30–300 GHz. As shown in Fig. 8.17, millimeter waves are located in the overlapping
wavelength range of microwave and far-infrared waves. Therefore, millimeter wave
has the advantages of two spectrums, and also has unique properties. Theory and
technology of millimeter wave are the extension of microwave to high frequency and
the development of light wave to low frequency. According to wave propagation
theory, higher frequency and shorter wavelength leads higher resolution and stronger
penetration ability, but also result in greater loss in the propagation process and
shorter transmission distance. Therefore, compared with microwave, millimeter
wave has high resolution, good directivity, strong anti-interference ability, and
good detection performance. Compared with infrared, millimeter waves have

lower atmospheric attenuation, better penetration of smoke and dust, and less influence by weather. These characteristics determine the millimeter-wave radar has the ability to work all-time and all-weather. Millimeter-wave radar can distinguish small targets and simultaneously recognize multiple targets; it has imaging capabilities, is small in size, has good mobility and concealment, and has strong survivability on the battlefield.

8.2.5.2   Millimeter Wave Radar Detection Principle

The most important task of millimeter wave radar is to find the target by radio and detect the distance, speed and direction of the target object.

The ranging principle of millimeter wave radar is very simple. Radar emits radio waves (millimeter waves) and then receives echoes. Measuring the position data and relative distance of the target through the time distance between sending and receiving. According to the propagation speed of the electromagnetic wave, the formula for the distance of the target can be determined as $s = ct/2$, where s is the target distance, t is the time from the electromagnetic wave emitted to target echo received, and c is the speed of light.

Millimeter wave radar speed measurement is based on the principle of Doppler effect (DopplerEffect). Doppler effect refers to the fact that when vibration sources such as sound, light, and radio waves move with the observer at a relative speed v, the frequency of vibration received by the observer is different from the frequency emitted by the vibration source. Because this phenomenon was first discovered by Austrian scientist Doppler, it is called the Doppler effect. In other words, when electromagnetic wave and detected target move relatively, the frequency of the echo is different from the frequency of the emitted wave. When the target approaches the radar antenna, the frequency of the reflected signal will be higher than the frequency of the transmitted signal; conversely, when the target moves away from the antenna, the frequency of the reflected signal will be lower than the frequency of the transmitted signal, as shown in Fig. 8.18. The frequency change formed by the

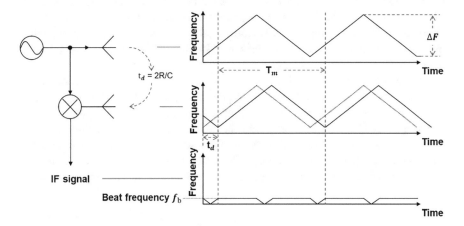

Fig. 8.18 FMCW radar sensor ranging principle

Doppler effect is called the Doppler frequency shift, which is proportional to the relative velocity v and inversely proportional to the frequency of vibration. In this way, by detecting this frequency difference, the moving speed of the target relative to the radar can be measured, that is, the relative speed of the target and the radar. According to the time distance between the transmitted emitting and receiving, the distance of the target can also be measured.

8.2.5.3   Application of Millimeter Wave Radar in ADAS

For vehicle safety, the most important judging basis is the relative distance and relative speed between the two vehicles. When the vehicle is running at high speed, it will easily lead to a rear-end collision if two vehicles is too close. With excellent ranging and speed measurement capabilities, millimeter wave radars are widely used in Vehicle ADAS such as adaptive cruise control (ACC), forward collision avoidance warning (FCW), blind spot detection (BSD), assisted parking (PA), assisted lane change (LCA).

Usually, in order to meet the detection needs of various distance ranges, vehicle will install multiple short-range, medium-range, and long-range millimeter-wave radars. Among them, the 24GHz radar system mainly realizes short-range detection (SRR, within 60 m), and the 77GHz radar system mainly realizes medium and long-distance detection (MRR, about 100 m; LRR, above 200 m). Different millimeter wave radars perform their duties and are installed distinct position of vehicle, as shown in Table 8.4.

According to the wave propagation theory, higher frequency and short wavelength reach better resolution and longer detection distance, but also cause the smaller detection angle (horizontal field of view). Therefore, the 77 GHz millimeter-wave radar can achieve a longer detection range and higher accuracy than 24 GHz. However, as the frequency increases, the corresponding chip design and manufacturing becomes more difficult and costly. The detection angle and distance are always contradictory. As shown in Fig. 8.19, it shows the comparison of the different detection distances and angles of the Continental Group 77 GHzARS 310 mm-wave radar in the short-range, medium-range, and long-range. Therefore, although 77 GHz can replace 24 GHz functionally, it is the mainstream in the future, but from the perspective of cost performance, 24 GHz millimeter-wave radar will be major appliance in the short range.

Figure 8.19 shows the detection range of Continental's ARS310 short-range, medium-range, and long-range radars. At present, the main standard configuration is one or two 77 GHzMRR/LRR plus four 24 GHzSRR. Although the detection distance of 24 GHzSRR is short, it has the advantage of large detection angle and low price. It can be equipped with multiple units to achieve close-range and all-round coverage of the vehicle body. The MRR and LRR functions are equivalent. The advantage of LRR is that its detectable distance is long, and the applicable speed

Table 8.4 Different millimeter wave radar performance and main functions

		Short-range radar (SRR)	Medium-range radar (MRR)	Long-range radar (LRR)
Working frequency		24GHz	76-77GHz	77GHz
Detection distance		< 60 m	100 m	>200 m
Function	Blind spot detection (BSD)	Rear	Rear	
	Lane change assist (LCA)	Rear	Rear	
	Rear cross traffic alert (RCTA)	Rear	Rear	
	Rear collision warning (RCW)	Rear	Rear	
	Automatic valet parking (AVP)	Rear		
	Reversing side warning system/ cross traffic alert (CTA)	Forward	Forward	
	Parking door opening assist (VEA)	Body		
	Active lane control (ALC)	Forward	Forward	
	Adaptive cruise control (ACC)		Forward	Forward
	Forward collision warning (FCW)		Forward	Forward
	Automatic emergency braking (AEB)		Forward	Forward
	Pedestrian detection system (PDS)	Forward	Forward	

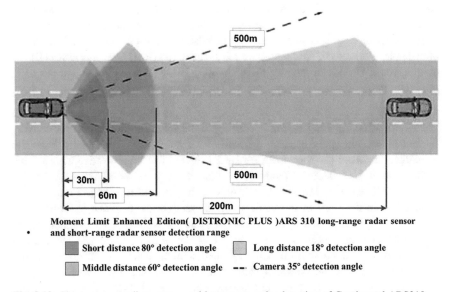

Moment Limit Enhanced Edition(DISTRONIC PLUS)ARS 310 long-range radar sensor
and short-range radar sensor detection range

■ Short distance 80° detection angle   ■ Long distance 18° detection angle

■ Middle distance 60° detection angle   --- Camera 35° detection angle

Fig. 8.19 Short-range, medium-range, and long-range radar detection of Continental ARS310

can achieve 250 km/h. However, in most countries with limited speed, Mid-range radar with 160 km/h applicable speed is always used to achieve adaptive cruise control (ACC) function.

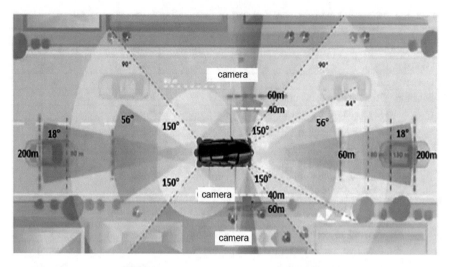

Fig. 8.20 Mercedes-Benz S-Class car radar signal

For example, the Mercedes-Benz S-Class uses 6 mm-wave radars (1 long +1 medium +4 short), as shown in Fig. 8.20, which are respectively distributed in the forward dual-mode long-range millimeter wave radar, and the backward medium and long-range millimeter wave radar, 4 short-range radars on the left and right of the front/rear bumper. "Short-range + medium-range + long-range" millimeter wave radars system complete adaptive cruise control (ACC), automatic emergency braking (AEB), front/rear collision warning (FCW/BCW), lane change assist (LCA), and blind spots. Various ADAS functions such as detection (BSD), reverse assist (BPA), parking assist (PA), etc. Among them, ACC, AEB, FCW, and LCA are the most important anti-collision warning functions in automobile ADAS.

1. Adaptive cruise: Adaptive cruise control (ACC) is a driving assistance function that can keep vehicle running according to the set speed or distance or actively control vehicle speed according to front vehicle speed, and finally keep the vehicle at a safe distance from forward vehicle. The biggest advantage of this function is that it can effectively liberate driver's feet and improve driving comfort. The realization principle of ACC is that the millimeter wave radar sensor installed in the front of the vehicle continuously scans the road in front of the vehicle, and the wheel speed sensor collects the vehicle speed signal during vehicle driving. When it is too close to front vehicle, the ACC system can properly take a breakthrough coordinated actions with the anti-lock braking system and the engine control system, and reduce the output power of the engine, so that it is always keep a safe distance from forward. When the ACC system taking a break, it usually limits the braking deceleration to a degree that does not affect the comfort. When a greater deceleration is required, the ACC system will send out an audible and light warning signal to inform the driver to take the initiative to take the brake operation.

Fig. 8.21 AEB sign

2. Automatic emergency braking: Automatic emergency braking (autonomous emergency braking, AEB) is an active safety assist function for automobiles. The AEB system uses millimeter-wave radar to measure the distance to the vehicle or obstacle in front, and then uses the data analysis module to compare the measured distance with the alarm distance and the safety distance. When it is less than the alarm distance, it will give an alarm prompt, and when it is less than the safety distance When the driver does not have time to step on the brake pedal, the AEB system will start to make the car brake automatically to ensure driving safety, as shown in Fig. 8.21.

  Studies have shown that 90% of traffic accidents are caused by drivers' inattention. AEB technology can reduce rear-end collisions by 38% in the real world, regardless of whether it is on urban roads (speed limit 60 km/h) or suburban areas. In the case of road driving, the effect is very significant. Therefore, the European New Car Safety Evaluation Association (EuroNCAP) took the lead to include the AEB system in the overall safety rating in 2014, and China also added AEB to the NCAP scoring system in 2018.

3. Forward collision warning function: Forward collision warning (FCW) uses millimeter-wave radar and front cameras to continuously monitor the forward automobiles to determine the distance, azimuth, and relative speed, and detect the potential collision hazard ahead. When no braking measures are taken, the instrument will display an alarm message and an audible alarm to warn the driver to take countermeasures. When it is judged that an accident is about to occur, the system will let the braking system automatically intervene in the work to avoid the accident or reduce the risk that the accident may cause, as shown in Fig. 8.22. AEB uses sensors to detect obstacles such as vehicles and pedestrians in front. If it finds that the distance is too close and there is a risk of collision, it will automatically brake. FCW can be understood as an early warning function before automatic braking. In fact, FCW and AEB systems are in a complementary relationship, and the purpose is to avoid or reduce the occurrence of collision accidents while driving.

4. Lane change assistance: Lane change assist (LCA) is to detect the lanes and the rear of the adjacent two sides through sensors such as millimeter wave radar and

Fig. 8.22 FCW schematic

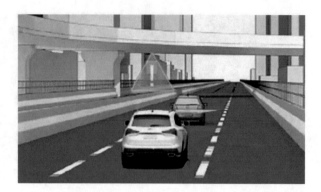

Fig. 8.23 Lane change
assistance signal

cameras, to obtain the movement information of the objects on the side and the rear, and to judge based on the current state of the vehicle. Finally, the driver is reminded by sound, light, etc., so that the driver can grasp the best time to change lanes, prevent traffic accidents caused by lane change, and also have a better preventive effect on rear collisions.

The lane change assist system includes three functions: blind spot detection (BSD), lane change warning (LCA), and rear collision warning (RCW), which can effectively prevent traffic accidents such as lane change, turning, and rear collision, greatly improving safety performance of lane changing, as is shown in Fig. 8.23.

Among them, the BSD judges the relative position and speed of moving object. When it is in vehicle blind area, BSD will promptly remind driver to pay attention to the risk of changing lanes. LCA detects that backward vehicle is approaching at a large relative speed in the adjacent area, and when the time distance between the two vehicles is less than a certain range, it reminds driver through sound and light. When RCW detects fast approaching moving object behind the same lane and the risk of collision, it will also promptly warn driver through sound, light, and other methods to reduce the damage caused by the collision by wearing seat belts and other methods.

5. Fusion of multiple sensors: In the process of implementing these driving assistance functions, it is not difficult to find that although the millimeter wave radar plays the core role of object detection, ranging, and speed measurement, the

whole process also requires the assistance of other sensors, such as LiDAR, camera, and ultrasonic radar, inertial sensors, etc. As more and more car manufacturers begin to integrate different sensors into automotive ADAS, the industry generally believes that sensor fusion is the key to the safety of highly automated driving. In environmental perception, each sensor has unique advantages and weaknesses. For example, millimeter-wave radar is not affected by the weather and can work all-weather and all-day, but the resolution is not high and cannot distinguish between people and objects; while the camera has a higher resolution and can perceive colors, but it is greatly affected by strong light; LiDAR can provide three-dimensional-scale sensing information, and has a strong ability to reconstruct the environment, but it is greatly affected by the weather. Sensors have their own advantages and disadvantages, and it is difficult to replace each other. In order to realize automatic driving in the future, vehicle perception system must include various sensors. With the development of autonomous driving from L2 to L5, the number and types of sensors integrated continue to increase. Only in this way can it be possible to ensure sufficient information acquisition and redundancy, and to achieve the safety standards required by OEMs.

Software is one of the cores of multiple sensor fusion. The algorithm is a barrier for the fusion of multiple sensors to more advanced autonomous driving technology, because the use of multiple sensors will increase the amount of processed information, and there may be conflicting information. How to ensure that the system quickly processes data and filters useless, wrong information, so as to ensure that the system finally makes timely and correct decisions. At present, many theoretical methods of sensor fusion have been proposed such as Bayes criterion method, Kalman filter method, D–S evidence theory method, fuzzy set theory method, artificial neural network method, and so on.

Therefore, in the case of using multiple sensors, if you want to ensure safety, you must perform information fusion on the sensors. The fusion of multiple sensors can significantly improve the redundancy and fault tolerance of the system, thereby ensuring the speed and correctness of decision-making. It is an inevitable trend for ADAS to move toward advanced automatic driving and ultimately realize unmanned driving at this stage.

8.2.6   Visual Perception System

In addition to radar ranging sensors (LiDAR, millimeter wave radar, ultrasonic radar, etc.), the environmental perception sensors used in autonomous driving technology also include vision sensors (monocular and binocular stereo vision, panoramic vision, and infrared cameras).

ADAS uses the camera as the main sensor because the camera resolution is much higher than other sensors, and it can obtain enough environmental details to help the vehicle to understand the environment. The on-board camera can describe the appearance and shape of the object, read signs, etc., which others cannot do. From the perspective of cost reduction, the camera is one of the powerful candidates for the identification sensor. The camera is the best choice in normal environment light, but it is greatly affected by weather and external factors, such as insufficient light in the tunnel, and weather, the narrowing of sight caused by factors, etc.

Collecting image information, road sign recognition, and lane line sensing can only be realized by the camera. At present, the main applications of cameras include monocular cameras, rear view cameras, stereo cameras or binocular cameras, and surround view cameras.

8.2.6.1   Camera Classification

1. Monocular camera: The model of the monocular camera can be approximated as a pinhole model, as shown in Fig. 8.24.

f the focal length of the camera.

   c the optical center of the lens.

   The light emitted by the object passes through the optical center of the camera, and then show image on the sensor. If the distance between the plane of the object and the plane of the camera is d, the actual height of the object is H, and the height on the sensor is h, then there is

$$\frac{d}{f} = \frac{h}{H} \tag{8.7}$$

According to formula (8.7), according to this idea, only the height h of the object on the sensor needs to be obtained. Once h is accurate enough, the accuracy of measured distance d can also be guaranteed.

The entire process includes sample collection and labeling. At the same time, large-scale training is performed on the labeled samples to extract features and models, and the models are classified and identified as actual image data, as shown in Fig. 8.25.

Another dimension needs to ensure the quality of the image source, and ensure the cleanness of the input data source through technologies such as wide dynamics,

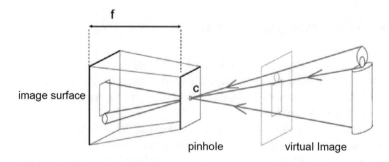

Fig. 8.24   Camera pinhole imaging model

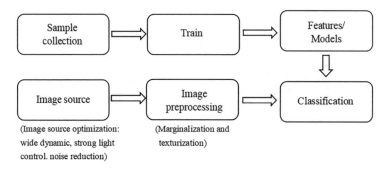

Fig. 8.25   Basic principles of visual ADAS

Fig. 8.26   Road and vehicle recognition

strong light suppression, and noise reduction. The clear data of the real environment is marginalized and textured and sent to the classifier for identification, such as shown in Fig. 8.26.

At the same time, in this link, we must pay great attention to the consistency of the model data and the image source data, that is, the sample labeled data and the actual image source must come from the same lens, image sensor, and the same ISP technology to ensure a high match between training and reality.

This part is currently difficult to achieve when testing data in the laboratory. Numerical public sample libraries may be used for training, and the cameras and lens angles used in the public sample libraries are not actually used.

2. Binocular camera: Binocular vision applied to outdoor scenes is rare. The bumblebee binocular is also applied to indoor scenes, and the ZED camera is also easier to implement indoors. It is recommended to build a binocular sensor for

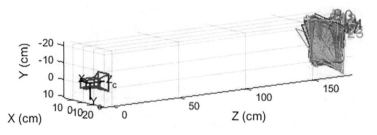

Fig. 8.27 Binocular stereo vision calibration

outdoor production of visual odometry or recognition algorithms, and determine the baseline length according to specific requirements while ensuring synchronous triggering. Binocular vision cannot bypass the disparity map and binocular calibration. The current more common binocular calibration method is to use the Zhang Zhengyou method and the Camera Calibration Toolbox for calibration. This method is also integrated in OpenCV, as shown in Fig. 8.27.

3. Panoramic camera: Panoramic cameras are divided into single-lens panoramic cameras and multi-lens splicing panoramic cameras. It is recommended to use a panoramic camera as a visual odometer, which has a large field of view and a high degree of correlation of feature points. It is a good choice to use panoramic vision with integrated navigation for high-precision map reconstruction, which can realize lane-level positioning of autonomous vehicles.

4. Infrared camera: Infrared camera belongs to another category of vision. Its night vision effect is better than daylight effect, and it can be applied to pedestrian and vehicle detection. After the appearance of LiDAR, infrared cameras were in an awkward position in autonomous driving applications. They were expensive and did not have the intuitive results of LiDAR. After using it, it was found that the laser discrete point cloud is still better than the recognition of obstacles (such as people). In addition, it was found in the test that the infrared camera can specifically capture the shadows on the glass. Infrared cameras can distinguish heating elements to a certain extent, such as roads, pedestrians, etc., but post-processing is required. LiDAR that uses absolute height or gradient to detect obstacles is more intuitive than that. Under night conditions, it can replace the color camera to detect and monitor forward-looking obstacles, as shown in Fig. 8.28.

8.2.6.2 Application of Computer Vision in Unmanned Driving

There are some more directly examples of computer vision in self-driving, such as the recognition of traffic signs and signal lights (Google), and the detection and positioning of highway lanes (Tesla). Some functional modules implemented based on LiDAR information can actually be implemented based on computer vision with

Fig. 8.28 Infrared vision imaging

a camera. The main problems that computer vision solves in the autonomous vehicle scene can be divided into two categories: object recognition and tracking and the positioning of the vehicle itself.

1. Object recognition and tracking: Through deep learning methods, objects encountered during driving can be identified, such as pedestrians, empty driving spaces, traffic signs on the road, traffic lights, and nearby vehicles. Since pedestrians and nearby vehicles and other objects are in motion, these objects need to be tracked to prevent collisions, which involves motion prediction algorithms such as OpticalFlow.

2. Vehicle positioning: Through the visual mileage calculation method based on topology and landmark algorithms or geometry, unmanned vehicles can determine their position in real time to meet the needs of autonomous navigation.

    (a) OpticalFlow and stereo vision: The recognition and tracking of objects and the vehicle positioning are inseparable from the underlying OpticalFlow and stereo vision technology. In the field of computer vision, OpticalFlow is a dense set of pixel-level correspondence relationships in a picture sequence or video. For example, a two-dimensional offset vector is estimated on each pixel, and the obtained OpticalFlow is represented by a two-dimensional vector field. Stereo vision establish a corresponding relationship from images obtained from two or more perspectives. Two issues are highly correlated. One is based on the images of a single camera at consecutive moments, and the other is based on the images of multiple cameras at the same moment. There are two basic assumptions when solving such problems: The corresponding points in different images are all from the same point in the physical world, so they have similar appearances. The spatial transformation of the set of corresponding points in different images basically satisfies rigid body condition, or the motion of the space divided into multiple rigid bodies. From this assumption, it can be concluded that two-dimensional vector field of OpticalFlow is flat and smooth.

(b) Object recognition and tracking: From color, offset, and distance information at the pixel level to the spatial position and movement trajectory at the object level are important functions of the self-driving vehicle's visual perception system. The perception system of autonomous vehicles needs to recognize and track multiple moving targets (Multi-Object Tracking, MOT) in real time, such as vehicles and pedestrians. Object recognition is one of the main problems of computer vision. In recent years, due to the revolutionary development of deep learning, CNN has been widely used in the field of computer vision, and the accuracy and speed of object recognition have been greatly improved. But, in general, the output of object recognition algorithms is generally noisy: The recognition of objects may be unstable, the objects may be occluded, and there may be short-term misrecognition. Naturally, the popular tracking-by-detection method in the MOT problem has to solve such a difficult point: How to obtain a robust object trajectory based on the noisy recognition result. At the ICCV2015 conference, researchers from Stanford University published an MOT algorithm based on Markov Decision Process (MDP) [7] to solve this problem. The tracking of moving targets is modeled by an MDP, as shown in Fig. 8.29.

1. The state of the moving target: s∈S=S_active∪S_tracked∪S_lost∪S_inactive, these subspaces each contain an infinite number of target states. The identified target first enters the active state, if it is misrecognized, the target enters the inactive state, otherwise it enters the tracked state. Target in the tracked state may enter the lost state, and target in the lost state may return to the tracked state, or remain in the lost state, or enter the inactive state after a long enough time.
2. Action $a \in A$, all actions are deterministic.
3. State change function T: S × A → S defines that the target state becomes s' under state s and action a.
4. The reward function R: S × A → R defines the instant reward for reaching the state s after action a. This function is learned from the training data.
5. The rule π: S → A determines the role a used in state s.

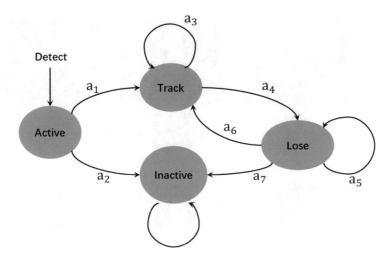

Fig. 8.29   DMM state diagram

As shown in Fig. 8.30, the state space of this MDP changes as follows.

In the active state, the object candidate box proposed by the object detecting algorithm is determined by an offline training support vector machine (SVM) to determine whether the next step is a_1 or a_2. The input of SVM is feature vector and spatial position size of the candidate object. It determines the MDP rule π_active in S_active.

In tracked state, an object appearance model based on the tracking–learning–detection tracking algorithm is used to determine whether the target object remains in the tracker state or enters the lost state. This appearance model uses bounding box of the target object in current frame as template, and all the object appearance templates collected in the tracked state are used to determine whether the target object is back to tracked state. In addition, in tracked state, object tracking uses the above-mentioned appearance model template, and the overlap ratio of candidate object and target object provided by the rectangular range OpticalFlow and the object recognition algorithm determines whether to remain in the tracked state. Update automatically.

In the lost state, if an object remains state for more than a threshold frame counts, it enters the inactive state; whether the object returns to the tracked state is determined by a classifier based on the similarity feature vector of the target object and the candidate object, corresponding to the π_lost in S_lost.

This MDP-based algorithm has reached an industry-leading level in the object tracking evaluation of the KITTI dataset.

Visual odometer method [8]: There are two categories of vision-based positioning algorithms: One is topology and landmark-based algorithms; the other is geometry-based visual odometer calculation method. Algorithm based on topology and

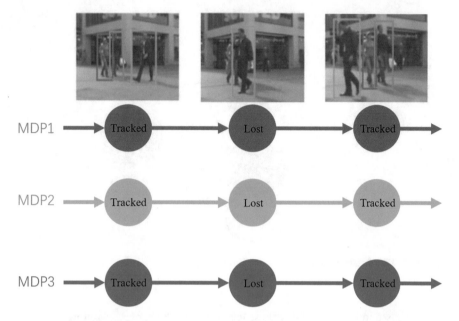

Fig. 8.30 State transition example

landmarks composes all landmarks into a topological map. When vehicle detects a landmark, it can roughly infer its own location. Algorithms based on topology and landmarks are easier than methods based on geometry, but require the establishment of accurate topological maps in advance, such as making landmarks at each intersection. The calculation of the visual odometer calculation method based on geometry is more complicated, but it does not need to establish a precise topological map in advance. This algorithm can expand the map while positioning. The following focuses on the visual odometer algorithm.

The visual odometer calculation method is mainly divided into two types: monocular and binocular. The problem of pure monocular algorithm is that the size of the observed object cannot be calculated, so the user must assume or calculate a preliminary size, or use it with other sensors (such as a gyroscope) for accurate positioning. The binocular visual odometer method calculates the depth of feature points by triangulation of the left and right images, and then calculates the size of object from depth information. Figure 8.31 shows the specific calculation process of the binocular visual mileage calculation method.

1. The binocular camera captures the left and right images.
2. The binocular image is triangulated to generate a disparity map of the current frame (DisparityMap).

Stereoscopic image   Visual channel

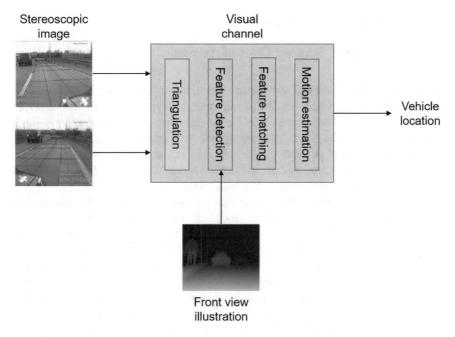

Vehicle location

Front view illustration

Fig. 8.31 The calculation process of binocular visual odometer calculation method

3. Extract the feature points of the current frame and the previous frame. If the feature points of previous frame have been extracted, it can be used directly. Harris corner detector can be used to extract feature points.
4. Comparing the feature points of current frame and previous frame, and finding the corresponding relationship between the feature points of two frames. Specifically, random sample consensus (RANSAC) algorithm can be used.
5. The movement of the vehicle between the two frames is calculated according to the corresponding relationship of the feature points between the frames. This calculation is achieved by minimizing the reprojection error between two frames.
6. Calculating the latest vehicle position based on the estimated vehicle movement between two frames and previous vehicle position.

Through the above visual odometer calculation method, the autonomous vehicle can locate in real time and conduct autonomous navigation. But the main problem of pure visual positioning calculation is that the algorithm is quite sensitive to light. Under different light conditions, the same scene cannot be recognized. Especially when the light is weak, the image will have a lot of noise, which greatly affects the quality of feature points. On reflective roads, this algorithm is also easy to fail, which is also a main reason that affects the popularization of visual odometer calculation method in unmanned driving scenes. A possible solution is to rely more on positioning based on the information returned by the wheels and radar when the light condition is awful.

8.3   Decision Planning Control

8.3.1   Decision Planning Control System

As an essential part of intelligent transportation, Intelligent vehicle is a comprehensive system involving multiple disciplines which is a typical cross-field research. In addition to the environmental perception system, the decision-control system is another important subsystem. They are equivalent to human eyes and brain, respectively. Environmental perception system is able to "see" the surrounding environment, and decision-control system makes decisions and judgments based on what it "thinks" according to the surrounding conditions.

The environment perception system uses sensors to perceive the surrounding environment information, and provide this information to decision-control system as input. The decision-control system makes decisions based on further processed information, then guides driving behavior. Similarly, the"brain"of intelligent vehicle obtains information from the "eyes" and direct the "hands and feet." For intelligent vehicle, the decision-control system is an essential part, which makes all decisions.

The task of decision-control system is to make the most reasonable decision and control based on predicted trajectory of surrounding objects as well as the routing intention and current location of the intelligent vehicle itself. According to solving different problem, the decision-control system could be divided into three layers, including decision-making, action planning, and feedback control [9].

The decision-making layer makes judgments through data fusion based on the results of environmental perception, and then sends control instructions to the control layer. The planning layer solves the planning problems of specific car actions. It means that the planning layer plan the route from A to B in a small space. The control layer controls the direction, throttle, and braking system of the intelligent vehicle according to the instructions from the decision-making layer and the planning layer. The decision-control system, with a high degree of intelligence and complexity, is the central part of the intelligent vehicle system.

8.3.2   Decision

Human driving ability not only requires the ability of properly manipulating the steering wheel, brake system, and throttle based on traffic rules, but also evaluating the social risks, health, legal, and life-threatening consequences of driving(such as "What should do if a pedestrian does not stop at a red light?"). Solving the driving problems needs to improve the human cognition, thus scientifically using the complex artificial intelligence systems to simulate them. The effect of decision-making system is interpreting the abstract information provided by the vehicle environment perception system, and then generating sustainable and safe driving decisions.

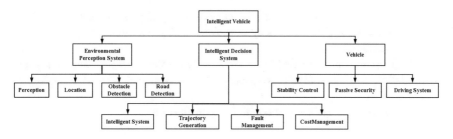

Fig. 8.32 Intelligent vehicle system

In order to operate reliably in the real world, intelligent vehicle should evaluate and determine the consequences of their potential actions by predicting the intentions of other traffic participants. The first decision-making system for intelligent vehicle appeared in the 2007 DARPA Urban Challenge Urmsonetal [10]. This system allowed intelligent vehicle to drive in the urban scenes they involved such as, simulating U-turns, crossroads, parking lots and real traffic. These early decision-making system based on finite state machine, decision tree, and heuristic algorithm. In recent years, the decision-making problem of intelligent vehicle can be solved by the lens of trajectory optimization. However, these methods cannot model the closed-loop interaction between vehicles and consider the potential results. Nowadays, real decision-making system cannot surpass human driver. The progress of decision-making system aims to improve the intelligence of the system. Cognitive systems, agent systems, fuzzy systems, neural networks, evolutionary algorithms, and rule-based methods constitute an intelligent decision-making system (IDMS). Figure 8.32 shows the location of IDMS from the perspective of the intelligent vehicle system.

In complex traffic scenarios, several intelligent vehicles could Inherent risk to human, which needs a real-time solution. In these circumstances, decision-making system should be reliability, safety and fault-tolerance. Thus, it is necessary to propose a real-time route planning algorithm for rapid random tree (RRT) method. This algorithm is the core of the planning and control software for the MIT team to participate in the 2007 DARPA Urban Challenge. In the event, this vehicle demonstrated the ability to complete a 60-mile (96.56 km) simulated military supply mission, and at the same time, it could experiment with other intelligent vehicles and human-driven vehicles for safe interaction. The proposed of Multi-Criteria Decision-Making (MCDM) and PetriNet addressed the problem of real-time autonomous driving. MCDM provides various benefits, such as the following:

1. The hierarchy is able to systematically and completely determine the destination of vehicle.
2. The utility function can be defined by heuristics, which reflect the choice of human driver. It can apply learning algorithms as well.

3. MCDM is used to integrate and evaluate of a large number of alternative driving program. The flexibility of policy reflects by defining attribute weights based on road conditions.
4. Driving actions are modeled as decisive finite automata, which additional goals, attributes, and alternative driving program can be added without changing the major system.
5. Decision-making system can be modeled as PetriNet.

8.3.3 Planning

In the past few decades, a large number of researchers have conducted extensive study on the motion planning of mobile robots and intelligent vehicles. Under special assumptions, different strategies are designed to meet various demands including kinematics, dynamics, and environmental constraints. In this part, the path and speed planning is proposed under a specific method, however other different strategies can also be considered.

8.3.3.1 Path Planning

The technology to obtain the best path can generally be divided into two categories: indirect technology and direct technology. Indirect technology discretizes state/control variables, and transforms the path planning problem into a parameter optimization problem, then solves the problem by using nonlinear programming or stochastic techniques. Direct technology use Pontryagin's maximum value principle to express the optimal condition as a boundary value problem [11]. The approximate solution is studied under a large number of possibilities and constraints. In the solution of direct technology, a local planner is encapsulated into a common process in order to deal with complex scene topology and obstacle avoidance.

The path planning system consists of several subsystems including cost map generation, global planning, and local planning, which run independently as separate processes. Cost map generation subsystem is responsible to generate the cost map and calculate trajectory to be used by the other two methods. It considers the security of different solutions (attention to obstacles in the environment and estimates of recent expected changes). Global planning is used to calculate the trajectory that the vehicle can travel between the current location to the target on the unstructured map. Local planning provides the mechanism of tracking to the system.

Cost map generation: Cost map saves the information about occupied/vacant areas in the map by the form of occupancy grid. This map uses the static map and obstacle information came from environmental perception system. The cost map calculates hierarchically, and is used to integrate different information sources into a single map. In each layer, the information about the occupied/vacant area around the vehicle is maintained in the form of an occupied grid, which uses different observation sources as input. Based on this information, both dynamic and static obstacles are marked on the map. For example, suppose that each cell in the map has 255 different cost values. Then, the cost map of each layer is represented as follows:

1. The value of 255 means that there is no information about a specific cell in the map.
2. The value of 254 indicates that the sensor has marked the specific unit as occupied. It is considered as a dangerous cell, so vehicles can never enter.
3. Considering the expansion method related to the size of the vehicle and its distance from obstacles, other cells are considered free, but their cost levels are different.

The cost value decreases with the distance to the nearest occupied cell, which can be expressed as follow:

$$C(i,j) = \exp\left(-1 \times \alpha \|c(i,j) - 0\| - \rho_{\text{inscribed}}\right) \times 253 \qquad (8.8)$$

where α is a proportion factor, the increase or decrease rate. $\|c(i,j) - 0\|$ is the distance between the cell and obstacles. $\rho_{\text{inscribed}}$ is the inner radius of vehicle limit.

In order to set different danger levels in the map, different distance thresholds are defined usually. For example, the following four thresholds can be defined:

ξ_{lethal}: There is a barrier in this cell, so the vehicle is in collision. The cost level is represented by 254.

$\xi_{\text{inscribed}}$: The distance between the unit and the nearest obstacle is less than $\rho_{\text{inscribed}}$ 9 k. If the center of the vehicle is in this unit, vehicles should avoid areas below the distance threshold. The cost level is represented by 253.

$\xi_{\text{circumscribed}}$: If the center of the vehicle is in this cell, it is very likely that the car is colliding with an obstacle, depending on the position of the vehicle. Vehicles should avoid obstacles below this threshold. However, There are also cases where no obstacles are encountered.

The other cells are considered safe (except for cells with unknown costs, because it is unknown whether they are occupied or not, so they are considered dangerous).

In the proposed method, only the routes that boundary cost is less than $\xi_{\text{circumscribed}}$ are considered. This cost is obtained by using Eq. (8.8) and other cost factors that will be explained later. The path of the cell passing the threshold will be cut off at the last safe point. In order to calculate the cost graph and the cost associated with each cell, ROS plugin COSTMAP2D is used, which implement some functions described in this section.

Now, we consider four different layers in the cost map.

The first layer represents obstacles in the previously captured static map. This map represents static obstacles in the entire area where the vehicle is traveling. This layer is the only layer based on non-primitive global planning. Because the non-primitive trajectory generation does not consider dynamic obstacles. Therefore, these should be avoided at the local planning level.

The second layer based on static map. For optimization reasons, in this and subsequent layers, each iteration does not calculate the cost map of the entire map.

In contrast, only a cell which in the center of the current car position is updated. Updating the entire map is not the target, because these layers are only used for local planning and manipulation. Static obstacles are included in the local plan. Because vehicles are not willing to pass through restricted areas, at the same time avoiding obstacles vehicles should know which areas are prohibited at the local planning level.

The third layer is used to represent dynamic obstacles detected by different sensors. Only use vertical obstacles that the vehicle may collide as input. The parameters in this layer are more dangerous than the obstacle calculated in the second layer. The obstacles detected in real time gain more priority than obstacles on a static map.

The last layer provides an estimate of the future movement of dynamic obstacles. The input point cloud is divided into voxel grids to reduce the dimensionality. For each voxel, an occupancy probability is assigned based on the number of points of the input point cloud from its neighborhood.

Using this probability, effective pixels (having a higher occupancy rate) are distinguished from noisy pixels (having a smaller probability). All layers are combined into one cost map. The cost map considering the movement of obstacles is very interesting, because the vehicle will try to avoid obstacles on the side that does not cross the track.

Global planning: Intelligent vehicle usually uses two global planners including the original global planner and the non-original global planner. These planners aim to obtain a feasible path from the current position of the vehicle to the determined target.

Global planner based on the original information. The purpose of these two methods in the system is completely different. Generally, non-original global planner is used for general navigation. However, the original global planner is used to recover the vehicle in the presence of obstacles for a long time such as parking.

Global planner based on the original information plan a route from the vehicle location to the target. The route is generated by combining "movement primitives," which is a short, kinematically feasible exercise. In order to conform to the curvature limit of the vehicle, these "movement primitives" are generated by the model of vehicle.

These "movement primitives" are calculated as follows: Considering a set of predefined directions, for each direction, the model is constantly evolving until it reaches the predetermined direction at different speeds. In this process, a set of trajectories that realize vehicle restrictions are obtained, which will be used as planning components.

ARA algorithm is used to search for a feasible path with these information. As each node expands, a new x, y coordinate and position is explored until find the best path or reach the end of exploration time. In this search, the backward cost is higher than the forward cost, in order to prevent the vehicle from using the reverse path without degrading performance. In addition, the original search algorithm can be improved by adding a new cost, which penalizes forward and backward routes.

Global planner based on the non-original information. The non-original global planner calculates the least cost path from the position of the vehicle to the target, such as Dijkstra algorithm. Considering the algorithm need the global target speed, this planner is used to roughly estimate the route that vehicle will follow. In order to plan a smooth path, static obstacles in the cost map will be overinflated.

If this route does not take into account the incomplete restrictions of the vehicle, the initial angle of global planning direction is much larger than the maximum angle required by the local planning. So as to generate a feasible path, This is the reason why non-primitive planning and partial planning state machines are used in combination.

Local planning: When the global path is defined, a method is needed that can calculate the steering and speed commands required to control the vehicle in order to track the path. This method should be able to avoid obstacles on the road, and it must be safe and effective.

The basic idea of local path generation is to define a set of feasible paths and choose the best path based on cost. The winning path defines the steering and speed commands used for the vehicle [12]. The choice in the local path helps to overcome the existence of unforeseen obstacles in the road.

The current Euclidean coordinate system is generally converted into a new system based on the Frenet space. The calculation of this space is as follows: The global path is considered as the basic frame of the curvilinear coordinate system. According to the basic framework, the feasible local path is defined as follows:

The closest point (where the distance is perpendicular to the distance of the global path) to the main trajectory will be the origin of the curved coordinate system.

The horizontal axis represents the distance on the global path.

The vertical axis is represented by a vector perpendicular to the origin, which points to the left in the direction of the path.

In this mode, the trajectory can be easily calculated in the curve space (that is, maneuvering information is generated), and then these are converted into the original Euclidean space, where obstacle information is added by assigning the cost to each path.

Based on this idea, the method can be divided into five stages:

1. Cost map generation: Using the information generated by the sensors or the methods described in the previous chapters, the system constructs a cost map related to the distance of obstacles.
2. Basic frame structure: Based on the global path constructed in the previous part, the basic frame of the curvilinear coordinate system is generated.
3. Candidate path generation: The candidate path is generated into the curve space, and then converted into Euclidean space.
4. Choice of winning path: Allocate the cost of all paths and choose the path with the lowest cost.
5. Calculation of vehicle instructions: The vehicle speed and steering angle are calculated based on the characteristics of the winning path.

8.3.3.2 Speed Planning

When the path planning is given the selected track, the follow-up problem that motion planning needs to address is to add speed-related information to this track, which is called speed planning.

In fact, any path in an unstructured environment can be decomposed and a series of rotations can be entered with the help of path planning algorithms arcs and arcs consisting of rings and straight lines. The swing line is chosen because the arc of a gyro has variable curvature at each point which is proportional to the arc length, and it provides the smoothest connection between a straight line and a circular curve, often used in road and rail designs: centrifugal forces actually change proportionally to time, returning at a constant rate from zero (along a straight line) to the maximum (along the curve).

This decomposition is useful for finding optimal velocity distributions for closed forms, as both straight segments and arcs can be associated with constant velocity. More precisely, when a turn starts, the maximum speed is limited by the comfortable lateral acceleration threshold, and when the straight segment is tracked, the maximum longitudinal velocity, acceleration, and lift bar will be the limit applied to the reference speed.

The speed curve can be defined as follows:

1. When the curvature curve is an arc or a gyration line in front of it, take the constant-speed curve at the minimum v_{min}.
2. From the minimum value v_{min} to maximum allowed speed v_{max}. v_{max} smooth transition and return again to v_{min} that meets acceleration and constraints.
3. A set of one or two smooth transition curves (more than 2 classes) from zero to maximum speed, and vice versa.

In order to obtain the closed form expression of the second type of curve, the velocity trajectory is divided into several intervals. Let's say that the seven intervals. The arc length Sr is shown below.

$$
\ldots s_r(t) = \begin{cases} \ldots s[t_0, t_1][t_6, t_7]_{r\,max} \\ 0, \qquad t \in [t_1, t_2] or \ t \in [t_5, t_6] \\ -\ldots s[t_2, t_3][t_4, t_5]_{r\,max} \end{cases}
$$

$$
\dddot{s}_r(t) = \dddot{s}_r(t_{i-1}) + \ldots s_r(t)(t - t_{i-1})
$$

$$
\ddot{s}_r(t) = \dot{s}_r(t_{i-1}) + \ddot{s}_r(t_{i-1})(t - t_{i-1}) + \frac{1}{2}\ldots s_r(t_{i-1})(t - t_{i-1})^2 \tag{8.9}
$$

$$
s_r(t) - s_r(t_{i-1}) + \dot{s}_r(t_{i-1})(t - t_{i-1}) + \frac{1}{2!}\ddot{s}_r(t_{i-1})(t - t_{i-1})^2
$$

$$
+ \frac{1}{3!}\ldots s_r(t_{i-1})(t - t_{i-1})^3
$$

The arc length will be from the initial point $(s_r(t_0))$ of the closed roundabout to the last point of the straight segment $(s_r(t_7))$, the initial and final velocity $(\dot{s}_r(t_0), \dot{s}_r(t_7))$

will be the minimum speed vmin, initial acceleration and final acceleration $(\ddot{s}_r(t_0), \ddot{s}_r(t_7))$ setting, $(\ldots s_r(t_0), \ldots s_r(t_7))$ will be equal to zero.

With regard to comfort constraints, the maximum speed will be V^*_{max}, and the maximum speed and acceleration will be determined by the design parameters γ_{max} and J_{max}.

Note that the value of V^*_{max} corresponds to the previously defined Vmax, provided that sufficient distance reaches the target. If the available arc length is less than a critical value, the maximum speed is set to the initial speed V_0 resulting in a constant velocity profile, otherwise the maximum speed between V_0 and V_{max} is calculated. The closed polynthic expression of the Eq. (8.9) allows the maximum speed to be calculated as follows.

$$V^*_{max} = \begin{cases} V_{max}, \text{When condition 1 is met} \\ V_0, \text{When condition 2 is met} \end{cases} \tag{8.10}$$

condition 1:

$$\Delta_s < \frac{V_0}{2} \Delta \left(\frac{V_0}{\gamma_{max} \frac{\gamma_{max}}{J_{max} 2J_{max} 2_{max}} \frac{1}{}} \right) \tag{8.11}$$

condition 2:

$$\Delta_s < \frac{V_0}{2} \Delta \left(\frac{V_0}{\gamma_{max} \frac{\gamma_{max}}{J_{max} 2J_{max} 2_{max}} \frac{1}{}} \right) \tag{8.12}$$

Among them: $\Delta_s = s(t_7) - s(t_0)$.

Another algorithm enables the total time required to cover the path by slightly impairing passenger comfort, rather than reducing the speed to V_{min} in each turn, but considering only non-simple and turning.

8.3.4 Vehicle Control

Vehicle control is based on the position of the vehicle to control the intelligent car, so that it plans to travel. Vehicle control is generally divided into vertical control and lateral control. Vertical control is the control of the gear, throttle, and brakes of the intelligent car, and lateral control is the adjustment control of the steering wheel.

Mathematical models are very important in the analysis and control of automotive dynamics. According to the captured physical phenomena, there are several mathematical models in the literature with varying degrees of complexity and accuracy [13]. In general, the motion of the vehicle is considered in the plane, mainly describing the movement of the vertical and horizontal vehicles. Different

Fig. 8.33 Nonlinear bike model

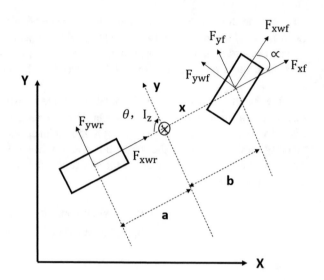

longitudinal and lateral power couplings must be considered when describing vehicle movement.

1. The coupling of power and motion is caused by the direction of the deflected plane caused by wheel steering.
2. The interaction between tires and road surfaces is another important source of coupling.
3. Longitudinal and lateral acceleration results in load transfer between the front and rear axles and the left and right wheels.

Complexity is used to balance complexity with accuracy. Complexity models can provide a good level of precision, but are still too complex for controller composition. Therefore, the nonlinear bicycle model is usually used for horizontal control, and the one-wheeled car model is used for vertical control design.

A nonlinear bicycle model considers longitudinal x, lateral y, and yaw motion θ. For this model, assuming that the mass of the vehicle is entirely on the rigid base of the vehicle, it considers the transmission of pitch loads and ignores the transverse load transfer caused by rolling motion.

In Fig. 8.33, the control angles α, a and b represent the distance between the wheels and the vehicle's center of gravity, and indicators f and r represent the front and rear.

The dynamic equation is

$$m(\ddot{x} - \dot{\gamma}\theta) = \sum i = f, r \ F_{xi} + F_r$$
$$m(\ddot{\gamma} + \dot{x}\theta) = \sum i = f, r \ F_{yi} \tag{8.13}$$
$$I_z\theta = F_{yf}a - F_{yr}b$$

where m is the vehicle mass, I_z is the inertial yaw torque, F_r is the force of resistance, F_{xi} and F_{yi} are the vertical and lateral tire forces along the x and y axes, respectively.

These forces can be associated with the longitudinal relationship between F_{xwi} tire force, lateral force F_{ywi}, and steering angle α.

$$\begin{cases} F_{xf} = F_{xwf} \cos(\alpha) - F_{ywf} \sin(\alpha) \\ F_{yf} = F_{xwf} \sin(\alpha) - F_{ywf} \cos(\alpha) \end{cases} \tag{8.14}$$

When the drive torque TD and brake torque TB are applied, the rotational motion can be exported as

$$I_z\dot{w}_{wi} = T_{di} - F_{xi}R - T_{bi}(i = f, r) \tag{8.15}$$

where, R is the radius of the wheel, and the $\dot{w}_{wi}$ is the velocity of the swing angle.

The trajectory of the vehicle's center of gravity is given in the absolute inertial coordinate system

$$\begin{cases} \dot{X} = \dot{x}\cos\theta - \dot{\gamma}\sin\theta \\ \dot{Y} = \dot{x}\sin\theta + \dot{\gamma}\cos\theta \end{cases} \tag{8.16}$$

8.3.4.1 Vertical Motion Control

For controller synthesis, the vertical model is usually based on a single-wheeled vehicle model. Therefore, the number of longitudinal forces acting on the vehicle's center of gravity is

$$m\dot{v} = F_P - F_r \tag{8.17}$$

FP is the thrust force and Fr is the sum of the resistance when the speed is. $v = \dot{x}$ Propulsion is a controlled input caused by brake and drive movements.

The equation that describes wheel dynamics is

$$I_z\dot{w} = T_d - F_xR - T_b \tag{8.18}$$

For the synthesis of the longitudinal controller, it is assumed that a non-slip roll is used,

$$v = Rw; \mathrm{F}_p = F_x \tag{8.19}$$

Therefore, longitudinal dynamics are

$$\left(m + \frac{i_z}{R^2}\right)\dot{v} = \frac{T_d - T_b}{R} - F_r \tag{8.20}$$

The longitudinal control is integrated by the Leapunov method, and the speed tracking error is considered

$$e = v_{ref} - v \tag{8.21}$$

Among them, v and v_{ref} are the actual and reference speed.
The guide of error is

$$\dot{e} = \dot{v}_{ref} - \dot{v} = \dot{v}_{ref} - \frac{1}{M_i}(T_d - (T_b + RF_r)) \tag{8.22}$$

where $M_t = (mR^2 + I_w)/R$, the expression of $\dot{v}$ is given by the nonlinear longitudinal model.

Note that Tb can be considered zero because the brakes are inactive when the throttle is active.

As we all know, in the Lyapunov method, in order to ensure that the trace error converges to zero, a Lyapunov candidate function is proposed, which validates two conditions: It must be positive, and its guide must be negative relative to time.

8.3.4.2 Horizontal Motion Control

The problem of transverse control is complex due to longitudinal and lateral coupling dynamics as well as tire behavior.

Nonlinear bike models capture these phenomena in a simplified way. The algorithm that selects the vehicle steering control task is fuzzy logic. Another algorithm that frequently chooses to perform vehicle steering control tasks is the prediction controller. When the reference track is a priori known, the prediction algorithm has many advantages over other algorithms and is easier to implement as a PID controller.

The principles of the control strategy included in the term predictive control are as follows:

1. This algorithm uses a clear plant model that can predict system output up to a given time (predictive horizon).
2. The future control signal obtained by the controller minimizes the target function to a certain number of steps (control layer).

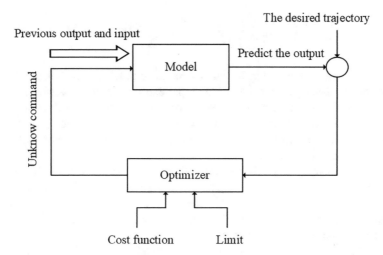

Fig. 8.34 model predicts the controller structure

3. The concept of sliding layer. Make predictions and minimize the target function to get input commands to the factory. The first control command obtained during minimization is applied, discards the rest, and moves the horizon to the future, repeating this step in each sampling cycle.

Different predictive control algorithms differ from models that describe systems and cost functions. Figure 8.34 shows the overall structure of the model prediction controller. The successful implementation of many prediction controllers has been proposed in the existing literature. For simplicity, we introduce dynamic matrix control algorithms in particular. The mathematical model used in this method to represent the system is the step response of the segmented linearization system, and the cost function used is designed to reduce future errors and control work, and the name of the algorithm derives from the fact that the dynamics of the system are represented by a single matrix of step response elements.

The mathematical expression for the prediction and cost functions is

$$\hat{\gamma} = Gu + f \tag{8.23}$$

The lateral torque that is positively related to u is
The lateral torque that is positively related to u is

$$J = \sum_{j=1}^{p} \left[\widehat{\gamma}\left(t+j\|t\right) - w(t+j)\right]^{2} + \sum_{j=1}^{m} \lambda[\Delta u(t+j-1)]^{2} \qquad (8.24)$$

Among them, $\widehat{\gamma}$ is a vector, its dimensionality is equal to the prediction horizon, the prediction layer p、 w is the future output expectation, u is a vector, its dimensional value is equal to the control layer m contains the future control action, G is the control of the dynamic matrix local linearization system, the free response is to predict how the system will run if the command remains unchanged and the last calculated command is equal.

The main methods of predicting controllers can be summarized as models of the information prediction process using past input signals, past control commands, and future control actions calculated by the optimizer. In order to calculate future control signals, the optimizer uses the cost function mentioned earlier. The model process is the basis for the correct operation of the system.

Note that if the limitations of the physical model are not included in the optimization process, you can get the next cost function minimized.

$$J = ee^{T} + \lambda uu^{T} \qquad (8.25)$$

Among them, e is the vector of the prediction error, u is the vector of the future signal control increment, and the mathematical expression of the future command is calculated by using the analog of J and equal to zero

$$u = \left(G^{T}G + \lambda I\right)^{-1} G^{T}(w - f) \qquad (8.26)$$

The optimizer will be able to calculate the steering angle by minimizing the difference between the free response and the desired trajectory. In other words, the optimizer calculates the steering angle to produce the best path tracking.

The software reads the sensor and sets the value of the internal state of the system in each iteration. These states are the position, direction, and speed of the vehicle, and these values are used to calculate the system's step response. The parameters of the step response form a dynamic matrix G. At the beginning of the iteration, the prediction of vehicle behavior is calculated using the sensor's read value. Compare the prediction of vehicle motion with the expected trajectory from the point closest to the prototype. The future error vector is the result of a previous comparison. Use this equation to get future commands, but apply only the first item of the future command vector to remember the concept of sliding the horizon, and finally update the algorithm's variables.

8.4   Application of New Technologies for Intelligent Vehicles

8.4.1   Deep Learning

Deep learning is a machine learning technology that uses neural networks with multiple hidden layers for image classification, speech recognition, language understanding, and other tasks. The concept of deep learning originates from the study of artificial neural network, and the multilayer perceptron with multiple hidden layers is a deep learning structure. Deep learning combines low-level features to form more abstract high-level representation attribute categories or features to discover distributed feature representations of data. Deep learning has been a great success in the field of computer vision, completely overturning traditional computer pattern recognition methods. Before it appeared, most recognition tasks had to go through two basic steps: manual feature extraction and classifier judgment, while deep learning can automatically learn features from training samples. There are two main reasons for the rapid application of deep learning:

1. Easy access to a large number of manually annotated data sets, such as ImageNet Large Scale Visual Recognition Challenge (ILSVRC).
2. The algorithm can process graphics in parallel on GPUS, which improves the learning efficiency and prediction ability.

By using the self-learning characteristic of deep neural network, the huge and complex neural network model is trained by high-performance GPUS, and then transplanted to the embedded development platform, so that real-time and efficient processing of image and video information can be realized.

There are five main types of deep learning: CNN, RNN, LSTM, RBM, and AutoEncoder. In the field of automatic driving, deep learning is mainly applied to image processing of camera data. Therefore, CNN is mainly used, which refers to convolutional neural network and has been proved to achieve good results in image processing. In addition to image processing, deep learning can also be used for radar data processing. However, due to the extremely rich information of images and the difficulty of manual modeling, deep learning can maximize its advantages.

8.4.1.1   Vehicle Application

Deep learning techniques can be applied to many use cases in the automotive industry. Computer vision, for example, is an area where deep learning systems have improved greatly in recent years. Using convolutional neural networks to detect vehicles and lanes, replacing expensive sensors (such as lidar) with cameras, using neural networks to train vehicles to drive automatically with observation cameras, laser rangefinders, and input from real drivers.

Unmanned vehicle to safely on the road, you need to know in what kind of environment, which driveway, the total number of several lanes, where is the road

Fig. 8.35 Upper left—pedestrian, upper right—traffic sign, lower left—vehicle, lower right—traffic light

edge, and the lane line, have the car around people or other moving object, is there any static obstacles, the position of the signal lights and traffic signs, and the instructions of information, etc.

The identification work is mainly divided into three parts: object detection, driving area identification and driving path identification.

1. Object recognition. As shown in Fig. 8.35, the perception system of unmanned vehicles needs to recognize and track multiple moving targets in real time, such as vehicles and pedestrians. Object recognition is the core problem of computer vision [14]. But the output of the traditional target recognition algorithm is usually noisy, which makes the recognition not stable or the object is blocked and misidentified. Deep learning, on the other hand, has the advantages of improving the recognition accuracy of partially occluded objects and reducing the influence of light changes on the object recognition accuracy through the training of large amounts of data, which greatly improves the recognition accuracy.

2. Driving area identification: Before deep learning, there are two methods to detect the driving area. One is based on binocular camera stereo vision or structure from motion, and the other is based on image segmentation such as local features and Markov fields. As shown in Fig. 8.36, green represents the drivable area, but notice that the green area on the left already covers the sidewalk.

   This problem can be improved by applying semantic segmentation in deep learning method, and a high precision driving region division can be obtained. Semantic segmentation determines the semantic information of each pixel in the

Fig. 8.36 Driving area identification in traditional method

Fig. 8.37 Semantic segmentation

image by extracting the features of the original image, and divides the driving area by restoring the semantic information. The effect is shown in Fig. 8.37.

3. Driving path identification: Driving path identification to solve the problem is mainly in no lane line or lane line condition is very poor situation how to drive the problem. If all traffic conditions are as shown in Fig. 8.38 then it can be viewed as a better result.

But sometimes lane lines are hard to detect due to road conditions or weather. Existing deep learning offers a solution: you can train a neural network with data from people driving in the absence of lane lines, and then train the neural network to approximate how the car will drive in the future even when there are no lane lines. The principle is also relatively easy to understand: Find a person to drive a car, and save the camera data during the whole driving process, the driving strategy of the person and the driving path of the vehicle. The neural network is trained with each frame of picture as input and the path of the vehicle in the future (a very short period of time) as output. Its application results are shown in Fig. 8.39. It can be seen that

Fig. 8.38 Lane identification

Fig. 8.39 No-lane driving path identification

the driving path provided by the neural network basically conforms to human judgment.

In addition to environmental perception, there are many applications for deep learning in smart cars, such as visual inspection in manufacturing. The increased deployment of mobile devices and IoT sensors has led to the generation of vast amounts of image and video data, often maintained using spreadsheets and manually. Further learning can help organize this data and improve the data collection process.

Social Media Analysis The application of computer vision can be extended to the analysis of social media. Publicly provided consumer-generated car image data through social media can provide valuable information. In-depth learning can help and improve data collection and analysis.

Driving Behavior Judgment Learn driving conditions and driver behavior by processing large amounts of sensor data (based on sensors such as cameras).

Conversational User Interface Existing intelligent vehicles already have a large number of service platforms. With the application of deep learning, in-car voice dialogue system will become more natural and interactive, and even allow hands-free interaction with vehicles.

8.4.1.2   Convolutional Neural Network

Convolutional neural network (CNN) [15] has made great contribution to the realization of the unmanned vehicle, the algorithm and application will be introduced below. Compared with the usual deep neural network DNN, the current CNN has the following characteristics.

1. A neuron at a higher level only receives input from some neurons at a lower level, which is a neighborhood in a two-dimensional space.
2. The input weight of different neurons in the same layer is shared, so the translation invariance in the visual input can be utilized, which can not only greatly reduce the number of parameters of the CNN model, but also accelerate the training speed.

Since CNN made a breakthrough in image classification in 2012, target detection has naturally become the next target applied by CNN. Object detection algorithms using CNN emerge in an endless stream. Here are some representative algorithms to be introduced. R-CNN series algorithm is a two-stage algorithm, which divides object recognition into two aspects.

1. The selection of the possible location of the object. Input a picture, because the position size of the object in it has too many possibilities, so it needs an efficient method to find them, the key is to find all the objects as far as possible under a certain upper limit of the number of areas, the key index is the recall rate.
2. Candidate region identification. Given a rectangular area in the image, identify the objects in it and correct the area size and aspect ratio, output the object category and the "smaller" rectangular box to focus on the recognition accuracy.

Faster R-CNN in the R-CNN series of algorithms will be introduced below, which can be divided into RPN and Fast R-CNN according to the above two steps.

1. **RPN:** The function of RPN is to efficiently produce a list of candidates. RPN selects CNN as the basis, image features are extracted through multiple convolutional layers, a 3×3 rolling window is used to connect a 256 or

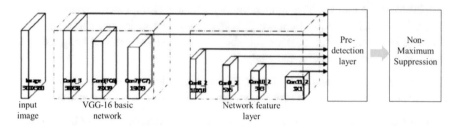

input   VGG-16 basic   Network feature
image   network   layer

Fig. 8.40 Network structure of SSD

512 dimension full connection layer on the feature map output by the last convolutional layer, and finally branches to two full connection layers, one outputs object category, the other outputs object position and size. In order to use different object sizes and aspect ratios, a total of nine combinations of three scales (128 × 128, 256 × 256, 512 × 512) and three aspect ratios (1:1, 1:2, and 2:1) need to be considered at each position. Finally, redundant candidate regions are removed according to the degree of spatial overlap to obtain the remaining possible regions of objects.

2. **Fast R-CNN** [16]: In the candidate region classification stage, the neural network based on full connection is used. For the feature extracted from the candidate box, the classifier is used to judge whether it belongs to a specific class. For the candidate box belonging to a certain feature, the regressor is used to further adjust its position.

3. **SSD** [17]: Since Faster R-CNN still fails to meet the real-time requirements, a Single Shot Detector (SSD) is created as an algorithm that can run in real time and has higher accuracy (as shown in Fig. 8.40). SSD follows the idea of sliding window. By discretizing the position, size, and aspect ratio of the object, SSD uses CNN to achieve the purpose of high-speed detection of the object.

SSD applies VGG-16 network to extract the underlying image features. By canceling the steps of generating candidate regions, picture scaling, and feature map sampling, the object position and classification can be judged in one step. On the basis of VGG network, SSD adds a decreasing convolutional layer, and the convolutional layers of different scales respectively use 3 × 3 convolutional kernel for object position shift and classification judgment, so that SSD can detect objects of different sizes.

The environmental perception of intelligent cars mainly relies on the problems related to computer vision, which has become an important stage for CNN to play its role. CNN uses a wide variety of networks in unmanned target detection. It is hoped that the above content can help readers to guide their further study.

8.4.2   Reinforcement Learning

8.4.2.1   Introduction to Reinforcement Learning

Reinforcement learning is the latest development in the field of machine learning in recent years [18]. The purpose of reinforcement learning is to learn how to take the optimal behavior in the corresponding observation by interacting with the environment. The quality of behavior can be determined by the reward given by the environment. There are different observations and rewards for different environments. For example, environmental observation in driving is the image and point cloud of the surrounding environment collected by cameras and LiDAR, as well as the output of other sensors, such as driving speed, GPS positioning, and driving direction. Depending on the driving task, the reward can be determined by the driving efficiency, comfort, and safety.

The biggest difference between reinforcement learning and traditional machine learning is that reinforcement learning is a closed-loop learning system, and the behavior selected by the reinforcement learning algorithm will directly affect the environment, and then affect the observations obtained from the environment after the algorithm. Traditional machine learning takes training data collection and model learning as two independent processes. For example, if a face classification model is needed to be learn, the traditional machine learning method first needs to hire an annotator to label a batch of face image data, and then train the model with these data. Finally, the trained face recognition model can be tested in real applications. If it is found that the test results are not ideal, then the problems in the model need to be analyzed to find the causes from the data collection or model training, and then solve these problems in some ways. For the same problem, the method of reinforcement learning is to try to predict in the face recognition system, and adjust the prediction based on the satisfaction of user feedback, so as to integrate the process of collecting training data and model learning. The interactive process of reinforcement learning and environment is shown in Fig. 8.41.

There are many challenges in reinforcement learning that traditional machine learning does not have. First of all, there is no priori information in reinforcement learning to determine which behavior to take at each moment, and it must explore all possible behaviors to determine the optimal behavior. How to effectively explore in

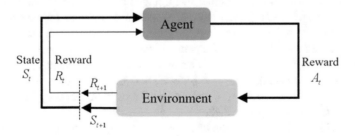

Fig. 8.41   The interactive process of reinforcement learning and environment

the context of a large number of possible behaviors is the most important problem in reinforcement learning. Second, in reinforcement learning, an action may affect rewards not only at the present moment, but also at all subsequent moments. In the worst case, a good action won't be rewarded at the present moment, but only after many steps have been performed correctly. So, it can be very difficult to determine the reward related to the behavior many steps ago.

Although reinforcement learning has many challenges, it can also solve many problems that traditional machine learning cannot solve. First, in view of the process that does not require labeling, reinforcement learning can more effectively solve the special cases that exist in the environment. For example, there may be special situations of pedestrians and animals jaywalking in driverless environment. As long as the simulator can simulate these special situations, reinforcement learning can learn how to make the correct behavior in these special situations. Secondly, reinforcement learning can take all the modules as a whole system, thus adding robustness to some of its modules. For example, the perception module in autonomous vehicle cannot be completely reliable, but reinforcement learning can do it, even if some modules fail. Finally, reinforcement learning can make it easier to learn a series of behaviors. Autonomous vehicles need to perform a series of correct behaviors in order to drive successfully. If there is only labeled data, when the learned model produces a little offset at each moment, it may end up with so much offset that it can lead to devastating consequences, but the advantage of reinforcement learning can learn to correct the offset automatically.

8.4.2.2 Application of Reinforcement Learning in Autonomous Driving

The decision-making of autonomous vehicle refers to how to control the behavior of the vehicle to achieve the driving goal given the environmental information parsed by the perceptual module. For example, vehicle acceleration, deceleration, left turn, right turn, lane change, and overtaking are the output of the decision module. The decision-making module not only needs to consider the safety and comfort of the vehicle to ensure that it can reach the target as soon as possible, but also needs to ensure the safety of passengers in the case of malicious driving of nearby vehicles. Therefore, on the one hand, the decision module needs to make long-term planning for the driving plan, and on the other hand, it also needs to predict the behavior of surrounding vehicles and pedestrians. Moreover, the decision module in autonomous driving has strict requirements for safety and reliability. Existing decision modules are established based on the manual rules generally and can cope with most driving situations, but the rule-based decision-making system cannot be enumerated to all situations, especially the emergency. Therefore, there is an urgent need to establish an adaptive system to cope with all kinds of emergency in the driving environment, as shown in Fig. 8.42.

In the decision-making process of automatic driving, simulator plays an important role in simulating common driving scenes, such as lane conditions, road

Fig. 8.42 Using reinforcement Learning in TORCS Simulator

conditions, obstacle distribution and driving behavior, weather, etc. Meanwhile, it has the function of playing back the data collected in the real scene.

Another important function of the simulator is training the reinforcement learning model to deal with various emergencies. Only by simulating enough emergencies, reinforcement learning can learn the corresponding processing method, instead of making rules for each emergency separately. Existing reinforcement learning algorithms have obtained promising results in the simulation environments of autonomous driving. However, many improvements are needed to make the reinforcement learning apply in the real scene of automatic driving.

The first direction of improvement is to enhance the adaptive ability of reinforcement learning. When the driving scene changes, the existing reinforcement learning algorithms need many trials and errors in order to learn the correct behavior, while human drivers only need to make a few trial and error to learn the correct behavior. How to learn the correct behavior with only a very small number of samples is an important condition to enhance the practicality of algorithms.

The second direction of improvement is the interpretability of the reinforcement learning model. At present, both strategy function and value function in reinforcement learning are represented by deep neural network with poor interpretability. It is difficult to find out the cause and troubleshoot when problems occur in actual use. In the automatic driving task involving human life, it is totally unacceptable that the cause cannot be found.

The third direction of improvement is reasoning and imagination. Human drivers need to have certain reasoning and Imagination ability in the process of learning. For example, when human drivers are driving, they don't have to try and know that dangerous behaviors will bring devastating consequences. This is because human drivers have a good enough model of the world to reason and imagine the possible consequences of corresponding actions. This ability is not only very important for the environment with dangerous behavior, but also can greatly accelerate the convergence speed in a safe environment.

Only by making a substantial breakthrough in these directions can reinforcement learning really be used in important tasks such as autonomous driving or robots.

8.4.3 Slam

The full term of SLAM is Simultaneous Localization and Mapping. That is, in a static unknown environment, the environment map is learned through the movement and measurement of a robot, and then the position of the robot on the map is determined, as shown in Fig. 8.43.

SLAM can be divided into visual SLAM (VSLAM) [19] and laser SLAM [20]. The realization of visual SLAM depends on the image information returned by the camera, while the laser SLAM depends on the point cloud information returned by the LiDAR.

Laser SLAM starts earlier than visual SLAM, and is relatively mature in theory, technology, and product implementation. At present, there are two main ways to realize visual SLAM scheme, one is the RGBD-based depth camera, such as Kinect; and the other is based on monocular, binocular, or fisheye camera.

At present, visual SLAM is still in the stage of further research, application scene expansion, and gradual implementation of production. The application scene of visual SLAM is much richer, and it can work both indoor and outdoor, but it is highly dependent on light and cannot work in dark places or some untextured areas. Laser SLAM is a mature positioning and navigation scheme, which is mainly used indoors. But visual SLAM is the mainstream research direction in the future.

Fig. 8.43 Effect of SLAM mapping

Therefore, in the future, multi-sensor fusion is an inevitable trend. Only by learning from each other's strong points and combining their advantages, can we create a truly easy-to-use SLAM scheme.

Initially, SLAM was proposed to solve the localization and mapping problems of mobile robots in unknown environments. Therefore, the significance of SLAM for autonomous driving is how to help vehicles sense the surrounding environment, so as to better complete advanced tasks such as navigation, obstacle avoidance, and path planning.

High-precision maps are available, regardless of the format of the map, the storage, and how to use it. First of all, can the constructed map really help autonomous vehicles to complete such tasks as obstacle avoidance or path planning? As the environment is dynamic, it is uncertain where there will be a car on the road and when there will be a pedestrian. Therefore, from the perspective of real-time perception of the surrounding environment, the map constructed in advance cannot solve this problem. In addition, the GPS positioning method is passive and dependent on signal source, which makes it unreliable in some special scenarios, such as GPS signal blocking in urban environment, weak signal in field environment, and signal interference and monitoring of unmanned combat vehicles during combat. Therefore, such an active and passive working mode as SLAM is a very necessary technology in the above scenarios.

References

1. Bishop R (2000) Intelligent vehicle applications worldwide[J]. IEEE Intell Syst Appl 15(1): 78–81
2. Shengchao LI (2019) Cognition of overall development of intelligent vehicles. Auto Engineer
3. Yan X, Zhu Z, Zhou K et al (2018) Application and analysis of artificial intelligence in vehicle intelligent driving system [J]. J Hubei Univ Autom Technol 32(01):40–46
4. Coombes M, Mc Aree O, Chen WH et al (2012) Development of an autopilot system for rapid prototyping of high level control algorithms[C]. In: Proceedings of 2012 UKACC international conference on control. IEEE, Piscataway, pp 292–297
5. Kim IY, Yang KS, Baek JJ et al (2013) Development of intelligent electric vehicle for study of unmanned autonomous driving algorithm[C]. In: 2013 world electric vehicle symposium and exhibition (EVS27). IEEE, Piscataway, pp 1–6
6. Janai J, Güney F, Behl A et al (2020) Computer vision for autonomous vehicles: problems, datasets and state of the art[J]. Foundations and Trends® in Computer Graphics and Vision 12(1–3):1–308
7. Veeramani B, Raymond JW, Chanda P (2018) DeepSort: deep convolutional networks for sorting haploid maize seeds[J]. BMC Bioinform 19(9):1–9
8. Mur-Artal R, Montiel JMM, Tardos JD (2015) ORB-SLAM: a versatile and accurate monocular SLAM system[J]. IEEE Trans Robot 31(5):1147–1163
9. Hubmann C, Becker M, Althoff D et al (2017) Decision making for autonomous driving considering interaction and uncertain prediction of surrounding vehicles[C]. In: 2017 IEEE intelligent vehicles symposium (IV). IEEE, Piscataway
10. Urmson C, Anhalt J, Bagnell D et al (2008) Autonomous driving in urban environments: boss and the urban challenge[J]. J Field Robot 25(8):425–466

11. Ahmed N, Pawase CJ, Chang KH (2021) Distributed 3-D path planning for multi-UAVs with full area surveillance based on particle swarm optimization[J]. Appl Sci 11(8):3417
12. Chu K, Lee M, Sunwoo M (2012) Local path planning for off-road autonomous driving with avoidance of static obstacles[J]. IEEE Trans Intell Transp Syst 13(4):1599–1616
13. Tapi MD, Bagny-Beilhe L, Dumont Y (2020) Miridae control using sex-pheromone traps. Modeling, analysis and simulations[J]. Nonlinear Anal Real World Appl 54:103082
14. Sun Z, Bebis G, Miller R (2006) On-road vehicle detection: a review[J]. IEEE Trans Pattern Anal Mach Intell 28(5):694–711
15. Zeiler MD, Fergus R (2014) Visualizing and understanding convolutional networks[C]. In: European conference on computer vision. Springer, Cham, pp 818–833
16. Ren S, He K, Girshick R et al (2016) Faster R-CNN: towards real-time object detection with region proposal networks[J]. IEEE Trans Pattern Anal Mach Intell 39(6):1137–1149
17. Liu W, Anguelov D, Erhan D et al (2016) Ssd: single shot multibox detector[C]. In: European conference on computer vision. Springer, Cham, pp 21–37
18. Lillicrap T P, Hunt J J, Pritzel A, et al. Continuous control with deep reinforcement learning [J]. arXiv preprint arXiv:1509.02971, 2015
19. Karlsson N, Di Bernardo E, Ostrowski J et al (2005) The vSLAM algorithm for robust localization and mapping[C]. In: Proceedings of the 2005 IEEE international conference on robotics and automation. IEEE, Piscataway, pp 24–29
20. Kohlbrecher S, Von Stryk O, Meyer J et al (2011) A flexible and scalable SLAM system with full 3D motion estimation[C]. In: 2011 IEEE international symposium on safety, security, and rescue robotics. IEEE, Piscataway, pp 155–160

Chapter 9
Intelligent Internet of Vehicles (IoV) and Vehicle Infrastructure Cooperative Technology

CHEN Peng, DING Chuan, ZHANG Junjie, LU Guangquan, DAI Rongjian, CAI Pinlong, WEI Lei, LIU Qian, and HAN Xu

9.1 Introduction

The rapid development of communication technology has promoted the process of informatization and intelligence of the transportation system. The development of wireless communication technology has provided basic conditions for high-speed moving vehicles to realize the interconnection between vehicles and vehicles, vehicles and roads, and vehicles and management centers, as well as promotes the rapid development of Intelligent IoV and Vehicle Infrastructure Cooperative Technology. This chapter focuses on the basic technologies of the IoV and vehicle-to-infrastructure cooperation.

9.1.1 From Driver Assistance to Connected Automated Vehicle

In order to avoid traffic accidents caused by improper operation, advanced driver assistance system (ADAS) has gradually been used in automobiles. ADAS is an active safety technology that collects data inside and outside the vehicles, performs technical processing such as identification, detection and tracking of static and dynamic objects, and provide warning or takeover of possible dangers based on various sensors installed on the car. ADAS contains Adaptive Cruise Control (ACC)

CHEN Peng · DING Chuan · LU Guangquan (✉) · DAI Rongjian · CAI Pinlong · WEI Lei · LIU Qian · HAN Xu
School of Transportation Science and Engineering, Beihang University, Beijing, China
e-mail: lugq@buaa.edu.cn

ZHANG Junjie
HefeiInnovation Research Institute, Beihang University, Hefei, China

© Tsinghua University Press 2022
W. Yunpeng et al. (eds.), *Intelligent Road Transport Systems*,
https://doi.org/10.1007/978-981-16-5776-4_9

System, Lane Departure Warning System (LDWS), Collision Warning/Collision Avoidance (CW/CA), Driver Condition Warning (DCW), Vision Enhancement (VE), Intersection Collision Avoidance (ICA), etc.

With the application of electronic technology in automobiles and the upgrading of industrial technology, automobiles have gradually changed from assisted driving to automated driving. Automated driving vehicles refer to the vehicles that can automatically perceive the surrounding environment and navigate independently without manual intervention. They integrate environmental perception, planning, and decision-making, and multilevel assisted driving, on the basis of high-precision perception, multi-source information fusion, machine learning, and artificial intelligence technology. Society of Automotive Engineers International (SAE International) divides automated driving into six levels:

Level 0: No Driving Automation. The user fully control vehicle.

Level 1: Driver Assistance. The system can assist the user to complete certain driving tasks, either the lateral or the longitudinal vehicle motion control.

Level 2: Partial Driving Automation. The system can automatically complete certain driving tasks, but the user needs to supervise the automation system and ensure to takeover at any time when necessary.

Level 3: Conditional Driving Automation. The system can automatically complete certain driving tasks, but the user is receptive to takeover request to intervene.

Level 4: High Driving Automation. The system can automatically complete driving tasks and monitor the driving environment under certain environments and specific conditions. User is not expected to respond to a request to intervene.

Level 5: Full Driving Automation. The system can automatically complete all driving tasks under all conditions. User is not expected to respond to a request to intervene.

The development of communication technology brings "Internet" into the traditional automobile world. It enables vehicle information sharing and cooperative control. Connected Automated Vehicle (CAV) is the integration of automated driving vehicle and modern communication and network technology, with the function of information interconnection and sharing between vehicle and people, vehicle, road, and cloud. It has the capabilities of complex environment perception, information interaction, intelligent decision-making, and cooperative control. CAVs can meet various travel needs of travelers, including improving driving safety, achieving energy conservation and emission reduction, and optimizing the allocation of transportation resources. Vehicle network is generally divided into three levels:

Level 1: Assisted Information Interaction Network. Vehicles will share their own perceived information with others.

Level 2: Cooperative Awareness Network. Vehicles will perceive the environment cooperatively.

Level 3: Cooperative Decision and Control Network. The cooperative control between vehicles can be carried out according to the vehicle state.

The control architecture of CAV can be divided into perception, transmission, decision-making, and control layers. The perception layer detects the vehicle's own

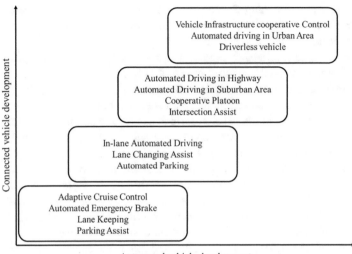

Fig. 9.1 Development of connected and automated vehicle

state and the surrounding environment information accurately in real time and carries out data fusion by using on-board sensors such as GPS, radar, camera, etc. The transmission layer realizes the information interconnection and sharing among people, vehicles, roads, and cloud by using long-distance and short-distance communication technology. The decision-making layer is the "central nerve" of CAV. and it makes decisions on vehicle behaviors based on environment perception. The control layer controls the vehicle motion state to realize the decisions. At present, the development of CAV technology is still in the stage of assistance driving. Driver has the fully control of the vehicle and decision information mainly plays the role of prediction warning. When the technology of CAV is relatively complete, the decision-making layer includes not only the on-board decision-making system, but also the cloud cooperative decision-making system, which can control the operation of the whole transportation system. Figure 9.1 shows the comprehensive application of CAVs in the process of integration of automated and connected at different stages.

9.1.2   From Internet of Vehicles to Vehicle Infrastructure Cooperative Technology

With the increasing intelligence of vehicles and roads, more and more vehicles and roadside infrastructure are equipped with communication technology. The development of IoV and its related applications is an inevitable trend. The IoV refers to the wireless communication equipment loaded on the vehicle, which can extract static and dynamic information of all vehicles on the information network platform, and supervise and provide comprehensive services. The IoV makes intelligent and safe driving realized by collecting and sharing information between vehicle-to-vehicle, vehicle-to-infrastructure, and vehicle-to-network.

Traffic safety and efficiency are two major challenges. With the rapid development of China's national economy in recent years, car ownership is growing and traffic problems are becoming more and more serious. However, the traditional solutions have great limitations. First of all, vehicles cannot perceive the surrounding environment and achieve active safety, which has high requirements for the driver's response. Second, incomplete traffic guidance and control is caused by the lack of real-time and accurate information. The development of IoV is an opportunity to solve these traffic problems. It is predicted that the application of IoV can reduce traffic congestion by about 60%, improve short distance transportation efficiency by nearly 70%, and increase the capacity of the existing road network by two to three times. In addition, IoV can also realize the effective use of the whole public resources, and greatly reduce the management costs for the government.

With advanced wireless communication and Internet technology, IoV technology has developed rapidly, including dynamic real-time information interaction between vehicle and vehicle, expanding the space–time scope of traffic information, enriching the information acquisition methods, and carrying out active safety control and road collaborative control in a large range. As the next generation of intelligent transportation technology, intelligent vehicle-to-infrastructure cooperation is the development of IoV technology. Based on the interaction of vehicle-to-vehicle and vehicle-to-infrastructure, the information sharing and cooperative operation between vehicles and road facilities is realized. It can transfer vehicle safety control from independent mode to cooperative mode. Establishing a Vehicle Infrastructure Cooperative System is of great significance to improve the efficiency, safety, and sustainable development of the traffic system.

Intelligent Vehicle Infrastructure Cooperative System is mainly composed of intelligent vehicle system, intelligent infrastructure system, and communication platform. The intelligent vehicle system is responsible for the perception, control, and delivery of vehicle's own state and the surrounding driving environment information. The intelligent infrastructure system is responsible for the monitoring of traffic flow information (such as traffic flow and average speed) and the perception, processing, and delivery of road surface and geometry condition and abnormal information. The communication platform is responsible for the communication of the whole system and realize the interaction between roadside infrastructure and vehicle units.

As one of the main directions of the development of intelligent transportation technology, Intelligent Vehicle Infrastructure Cooperative System is widely used. Firstly, concerning vehicle itself, the system can realize the cooperative safety of vehicles, such as active obstacle avoidance by people and vehicles, warning and control of dangerous road sections. Second, the system can achieve traffic cooperative control in a wide range, such as traffic signal cooperative control, real-time route guidance, and public transit priority control. Third, the system can also provide comprehensive information services according to the needs of users, such as traffic demand management and real-time traffic information query.

9.2   Intersection Signal Control Technology Based on Internet of Vehicles

For the urban road traffic network, the intersection is the most important node of the network. It realizes the traffic flow conversion of different sections. The performance of the intersection directly restricts the capacity of the road traffic network. Traffic flow in different directions will form multiple serious conflict points when passing through the intersection. Signal control can assign different rights of way to avoid the conflicts of traffic flows and separate conflicting traffic flows in time. Therefore, signal control is a key method to improve intersection efficiency and traffic safety. In recent years, with the rapid development of wireless communication technology, Internet technology, and sensor detection technology, vehicles conduct real-time information interaction and sharing through vehicle-to-vehicle and vehicle-to-road communication, forming a huge vehicle network, providing a new opportunity to improve the service level of the intersection signal control system, and the signal control of intersections in the networked environment has become another key issue in the transportation field.

9.2.1   Prediction of Traffic Flow Parameters Based on Data of Internet of Vehicles

Signal control at intersections is inseparable from traffic flow parameter detection. Traffic survey is defined as collecting relevant data of vehicle operation in some selected sections of the road. The specific objects of the traffic survey vary with the purpose of the research, but mainly the traffic flow parameters. The related technology used in traffic survey is called traffic detection technology. The traditional traffic detection technology mainly includes manual survey, loop induction coil, video detection, and infrared detection. However, limited by the detection range, high cost, high damage rate, and other practical disadvantages, the traditional detection equipment has poor operating efficiency. With the gradual deployment of IoV technology, V2X communication can provide a large amount of IoV data, such as the real-time speed and location information of vehicles. Direct application of these data to traffic flow parameter prediction can alleviate the inherent shortcomings of traditional detection methods. This section will focus on the traffic flow parameter prediction method based on the data of the IoV.

9.2.1.1 Traffic Flow Parameters Based on Internet of Vehicles Data

1. Traffic volume: Traffic volume, also known as traffic flow, refers to the number of vehicles passing through a certain place or a certain section of the road within a unit time, and the unit is veh/h. In the IoV environment, N_1 represents the number of all vehicles detected by the roadside detection equipment at the observation start time t_1. N_2 represents the number of all vehicles detected by the roadside detection equipment at the observation end time t_2. The traffic volume detected by the roadside equipment, from t_1 to t_2, is

$$q = \frac{N_2 - N_1}{t_2 - t_1} \tag{9.1}$$

   where t_1 represents the start time of observation; t_2 represents the end time of observation.

2. Density: Density refers to the number of vehicles per unit length of road and intuitively reflects the degree of traffic density. For single lane, the unit is veh/h/ln, and for multilane, the unit is veh/h. In the IoV environment, the roadside detection equipment can directly perform real-time statistics on the number of road vehicles, so this parameter can be easily detected. The calculation formula is

$$\rho = \frac{N_{ij}^t}{L} = \frac{N_{ij}^{t-1} + N_i^t - N_j^t}{L} \tag{9.2}$$

   where L represents the length of the road section ij; N_{ij}^t represents the number of vehicles passing the road section ij in the detection period t; N_{ij}^{t-1} represents the number of vehicles passing the road section ij in the previous detection period $t-1$; N_i^t represents the number of vehicles passed by the upstream road section detection device i in the detection period t; N_j^t represents the number of vehicles passed by the downstream road section detection device j in the detection period t.

3. Speed: Speed includes instantaneous speed and average speed. The instantaneous speed is the speed of the vehicle passing a certain point on the road at a certain moment. Average speed is divided into time average speed and space average speed.

   (a) Time average speed: The time average speed, also called arithmetic average speed, simply calculated as the average of the vehicle speed detected by the roadside equipment. The calculation formula is

$$v_s = \frac{1}{N} \sum_{i=1}^{N} v_i \tag{9.3}$$

   where N represents the number of vehicles detected; v_i represents the detection speed of the ith vehicle.

(b) Space average speed: The space average speed, also called the harmonic average speed, is the quotient of a certain observation distance passed by the vehicle and the average travel time used. In the case of free flow, there is no difference between the time average speed and the space average speed. However, when traffic congestion, traffic accidents, or road signal control occur, since the space average speed takes into account the different delays, it can better reflect the driving situation of the vehicle, so it is closer to the real situation. In the IoV environment, because the roadside detection equipment can automatically identify vehicles and record the exact time when each vehicle enters and exits a certain road section, the space average speed can be directly obtained. The calculation method is as follows:

$$v_s = \frac{L}{N} = \frac{L}{\frac{1}{N}\sum_{i=1}^{N}\frac{L}{v_i}} \tag{9.4}$$

where L represents the length of the road section; N represents the number of vehicles passing through the detection time; v_i represents the detection speed of the ith vehicle.

4. Occupancy rate: Occupancy rate includes time occupancy rate and space occupancy rate. The time occupancy rate is the ratio of the time the detector is occupied by the vehicle to the observation time. The space occupancy rate is the ratio of the total length of all vehicles on the road to the length of the road section. Compared with the time occupancy rate, the space occupancy rate can better reflect the occupation of the road by the vehicle. In the IoV environment, the traffic volume can be directly detected and the vehicle length can be accurately obtained through V2X communication, so that the space occupancy rate of the road can be accurately calculated. The calculation formula is

$$o = \frac{1}{L} = \sum_{i=1}^{N} l_i \tag{9.5}$$

where l_i represents the length of the ith vehicle; L represents the length of the road section; N represents the number of vehicles passing through the detection time.

5. Road delay time: Due to the existence of traffic signal control, traffic accidents, traffic congestions, etc., vehicles on the road will be disturbed to a certain extent, resulting in delays. The average value of the difference between the driving time of a vehicle on a certain road and the ideal time in a period of time is called the road delay time. In the IoV environment, the travel time of the vehicle can be directly obtained, making the parameter measurable. The calculation formula is

$$T_{\mathrm{D}} = \frac{1}{N} = \sum_{i=1}^{N} (t_i - t_s) \tag{9.6}$$

where N represents the number of vehicles passing through the detection time; t_i represents the actual travel time of the ith vehicle; t_s represents the ideal travel time of the road.

9.2.1.2 Data Filtering

Signal control at intersections in a networked environment and formulating reasonable and feasible signal timing plans are not enough to rely only on real-time traffic flow parameter detected. In practice, it is often necessary to use the detected real-time data to predict future traffic flow parameters. In the IoV environment, due to the widespread existence of other uncertain factors such as driving speed, channel quality, and environmental changes, wireless signal transmission might be interfered, which can cause data loss or errors. Therefore, it is necessary to further analyze the completeness, rationality, and validity of the collected raw data, and filter the raw data to eliminate unreasonable data.

The IoV can directly detect the time when the vehicle passes the upstream and downstream detectors of the road section. Therefore, it can be judged whether the raw data collected on the road section is correct according to the travel time of the vehicle. The minimum value of vehicle travel time can be regarded as the time taken by the vehicle to pass through the section at the maximum driving speed under the condition of free flow. The maximum travel speed is related to many factors such as road conditions, weather conditions, and traffic control. For expressways and urban arterial roads, the range of road travel time is as follows:

$$\frac{L}{v_{\max}} \leq t_i \leq \frac{L}{v_{\min}} \tag{9.7}$$

$$\frac{L}{\alpha v_{\max}} \leq t_i \leq \frac{L}{Cl_q} + T_{\mathrm{d}} \tag{9.8}$$

where L represents the length of the road section; $v_{\max}$ represents the road speed limit; $v_{\min}$ represents the average driving speed of the road when the road downstream is blocked; α represents the correction coefficient, and the value is generally 1.3~1.5; C represents the average queue length, that is, the ratio of the queuing length to the number of queuing vehicles; if there is signal control, T_{d} takes the maximum red light time, and if there is no signal control, it takes the average delay time at the intersection.

When filtering the raw data, we consider that only the data that satisfy the formula (9.7) or formula (9.8) will be kept and the others will be dropped.

9.2.1.3   Data Repair

In order to ensure data integrity, missing data can be further repaired. For commonly used data, please refer to Chap. 3.

9.2.1.4   Traffic Flow Parameter Prediction Based on IoV Data

Common methods of traffic flow parameter prediction include prediction methods based on parameter models, prediction methods based on traffic simulation, prediction methods based on machine learning, and prediction methods based on deep learning. Please refer to Chap. 5 for related methods.

9.2.2   Signal Timing Identification Based on Data of Internet of Vehicles

An important task of urban traffic management and control is to formulate reasonable and effective intersection signal timing scheme and to adjust and optimize it in time according to the actual situation of the intersection. When evaluating the timing scheme, it needs to rely on signal parameters, such as split, offset, cycle, etc. Therefore, the signal timing scheme of intersections is particularly critical. At present, in China's first-tier cities (such as Beijing, Shanghai, and Guangzhou), a unified traffic management system is used in all intersections, and most of the intersection timing schemes can be acquired and updated in real time. However, in most small and medium-sized cities in China, due to the outdated traffic control system, the intersection timing plan is not archived online, nor can it be updated in real time. It is difficult to obtain the timing plan and the data reliability is poor. Therefore, it is particularly important to tackle how to accurately estimate the signal timing scheme at the intersection when the signal timing is not easy to obtain, and how to effectively identify the signal timing.

9.2.2.1   Basic Concepts of Urban Traffic Signal Control

1. Basic time parameter. The basic time parameters of signal timing at urban intersections include all-red time, yellow light time, green time, minimum green time of phase, maximum green time of phase, loss time, effective green time of phase, etc. Among them, yellow light time, red light time, minimum green time of phase, maximum green time of phase, loss time and so on are usually constant parameters, while green time is a variable parameter.

   All-red time, also known as red-light clearance time, refers to the duration when all the traffic signal lights appear red light in the process of switching between two phases. The function of all-red time is to clear all vehicles in the intersection, and the time of yellow light and all-red time are collectively called clearance time or interval time between green lights. In the practical application process, the intersection all-red time is generally set to less than 3 s.

The yellow light alerts drivers to an impending change in the right of way ahead. With regard to the current road traffic flow, the main function of yellow light time is to allow drivers to make decisions and eliminate the dilemma area. The duration of yellow light is related to intersection speed limit, vehicle deceleration, and driver reaction time, etc. In practical application, it is generally set at 2~3 s. If the signal light has a countdown reminder, the driver can judge the change of traffic signal in advance, and in principle, the duration of yellow light can be reduced or even canceled.

The green time refers to the duration of the right of way. The green time is one of the core control elements of urban traffic signal control, and its length directly affects the signal timing parameters, thus affecting the traffic signal control effect. In practical application, the setting of green time must be determined according to the actual traffic flow and saturated traffic flow.

The maximum phase green time is the longest time that a phase green can last, mainly to prevent the infinite extension of phase green time from affecting the normal release of traffic flow in other phases, thus causing traffic congestion and traffic accidents. In practical application, the phase maximum green time is generally set at about 80 s.

Loss time refers to the signal time that cannot be effectively used within a signal cycle, including vehicle starting loss time, vehicle braking loss time, etc.

The phase effective green time refers to the actual time used for vehicle passage within a signal cycle. The calculation method is the sum of green time and yellow light time divided by loss time. In practical application, it is generally not clear to distinguish the phase effective green time from the phase green time.

2. Signal timing parameters: Two-phase signal timing at urban intersections is shown in Fig. 9.2, where G represents green time, R represents red time, r represents all-red time, and C represents signal cycle.

$$C = \frac{L}{1 - Y} \tag{9.9}$$

Cycle refers to the time required for the semaphore phase to display a cycle in a specified order. The signal cycle length can be calculated as above.

The parameters in the formula are described as follows.

L represents the total signal loss time.

$$L = \sum_k (L_s + I - A)_k \tag{9.10}$$

where L_s represents the starting loss time, and 3 s can be taken if there is no measured data. A represents the time of yellow light in seconds. I represents the

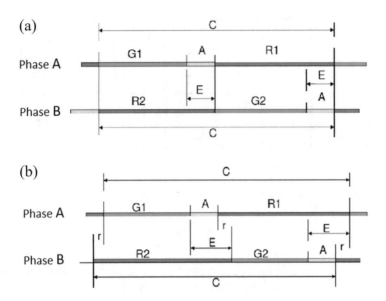

Fig. 9.2 The indication of two phase signals timing. (**a**) With all-red time. (**b**) Without all-red time

time between green lights in seconds. k represents the number of green light intervals in a period.

Y is the flow ratio summation.

$$Y = \sum_j \max\left[y_j, y_j', \ldots\right] = \sum_j \max\left[\left(\frac{q_d}{s_d}\right)_j, \left(\frac{q_d}{s_d}\right)_j', \ldots\right] \qquad (9.11)$$

where j represents the number of phases in a cycle. y_j, y_j' represents the flow ratio of the jth phase, and the unit is pcu/h. s_d represents the designed saturation flow, unit is pcu/h. q_d represents the designed traffic volume, which can be estimated according to Eq. (9.12).

$$q_d = \frac{Q}{\text{PHF}} \qquad (9.12)$$

where Q represents the peak hour traffic volume. PHF is the peak hour coefficient and the main inlet is 0.75 and the secondary inlet is 0.8.

The cycle is directly related to the intersection saturation. Generally, The higher the intersection saturation, the longer the signal cycle. In practical application, the value is generally 40~120 s, and for the intersection with large flow, the value can be about 180 s.

The split refers to the ratio of the total effective green time to the signal cycle, and represents the ratio of the time that can be used for vehicle traffic within a signal cycle. Its calculation method is as follows.

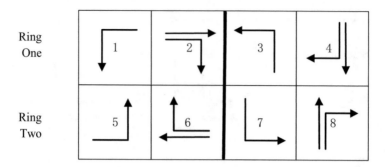

Fig. 9.3 Double ring eight phase structure at the intersection

$$\lambda = \frac{G_e}{C} \tag{9.13}$$

where C represents the signal cycle. G_e represents the total effective green time, which can be calculated according to Eq. (9.14).

$$G_e = C - L \tag{9.14}$$

where L represents the total signal loss time.

According to the total effective green time, the effective green time of each phase can be further obtained.

$$g_{ej} = G_e \frac{\max \left[y_j, y'_j, \ldots \right]}{Y} \tag{9.15}$$

The offset is also known as green time difference or green start time. For two intersections, offset refers to the difference between the starting time of green light at the same phase at adjacent intersections.

3. Signal structure parameters: At present, the definition of phase mainly includes two kinds, among which domestic scholars generally give the following definition: A flow direction or a combination of several nonconflicting traffic flows to obtain the right of passage for a period of continuous time. According to the above definition, a complete signal cycle is composed of multiple phases, and the signal cycle time is equal to the sum of the green time and the all-red time of each phase. In foreign countries, the phase is usually defined by the American Electrical Manufacturers Association as the green time assigned to any independent traffic flow in a cycle. For intersections, NEMA also defines a Dual-Ring eight-phase structure, as shown in Fig. 9.3.

As shown in Fig. 9.3, phases 2, 4, 6, and 8 are defined as straight, and phases 1, 3, 5, and 7 are defined as left turn. No separate indicator light is set for right turn. First set the phase with the largest straight flow as phase 2 then rotate

Fig. 9.4 Schematic
diagram of phase number

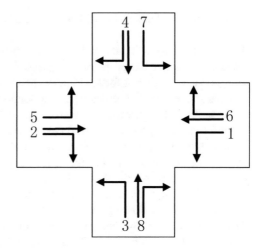

Table 9.1 All possible phase combinations in a double-ring structure

Combination number	Number of phase member 1	Number of phase member 2	Combination number	Number of phase member 1	Number of phase member
1	1	5	5	3	7
2	2	5	6	4	7
3	1	6	7	3	8
4	2	6	8	4	8

clockwise to get 4, 6, 8 phase. Then, set phase 1 as the left turn in the opposite
direction of phase 2, and rotate clockwise to get phase 3, 5, 7.

The phase number of the intersection is shown in Fig. 9.4.

The double-loop structure has two phase groups corresponding to all phase
combinations in the east–west direction and north–south direction respectively.
Phases 1–4 belong to one ring, and phases 5–8 belong to another ring. Because of
the conflict between traffic flow combinations in different directions, there is a
separation line between the two phase groups, and the phases on both sides
cannot cross the separation line. All possible phase combinations in the double-
loop structure shown in Fig. 9.3 are shown in Table 9.1.

The display order of phases is called phase sequence. The phase sequence can
be arranged in accordance with the order in which the traffic flow obtains the right
of way. Taking the double-loop structure shown in Fig. 9.4 as an example,
according to all possible phase combinations, it can be seen that there are three
possible phase sequences in a phase group and nine possible phase sequences in a
signal cycle.

9.2.2.2 Signal Timing Identification Based on Internet of Vehicles Data

At present, many scholars at home and abroad have done related research on intersection signal timing based on traffic flow characteristics. Li Li et al. from Tsinghua University put forward intersection traffic flow prediction method, timing modeling based on time series trend and random intersection signal timing model. Jeff et al. from the University of Washington, USA, based on the traffic flow theory and machine learning related methods, estimated the single-lane driving condition of vehicles without overtaking. According to the start time of the track signal cycle, a specific method for estimating the effective green time and red time is proposed, which is an extended judgment method of the delay model. In addition, some experts and scholars reversely deduced fixed signal intersection timing based on high-frequency updated floating car data. Kerper et al. from the German Institute of Driver Information System reversely deduced fixed signal intersection timing based on 1 Hz updated floating car data, and the results showed high reliability. Fayazi from Clemson University, USA, based on the research of this method, made signal timing estimation based on the sparsely updated track data provided by the San Francisco bus system.

The above research methods are all based on continuous time series traffic flow to carry out signal timing inversion, which requires observation of the effective trajectory of continuous period, and most of the studies require very high data sampling rate. However, the existing domestic floating vehicle data and mobile navigation data still cannot meet the data sampling rate requirements of such methods. Therefore, this part will introduce how to reverse the signal cycle based on the limited vehicle trajectory data.

Since vehicle trajectory can reflect the real appearance of traffic flow and provide a basis for the design and optimization of traffic control system, vehicle trajectory can also reflect the control results of traffic signal control system. First of all, information can be extracted from vehicle trajectory data to accurately classify vehicles. Then, signal timing can be reversely deduced from the extracted trajectory data. By analyzing the error, accurate intersection signal timing can be backward deduced.

The identification steps are as follows:

1. Vehicle steering matching.

   Step 1: Number the direction of the intersection according to the channelization of the intersection and list all possible directions of the vehicle.
   Step 2: Change determining vehicle direction to finding vehicle starting and ending direction number based on intersection direction number.
   Step 3: Find the driving track in all directions during each time period.
   Step 4: Match the starting and ending points of the vehicle with the intersection to determine the driving direction, and then aggregate the different steering trajectories at different periods.

2. Effective trajectory extraction. According to the intersection situation, the effective track can be defined as the traveling track of the vehicle which is 300~400 m away from the stop line at the intersection.

   Step 1: According to Eqs. (9.7) and (9.8), judge whether the travel time is within the reliable range, and exclude the trajectory data outside the reliable range.

   Step 2: Judge whether there is a round-trip folding point in the trajectory. If there is a round-trip folding point, it indicates that there is a round-trip situation in the vehicle running. At this time, the trajectory data should be eliminated.

   Step 3: Determine whether the screening requirements of direction characteristics are met. Under the premise of reliable travel time, the number of vehicle traveling direction and its change should be consistent with the direction of track traveling.

   Step 4: Judge whether the driving distance meets the requirements, that is, judge whether the distance between the starting point and the end point of the driving track and the intersection meets the set requirements.

   Step 5: Further eliminate the vehicle tracks that do not meet the set requirements, that is, the vehicle track length is between 300 and 400 m from the stop line at the intersection.

3. Signal identification.

   Step 1: Determine the trajectory stop point and nonstop point according to the periodic effective trajectory data. The stop point is defined as a series of points where the position of the vehicle is unchanged in the trajectory of the vehicle stopping, and the first point is taken as the stop point. Nonstop point is to select the point closest to the stop line in the trajectory of the vehicle not stopping.

   Step 2: Calculate the time difference between all stop points and nonstop points as follows.

   $$td_i^{(\text{spi,nsi})} = t_{\text{spi}} - t_{\text{nsi}} \qquad (9.16)$$

   where sp_i represents the stop point and ns_i represents nonstop point. t_{spi} represents the time corresponding to the stop point and t_{nsi} represents the time corresponding to the nonstop point.

   Step 3: Calculate the variance and mean of the time difference under different cycle estimates.

   Step 4: Output the cycle length estimated value with smaller mean variance of time difference, and obtain the cycle length estimated value.

Example 9.1 Take the trajectory data of 100 m before and after the intersection of Huanggang Road and Hongli Road in Shenzhen City as the research sample, please carry out signal timing identification. Some initial data are shown in Table 9.2.

The data are illustrated as follows. The flow direction 1 is from east to west, 2 is from west to east, 3 is from south to north, 4 is from north to south. Intersections ID1019727 stands for Huanggang Road–Fuzhong Road, 984921 stands for Huanggang Road–Hongli Road, and 1013582 stands for Huanggang Road–Sungang West Road. The intersections are arranged from south to north. S is the distance from the starting point of the vehicle.

Firstly, classify trajectories to determine the nonstop point and stop point. There will be 1~3 track errors when points are directly extracted according to their positions, so it is necessary to calculate the speed of each point, identify the stop point through the speed, and then compare all points in the non-stop track to screen out the nearest point to the stop line. The screening process is shown in Fig. 9.5.

Carry out data visualization, and the trajectory data set is counted into the estimated period to obtain the spatiotemporal map of trajectory distribution. Further eliminate the stopping tracks with less than 40 m and the non-stopping tracks with less than 100 m. In the case of a relatively correct period, there are about 20 tracks, 16 non-stopping tracks, and only 3–4 stopping tracks, as shown in Fig. 9.6.

Calculate the time difference between all stop points and nonstop points according to Eq. (9.10), then calculate the variance of the time difference under different cycle estimates, and then take the average value. The number of stop points and nonstop points is different, so the variance of the time difference between stop points and each nonstop point is solved by calculating the mean value of the stop point. The smaller the mean variance of time difference, the more accurate the estimation result of cycle length. The calculation result of variance of time difference is shown in Fig. 9.7.

It can be clearly seen from Fig. 9.7 that the variance of time difference under different cycle lengths varies greatly. When the cycle length is 170 s, the variance of time difference is the smallest. Therefore, the signal cycle length with the highest possibility is 170 s, and then it is the estimated signal cycle with the highest accuracy.

Table 9.2 Some initial data

	ID	ID of entering intersection	Into the flow	ID of exiting intersection	Out of the flow	UNK time stamp	Longitude	Latitude	Link	Link distance	Link length	S
0	7360	1019727	3	1013582	3	1493090064	114.0732	22.54325	9906221	5	189	0
1	7360	1019727	3	1013582	3	1493090067	114.0732	22.54332	9906221	13	189	7.783645
2	7360	1019727	3	1013582	3	1493090070	114.0732	22.54332	9906221	13	189	7.783645
3	7360	1019727	3	1013582	3	1493090073	114.0732	22.54338	9906221	20	189	14.49178
4	7360	1019727	3	1013582	3	1493090076	114.0732	22.54347	9906221	30	189	24.48443
5	7360	1019727	3	1013582	3	1493090079	114.0732	22.54363	9906221	47	189	42.30396
6	7360	1019727	3	1013582	3	1493090082	114.0732	22.54383	9906221	70	189	64.50123
7	7360	1019727	3	1013582	3	1493090085	114.0732	22.5441	9906221	100	189	94.52127
8	7360	1019727	3	1013582	3	1493090088	114.0732	22.54438	9906221	131	189	125.6465
9	7360	1019727	3	1013582	3	1493090091	114.0732	22.54473	9906221	170	189	164.5813
10	7360	1019727	3	1013582	3	1493090094	114.0731	22.54508	10153871	11	9	203.4893
11	7360	1019727	3	1013582	3	1493090094	114.0731	22.54508	10153761	11	37	203.4893
12	7360	1019727	3	1013582	3	1493090097	114.0731	22.54542	1.69E+08	12	154	241.3017
13	7360	1019727	3	1013582	3	1493090100	114.0731	22.54581	1.69E+08	55	154	284.6887
14	7360	1019727	3	1013582	3	1493090103	114.0731	22.5462	1.69E+08	99	154	328.0829
15	7360	1019727	3	1013582	3	1493090106	114.0731	22.54657	1.69E+08	140	154	369.2371
16	7360	1019727	3	1013582	3	1493090109	114.0731	22.5469	1.69E+08	22	94	405.9083
17	7360	1019727	3	1013582	3	1493090112	114.0731	22.54712	1.69E+08	46	94	430.355
18	7360	1019727	3	1013582	3	1493090115	114.0731	22.54739	1.69E+08	76	94	460.3573
19	7360	1019727	3	1013582	3	1493090118	114.0731	22.54769	10416181	7	8	493.7151
20	7360	1019727	3	1013582	3	1493090118	114.0731	22.54769	10752581	7	71	493.7151
21	7360	1019727	3	1013582	3	1493090121	114.0731	22.54787	10752581	27	71	513.7298
22	7360	1019727	3	1013582	3	1493090124	114.0731	22.5479	10752581	31	71	517.0656
23	7360	1019727	3	1013582	3	1493090127	114.0731	22.5479	10752581	31	71	517.0656

(continued)

Table 9.2 (continued)

	ID	ID of entering intersection	Into the flow	ID of exiting intersection	Out of the flow	UNK time stamp	Longitude	Latitude	Link	Link distance	Link length	S
24	7360	1019727	3	1013582	3	1493090130	114.0731	22.5479	10752581	31	71	517.0656
25	7360	1019727	3	1013582	3	1493090133	114.0731	22.5479	10752581	31	71	517.0656
26	7360	1019727	3	1013582	3	1493090136	114.0731	22.5479	10752581	31	71	517.0656
27	7360	1019727	3	1013582	3	1493090139	114.0731	22.5479	10752581	31	71	517.0656
28	7360	1019727	3	1013582	3	1493090142	114.0731	22.5479	10752581	31	71	517.0656
29	7360	1019727	3	1013582	3	1493090145	114.0731	22.5479	10752581	31	71	517.0656
30	7360	1019727	3	1013582	3	1493090148	114.0731	22.5479	10752581	31	71	517.0656
31	7360	1019727	3	1013582	3	1493090151	114.0731	22.5479	10752581	31	71	517.0656
32	7360	1019727	3	1013582	3	1493090154	114.0731	22.5479	10752581	31	71	517.0656
33	7360	1019727	3	1013582	3	1493090157	114.0731	22.5479	10752581	31	71	517.0656
34	7360	1019727	3	1013582	3	1493090160	114.0731	22.5479	10752581	31	71	517.0656
35	7360	1019727	3	1013582	3	1493090163	114.0731	22.5479	10752581	31	71	517.0656
36	7360	1019727	3	1013582	3	1493090166	114.0731	22.5479	10752581	31	71	517.0656
37	7360	1019727	3	1013582	3	1493090169	114.0731	22.5479	10752581	31	71	517.0656
38	7360	1019727	3	1013582	3	1493090172	114.0731	22.5479	10752581	31	71	517.0656
39	7360	1019727	3	1013582	3	1493090175	114.0731	22.5479	10752581	31	71	517.0656
40	7360	1019727	3	1013582	3	1493090178	114.0731	22.5479	10752581	31	71	517.0656

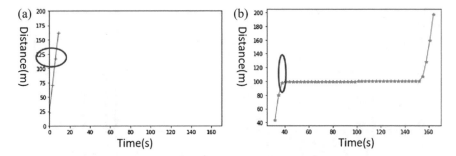

Fig. 9.5 Nonstop point and stop point. (**a**) Nonstop point. (**b**) Stop point

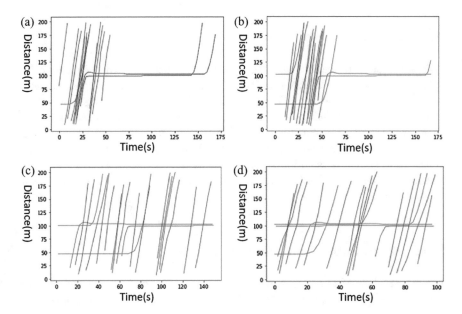

Fig. 9.6 The distribution of trajectory sets after screening. (**a**) The signal cycle length is 171 s. (**b**) The signal cycle length is 160 s. (**c**) The signal cycle length is 150 s. (**d**) The signal cycle length is 100 s

9.2.3   Signal Timing Optimization Based on Data of Internet of Vehicles

In the traditional traffic environment, detection equipment near intersections such as video surveillance and loop coils are widely used in the collection of traffic information. However, once these traditional devices encounter bad weather or obstructions, the quality of the data they collect is difficult to guarantee. At the same time, limited by the inherent characteristics of the sensor, the detector can only detect

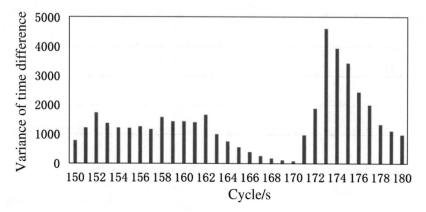

Fig. 9.7 Time difference variance calculation results

limited vehicle information. For example, the coil detector can only obtain the number of vehicles but cannot identify the type of the vehicle, while the magnetic sensor cannot perceive a stationary vehicle. Different from image and video processing and sensor network technology, V2X communication technology based on dedicated short-range communication (DSRC) can obtain multiple information of the vehicle, such as the lane of the vehicle, vehicle speed, vehicle type, vehicle priority, and so on. These information can be used to design more reasonable and effective intersection signal control methods, so as to achieve signal timing optimization.

9.2.3.1 Traffic Signal Control Evaluation Index

The optimization of intersection signal timing needs to evaluate the optimization effect, and two traffic signal control evaluation indexes are introduced below.

1. Delay. The average delay of vehicles at intersections mainly consists of two parts, one is consistent delay du, and the other is random delay dr. The calculation method is as follows:

$$d_{ui} = \sum_j \frac{c(1 - g_{ei}/c)}{2(1 - y_{ij})} \tag{9.17}$$

$$d_{ri} = \sum_j \frac{x_{ij}^2}{2q_{ij}(1 - x_{ij})} \tag{9.18}$$

$$d_i = d_{ui} + d_{ri} \tag{9.19}$$

where d_i represents the average delay time of the ith phase; d_{ui} represents the average consistent delay time of the ith phase; d_{ri} represents the average random delay time of the ith phase; c represents the cycle duration; g_{ei} represents the effective green light time of the ith phase; q_{ij} represents the actual arrival equivalent traffic volume of the jth entrance of the ith phase; y_{ij} represents the flow ratio of the jth entrance of the ith phase; x_{ij} represents the saturation of the jth entrance of the ith phase. Therefore, the average delay time of vehicles at an intersection in a cycle is the weighted average of the delays of each phase, namely

$$d = \frac{\sum_i d_i q_i}{\sum_i q_i} \tag{9.20}$$

2. Number of stops. The number of stops is an effective index reflecting the traffic operation status of the road network interval, which can be used to evaluate the traffic signal control effect of the measured road network interval. The average number of stops in the ith phase is represented by h_i, and the calculation method is as follows:

$$h_i = \sum_j 0.9 \frac{(c - g_{ei})}{1 - y_{ij}} \tag{9.21}$$

The average number of vehicle stops at an intersection in a cycle is the weighted average of the number of stops in each phase, namely

$$h = \frac{\sum_i h_i q_i}{\sum_i q_i} \tag{9.22}$$

9.2.3.2   Signal Timing Optimization Method Based on IoV Data

Traffic Signal Timing Optimization with Dynamic Adjustment in the Environment of IoV

In traditional traffic control systems, the green light time allocated to each phase of the intersection is often fixed. However, the fixed green light time cannot meet the time-varying traffic flow. Therefore, this part introduces a traffic signal timing optimization method with dynamic adjustment in the environment of IoV, which can dynamically allocate reasonable green light time for traffic flows in different directions on the basis of meeting traffic demand.

The basic idea. For vehicles, the quality of driving through the intersection is closely related to the waiting time and the number of stops. If the waiting time is too long, the number of vehicles passing through the intersection will be reduced; if the number of stops is too much, it is easy to cause waste of energy consumption and environmental pollution. Under the environment of IoV, the detection equipment can accurately record the real-time information of vehicles entering and leaving the road, and allocate reasonable green time for each signal phase according to the change of traffic volume. In other words, when the traffic volume is low, a shorter green light time can be allocated to reduce the waiting time of the vehicle; when the traffic volume is large, a longer green light time can be allocated to reduce the number of stops.

Optimization steps. The allocation of green light time is actually the allocation of right of way. When vehicles in a certain direction are about to obtain the right of way, the control system will calculate the green light time according to the real-time traffic flow of the current road. After the allocated green light time, the right of way will be transferred to the road in the next direction.

Taking into account that different types of vehicles have different demands for green light time, the number of vehicles should not only be considered in the process of signal timing optimization, but also the type of vehicle. Vehicles can be divided into three categories: large, medium, and small. Large vehicles include large trucks; medium vehicles include ordinary trucks and buses; small vehicles mainly refer to cars, taxis, etc. Different weights W_1, W_2, and W_3 are assigned to the three types of vehicles, respectively, and the final weight value that affect the green light time distribution are obtained by accumulating the weights of road vehicles.

Assume that the total number of vehicles detected on the current road is N, in which the number of vehicles turning left, straight and right is N_l, N_s, and N_r, and the weight of a single vehicle is W_i, then

$$W_{\text{sum}} = \sum_{i=1}^{N} \text{flag} \cdot W_i \qquad (9.23)$$

In the process of signal timing optimization, the influence of right-turning vehicles can be ignored, and only consider straight and left-turning vehicles. Therefore, the calculation method of *flag* and W_i is

$$\text{flag} = \begin{cases} 1 & \text{Vehicles Turning Left} \\ 1 & \text{Straight Vehicles} \\ 0 & \text{Vehicles Turning Right} \end{cases} \qquad (9.24)$$

$$W_i = \begin{cases} W_1 & \text{Large Vehicle} \\ W_2 & \text{Medium Vehicle} \\ W_3 & \text{Small Vehicle} \end{cases} \qquad (9.25)$$

In order to improve the traffic efficiency of the intersection, it is necessary to transfer the green light control right to the next signal phase when the total vehicle weight of the current signal phase is less than a certain threshold. Assuming that the weight threshold is W_t, the specific steps for optimizing the green light time are as follows.

Step 1: When the vehicle enters the road, an arrival message AM_i containing its specific identifier, driving lane, vehicle type, and location is sent to the RSU_1 via V2I communication. When the vehicle leaves the road, it sends a departure message DM_i containing only its identifier to the RSU_2.

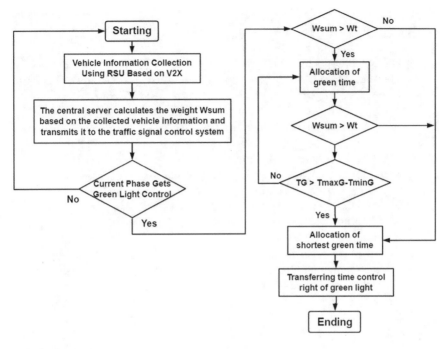

Fig. 9.8 Green light time optimization process

Step 2: The central server calculates the total weight W_{sum} corresponding to each signal phase through the collected vehicle information, and forwards it to the corresponding traffic signal control system of the road.

Step 3: The traffic signal control system determines whether the current signal phase has the green light control right. If yes, go to Step 4; otherwise, go to Step 1.

Step 4: The traffic signal control system compares W_{sum} and W_t. If $W_{sum} > W_t$, it means that the road is relatively congested, the next step is to execute Step 5, otherwise, to execute Step 8.

Step 5: The signal control system assigns green light time of a time step to the current phase.

Step 6: The traffic signal control system continues to compare W_{sum} and W_t. If $W_{sum} > W_t$, it means that the road is still congested, the next step is to execute Step 7, otherwise, to execute Step 8.

Step 7: Judge whether the current cumulative allocated green light time is longer than the difference between the longest green light time T_G and the shortest green light time T_{maxG}. If so, execute Step 8; otherwise, switch to Step 5.

Step 8: The traffic signal control system allocates the shortest green time T_{minG} to the current signal phase, which means it enters the signal phase countdown stage.

Step 9: The traffic signal control system transfers the green light time controller to the next signal phase, and the process ends.

The specific process of the optimization of green passage time above is shown in Fig. 9.8.

Example 9.2 Taking an intersection in Qinhuangdao City as an example, the lane group of the intersection is shown in Fig. 9.9. There are 8 lane groups at the intersection, among which, lane group 1 is the east entrance and goes straight; lane group 2 is the east entrance and turns left; lane group 3 is the south entrance and goes straight; lane group 4 is the south entrance and turns left; lane group 5 is the west entrance and goes straight; lane group 6 is the west entrance and turns left; lane group 7 is the north entrance and goes straight; lane group 8 is the north entrance and turns left. Since there is a dedicated entrance lane for right-turning and it can turn right during the red light period, signal control for right-turning is not carried out. The intersection phase is shown in Fig. 9.10. The solid line is the red traffic lane, while the dotted line is the green traffic lane. The phase 1–4 are east–west straight, east–west left, north–south straight, and north–south left, respectively. The actual traffic data of this intersection obtained from license plate recognition data are shown in Table 9.3. In the simulation period of 40 min, the first period is 20 min, the saturation of flow is 0.89, and the traffic flow is in an unsaturated state, and the actual volume at the intersection is calculated according to the license plate recognition data; the second period is 20 min, the saturation is 0.95, and it is in the saturation state, which is the artificial assumed flow. Considering the randomness of the traffic flow, it is considered that the saturation state is reached when the saturation is 0.95. The control effect of the dynamic programming timing scheme on the state of each traffic flow is investigated through the change of flow.

The number of entrance lanes in each lane group is

$$\mathbf{M} = (3, 1, 3, 1, 3, 13, 1)^{\mathrm{T}}$$

The initial queue length is

$$\mathbf{L_0} = (12, 10, 5, 12, 10, 10, 5, 9)^{\mathrm{T}}$$

The dissipation rate of queued vehicles in each lane group is

$$\boldsymbol{\theta_t} = (0.33, 0.30, 0.33, 0.30, 0.33, 0.30, 0.33, 0.30)^{\mathrm{T}}$$

The initial signal state is the east–west straight green light, and the other directions are red light. The sequence is 1 (east–west straight), 2 (east–west left), 3 (north–south straight) and 4 (north–south left). Vehicle arrival rate is

$$\Gamma_t = \begin{matrix} (0.31, 0.04, 0.11, 0.06, 0.28, 0.04, 0.1, 0.03) & 0 < t \le 1200 \text{ s} \\ (0.42, 0.04, 0.11, 0.06, 0.39, 0.04, 0.1, 0.03) & 1200 < t \le 2400 \text{ s} \\ (0.36, 0.04, 0.11, 0.06, 0.34, 0.04, 0.1, 0.03) & 2400 < t \le 3600 \text{ s} \end{matrix}$$

The minimum green time constraint is

(continued)

Example 9.2 (continued)
$$X = (10, 10, 10, 10, 10, 10, 10, 10)^T$$

The maximum green time constraint is

$$H = (80, 60, 60, 60, 80, 60, 60, 60)^T$$

The maximum queue length constraint is

$$F = (70, 60, 50, 40, 70, 60, 50, 40)^T$$

The maximum queue length dissipation constraint is

$$D = (8, 5, 8, 5, 8, 5, 8, 5)^T$$

The output result of the model is the number of times that the nth lane group $s_t(n)$ lasts 1, which can be converted into the phase effective green time. Taking into account the tendency of traffic flow at intersections to increase first and then decrease, determine the two situations of unsaturated state (saturation of 0.8) and saturated state (saturation of 0.95). According to the green light time optimization process shown in Fig. 9.7, through iterative calculations obtain the signal timing schemes under two traffic conditions, as shown in Fig. 9.11.

Fig. 9.9 The lane group of the intersection

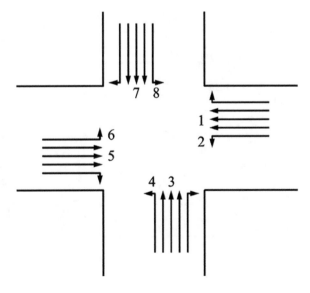

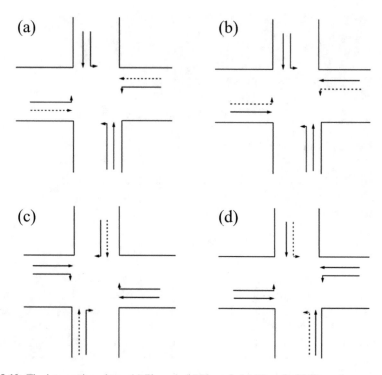

Fig. 9.10 The intersection phase. (a) Phase 1. (b) Phase 2. (c) Phase 3. (d) Phase 4

Table 9.3 Traffic flow and saturation

Traffic state	Unsaturated state		Saturation state	
	Traffic flow (veh/h)	Saturation	Traffic flow (veh/h)	Saturation
East entrance right	150	0.1	150	0.1
East entrance straight	1103	0.85	1234	0.95
East entrance left	173	0.51	199	0.59
South entrance right	150	0.1	150	0.1
South entrance straight	514	0.51	523	0.52
South entrance left	213	0.89	208	0.87
West entrance right	150	0.1	150	0.1
West entrance straight	1016	0.78	1111	0.86
West entrance left	187	0.56	225	0.67
North entrance right	150	0.1	150	0.1
North entrance Straight	550	0.54	573	0.57
North entrance left	194	0.81	187	0.78
Intersection	4550	0.89	4860	0.95

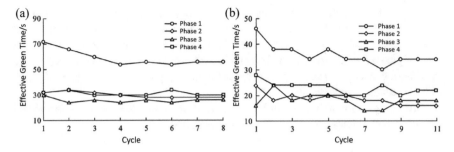

Fig. 9.11 The optimized timing scheme under unsaturated and saturated traffic conditions. (a) Unsaturated state. (b) Saturation state

Traffic Signal Timing Optimization of On-Demand Distribution in IoV Environment

The basic idea. Another method available for signal timing optimization in IoV environment is of on-demand distribution. This method takes into account that the vehicle queue length can directly reflect the effect of signal control at intersections. Therefore, the expected queue length of vehicles at the intersection can be directly used as the research object. According to the reliability of phase clearing at the intersection and assuming the traffic flow arrival rate as a random variable, an optimization model of intersection signal timing can be constructed by allocating green time according to demand.

Optimization model. Suppose a signal intersection has n phases, v_i, s_i and g_i represent the vehicle arrival rate (pcu/s), saturated flow rate (pcu/s) and green light time (s) of the key entrance lane of phase i, respectively; C represents the signal period (s); b_i represents the target value of the expected queue length of motor vehicle in phase i, and the value is related to the length of each entrance lane corresponding to phase i and the requirements of the signal control system. The value is

$$b_i \leq \min \left(q_1^i, q_2^i, \ldots, q_n^i\right) \tag{9.26}$$

where q_n^i represents the allowed queue length of the nth entrance lane corresponding to phase i.

The positive and negative deviations of the expected queuing length of vehicles in phase i from b_i are d_i^+, d_i^-, namely

$$d_i^+ = \max \left\{ \max \left[E(Cv_i - g_i s_i), 0 \right] - b_i \right), 0 \right\} \tag{9.27}$$

$$d_i^- = \max \left\{ \max \left[b_i - E(Cv_i - g_i s_i), 0 \right] \right), 0 \right\} \tag{9.28}$$

where $E(Cv_i - g_i s_i)$ represents the expected queue length of vehicles in a certain period of phase i. If the vehicle arriving within the cycle is less than the maximum capacity of phase i, this value is negative, indicating that there is redundancy in the green time of phase i; if the vehicle arriving within the cycle is more than the maximum capacity of phase i, this value is positive, indicating that the green light time of phase i is insufficient; if the vehicle arriving within the cycle is equal to the maximum capacity of phase i, this value is 0, indicating that the green time of phase i just meets the requirements.

According to the above definition, an optimization model of intersection signal timing when vehicles arrive at random is established.

$$\min Q = \sum_{i=1}^{n} \left(u_i^+ d_i^+ + u_i^- d_i^- \right) \tag{9.29}$$

$$\text{s.t.} \quad \max \left[E(Cv_i - g_i s_i), 0 \right] + d_i^- - d_i^+ = b_i \tag{9.30}$$

$$P(Cv_i \le g_i s_i) = \alpha_i \tag{9.31}$$

$$\sum_{i=1}^{n} g_i + L = C \tag{9.32}$$

$$L = \sum_{i=1}^{n} (I_i - A + H) \tag{9.33}$$

$$0 \le z_i \le g_i \tag{9.34}$$

where Q represents the objective function; u_i^+ and u_i^- represent weight factors of positive and negative deviation of queue length respectively; $P(Cv_i \le g_i s_i)$ represents the probability that all arriving traffic flows in the period can pass during the green light time of phase i; α_i represents the confidence level of phase i, namely the reliability of phase clearing; L represents total loss time per cycle(s); A represents the yellow light duration of each phase(s); H represents the starting loss time per phase(s); I_i represents the green light interval of phase i(s); z_i represents the shortest green light time of the phase that meets the safe passage of pedestrians and vehicles (s).

In this model, the objective function minimizes the deviation between the queuing length of vehicles in each phase at the intersection and the target queuing length of vehicles. Constraints (9.43) indicates that the signal timing parameters should meet the expected queue length target value of each phase. Equation (9.44) shows the phase clearing reliability constraint that phase i should satisfy. Equation (9.45) shows that the sum of green light time and cycle loss time of each phase is equal to cycle length. Equation (9.46) represents the total loss time of the cycle composed of the loss time of each phase. Equation (9.47) indicates the minimum green time constraint that the phase green time should meet for the safe passage of pedestrians and motor vehicles.

Therefore, the N-phase intersection signal timing optimization model is obtained. This model is also applicable to three-phase and multiphase signal intersections. The optimization and evaluation of intersection signal timing can be carried out by using this model in the environment of IoV.

Example 9.3 Taking the two-phase signalized intersection as an example. Under the conditions that s_i, L, z_i, and b_i are known, calculate the cycle length, green light time, and phase clearing reliability of the key entrances of each phase at the signalized intersection under different arrival rates.

Assuming that the two phase signals have no primary and secondary distinction, the arrival rate of the traffic flow in the intersection phases obey the normal distribution. The expected value is 700 pcu/h, and the saturation flow rate of both phases is 1800 pcu/h. The shortest green light time of the phase is 20 s, the loss time is 3 s, and the total loss time of the cycle is 6 s. The deviation weight factor is 1. The expected queue length of two phases is equal, and the phase clearing reliability of two phases is equal. When b changes, the arrival rate of vehicles in phase 2 is a constant. When the arrival rate of vehicles in phase 1 is a random variable, the optimization results of cycle duration C, green light time g_1 and g_2 of each phase, phase clearing reliability, namely confidence level α, are shown in Table 9.4.

When b changes, the arrival rates of vehicles in phase 1 and phase 2 are random variables. The calculation results of cycle duration C, green time g_1 and g_2 in each phase, and phase clearing reliability, namely confidence level α, are shown in Table 9.5.

Combining Tables 9.4 and 9.5, it can be seen that when the target value of the expected queue length and cycle duration are determined, in order to control the queue length of the intersection as much as possible, the phase with the larger vehicle arrival rate needs to allocate more green light time, and the greater the variance of the two-phase vehicle arrival rate, the more obvious this feature is.

Table 9.4 Timing optimization with vehicle arrival rate as random variable in phase 1

| b/pcu | $\sigma_1^2 = 70, \sigma_2^2 = 0$ | | | | $\sigma_1^2 = 210, \sigma_2^2 = 0$ | | | | $\sigma_1^2 = 280, \sigma_2^2 = 0$ | | | |
	C/s	g_1/s	g_2/s	α	C/s	g_1/s	g_2/s	α	C/s	g_1/s	g_2/s	α
4	50	24	20	0.995	50	24	20	0.8	52	26	20	0.758
6	52	26	20	0.997	59	30	23	0.848	59	30	23	0.782
10	80	43	31	0.999	81	44	31	0.898	82	44	32	0.832
16	–	–	–	–	127	72	49	0.933	129	73	50	0.871
30	–	–	–	–	210	122	82	0.952	216	126	84	0.894

Table 9.5 The results of timing optimization when the arrival rates of phase 1 and 2 vehicles are all random variables

b/pcu	$\sigma_1^2 = 70, \sigma_2^2 = 70$				$\sigma_1^2 = 70, \sigma_2^2 = 140$				$\sigma_1^2 = 70, \sigma_2^2 = 210$			
	C/s	g_1/s	g_2/s	α	C/s	g_1/s	g_2/s	α	C/s	g_1/s	g_2/s	α
4	46	20	20	0.871	56	24	26	0.841	59	24	29	0.771
6	62	28	28	0.945	64	28	30	0.864	70	30	34	0.811
10	80	37	37	0.971	106	47	53	0.922	110	47	57	0.860
16	144	69	69	0.990	158	71	81	0.943	168	73	89	0.885
30	260	127	127	0.995	278	127	145	0.957	300	132	162	0.903

9.3 Travel Path Planning Technology in the Environment of Internet of Vehicles

The Internet of Vehicles (IoV) is a vast interactive network that contains a large amount of information (e.g., location, speed, and path of vehicles and traffic signals). By the aid of GPS, RFID, sensors, cameras, radar, and other information collection equipment, the vehicle can collect the information of its own operation state and the surrounding driving environment. Through the Internet technology, the information processing system of IoV can received a wide variety of information of vehicles; and these large amounts of information about vehicles and traffic signals can be analyzed and processed through computer technology, thus the operation state of the dynamic road traffic network can be predicted and the optimal paths for individual vehicles can be calculated. In the environment of IoV, the real-time operation status of road network can be obtained; moreover, the operation of road network can be predicted accurately according to the historical and real-time traffic data. Based on this, travelers can choose departure time reasonably and make the path planning according to operation status of road network at the departure time, and can even adjust travel path dynamically. In addition, with the development of sharing economy under the internet environment, autonomous vehicles can provide travel services for passengers. Therefore, how to optimally determine the required size of autonomous vehicle fleet and design the vehicle paths under the given conditions of demand to minimize the total cost are key issues to be concerned about for the dial a ride problem (DARP).

9.3.1 Short-Time Forecasting of Road Network Operation Condition Based on Internet of Vehicles Data

In the environment of IoV, the current operating status of the road network can be collected by various detectors, but a reasonable vehicle path planning requires short-term prediction of the road network operating status based on historical and current data, thus travelers can avoid congestion in time. It not only saves the travel time of travelers, but also avoids the aggravation of congestion in the congested road section. Based on that, traffic streams on can operate smoothly in the transportation network.

Short-term traffic prediction tries to predict the future traffic state of the road network based on the historical and current data collected by various detectors in the road network. To achieve this objective, we should first analyze the regularity of traffic data and then choose the appropriate prediction model. At present, many models have been applied to short-term traffic prediction, which are mainly divided into statistical model, artificial intelligence model and combination prediction model. Refer to Chap. 5 for related methods.

9.3.2 Optimal Path Problem and Algorithm for Dynamic Road Network in Internet of Vehicles

In the environment of the IoV, the changes of travel time can be collected in real time, in other words, the travel time of a road is a time-dependent function rather than a static value. The optimal path problem of dynamic road network is to get the shortest path in the road network where travel time varies with time, and the deterministic time-dependent function is usually used to express the road section impedance. The time-dependent function can be formulated based on travel time or travel speed. Malandraki et al. from Northwestern University in the United States used a time-dependent function for the first time to express the travel time in the form of a piecewise function. However, this method requires the assumption that the vehicles wait for a certain time at the nodes to meet the First-In, First-Out (FIFO) characteristics of the road network, which is not consistent with the actual situation. Ichoua et al. of the University of Montreal used a time-dependent function to represent the travel speed, and calculated the travel time according to the travel speed, thus the road network can meet the FIFO characteristics and overcomes the shortcomings of the models such as Malandraki.

9.3.2.1 Definition of Dynamic Road Network

For a dynamic network $G = (V, A, C)$, $V = \{1, 2 \ldots, n\}$ is the set of nodes and $A = \{1, 2 \ldots, m\}$ is a collection of sections. $C = \{c_{ij}(t) | (i, j) \in A\}$ is the set of travel time–dependent functions of the road segment (i, j). The form of time-dependent function can be discrete step function or continuous linear piecewise function. If a piecewise function or a step function is used to define M departure periods: $T = \{0, 1, \ldots, M - 1\}$, then the travel time corresponding to the road segment(i, j) is $c_{ij}(t)$ in the period $t(t \in T)$.

9.3.2.2 Modeling of Optimal Path Problem for Dynamic Road Network

According to the departure time, the optimal path problems of time-varying network can be divided into two categories: (1) Given the departure time, the shortest path of travel time between two points can be searched; (2) Considering all possible departure times throughout the day, the shortest path of travel time between two points is solved, in which, the first problem is the basis of the second problem, and the first problem should be defined at first.

The nodes $v0 \in V$ and $vd \in V$ represent the start point and the end point respectively. X is a set of feasible solutions. The path $\lambda(\lambda \in X)$ is a feasible path between the starting point $v0$ and the end point vd. Cost $(\lambda, t0)$ is the travel time of the path λ which starts from the starting point $v0$ at the moment $t0$. Then, the mathematical model of the optimal path problem of dynamic transportation network is

$$Z = \min_{\lambda \in X} (\text{Cost}(\lambda, t_0)) \tag{9.35}$$

According to the definition of dynamic road network, Eq. (9.35) can be expressed as follows:

$$Z = \min \left(\sum_{(i,j) \in A} \sum_{t=1}^{T} c_{ij}(t) \cdot x_{ij}^{t} \right) \tag{9.36}$$

$$\text{s.t. } x_{ij}^{t} = \begin{cases} 1, & (i,j) \in A \\ 0, & \text{others} \end{cases} \tag{9.37}$$

$$\sum_{(i,j) \in A} \sum_{t=1}^{T} x_{ij}^{t} = 1 \tag{9.38}$$

$$\sum_{t=1}^{T} \sum_{\{j:(i,j) \in A\}} x_{ij}^{t} - \sum_{t=1}^{T} \sum_{\{j:(i,j) \in A\}} x_{ji}^{t} = \begin{cases} 1, & i = v_0 \\ -1, & i = v_d \\ 0, & \text{others} \end{cases} \tag{9.39}$$

where, as a constraint condition, Eq. (9.37), indicates that section (i, j) is used by path λ in time period t, then $x_{ij}^{t} = 1$, otherwise $x_{ij}^{t} = 0$; Eq. (9.38) indicates that the road section can only be used once; Eq. (9.39) indicates that the starting point only starts but does not arrive, the end point only arrives but does not start, and the arrival and departure times of the intermediate node are equal.

9.3.2.3 Optimal Path Algorithm for Dynamic Transportation Network

Static shortest circuit algorithm includes label setting method, label modification method, A* algorithm, etc., which can be extended to solve the optimal path problem of dynamic road network satisfying FIFO condition. The following are several optimal path algorithms of dynamic road network, which are respectively the improved Dijkstra algorithm, A* algorithm based on Euclidean distance and the improved A* algorithm.

The symbol of the algorithm is defined as follows: $t0$ is the given departure time; $v0$, vd are the starting point and the ending point; V,A is the collection of all nodes and road sections; S, W are respectively the node set of the shortest path which have been found and the node set of the shortest path which have not been found; li and pi are the numbers of node i and its precursor nodes respectively.

9.3.2.4 Improve Dijkstra Algorithm

The improved Dijkstra algorithm is to extend the traditional Dijkstra algorithm to the dynamic road network. The optimal path algorithm in two cases is given below.

1. Given the departure time $t0$, solve a node to get the optimal path of all other nodes

Step 1: Initialization. Set $i = v0$, $li = t0$, $pi = 0$; for $\forall j \neq i, l_j = +\infty, p_i = 0; S = \{i\}$, $W = \Phi$

Step 2: Node label update. For all successor nodes j of i, if $lj > li + cij(li)$, then $lj = li + cij(li)$, $pj = i$, and if $j \notin W$, then $W = W \cup \{j\}$.

Step 3: Node selection. Let $v*$ be the node with the smallest label in W, that is, $lv* = \min(lj), j \in W$, let $i = v*$, $S = S \cup \{i\}$, $W = W - \{i\}$.

Step 4: Judgment of the termination conditions. If $W = \Phi$, stop the calculation, otherwise go to Step 2.

2. Given the departure time $t0$, find the optimal path between two points.

Now, the algorithm is also composed of four steps: initialization, updating node label, node selection, and judgment of termination conditions. The first three steps are the same as in the previous case, except for judging the termination condition. In this case, the judgment condition is: if $i = vd$, then the calculation is stopped; otherwise, go to Step 2.

$(li - t0)$ is the shortest path travel time from $v0$ to vd. It can be traced back to the shortest path by the precursor node pi of node i.

Example 9.4 Consider the simple network shown in Fig. 9.12, which consists of 5 nodes and 7 road sections (the data marked on the sections in Fig. 9.12 is the corresponding section travel time when $t_o = 0$), and use Dijkstra algorithm to find the shortest path from node 1 to node 5.

(continued)

Example 9.4 (continued)

First, initialization. Let the current node i be 1, the label is 0, and all nodes except node 1 are labeled $+\infty$. The precursor node pi is 0, at this moment, the $S = \{1\}$, $W = \Phi$.

Search for the successor node 2 and node 3 starting from node 1, $l_2 = l_3 = +\infty$, $l_1 + c_{12}(l_1) = 0 + 1 = 1$, $l_1 + c_{13}(l_1) = 0 + 2 = 2$, $l_2 > 1$, $l_3 > 2$, $l_2 = 1$, $l_3 = 2$, $p_2 = p_3 = 1$, then $W = W \cup \{2, 3\}$. Select the node with the smallest label in W, $i = v^* = 2$, add it to S and remove it from W, $S = S \cup \{2\} = \{1, 2\}$, $W = W - \{2\} = \{3\}$.

Search for the successor nodes starting from node 2 (except the node included in S) and find node 3 and node 4. $l_2 + c_{23}(l_2) = 1 + 2 = 3 > 2$, therefore l_3 and p_3 remain unchanged. $l_2 + c_{24}(l_2) = 1 + 2 = 3 < +\infty$, so change l_4 to 3 and p_4 to 2. After comparing l_3 and l_4, it is found that l_3 is the smallest, so node 3 is placed in S and removed from W, $i = v^* = 3$, $S = S \cup \{3\} = \{1, 2, 3\}$, $W = W - \{3\} = \{4\}$.

Search for the successor nodes starting from node 3 (except the nodes included in S) and find node 5. $l_3 + c_{35}(l_3) = 2 + 1.5 = 3.5 < +\infty$, so l_5 is modified to 3.5, and p5 is modified to 3. Set node 5 in S, $i = v^* = 5$, $S = S \cup \{5\} = \{1, 2, 3, 5\}$. Only the shortest path from node 1 to node 5 is found here, so the calculation is stopped when $i = 5$.

Therefore, the shortest path between node 1 and node 5 is $1 \rightarrow 3 \rightarrow 5$. The travel time of the shortest path is 3.5.

9.3.2.5   A* Algorithm Based on Euclidean Distance

A* algorithm is a heuristic-based search algorithm, which evaluates the unextended nodes by introducing an evaluation function, and selects the most promising nodes to be extended until the target node is found. Compared with Dijkstra algorithm, the number of nodes searched by A* algorithm is greatly reduced, so it can greatly improve the computational efficiency.

In the A* algorithm, the evaluation function $f(i)$ is defined by formula (9.40), that is:

Fig. 9.12 Simple transportation network

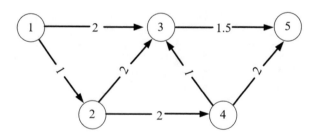

$$f(i) = g(i) + h(i) \tag{9.40}$$

where, $g(i)$ represents the actual cost from the starting point to node i; $h(i)$ is a heuristic function that represents the estimated cost of the optimal path from node i to the target node.

It can be proved that the necessary and sufficient strip with the shortest path can be found by the A* algorithm, which is:

$$h(i) - h(j) \le e(i,j) \tag{9.41}$$

where, $e(i,j)$ represents the actual shortest path cost from node i to j. Equation (9.41) represents the difference of the estimated cost of two nodes. It cannot be higher than the actual shortest path cost between the two nodes. It can be proved that if the compatibility condition is true, then the evaluation function $f(i)$ is monotonically increasing, that is, the sequence of nodes extended by the A* algorithm with $f(i)$ is non-decreasing. The choice of estimated cost $h(i)$ has a great influence on the efficiency of A* algorithm. The closer the estimated value is to the actual shortest path cost, the higher the efficiency of A* algorithm will be. It can be proved that when the estimated value $h(i)$ is equal to the actual cost of the shortest path, A* algorithm only searches nodes on the shortest path. When $h(i) = 0$, A* algorithm is transformed into Dijkstra algorithm. Therefore, the design of heuristic function $h(i)$ is the key to A* algorithm.

For dynamic road network, set $v_{ij}(t)$, $t \in \{0, 1, ..., M - 1\}$, which represents the travel speed of road section (i, j) at time period t, and $v_{\max}$ represents the maximum travel speed of all sections in the road network at all time periods; $\mathrm{dis}(i,j)$ represents the Euclidean distance from node i to j. So, we can define the heuristic function $h(i)$ as

$$h(i) = \frac{\mathrm{dis}(i, v_d)}{v_{\max}} \tag{9.42}$$

The following proves that $h(i)$ of formula (9.42) satisfies the compatibility condition.

Given that $d(i,j)$ represents the actual shortest path length of nodes i to j and $e(i,j)$ represents the travel time of the actual shortest path from nodes i to j, then

$$h(i) - h(j) = [\mathrm{dis}(i, v_d) - \mathrm{dis}/(j, v_d)]/v_{\max} \tag{9.43}$$

$$e(i,j) = \frac{d(i,j)}{v(t)} \ge \frac{\mathrm{dis}(i,j)}{v(t)} \ge \frac{\mathrm{dis}(i,j)}{v_{\max}} \tag{9.44}$$

According to the trigonometric inequality relationship, $\mathrm{dis}(i,v_d) - \mathrm{dis}(j,v_d) \le \mathrm{dis}(i,j)$ holds, so $h(i) - h(j) \le e(i,j)$, and the compatibility condition holds.

1. Given the departure time $t0$, find the optimal path between two nodes

Step 1: Pretreatment. First, the maximum travel speed $v_{\max}$ in the road network is calculated, and then the estimated path travel time $h(i) = \mathrm{dis}(i, v_d)/v_{\max}$ from any node i to the target node v_d is calculated.

Step 2: Initialization. Set $i = v0$, $li = t0$, $pi = 0$; for $\forall j \neq i$, $l_j = +\infty$, $p_i = 0$, $S = \{i\}$, $W = \Phi$.

Step 3: Node labels update. With respect to all successor nodes j of i, if $l_j > l_i + c_{ij}(l_i) + h(i)$, then $l_j = l_i + c_{ij}(l_i) + h(i)$, $p_j = i$, and if $j \notin W$, then $W = W \cup \{j\}$.

Step 4: Node selection. Let v^* be the node with the smallest label in W, that is

$$l_{v*} = \min(l_j), j \in W, \text{ set } i = v^*, S = S \cup \{i\}, W = W - \{i\}.$$

Step 5: Judgment of termination conditions. If $i = v_d$, stop the calculation; otherwise go to Step 2.

$(li - t0)$ refers to the shortest path travel time from node $v0$ to node v_d. It can be traced back to the shortest path through the precursor node pi of node i.

9.3.2.6 Improved A* Algorithm

Chabini et al. proposed an improved A* algorithm and constructed a new heuristic function $h(i)$. Compared with the Euclidean A* algorithm, the estimated value of $h(i)$ is closer to the actual cost of the shortest path, so it has higher efficiency. The heuristic function $h(i)$ is constructed as follows.

For any $(i,j) \in A$, and $c_{ij}^{\min}(t) = \min c_{ij}(t)$, $t = 0, 1, \ldots, M - 1$, a static network G' can be constructed by $c_{ij}^{\min}(t)$, and $e^{\min}(i,j)$ is expressed in G'. $h(i)$ is defined as shown in formula (9.45), which can prove that $h(i)$ satisfies the compatibility condition.

$$h(i) = e^{\min}(i, v_d) \tag{9.45}$$

2. Given the departure time t0, find the optimal path between two nodes

Step 1: Preprocessing. The static network G' is constructed, and the shortest path travel time $e^{\min}(i, v_d)$ from any other node i to the target node v_d is calculated by Dijkstra algorithm.

Step 2: Set $i = v_0$, $l_i = t_0$, $p_i = 0$; for $\forall j \neq i$, $l_j = +\infty$, $p_i = 0$; $S = \{i\}$, $W = \Phi$

Step 3: Node labels update. For all successor nodes j of i, if $l_j > l_i + c_{ij}(l_i) + h(i)$, then $l_j = l_i + c_{ij}(l_i) + h(i)$, $p_j = i$,if $j \notin W$,then $W = W \cup \{j\}$.

Step 4: Node selection. Let v^* be the node with the smallest label in W, that is, $l_{v*} = \min(l_j)$, $j \in W$ $lv^* = \min(lj)$, $j \in W$. Let $i = v^*$, $S = S = S \cup \{i\}$, $W = W - \{i\}$.

Step 5: Judgment of termination conditions. If $I = v_d$, stop the calculation; otherwise go to Step 2.

$(l_i - t_0)$ is the shortest path travel time from $v0$ to node v_d, which can be traced back to the shortest path through the precursor node pi of node i.

Example 9.5 Considering a simple network as shown in Fig. 9.13, the numbers on the arrows indicate the travel time and the maximum travel speed between nodes in turn, and the Euclidean distance from each node to node 5 is dis(1, 5) = 2.18, dis(2, 5) = 2, dis(3, 5) = 1.22, dis(4, 5) = 1. Please use A* algorithm to calculate the shortest path from node 1 to node 5 at $t0 = 0$.

Firstly, preprocessed to obtain the maximum travel speed $v_{max} = 1.5$ in the road network. Then calculate the estimated path travel time from any node i to the target node5, $h(i) = $ dis$(i, v_d)/v_{max}$, $h(1) = 2.18/1.5 = 1.45$, $h(2) = 2/1.5 = 1.33$, $h(3) = 1.22/1.5 = 0.81$, $h(4) = 1/1.5 = 0.67$, and initialize; let the current node i be 1,the label is 0, all nodes except node 1 are labeled $+\infty$, and the predecessor node pi is 0, at this time, $S = \{1\}$, $W = \Phi$.

Search for the successor nodes 2 and 3 of the road segment starting from node 1. $l_2 = +\infty$, $l_3 = +\infty$, $l_1 + c_{12}(l_1) + h(1) = 0 + 1 + 1.45 = 2.45$, $l_1 + c_{13}(l_1) + h(1) = 0 + 2 + 1.45 = 3.45$, $l_2 > 2.45$, $l_3 > 3.45$, $l_3 = 3.45$, $p_2 = p_3 = 1$, therefore, $W = W \cup \{2, 3\}$. Select the node $i = v^* = 2$ with the smallest label in W, add it to S and remove it from W, $S = S \cup \{2\}$, $W = W - \{2\} = \{3\}$.

Search for the successor nodes of the road section starting from node 2 (except the nodes included in S), then find node 3 and node 4. $l2 + c23 + h(2) = 2.45 + 2 + 1.33 = 5.78 > 3.45$.

So the values of $l3$ and $p3$ remain unchanged, $l2 + c24 + h(2) = 2.45 + 1 + 1.33 = 4.78 < +\infty$, so change l4 to 4.78 and p4 to 2. After comparing $l3$ and $l4$, it is found that $l3$ is the smallest, so node 3 is placed in S and removed it from W, $i = v^* = 3$, $S = S \cup \{3\} = \{1,2,3\}$, $W = W - \{3\} = \{4\}$.

Search for the successor nodes of the road section starting from node 3, and find node 4 and node 5. $l_3 + c_{34}(l_3) + h(3) = 3.45 + 1 + 0.81 = 5.76 > 4.78$ so the values of $l4$ and $p4$ remain unchanged. $l_3 + c_{35}(l_3) + h(3) = 3.45 + 1.5 + 0.81 = 5.76 < +\infty$, so change l5 to 5.76 and p5 to 3. After comparing $l4$ and $l5$, it is found that l4 is the smallest, and node 4 is placed in S, $i = v^* = 4$, $S = S \cup \{4\} = \{1, 2, 3, 4\}$, $W = W - \{4\} = \{5\}$.

Search for the successor nodes of the road section starting from node 4, and there is only node 5. $l_4 + c_{45}(l_4) + h(4) = 4.78 + 2 + 0.67 = 7.45 > 5.76$, so the values of $l5$ and $p5$ remain unchanged, place node 5 in S, $i = v^* = 5$, $S = S \cup \{5\} = \{1, 2, 3, 4, 5\}$, and stop the calculation because of $i = 5$. Therefore, the shortest path between node 1 and node 5 is $1 \rightarrow 3 \rightarrow 5$, and the shortest path travel time is 5.76.

Because of the small scale of the road network in this case, the advantages of A* algorithm cannot be reflected here. However, with the increase of the scale of the road network, the advantages of this algorithm over Dijkstra algorithm in computational efficiency will be highlighted.

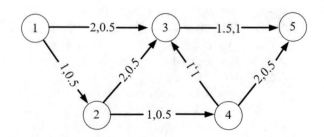

Fig. 9.13 Simple transportation network

9.3.3 Vehicle Service Routing Problem and Algorithm in the Environment of Internet of Vehicles

The sharing economy, born under the background of "Internet Plus," shows a strong development trend and potential. Sharing economy has brought a new development mode, consumption mode, and business operation mode, which has become a global economic development trend that cannot be ignored. The development of IoV and autonomous driving has laid a foundation for the development of shared cars, and promoted the transformation of automobile consumption pattern from the purchase of cars to the purchase of travel services.

Users book their travel needs in advance through the Internet, and service providers provide car-hailing services for users. As a provider of car-sharing service, how to determine the fleet size and optimize the vehicle service path on the basis of ensuring the service quality, so as to seek the minimum total cost is a problem worthy of further study. In addition, the distribution path optimization in the vehicle distribution problem is also similar to it.

9.3.3.1 Problem Description

The DARP problem can be described as follows: If all the vehicles are autonomous vehicles (AV), the vehicles in the fleet start from the station, provide ride-hailing services to multiple users, and then all of them arrive at the terminal within a day, the scheduling and service routing of the vehicles need to be properly arranged.

The known conditions. There are already conditions for the DARP problem:

1. Users' travel needs need to be booked in advance, that is, all users' needs have been completed before the service starts one day.
2. Each car only serves one or one group of passengers (starting from the same point and leaving at the same time) at a time, and carpooling is not considered. The vehicle will not pick up other passengers, so a service will not be interrupted.
3. The ride-hailing service has a hard time window, that is, the latest time for picking up passengers stipulated by the user. Vehicles are not allowed to arrive later than this time. If they arrive in advance, they need to wait. This time can be estimated by the chosen path.

4. All AVs must depart from the departure station and return to the terminal station. If there are participating vehicles in the fleet, they must directly arrive at the terminal station from the departure station and take a rest.

9.3.3.2 The optimization Goals

For DARP problems, there can be multiple optimization goals, including the minimum total cost, the maximum total revenue, the maximum user satisfaction, the minimum fleet size, etc. These goals are mainly divided into two categories, one is the cost of service (benefit), and the other is the quality of service. Generally speaking, these goals are in conflict with each other and cannot be optimal at the same time. Sometimes in practice, it is generally necessary to find a balance between cost (benefit) and service quality. At this time, several comprehensive goals need to be considered.

The time window of user needs is set as a hard time window, and user satisfaction is no longer considered in the optimization goal. Therefore, the optimization goal of this book is to minimize the overall cost of the system.

9.3.3.3 Problem Modeling

For convenience, firstly, the main parameters and variables used in the model are introduced, as shown in Table 9.6.

The trip demand of users are distributed within the study area, and the demand number is denoted by i. In this problem, a demand network is established from the point of view of demand, as shown in Fig. 9.14. For modeling purposes, we set the virtual starting point o and virtual ending point d, and all AVs should start from point o and finally arrive at point d. There are two types of nodes in the network, one is the pick-up point $i^- \in I^-$ and the drop-off point $i^+ \in I^+$ for each travel demand $i \in I$, and the other is the starting point and the ending point. Therefore, the set of all points in the network is $N = \{i^-, i^+ | \forall i \in I\} \cup \{o, d\}$.

Path set in the network is $A = A_1 \cup A_2 \cup \{(o, i^-)\}_{i \in I} \cup \{(i^+, d)\}_{i \in I} \infty \{(o, d)\}$, where $A_1 = \{(i^-, i^+) | \forall i \in I\} A_2 = \{(i^+, j^-) | \forall (i,) \in A^r\}$. It can be seen there are five types of connections. The dispatch path (o, i^-) is the first time AVs are dispatched from the starting point to the trip demand i through the path; The service path (i^-, i^+) is the path that each demand i is served; The reassignment path (i^+, j^-) is the path of AVS from the drop-off point i^+ of current demand I to the pick-up point j^- of another demand j after the service demand i; the destination path . refers to the last demand of AVS service i: the path from the drop-off point i^+ to the destination d; The virtual path (o, d) is the path from the starting point o to the ending point d for the undispatched AVs, and the path does not generate costs.

It should be pointed out that the re-dispatch path needs to meet the reachable condition, that is to say the travel time required by AVS on the re-dispatch path

Table 9.6 Parameters and variables

Parameter	Explanation	Parameter	Explanation
$i \in I$	Trip demand number $i \in I$	f	Fixed cost per vehicle
$i^- \in \Gamma$	Pick-up points for demand $i \in I$	m_{ij}	Capacity of path (i,j)
$i^+ \in I^+$	Drop-off points for demand $i \in I$'s	c_{ij}	The cost of using path (i,j)
I	Trip demand set; $|I| = N$	t_{i-}, t_{i+}	Start and end time for service demand i
Γ, I^+	Collection of pick-up points and drop-off points for all trip demand I	$v(a,b,t)$	At time t, the speed of the vehicle from a to b, a and b are pick-up and drop-off points
(i^-, i^+, t_i)	Travel at time t_i, demand i starts from the starting point and arrives at the end point i^+	$TD(a,b)$	The actual driving distance of the vehicle from a to b
A	The reachable demand set $A' = \{(i,j) \| i,j \in I, i$ can reach $j\}$	θ	Buffer time for re-dispatch of AV
l_i	Losses due to unmet needs i	μ	Buffer distance for re-dispatch of AV
d_{ij}	Cost of travel from demand i to demand j	Decision variable	
p_{ij}	Parking costs in the process of seeing off customers in demand i to picking up customers in demand j	x_{ij}	Integer: number of AVS using path(i,j); $x_{ij} \geq 0$
d_i	Fixed costs for dispatching and returning vehicles due to demand i		
F	Fleet size (total number of autonomous vehicles)		

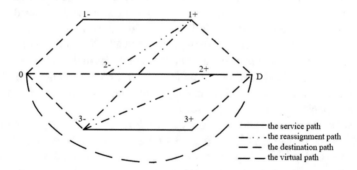

Fig. 9.14 An example network of the three requirements

should be less than the end time of demand i and the start time of demand j, as shown in Eq. (9.46).

$$t_{j^-} - t_{i^+} \geq \frac{TD(i^+, j^-)}{v(i^+, j^-, t_{i+})} \qquad (9.46)$$

A path in a network has several attributes, including path distance, travel time, cost, and capacity. The capacity of the path is defined by Eq. (9.47). Except for the virtual path, the capacity of all paths is 1, because each demand only needs to be serviced once. The capacity of the virtual path is the total number of AVs that can be dispatched, that is, fleet size F. However, the actual number of AVs required through the model is not equal to fleet size F.

$$m = \begin{cases} 1, & \text{if}(i,j) \in A \backslash \{o,d\} \\ F, & \text{if}(i,j) = \{o,d\} \end{cases} \qquad (9.47)$$

Here, the objective function of the model is to minimize the overall cost of the system. The influencing factors include fleet establishment, vehicle use, vehicle dispatch return to station, parking, and the penalty brought by unserved demand. Therefore, different paths have different cost calculation methods, as shown in Eq. (9.48)

$$c_{ij} = \begin{cases} 0, & \text{if } i = 0, j = d \\ d_j, & \text{if } i \neq 0, j = d \\ f + d_i, & \text{if } i = 0, j \neq d \\ -l_i, & \text{if } (i,j) \in A_1 \\ d_{ij} + p_{ij}, & \text{if } (i,j) \in A_2 \end{cases} \qquad (9.48)$$

d_i and d_j are the cost of dispatching and returning path respectively; f is the additional cost of vehicle maintenance and refueling when dispatching the vehicle. It is assumed that there is no operating cost on the service path, and the unserved demand will bring revenue reduction and user complaints. It can be understood that the negative profit of the demand is $-li$. In addition, the vehicle redistribution will produce travel cost d_{ij} (such as fuel consumption) and parking cost p_{ij}.

For the redispatch path (i^+, j^-), the time difference between the ending time of demand i and the beginning time of demand j should be greater than the actual travel time from i^+ to j^-, as shown in Eq. (9.49). However, considering the uncertainty of the time of AV on the redispatch path, a buffer time θ needs to be added. Similarly, the buffer distance μ is the maximum distance between the drop-off points and pick-up points of two consecutive requirements. The buffer distance is a relatively relaxed condition for the optimization results of the model, but the network can be simplified to improve the efficiency of the model. Especially for a large demand network, this parameter requires AV to serve the demands with closer distances as much as possible, rather than serve the demands of farther, to avoid the waste of resources on the redispatch path.

$$t_{j^-} - t_{i^+} \geq \theta + \frac{TD(i^+, j^-)}{v(i^+, j^-, l^+)} \tag{9.49}$$

According to the above demand network architecture, the following linear optimization model is established, as shown in Eqs. (9.50) to (9.55).

$$\min_{\{x_{ij}\}} \sum_{(i,j) \in A} c_{ij} x_{ij} \tag{9.50}$$

$$\text{s.t. } x_{ij} \leq m_{ij}, \forall (i,j) \in A \tag{9.51}$$

$$\sum_j x_{ji} = \sum_j x_{ij}, \forall i \in I \setminus \{o, d\} \tag{9.52}$$

$$\sum_{j \in I \setminus \{o, d\}} x_{ij} = F, \quad i = o \tag{9.53}$$

$$\sum_{j \in I \setminus \{o, d\}} x_{ji} = F, \quad i = d \tag{9.54}$$

$$x_{ij} \geq 0, \forall (i,j) \in A \tag{9.55}$$

Equations (9.51) to (9.53) are flow balance limits. Equations (9.54) and (9.55) are path capacity limits. When the travel demand is known, the maximum number of AVs and service paths needed to serve these demands can be obtained.

Example 9.6 Here, an 8 × 8 network with 64 nodes is constructed as the generation road network for travel demand, as shown in Fig. 9.15. Each node can be the starting point i^- or the end point i^+ of travel demand. The coordinates of the starting point $i^-(O\_x, O\_y)$ and the end point $i^+(D\_x, D\_y)$ are randomly generated, and the starting point and the end point are not the same node. As shown in Fig. 9.15, the starting point of demand 1 is (1, 2) and the end point is (4, 5).

The analysis cycle of the service path is 4 h, that is, 246 min, and the departure time ti-of each travel demand is required to be generated in the range [0,240]. In order to simplify the calculation, the influence of other vehicles and congestion on the road section is not considered. It is assumed that the AVs travel at the same speed on each section of the network. It is considered that it takes 5 min for the vehicle to travel a unit length of distance of one unit length of distance, then the end time of each trip demand is t_{i^+} as shown in Eq. (9.56).

(continued)

Example 9.6 (continued)

$$\mathrm{TD}(i^+, j^-) = |D\_x - O\_x| + |D\_y, D\_y| \tag{9.56}$$

$$t_{i^+} = t_{i^-} \times (\mathrm{TD}(i^+, j^-)) \tag{9.57}$$

According to the above method, we randomly generated 10 trip demands in the road network, as shown in Table 9.7.

The parameters in the travel cost are set as: $di = dj = 30$ per veh, $dij = 5 \cdot \mathrm{TD}$ (i, j), $li = 100 \cdot \mathrm{TD}(i)$, $f = 30$ per veh, $pij = 1$ per hour. The fleet size is set to $F = 15$

The algorithm is programmed with Java language. The test environment is a common desktop computer. The configuration is as follows: CPU: Intel Core i7 (4 cores), 3.41 GHz; memory: 8 GB; operating system: Windows 10 Professional. After calculating the test case, the total system cost $\sum\limits_{(i,j) \in A} c_{ij} x_{ij} =$ -5695, and the negative sign represents profit, that is, the maximum profit is 5695. The result of solution xij obtained by calculation is shown in Table 9.8. Numbers indicate the number of AVs using the path, and 0 indicates that no vehicles use the path. The service path of AVs obtained from the solution is as follows:

$1o\tilde{0} - \tilde{0} + \tilde{2}\text{-}\tilde{2} + \tilde{9} - \tilde{9} + \tilde{d};$

$2o\tilde{1}\tilde{1} + \tilde{7}\text{-}\tilde{7} + \tilde{d};$

$3o\tilde{3} - \tilde{3} + \tilde{d};$

$4o\tilde{5} - \tilde{5} + \tilde{d};$

$5o\tilde{6} - \tilde{6} + \tilde{8} + \tilde{4} - \tilde{4} + \tilde{d};$

$6o\tilde{d}$

The first car serves demand 0, 2, and 9; the second car serves demand 1 and 7; the third and fourth cars only serve demand 3 and 5, respectively; and the fifth car serves demand 6, 8, and 4 in turn. It can be seen that in this test case, only five AVs can meet ten travel needs, and the total cost at this time is the smallest, while the remaining ten vehicles directly arrive at the terminal station from the departure station.

It can be seen that on the basis of obtaining travel demand information under the environment of IoV, this model can not only obtain the service path of AVS, but also obtain the total number of vehicles required to meet all demands, so as to pursue the minimum overall cost or maximum overall profit.

Fig. 9.15 Test road network

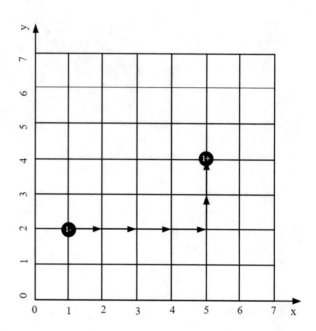

Table 9.7 Demand cases

Demand	O_x	O_y	D_x	D_y	The start time	The end time	Duration time
0	1	7	5	5	183	213	30
1	4	0	5	0	183	188	5
2	6	6	5	6	189	194	5
3	5	7	0	7	164	189	25
4	1	4	0	4	112	117	5
5	6	2	7	2	42	47	5
6	1	5	5	4	225	250	25
7	6	5	4	5	74	84	10
8	6	6	0	7	174	209	35
9	5	2	5	0	88	98	10

9.4 Intelligent Control Technology in Vehicle-to-Infrastructure Cooperation Environment

With the help of the information obtained from vehicle-to-vehicle/vehicle-to-infrastructure communication, the cooperative operation relationship between vehicles can further enhance vehicle operation safety and improve traffic efficiency to a certain extent. Perception range, model, and content of vehicles have changed in the vehicle-to-infrastructure cooperation environment under the support of

Table 9.8 Test case solution

	0-	1-	2-	3-	4-	5-	6-	7-	8-	9-	0+	1+	2+	3+	4+	5+	6+	7+	8+	9+	d
0-	0	0	0	0	0	0	0	0	0	0	0	1	0	0	0	0	0	0	0	0	0
1-	0	0	0	0	0	0	0	0	0	0	0	0	1	0	0	0	0	0	0	0	0
2-	0	0	0	0	0	0	0	0	0	0	0	0	0	1	0	0	0	0	0	0	0
3-	0	0	0	0	0	0	0	0	0	0	0	0	0	0	1	0	0	0	0	0	0
4-	0	0	0	0	0	0	0	0	0	0	0	0	0	0	0	1	0	0	0	0	0
5-	0	0	0	0	0	0	0	0	0	0	0	0	0	0	0	0	1	0	0	0	0
6-	0	0	0	0	0	0	0	0	0	0	0	0	0	0	0	0	0	1	0	0	0
7-	0	0	0	0	0	0	0	0	0	0	0	0	0	0	0	0	0	0	1	0	0
8-	0	0	0	0	0	0	0	0	0	0	0	0	0	0	0	0	0	0	0	1	0
9-	0	0	0	0	0	0	0	0	0	0	0	0	0	0	0	0	0	0	0	0	0
0+	0	0	1	0	0	0	0	0	0	0	0	0	0	0	0	0	0	0	0	0	0
1+	0	0	0	0	0	0	0	1	0	0	0	0	0	0	0	0	0	0	0	0	0
2+	0	0	0	0	0	0	0	0	0	1	0	0	0	0	0	0	0	0	0	0	0
3+	0	0	0	0	0	0	0	0	0	0	0	0	0	0	0	0	0	0	0	0	1
4+	0	0	0	0	0	0	0	0	0	0	0	0	0	0	0	0	0	0	0	0	1
5+	0	0	0	0	0	0	0	0	0	0	0	0	0	0	0	0	0	0	0	0	1
6+	0	0	0	0	0	0	0	0	1	0	0	0	0	0	0	0	0	0	0	0	0
7+	0	0	0	0	0	0	0	0	0	0	0	0	0	0	0	0	0	0	0	0	1
8+	0	0	0	0	1	0	0	0	0	0	0	0	0	0	0	0	0	0	0	0	0
9+	0	0	0	0	0	0	0	0	0	0	0	0	0	0	0	0	0	0	0	0	1
O	1	1	0	1	0	1	1	0	1	0	0	0	0	0	0	0	0	0	0	0	10

communication technology. Therefore, cooperative control of vehicles is different from the intelligent control based on sensors of vehicle self. This section focuses on the platoon control, lane change control, and collaborative traffic control technology at intersections in vehicle-to-infrastructure cooperation environment.

9.4.1 Platoon Cooperative Control of Car-Following Based on Vehicle-to-Vehicle Communication

9.4.1.1 Car-Following Process and Platoon Control Based on Vehicle-to-Vehicle Communication

Car-following is the most common driving behavior. Based on car-following behavior in micro driving, the car-following models study the interaction between two adjacent cars in traffic flow. The change of the lead vehicle's motion state causes the corresponding behavior of the following vehicle. Car-following model explains the traffic flow characteristics of a single lane by analyzing car-following behavior by each vehicle, and links the micro behavior of the driver with the macro phenomenon of traffic flow. Many scholars have conducted in-depth research on car-following models and obtained rich research results, including stimulus–response model, safety distance model, physiological and psychological model, artificial intelligence model, optimal velocity model, intelligent driving model, and cellular automata model.

In the traditional driving environment, N vehicles move on the single lane as shown in Fig. 9.16. The driver of the second vehicle perceives the brake information of the front vehicle if the lead vehicle makes an emergency brake. Then it takes the brake. The driver of the third vehicle takes the brake until the second vehicle makes an emergency brake in the platoon. In turn, rear-end collisions may occur because of the limitations of driver's perception in this situation.

In vehicle-to-infrastructure cooperation environment, the vehicle can not only obtain its own speed, direction, position, driving route, and other motion state information, but also its motion state information can be obtained by other surrounding vehicles. Then information interaction can be realized among vehicles. In addition, the vehicle will carry out effective information interaction with the roadside unit to realize the information exchanging and sharing of vehicle-to-vehicle and vehicle-to-infrastructure, as shown in Fig. 9.17. In this situation, the warning information will be sent directly to the $N - 1$ vehicles behind. Then the vehicle of

Fig. 9.16 Car-following process of N vehicles

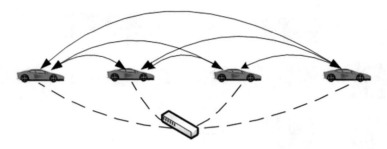

Fig. 9.17 Car-following of the platoon in the vehicle-to-infrastructure cooperation environment

platoon will acquire motion state information of front vehicles, and there will be enough response time to take measures to avoid rear-end collisions.

Therefore, platoon control based on vehicle-to-vehicle communication is to form a fleet of vehicles entering the road system, and maintain a certain platoon. The state of movement of lead vehicle and other vehicles of platoon are acquired using the wireless networks and vehicular sensors. Then all vehicles of platoon can drive automatically based on those motion state information, and also can keep a safe following distance. Given the platoon control system based on the vehicle-to-vehicle communication, it can effectively improve the safety of car-following, string stability, and reducing traffic accident. Therefore, it is one of the hot areas of future advanced intelligent driving assistance system and automatic driving system and their related technology researches.

9.4.1.2 Structure of Platoon Control System in Car Following Process

Platoon control based on the vehicle-to-vehicle communication can be regarded as an extension of the conventional adaptive cruise control system. It connects multiple vehicles to form a platoon through the wireless communications. Information sharing among vehicles can improve the perception ability of the vehicle, which make active control time more accurately. Platoon control based on the vehicle-to-vehicle communication can be divided into the control target generation and control algorithm, its logical structure is shown in Fig. 9.18. According to all the relevant information obtained, desired spacing headway is calculated by the cooperative control module based on vehicle-to-vehicle communication, and the control target is generated. Then the control algorithm realizes the control target by adjusting acceleration and deceleration according to the transfer function or performance parameters of the actuator.

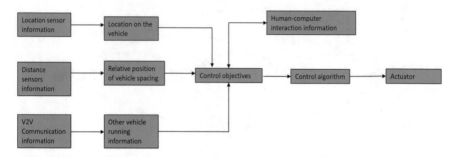

Fig. 9.18 Logical structure of platoon cooperative control based on vehicle-to-vehicle communication

9.4.1.3 Goal of Platoon Cooperative Control in Car Following Process

The purpose of platoon control is to design a longitudinal control system to make all vehicles of platoon move at the same velocity and maintain the same desired safe spacing headway between the vehicles. If the spacing headway between vehicles can be maintained at a safe and reliable distance, the rear-end collision can be effectively avoided, and the traffic capacity can be improved.

Vehicle-to-vehicle distance control refers to the longitudinal spacing headway control of platoon. At present, there are mainly two kinds of vehicle-to-vehicle distance strategies, namely fixed time headway and fixed spacing headway.

The fixed time headway strategy is defined as a function between the desired vehicle-to-vehicle distance and the velocity, which can be expressed as

$$S_i = L + t_h v_i$$

where S_i is the desired spacing headway of the ith vehicle; v_i is the speed of the ith vehicle; L is a constant, and it contains the length of the $i - 1$th vehicle as l_{i-1}; t_h is the time headway.

The error of the fixed time headway strategy can be expressed as

$$\delta_i = \sigma_i + L + t_h v_i$$
$$\sigma_i = x_i - x_{i-1} \tag{9.58}$$

$\dot{\delta}_i = -\lambda \delta_i$, λ is a positive control parameter, which is in the range of around 0.5. It can guarantee that δ_i is close to 0. The control law of desired acceleration of the controlled vehicle can be obtained by differentiating the above formula (9.58) as

$$\ddot{x}_i = -\frac{1}{t_h}(\sigma_i + \lambda \delta_i) \tag{9.59}$$

The choice of fixed time headway is related to the response time of vehicle control system and the acceptance degree of the driver or passenger and so on.

The fixed spacing headway strategy means that the vehicle adopts a fixed safe distance in the car-following process. Generally, the safe distance model under the relative distance braking mode is adopted. In the safe distance model, it is set τ_d as the difference between the starting acceleration or braking time of the preceding vehicle and the following vehicle, and the following vehicle continues to drive in the current state during that period of time. Using the deceleration of preceding vehicle as the initial moment, and the safety vehicle-to-vehicle distance under the relative distance braking mode was

$$\Delta d = D_b - D_f + \Delta D$$

where D_b and D_f are the driving distances of the preceding and following vehicles after braking, respectively; Δd is the safe distance between vehicles; ΔD is a safety margin in ensuring the vehicle's safe running. Their specific calculation methods are as follows:

$$D_f = \int_0^{T_f} v_f dt \qquad (9.60)$$

$$D_b = v_b(0)\tau_d + \int_0^{T_b} v_b dt \qquad (9.61)$$

where $v_f(0)$ and $v_b(0)$ are the initial velocity at the deceleration moment of the preceding and following vehicles, respectively; T_f and T_b are the running time of the preceding vehicle and following vehicle at the deceleration moment, respectively. τ_d is related to the perceived response time of the system, the response time of the vehicle control actuator and the delay of V2V information and so on.

9.4.1.4 Stability Analysis of the Platoon Cooperative Control in Car Following Process

The main emphasis of the spacing headway control is to maintain the string stability. Linear stability of the platoon mainly refers to the fluctuation of the spacing headway will or not magnify, and the string stability will or not weaken with the increasing number of vehicles in the platoon.

At present, many research organizations have designed many control strategies from different perspectives to achieve the string stability on the basis of ensuring the car following safety. Compared with the traditional ACC system, platoon control system based on the vehicle-to-vehicle communication can greatly reduce the car-following safety distance and improve the traffic capacity.

9.4.1.5 String Stability Analysis Based on Fixed Time Headway

As shown in Fig. 9.19, it is assumed that the gap between the i vehicle and the $i-1$th vehicle as $\xi_i(t)$, which denotes $\xi_i(t) = x_{i-1}(t) - x_i(t) - l_{i-1}$, where l_{i-1} is the $i-1$ vehicle length. According to the fixed time headway strategy, the desired gap between the ith vehicle and the $i-1$th vehicle is calculated as $t_h v_i(t)$, then the main control objective is to adjust the distance error as

$$\delta_i = x_{i-1}(t) - x_i(t) - l_{i-1} - t_h v_i(t) \tag{9.62}$$

So that the distance error is 0, and $\forall t \in (0, \infty], a_1(t) = 0 \Rightarrow \forall i, \lim_{t \to \infty} \delta_i(t) = 0$.

String stability means that the distance error does not magnify when it propagates to the rear of the platoon. As for the linear system, let $\phi_i(s)$ and $\phi_{i-1}(s)$ are frequency domain representation of distance errors of two adjacent vehicles in the platoon, respectively. $H(s)$ is the transfer function of the distance error of successive vehicles, which can be written as $H(s) = \phi_i(s)/\phi_{i-1}(s)$. Then the condition of the string stability is

$$\|H(s)\|_\infty = \sup \left| \frac{\phi_i}{\phi_{i-1}} \right| \le 1 \tag{9.63}$$

When the infinite norm of the transfer function is less than 1, it means that the car-following error of the preceding vehicle will not be transmitted and amplified to the rear of the platoon, and then the string stability can be guaranteed.

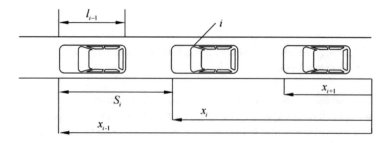

Fig. 9.19 Vehicle platoon in ACC system

Example 9.7 A linearized vehicle dynamics model be expressed as

$$\dot{X}_i(t) = A_i X_i(t) + B_i u_i(t - \varsigma) \tag{9.64}$$

where $\dot{X}_i(t) = \begin{bmatrix} x_i \\ v_i \\ a_i \end{bmatrix}$, $A_i = \begin{bmatrix} 0 & 1 & 0 \\ 0 & 0 & 1 \\ 0 & 0 & -1/\tau \end{bmatrix}$, $B_i = \begin{bmatrix} 0 \\ 0 \\ 1/\tau \end{bmatrix}$, $u_i(t)$ is the con-

troller model, τ is delay, ς is lag, a_i represents the acceleration of the ith vehicle. The PID controller is established to analyze the conditions of the string stability based on the control algorithm of fixed time headway.

Since the vehicle actuators and sensors exist hysteresis effect, the data of establishing the controller is not the current data, but the data with a certain delay time. That is to say, the vehicle spacing headway deviation is $\delta_i(t - \varsigma)$, The PID controller can be set as

$$u_i(t - \varsigma) = k_P \delta_i(t - \varsigma) + k_I \int \delta_i(t - \varsigma) dt + k_D \dot{\delta}_i(t - \varsigma) \tag{9.65}$$

According to the above method, the vehicle spacing headway deviation model of two consecutive vehicles in the time domain can be obtained as

$$\tau \ldots \delta_i(t) + \ldots \delta_i(t) + h k_D \ldots \delta_i(t - \varsigma) + (h k_P + k_D) \ddot{\delta}_i(t - \varsigma)$$
$$+ (h k_I + k_P) \dot{\delta}_i(t - \varsigma) + k_I \delta_i(t - \varsigma) = k_D \delta_{i-1}(t - \varsigma) + k_P \dot{\delta}_{i-1}(t - \varsigma) + k_I \delta_{i-1}(t - \varsigma) \tag{9.66}$$

The model of vehicle spacing headway deviation between two consecutive vehicles in the frequency domain is

$$H(s) = \frac{\left[(k_D s + k_P s + k_I)e^{-\varsigma s}\right]}{\left[\tau s^4 + s^3 + (h k_D s^3 + (h k_P + k_D)s^2 + (h k_I + k_P)s + k_I)e^{-\varsigma s}\right]} \tag{9.67}$$

According to the criterion of string stability, it can be known that $|H(i\varpi)|^2 < 1, \forall \varpi > 0$. The parameter k_I will slow down the convergence speed that the vehicle spacing headway deviation tends to zero. And the sign of the vehicle spacing headway deviation also change, thereby reducing the comfort of the vehicle. Therefore, it is assumed that $k_I = 0$. Then Eq. (9.67) can be simplified as

$$H(s) = \frac{\left[(k_D s + k_P s)e^{-\varsigma s}\right]}{\left[\tau s^4 + s^3 + (h k_D s^3 + (h k_P + k_D)s^2 + k_P s)e^{-\varsigma s}\right]} \tag{9.68}$$

(continued)

Example 9.7 (continued)

According to the string stability analysis method, the string stability condition can be calculated when the PD controller is used as

$$
\begin{cases}
0 < k_D < \tau/2h\varsigma, \\
2/h^2 < k_P, \\
2(\varsigma + \tau) < \left((1 + hk_D)^2 + 2\varsigma\tau k_P\right)/(hk_P + k_D).
\end{cases} \tag{9.69}
$$

When the hysteresis phenomenon is ignored, i.e. $\varsigma = 0$ and $\tau = 0$. The string stability condition is $2/h^2 < k_P$. That is to say, the parameter k_D can take any value. Due to the hysteresis of actuators and sensors, the platoon is not stable if the value of k_D is greater than $\tau/2h\varsigma$.

9.4.1.6 String Stability Analysis Based on Fixed Spacing Headway

We consider a group of $N + 1$ vehicles on a single-lane, as shown in Fig. 9.20. Without loss of generality, we use 0 to mark the leading vehicle in the platoon, and 1, 2, ..., N are marked for the following vehicles of platoon. Each vehicle's speed, acceleration, and position can be transmitted among each other via wireless networks.

The distance error of the i vehicle is defined as follows:

$$
\Delta\delta_i = x_{i-1} - x_i - \delta_d - l_i \tag{9.70}
$$

where $x_i(i = 1, 2, \ldots, N + 1)$ represents the position of the ith vehicle; l_i is the length of the ith vehicle; δ_d is the minimum safe distance between two vehicles; $\Delta\delta_i$ represents error between the desired vehicle spacing and the actual vehicle spacing.

The condition of the string stability can be described as

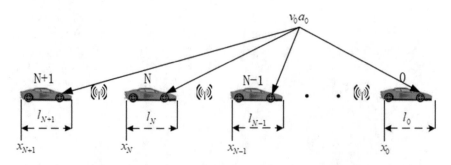

Fig. 9.20 Vehicle platoon structure

$$\|\Omega(s)\|_\infty = \sup \left| \frac{\Delta \delta_i(s)}{\Delta \delta_{i-1}(s)} \right| \leq 1 \tag{9.71}$$

where $\Omega(s)$ is the transfer function of the error. When the infinite norm of the transfer function is less than 1, it means that the error of the preceding vehicle will not be transmitted and amplified to the rear of the platoon. Then the string stability can be guaranteed.

Example 9.8 Considering a platoon of $n + 1$ vehicles, the first vehicle is set as the lead vehicle, and the other n vehicles are all the following vehicles. All vehicles use the same adaptive cruise control system, and the fixed spacing headway strategy is used to analyze the string stability.

We suppose the control law of the vehicle as $\ddot{x}_i = u_i$, x_i is the position of the ith vehicle (x_0 is the position of the lead vehicle), u_i is the acceleration of the ith vehicle, $\dot{x}_i = v_i$ is the velocity of the ith vehicle, l_i is the length of the ith vehicle, σ_d is the minimum safe distance between two vehicles, and the acceleration of the ith vehicle can be expressed as

$$\ddot{x}_i = -k_v \Delta \delta_i - k_p \Delta \dot{\delta}_i - k_d (\dot{x}_i - \dot{x}_0) \tag{9.72}$$

In the fixed spacing headway control strategy, the spacing error $\Delta \delta_i$ between the ith vehicle and the $i-1$th is represented as

$$\Delta \delta_i = x_{i-1} - x_i - \delta_d - l_i \tag{9.73}$$

The derivative of both sides of Eq. (9.73) can be obtained as

$$\Delta \ddot{\delta}_i = \ddot{x}_{i-1} - \ddot{x}_i \tag{9.74}$$

We can obtain the following Eq. (9.75) according to Eqs. (9.72), (9.73), and (9.74).

$$\Delta \ddot{\delta}_i + (k_v + k_d) \Delta \dot{\delta}_i + k_p \Delta \delta_i = k_v \Delta \dot{\delta}_{i-1} + k_p \Delta \delta_{i-1} \tag{9.75}$$

We take the Laplace transform of Eq. (9.75), we can obtain the transfer function $\Omega(s)$ as

$$\Omega(s) = \frac{k_v s + k_p}{s^2 + (k_v + k_d)s + k_p} \tag{9.76}$$

(continued)

Example 9.8 (continued)

According to the criterion of the string stability, it can be known that the vehicle spacing error will not be magnified from the front to the rear of the platoon with the same adaptive cruise control system and spacing headway strategy when $\|\Omega(i\varpi)\|_\infty \leq 1$. It shows that the platoon is stable.

Therefore, we only need $\sqrt{k_p} > k_v$ and $k_v + k_d > \sqrt{2k_p}$, then the transfer function $\Omega(s)$ is less than 1. In other words, we need the parameter k_v, k_d, and k_p should satisfy Equation (9.77) to keep $\|\Omega(i\varpi)\|_\infty \leq 1$.

$$\begin{cases} \sqrt{k_p} > k_v, \\ k_v + k_d > \sqrt{2k_p}. \end{cases} \tag{9.77}$$

Then the string stability can be guaranteed.

9.4.2 Lane-Changing Control Under the Vehicle-to-Infrastructure Cooperation Environment

9.4.2.1 Lane-Changing Behavior

Lane-changing is the process that vehicles are moving in one lane change to another. Lane-changing behavior refers to driving behavior that drivers take to avoid the lane and change to the adjacent lane for satisfying their driving comfort and driving intention. Lane-changing behavior often happen when queuing, congestion, the change of traffic flow. With the mature and pervasion of vehicle-to-infrastructure cooperation technology, multi-vehicle's coordination will become an important way to further improve the safety and efficiency of lane changing.

In the vehicle-to-infrastructure cooperation environment, vehicular sensors can acquire the current driving state and positioning information of vehicles, and obtain the driving state and location information of surrounding vehicles through vehicle-to-vehicle information interaction. The movement of each vehicle is real-time optimized and controlled in the lane-changing process through the collaborative decision and action of each vehicle to avoid the risk of lane-changing and improve the success rate of lane-changing, as well as to reduce the negative impact of lane-changing behavior on vehicles of the target lane.

9.4.2.2 Objective of Cooperative Lane-Changing Control

The objectives of cooperative lane-changing control mainly include safety and comfort.

1. Safety. The safety of lane-changing process mainly embodies three aspects. The first is to ensure that the car-following safety of this lane in the lane-changing process; The second is to ensure that the interference of target lane should be possibly small to guarantee the safety of the target lane in the lane-changing process; The third is to ensure the vehicle's driving safety of lane-changing process.
2. Comfort. In the cooperative lane-changing process, the desired control input of each vehicle should be as close as possible to the desired acceleration of the driver for reducing acceleration's mutation because of operating vehicle of the driver after the end of lane-changing control. This situation can cause safety risks, decrease in traffic efficiency, and lead to driver discomfort. In the lane-changing process, vehicle's acceleration is divided into the lateral acceleration and the longitudinal acceleration. To ensure the comfort of lane-changing process, the acceleration constraint is established as shown in Eq. (9.78).

$$
\begin{cases}
h_{min} \leq a_{sh} \leq ch_{max} \\
v_{min} \leq a_{sv} \leq cv_{max} \\
a_{sv} \leq \min\left(a_{v\text{-ssafe}}, a_{v\text{-rsafe}}\right)
\end{cases}
\tag{9.78}
$$

where the lower limit of the acceleration constraint is set as the maximum longitudinal braking deceleration a_{hmin} and the maximum lateral braking deceleration a_{vmin} in normal driving process; the lower limits of the longitudinal and lateral acceleration are set as a_{chmax} and a_{cvmax}, respectively, and the maximum lateral acceleration is not greater than the critical values of lateral acceleration as $a_{v\text{-ssafe}}$ and $a_{v\text{-rsafe}}$, which can cause sideslip and rollover.

9.4.2.3 Cooperative Lane-Changing Control

The lane-changing process consists of the generation of lane-changing intention, the choice of lane-changing, the planning of lane-changing trajectory, and the trajectory control of vehicle's lane-changing, as shown in Fig. 9.21.

1. Generation of lane-changing intention. The lane-changing behavior can be divided into the mandatory lane-changing (MLC) and discretionary lane-changing (DLC) based on different interest motivations. MLC is refers to the vehicle must change the lane due to turning, obstacle, and parking in driving process. Therefore, the vehicles have been seeking a variety of suitable opportunities through deceleration and acceleration to change the lane before the latest lane-changing position. If the driver cannot complete the lane-changing in current position, then the vehicle will be stopped in this position to wait a suitable opportunity to change lane. DLC is carried out to achieve the desired driving speed without the constraint that lane-changing must be done. The difference between MLC and DLC is that the purpose of DLC is to get faster speed or reach the destination faster. This lane-changing behavior is not necessary to complete.

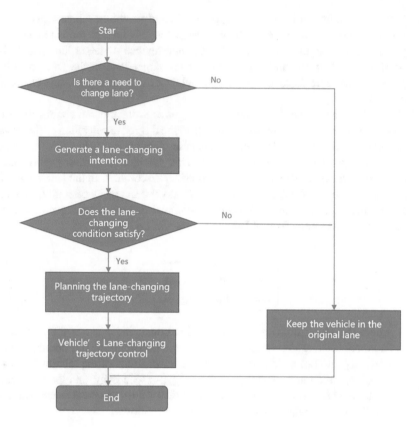

Fig. 9.21 Lane-changing process

In the generation stage of lane-changing intention, different reasons may lead to the generation of lane-changing intention. In the vehicle-to-vehicle cooperation environment, it can objectively judge whether to change lanes according to the actual traffic flow, thereby operating efficiency of traffic. The generation process of lane-changing intention is shown in Fig. 9.22. The generation mechanism of the DLC is different from that of the MLC.

Let the speed of the ith vehicle as $V_{S,\,i}$, and the velocities of the preceding vehicles in the current lane are denoted as $V_{S,\,i\,-\,1}$, $V_{S,\,i\,-\,2}\ldots V_{S,\,1}$, and the velocities of the following vehicles are denoted as $V_{S,\,i\,+\,1}$, $V_{S,\,i\,+\,2}\ldots V_{S,\,n}$, respectively. The average speed and variance in the current lane are V_S and σ_S^2, respectively. The velocities of the preceding vehicles in the target lane are denoted as $V_{T,\,j\,-\,1}$, $V_{T,\,j\,-\,2}\ldots V_{T,\,1}$, and the velocities of the following vehicles are denoted as $V_{T,\,j\,+\,1}$, $V_{T,\,j\,+\,2}\ldots V_{T,\,n}$, respectively. The average speed and variance in the target lane are V_T and σ_T^2, respectively.

The intention of MLC mainly includes the following two ways.

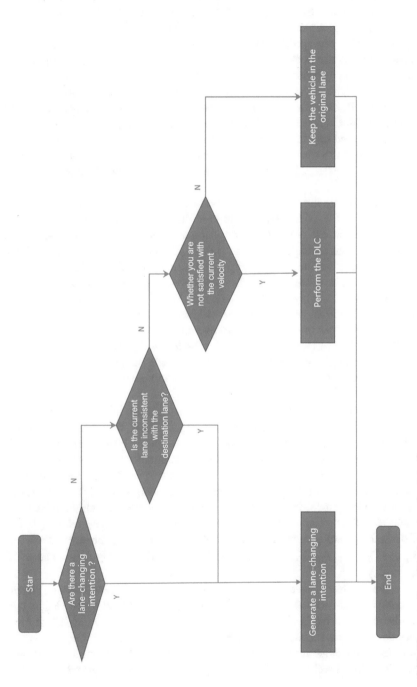

Fig. 9.22 The generation of the lane-changing intention

There are obstacles in front of the original lane, such as accidents, road construction, etc. Then a section of the original lane cannot be used, but the vehicle can run normally in the target lane, and the driver will have the intention to change lane. The specific forms are shown as $V_{S,\,i-1} \leq \Delta V_S$, where ΔV_S is a threshold value that is close to 0, and $V_{T,\,i-1} \geq \Delta V_T$, ΔV_T can be determined by the simulation experiments.

When the lane of the target vehicle is not consistent with the destination lane, the driver will have the intention to change lane. Such as the vehicle turns left/ right in the next intersection, this vehicle needs to early entry into the left/right turn lanes based on the lane guidance sign before entering the intersection. The specific forms are shown as $x \leq \Delta x$, where x is distance between the current position of vehicle and intersection, then the vehicle will have lane-changing intention.

Compared with the MLC, the DLC takes into account more information about the vehicles on the road. In the vehicle-to-infrastructure cooperative environment, the target vehicle can obtain the information of other vehicles. And the velocities of the current lane and the adjacent lane will affect the generation of the vehicle lane-changing intention. The velocity of the current lane is lower than the velocity of the adjacent lane. The speed difference is greater than a certain threshold, then the vehicle will have lane-changing intention. When there is a slow-moving large vehicle in the current lane, the vehicle will have lane-changing intention to drive in the adjacent lane or change lanes to overtake.

The intention of DLC mainly includes the following two ways.

There is a large speed difference between the current lane and the target lane, and the vehicle will have lane-changing intention when $V_T - V_S \geq \Delta V$ and $\sigma_S^2 \leq \Delta\sigma_S^2, \sigma_T^2 \leq \Delta\sigma_T^2$.

All vehicles in front of the target vehicle have different influences on the target vehicle. The greater the distance from the target vehicle, the smaller the impact. The preceding vehicle's speed of the target vehicle is lower than the average speed of the traffic flow, then there is a big impact for the target vehicle. For example, the velocity of the concrete-mixer truck is lower than the traffic speed in the current lane when the preceding vehicle of the target vehicle is a low-speed concrete-mixer truck, then it exerts a great influence on the target vehicle. In contrast, the target vehicle will have the intention to change lane if it is not disturbed by large vehicles. Specifically, the vehicle will change lane when $V_{S,\,i} - V_{S,\,i-1} \geq \Delta V_{S,\,i}, \ V_{S,\,i-1} < V_S, \ V_T - V_{S,\,i-1} \geq \Delta V_{T,\,i}$ and $\sigma_S^2 \leq \Delta\sigma_S^2, \sigma_T^2 \leq \Delta\sigma_T^2$.

In the process of lane-changing, different lane change intents have different priorities. The first level is the lane-changing to avoid obstacles. the vehicle will not be able to drive in the original lane when the obstacles occur, and it must change lane. The second level is the lane-changing in intersection, that is, the vehicle must change lane according to lane guide direction to pass the intersec-tion. The third level is the lane-changing to satisfy driver's desired speed. The driver is not satisfied with the current driving speed, but it needs to meet the lane-changing conditions before safety lane change.

2. The timing of lane-changing. The following points should be considered when the vehicle change the lane.

In the vehicle-to-infrastructure cooperation environment, the target vehicle can obtain the vehicle traveling information of its own and surrounding vehicles, and it make a reasonable judgment based on the information obtained.

The safety gap between vehicles must be considered to avoid collision in the limit cases when the vehicle wants to change lane. This mainly includes two aspects: In the initial stage of lane-changing, the target vehicle and its preceding vehicle in the original lane should keep a safety car-following process. Then the target vehicle and its preceding vehicle ensure a certain safe gap. In addition, there will be no collision between the target vehicle and the front and rear vehicles in the target lane in the process of lane-changing.

In the process of lane-changing, the target vehicle should not have a big impact on the driving state of the front and rear vehicles in the target lane, which is mainly considered from the perspective of time headway (TH) and time-to-collision (TTC). The TH and TTC of the target vehicle and the front vehicle of the target lane should be kept in a reasonable range after lane-changing. And the TH and TTC of the target vehicle and the rear vehicle of the target lane also should be kept in a reasonable range.

Example 9.9 The vehicle M runs in the lane 1. The vehicle M is not satisfied with the current driving speed, so the vehicle M changes the lane, as shown in Fig. 9.23. The M vehicle started to change lane at $t = 0$. The time headways must be more than above 2 s between the vehicle M and vehicle L (and vehicle F) if the lane-changing behavior not have an impact on the vehicles of lane 2, and the time-to-collision must be more than above 5 s between the vehicle M and vehicle L (and vehicle F). Under what conditions will the vehicle M can change lane?

Vehicles L, F, Q, and M are the preceding vehicle of the target lane, the following vehicle of the target lane, the preceding vehicle of the original lane, the following vehicle of the original lane, respectively. D_{MF} and D_{ML} are distances between the vehicle M and the preceding and following vehicle in the lane 2, respectively. In the process of lane-changing, it should be guaranteed that the vehicle M will not have a significant impact on the driving state of the vehicle F and L.

The motion state of each vehicle changes after the T of lane-changing time, as shown in Fig. 9.24. The vehicle M starts to change lane at the time $t = 0$ with an acceleration of $a_M(t)$, and the vehicle M begins to slow down with a deceleration of $a_F(t)$ after the reaction time of τ_1. The vehicle L drives at a constant speed within the T of lane-changing time.

(continued)

Example 9.9 (continued)

According to the kinematics formula, the displacements of vehicle F, L, and M can be obtained within the T of lane-changing time. The T of lane-changing time includes driver's reaction time τ_1 of the vehicle F and deceleration time τ_2, that is, $T = \tau_1 + \tau_2$.

$$
\begin{cases}
x_F(t) = v_F(t) \cdot (\tau_1 + \tau_2) + \dfrac{1}{2} a_F(t) \cdot \tau_2^2 \\[2mm]
x_L(t) = v_L(t) \cdot T \\[2mm]
x_M(t) = \int\limits_0^T \int\limits_0^\lambda [a_M(t)] dt d\lambda + v_M(t) \cdot T
\end{cases}
\tag{9.79}
$$

where $V_F(t)$ is the speed of the vehicle F, $V_L(t)$ is the speed of the vehicle L, and $V_M(t)$ is the speed of the vehicle M.

According to the requirements of this question, THs and TTCs should satisfy the following equations between the vehicle M and vehicle L (and vehicle F) after the lane-changing as

$$
\text{TH}_{M,L}(t) = \frac{D_{ML}(t) + L_L}{V_M(t)} > 2\,\text{s}, \quad \text{TC}_{M,L}(t) = \frac{D_{ML}(t)}{V_M(t) - V_L(t)} > 5\,\text{s} \tag{9.80}
$$

$$
\text{TH}_{F,M}(t) = \frac{D_{MF}(t) + L_M}{V_F(t)} > 2\,\text{s}, \quad \text{TTC}_{F,M}(t) = \frac{D_{MF}(t)}{V_F(t) - V_M(t)}
$$
$$
> 5\,\text{s} \tag{9.81}
$$

where $D_{ML}(t)$ is gap between the vehicle M and the vehicle L, and L_L is the length of the vehicle L; $D_{MF}(t)$ is gap between the vehicle M and the vehicle F, and L_M is the length of the vehicle M.

Equations (9.80) and (9.81) can be expressed as

$$
\begin{aligned}
D_{MF}(t) &> \max \{5(V_F(t) - V_M(t)), 2V_F(t) - L_M\} \\
D_{ML}(t) &> \max \{5(V_M(t) - V_L(t)), 2V_M(t) - L_L\}
\end{aligned}
\tag{9.82}
$$

According to Fig. 9.24, the gaps x_{MF} and x_{ML} between the vehicle M and the front and rear vehicles in the lane 2 before the lane-changing can be shown as Eq. (9.83):

$$
\begin{cases}
x_{ML} = x_M(t) - x_L(t) + D_{ML} \\
x_{MF} = x_F(t) + L_M - x_M(t) + D_{MF}
\end{cases}
\tag{9.83}
$$

(continued)

Example 9.9 (continued)

The lane-changing behavior not have a large impact on the vehicle F and vehicle L when the gaps between the vehicle M and the front and rear vehicles in the lane 2 satisfy Eq. (9.83), then the vehicle M can change lane.

1. Trajectory planning of lane-changing. The following points should be considered when the vehicle change the lane.

   In the process of lane-changing, the vehicle runs the lateral and longitudinal motion. The trajectory of lane-changing refers to the vehicle position, the lateral and longitudinal acceleration, and the lateral and longitudinal velocity from the beginning of lane-changing to the end of lane-changing. Lane-change trajectory is closely related to vehicle dynamic characteristics.

   The trajectory model of lane-changing mainly divided into two categories. The first class of the trajectory model of lane-changing: We set the vehicle's lateral and longitudinal displacement equations and the undetermined coefficients of the equations, and then solved the undetermined coefficients to obtain the lateral and longitudinal displacement equations through the acceleration and deceleration constraints and geometric constraints of crash avoidance in the process of lane-changing. Finally, we found the corresponding controller output. For example, Yugong Luo et al. used a time-based fifth-order polynomial to describe the desired lane-changing trajectory, which has the advantages of the closed form, continuous third derivative, and smooth curvature. First, the lateral and longitudinal displacement equations of lane-changing are set as shown in Eq. (9.84).

$$\begin{cases} x(t) = a_5t^5 + a_4t^4 + a_3t^3 + a_2t^2 + a_1t + a_0 \\ y(t) = b_5t^5 + b_4t^4 + b_3t^3 + b_2t^2 + b_1t + b_0 \end{cases} \tag{9.84}$$

where the polynomial coefficients can be obtained based on the duration of lane-changing, the position of the beginning and the end of lane-changing, the information of speed, collision avoidance, and so on. High order can make the acceleration to become continuous and smooth in the lane-changing process, and also ensure the comfort of lane-changing process. This method can plan the feasible lane-changing trajectory in real time.Overall, 12 unknown coefficients need to determine in the above function. Considering the boundary conditions of lane-changing process, that is, the initial state of lane-changing corresponds to the current state of the vehicle, and the final state is the same as the desired state of the vehicle, as shown in in Eqs. (9.85) and (9.86)

(continued)

Example 9.9 (continued)

$$\begin{cases} x(0) = x_0, & \dot{x}(0) = v_{x,0}, & \ddot{x}(0) = a_{x,0} \\ y(0) = y_0, & \dot{y}(0) = v_{y,0}, & \ddot{y}(0) = a_{y,0} \end{cases} \tag{9.85}$$

$$\begin{cases} x(t_f) = x_f, & \dot{x}(t_f) = v_{x,f} & \ddot{x}(t_f) = a_{x,f} \\ y(t_f) = y_f, & \dot{y}(t_f) = v_{y,f} & \ddot{y}(t_f) = a_{y,f} \end{cases} \tag{9.86}$$

where x_0 and y_0 are the longitudinal and lateral coordinates of the vehicle in the initial state, respectively; $v_{x,0}$ and $v_{y,0}$ are the longitudinal and lateral speeds of the vehicle in the initial state, respectively; $a_{x,0}$ and $a_{y,0}$ are the longitudinal and lateral accelerations of the vehicle in the initial state, respectively; x_f and y_f are the longitudinal and lateral coordinates of the vehicle in the final state, respectively; $v_{x,f}$ and $v_{y,f}$ are the longitudinal and lateral speeds of the vehicle in the final state, respectively; $a_{x,f}$ and $a_{y,f}$ are the longitudinal and lateral accelerations of the vehicle in the final state, respectively; t_f is the time of lane-changing. The second class of the trajectory model of lane-changing: It directly determines the general form of lane-changing controller output. And we mainly determine the variation law of the lateral acceleration in the lane-changing process, thereby obtaining the variation law of the lateral velocity and lateral displacement of the lane-changing process. The specific form of lane-changing trajectory is determined according to the physical constraints and the geometric constraint conditions of collision avoidance. Based on this specific form, driving controller output was derived. The common variation laws of lateral acceleration include sine functions, positive and negative trapezoids, and so on. Among them, the lateral acceleration as a sine function, which can be expressed by a general formula. Lane-changing trajectory can be determined only by the lane-changing time and the maximum lateral acceleration, which is completely in accord with the variation laws of lateral acceleration of most vehicles in the lane-changing. Shladover et al.'s general model of lateral acceleration variation is shown in Eq. (9.87), which can accurately describe most simple lane-changing processes.

$$a_{SV}(t) = \frac{2\pi H}{T_{LC}^2} \sin\left(\frac{2\pi}{T_{LC}}\left(t - T_{SV,delay}\right)\right) \quad T_{SV,delay} \le t$$

$$\le T_{SV,delay} + T_{LC} \tag{9.87}$$

where $T_{SV, delay}$ is the sum of time delays of the target vehicle before lane-changing, H is the total lateral displacement of lane-changing of the target vehicle, which is equal to the lane width, and its value is set as 3.75 m. It is the width of the standard lane. We carry out two integrations for Eq. (9.87), then

(continued)

Example 9.9 (continued)

the expressions of the lateral velocity and displacement of the target vehicle in the process of lane-changing can be obtained, as shown in Eqs. (9.88) and (9.89)

$$v_{SV}(t) = \frac{H}{T_{LC}} \left(1 - \cos\left(\frac{2\pi}{T_{LC}}\left(t - T_{SV,delay}\right)\right)\right) \quad T_{SV,delay} \leq t$$
$$\leq T_{SV,delay} + T_{LC} \tag{9.88}$$

$$y_{SV}(t) = \frac{H}{T_{LC}}\left(t - T_{SV,delay}\right)$$
$$- \frac{H}{2\pi} \sin\left(\frac{2\pi}{T_{LC}}\left(t - T_{SV,delay}\right)\right) \quad T_{SV,delay}$$
$$\leq t \leq T_{SV,delay} + T_{LC} \tag{9.89}$$

2. Vehicle movement control in the lane-changing process. Vehicle movement control is the last step of the whole process of lane-changing. The target vehicle must drive along the planned lane-changing trajectory to complete the expected lane-changing behavior. Control input is the velocity, acceleration, and transverse swing angular velocity, and output is torque and steering wheel angle. Control target is the vehicle driving trace and the planned lane-changing trace coincide.

   In the design process of trajectory tracking controller, the target vehicle is regarded as a rigid object with the same size. The difference between the actual trajectory and the reference trajectory is represented by the longitudinal and lateral coordinates of the vehicle and its heading angle. The control goal of the controller is to calculate the control input to minimize the error.

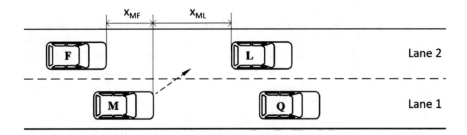

Fig. 9.23 Diagram of the lane-changing of the M vehicle

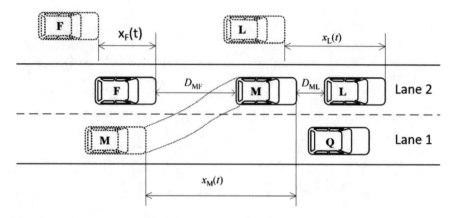

Fig. 9.24 Displacement changes during the lane-changing process

9.4.3 Traffic Control Method of Intersections Under the Vehicle-to-Infrastructure Cooperation Environment

9.4.3.1 Classification of Main Methods and Basic Principles of Cooperative Control at Intersections

Scholars from home and abroad have done a lot of researches for cooperative traffic control algorithm at unsignalled intersections. The main methods is divided into two categories: One is the centralized control method based on vehicle-to-infrastructure communication, and the other is the distributed control method based on vehicle-to-vehicle communication.

1. Centralized control method. Centralized control method is the control method based on the vehicle-to-infrastructure communication. This method needs to be placed a central controller with the communication system at an intersection. The central controller can response to traffic requirements of all entering vehicles at intersection and control efficient movement of the vehicles through the direct communication and control on the vehicles. This control method is equivalent to set a multiphase intelligent traffic light at the intersection. The basic principle of the centralized control method is shown in Fig. 9.25. All vehicles entering the intersection will send information such as position, speed, direction, and steering to the centralized controller on the side of the road. The centralized controller optimizes the traveling sequence of all vehicles and estimates the motion trajectory of the vehicles, and finally feeds back to all vehicles.
2. Distributed control method. The distributed control method does not need to set a controller on the roadside. Instead, all vehicles passing the intersection enter the intersection, risk analysis is carried out based on the vehicle-to-vehicle communication, and all vehicles yield to them according to the same rules as shown in Fig. 9.26. The distributed control methods mainly focus on vehicle behavior, the

Fig. 9.25 The basic
principle of the centralized
control method

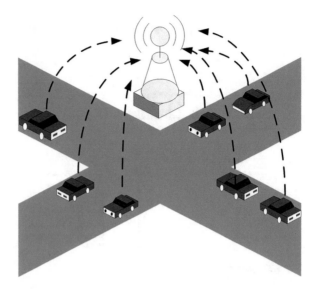

Fig. 9.26 The basic
principle of the distributed
control method

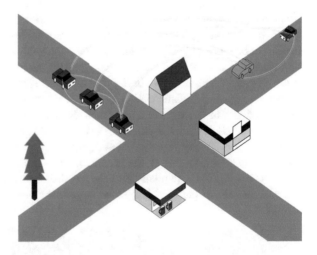

gap acceptable model, the control algorithm based on dynamic game theory, and collision avoidance decision model based on preemptive level and so on.

9.4.3.2   Distributed Traffic Control Model at Intersections

Gap acceptable model

The main idea of the gap acceptable model is that the vehicle will accept gap and pass through the traffic flow in the main lane, and then pass the intersection when the

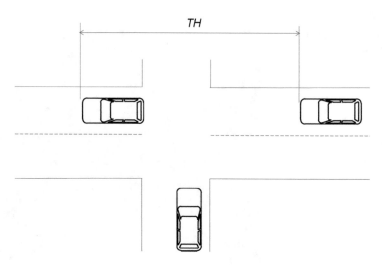

Fig. 9.27 Gap acceptable model

vehicle's gap in the secondary lane is larger than the prescribed critical gap. Otherwise, the vehicles will continue to wait and look for a more suitable gap. The gap acceptable model has an undetermined parameter as the prescribed critical gap, which can be represented by distance or time. The vehicle gap of the main lane must be greater than or equal to the critical gap, then the vehicles can enter the intersection in the secondary lane as shown in Fig. 9.27. Gap acceptable model is mainly used for cooperative control at intersection of primary and secondary lanes. The size of acceptance gap of the vehicle depends on the vehicle type and intersection width. Different vehicle types have different the acceptable gap through the same width of the intersection. The large vehicle's acceptable gap is higher than that of the small vehicle. The determination of the acceptable gap is also related to the traffic flow speed. According to the regulations of the Swedish Traffic Conflict Organization, the value of the critical gap should also be large if the velocity is high, otherwise, the reverse.

Example 9.10 According to the statistical analysis results of traffic flow, the time headway follows the distribution of negative exponential, that is, P-$(t < h) = 1 - e^{-qt}$ is the probability that the time headway is greater than t, q is the traffic flow. We assume that the traffic flow of the main lane is 500 vehicles/h. One vehicle pass the secondary lane that the time headway needs to 5 s, and that of two vehicles is 10 s, and so on, how many vehicles can pass in the secondary lane within one hour?

The number of vehicles that can pass through each time headway as

(continued)

Example 9.10 (continued)

$$N = \sum_{k=1}^{n} k \cdot (P(t \geq 5k) - P(t \geq 5k + 5))$$
$$= 1 \cdot (P(t \geq 5) - P(t \geq 10))$$
$$+ 2 \cdot (P(t \geq 10) - P(t \geq 15))$$
$$+ 3 \cdot (P(t \geq 15) - P(t \geq 20))$$
$$+ \cdots$$
$$= 0.7554$$

Therefore, the main lane can pass 500 vehicles, and the number of spacing headway have 499. Then the secondary lane can pass 377 vehicles, that is, $499 \times 0.7554 \approx 377$.

Collision avoidance decision model based on preemptive level

Collision avoidance decision model based on preemptive level determines the priority of vehicles of an intersection crossing according to the state of the predicted collision vehicles at the potential collision point. In order to implement this model, this algorithm provides a hypothetical description of the whole process of the driver from the discovery of the conflict to the avoidance of collision and finally to the disappearance of the conflict.

Hypothesis 1: In the running process of a vehicle, the driver will often estimate the future trajectory of his own vehicle and the potential collision vehicle to judge whether there is a conflict between the two vehicles.

Hypothesis 2: The driver will make a collision avoidance decision according to the actual situation of the conflict when the conflict exists.

Hypothesis 3: In the process of making the collision avoidance decision, the driver independently evaluates the dominant position of preemptive level for both sides in the conflict and makes a decision based on the evaluation results. The preemptive level is defined as the ratio between the total length of the vehicle and the length of the vehicle of passing through the potential collision point.

After obtaining the preemptive level of the vehicle, the rule of decision-making can be expressed as who thinks he is led all the others, it is man who will seize the right of way. The algorithm also sets up a safety critical preemptive level for each vehicle to ensure the vehicle safety. If the vehicle's preemptive level is greater than the safety critical preemptive level, the vehicle will take the lane-occupying behavior; otherwise, the vehicle will yield the lane. The lane-occupying behavior here refers to that the main vehicle expresses its possession of the right of way by accelerating or maintaining the original speed. While the behavior of yielding the lane refers to that the main vehicle expresses its acceptance of the occupied road by decelerating or other modest behavior.

Example 9.11 In Fig. 9.28, it is assumed that the length of vehicle A and vehicle B are both 4 m. They exist trajectory conflict through the communication technology. The speed of the vehicle A is 16 m/s, and there is a distance of 24 m from the conflict point. The speed of the vehicle B is 20 m/s, and there is a distance of 40 m from the conflict point. The vehicle A reaches the conflict point to cost 1.5 s, and the vehicle B needs to 2 s. Therefore, the vehicle B arrives at the potential collision point later. When the vehicle B reaches the conflict point, the vehicle A has passed 8 m at the conflict point. Therefore, the preemptive level of the vehicle A is $8/4 = 2$, and the preemptive level of the vehicle B is -2. Therefore, the vehicle A can accelerate or maintain the original speed to pass the conflict point, while the vehicle B needs to decelerate to ensure that the distance between the vehicle A and the vehicle B is greater than the safe distance.

Control algorithm based on dynamic game theory

Control algorithm based on dynamic game theory quantifies the vehicle risk to make decisions. It considers the process of vehicle's decision behavior on a smaller spatiotemporal scale. Then it is more reasonable for the decision-making of the autonomous vehicles to study the traffic behavior model of intersection by using repeated game theory in the vehicle-to-infrastructure cooperation environment.

According to the definition of dynamic repeated game theory, $\{c_1, c_2\}$ is set as a set of game participants.

In the process of the vehicle passing the intersection, the factors affecting their behavior decisions for c_1 and c_2 can be summarized into two aspects: safety factor and speed factor. Among them, the safety factor mainly refers to the factor that would lead to the collision or increase the seriousness of the conflict between two vehicles. The speed factor refers to that the vehicles are expected to cross the intersection at a faster speed to avoid slowing down or waiting. The control algorithm based on dynamic game theory can obtain the safety factor and speed factor. According to the changes of the safety and speed factors, the utility functions of c_1 and c_2 can be obtained. It can be determined that the process of vehicle traveling is actually a process of finite repeated games with complete information between c_1 and c_2 through the analysis of the utility function. c_1 and c_2 act at the same time in each stage, and c_1 and c_2 can calculate the action strategy taken by the other party at the end of each stage or the beginning of the next stage. This algorithm can well describe the driving behavior at the intersection.

The utility of risk factors can be quantified by the time difference between two vehicles passing the conflict point for the game behaviors of two vehicles passing the same intersection, that is, the smaller the time difference is, the more dangerous it is. In addition, the utility of each vehicle's speed factor can be quantified by the length of time from its current position to fully pass the

(continued)

Example 9.11 (continued)

intersection for each vehicle, that is, the shorter the time is, the higher the utility is. Then the combinational functions of risk factors and utility factors can be used to quantify the total utility.

It is necessary to quantify the influence of vehicle risk on decision-making when the game decision making model in the intersection is established. Figure 9.29 shows the schematic diagram of straight traffic on the left and straight traffic on the right at an unsignalled intersection. It is assumed that both east–west and north–south directions of the intersection is the lane 2. C_R and C_L are the straight vehicles of the right side and the left side, respectively, and C is the intersection point of the two vehicles. v_R^i and v_L^i are speeds of the two vehicles at the ith decision time point, respectively, and their accelerations are denoted as a_R^i and a_L^i, respectively. L_R^i and L_L^i are respectively distances that two vehicles arrive at the conflict point. Therefore, the vehicle j takes the ith decision-making time point to pass the conflict point, which needs to T_j^i,

that is, $T_j^i = \sqrt{\left(v_j^i/a_j^i\right)^2 + 2L_j^i/a_j^i} - v_j^i/a_j^i, i = 1, 2, 3, \ldots, N, j = R, L.$ Then

the efficiency benefit function is defined as $f_j^i = T_j^{i+1} + \left(T_j^{i+1} - T_j^i\right)/2.$ Two vehicles take the velocity and acceleration of the ith decision-making time point to pass the conflict point, which will generate the time difference, that is,

$\Delta T_j^i = \left| \left(\sqrt{(v_R^i/a_R^i)^2 + 2L_R^i/a_R^i} - v_R^i/a_R^i \right) - \left(\sqrt{(v_L^i/a_L^i)^2 + 2L_L^i/a_L^i} - v_L^i/a_L^i \right) \right|.$ The

safety benefit function is defined as $g_j^i = \Delta T_j^{i+1} + \left(\Delta T_j^{i+1} - \Delta T_j^i\right)/2$. The benefit function of C_R and C_L is expressed as

$$F_j^i = g_j^i/\left(f_j^i + g_j^i\right) \tag{9.90}$$

In order to analyze the game behavior of two vehicles at an unsignalized intersection, it is necessary to solve the game model at each decision time point in the crossing process. The specific solving steps of the game model in the intersection are shown as follows.

1. According to the motion information of the vehicle and the quantified model of risk value of the vehicle, the benefit function values of C_R and C_L are obtained under different combinations of behavior strategies ({accelerate, accelerate}, {accelerate, decelerate}, {decelerate, accelerate}, {decelerate, decelerate}) at the beginning of the first decision time step, respectively.
2. The Nash equilibrium point of the game between C_R and C_L on the first decision time step was found out based on the value of benefit function, and

(continued)

Example 9.11 (continued)
  the behavioral strategies of the two vehicles were obtained. Then, the
  motion information of the two vehicles at the beginning of the second
  decision time step was calculated according to the obtained strategies.
3. The benefit function values of C_R and C_L are respectively obtained at the
  beginning of the second decision time step according to the motion information of two vehicles and the quantified model of risk value. Repeat Step
  (2), the decision-making behavior of the vehicle can be obtained in the
  second decision time step, and the motion information of two vehicles are
  also obtained at the beginning of the third decision time step.
4. As an analogy, the decision behaviors of two vehicles are obtained in the
  1st to the Nth decision time steps, respectively.

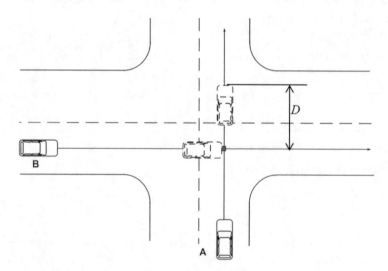

Fig. 9.28 Collision avoidance decision model based on preemptive level

Fig. 9.29 Diagram of two
vehicles crossing

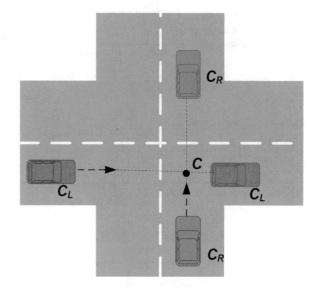

Example 9.12 In Fig. 9.29, it is assumed that the accelerations of both C_R and
C_L are 2 m/s^2. At the beginning of the first decision time step, the motion states
of the vehicles C_R and C_L are shown in the game theory as: $L_R = 10$ m,
$L_L = 13$ m, $v_R = 2$ m/s, $v_L = 2.5$ m/s. It is assumed that the coefficients of
speed and risk expectation are both 0.5, and the decision time step is 0.2 s. The
decision table of the first time step is shown in Table 9.9.

There is a Nash equilibrium in a single game in the first decision time step
as shown in Table 9.9. The behavioral strategies of C_R and C_L are acceleration
and deceleration, and the benefit function values of the two vehicles are (0.25,
0.20), respectively. According to the behavioral strategies of C_R and C_L, the
motion information of the two vehicles at the beginning of the second decision
time step can be obtained as $L_R = 9.56$ m, $L_L = 12.54$ m, $v_R = 2.4$ m/s and
$v_L = 2.1$ m/s, respectively. Similarly, the Nash equilibrium of the single game
of two vehicles in the second decision time step can be obtained according to
the above method as: the vehicle C_R decelerates, and the vehicle C_L acceler-
ates. And the benefit function values of the two vehicles are (0.35, 0.26),
respectively. Likewise, the behavioral decisions of two vehicles C_R and C_L can
be obtained in subsequent decision time steps when C_R and C_L achieved Nash
equilibrium, respectively, as shown in Table 9.10.

The speed changes of the two vehicles in the decision-making are shown in
Fig. 9.30.

Table 9.9 Nash equilibrium in the process of single game of the intersection

		C_L	
		Acceleration	Deceleration
C_R	Acceleration	(0.11, 0.10)	(0.25, 0.20)
	Deceleration	(−0.03, −0.03)	(0.09, 0.08)

Table 9.10 Decision table

	1	2	3	4	5	6	7	8	9	10	11
C_R	Acc.	Acc.	Acc.	Acc.	Acc.	Acc.	Acc.	Acc.	Acc.	Acc.	Acc.
C_L	Dec.	Dec.	Dec.	Dec.	Dec.	Dec.	Acc.	Acc.	Acc.	Acc.	Acc.

Acc. acceleration, *Dec.* deceleration

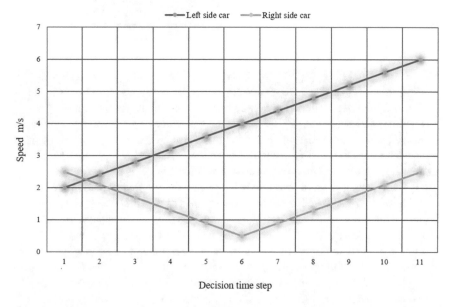

Fig. 9.30 Variations of velocity in decision-making

The distances between the two vehicles the intersection in the decision-making are shown in Fig. 9.31.

Cooperative collision avoidance method based on the rule-base in the intersection

A rule-base of vehicle traffic based on non-signalized intersections is adopted to achieve collision avoidance of vehicles at intersections through the vehicle-to-

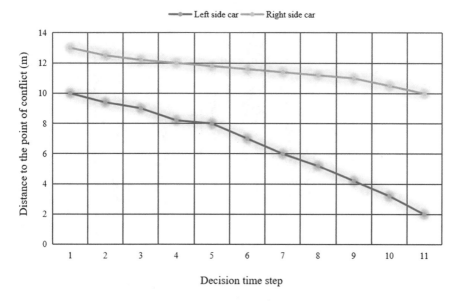

Fig. 9.31 Change of distance between two vehicles to the intersection in the decision

vehicle information interaction, as shown in Fig. 9.32. The whole process of collision avoidance technology includes four steps as vehicle-to-vehicle conflict detection, quantification of conflict severity, determination of priority, and resolving the conflict by avoiding and giving way. Each vehicle continuously sends out the motion state information of its own vehicle and receives the motion state information of surrounding vehicles in the process of driving. Each vehicle carries out conflict detection according to the motion state of its own vehicle and the motion state of other vehicles to judge whether there is conflict with other vehicles on its driving route. In the case of conflict, on the one hand, the severity of conflict is judged to provide a basis for the early warning and resolution of conflict. On the other hand, if the vehicle needs to avoid collision according to the rule-base of vehicle traffic, then the deceleration of the vehicle is calculated to achieve deceleration or stop to avoid collision.

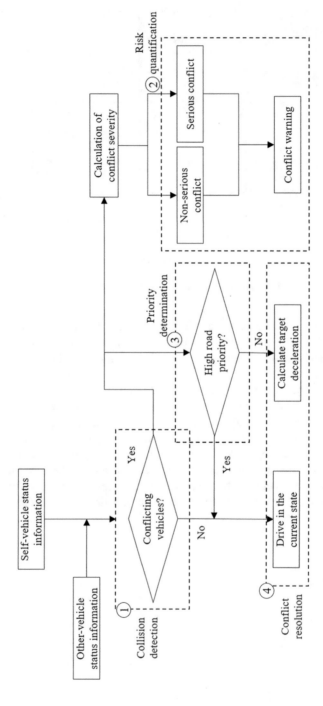

Fig. 9.32 Flow chart of conflict identification and collision avoidance technology in the intersection

Example 9.13 Two-way single-lane unsignalled intersection as the research example is shown in Fig. 9.33. A set of unsignalled intersection traffic rule-base with clear priority is formulated to determine the traffic order of conflicting vehicles, which is very necessary to improve the traffic efficiency and traffic safety at unsignalled intersections.

As for the example of the two-way single-lane unsignalled intersection, set the SV of the entrance 1 as the main research object, its traveling trace might have three scenarios as turn left (1A), go straight (1B), and turn right (1C). Traveling trace of POV can select any one of them (iA, iB, iC (i = 1, 2, 3, 4), and 12 of them in all). Therefore, the relative motion relationship between SV and POV has 36, that is, 3 × 12 = 36. The cases of 36 can be divided into four categories: no conflict, traffic diversion conflict, confluence conflict, and cross conflict.

Table 9.11 reflects the types of conflicts, where 0 represents no conflict, F represents traffic diversion conflict, H represents confluence conflict, and J represents cross conflict. Conflicts exist in 20 cases and no conflict exists in 16 cases. Table 9.12 records the priority status of the 20 conflicting cases.

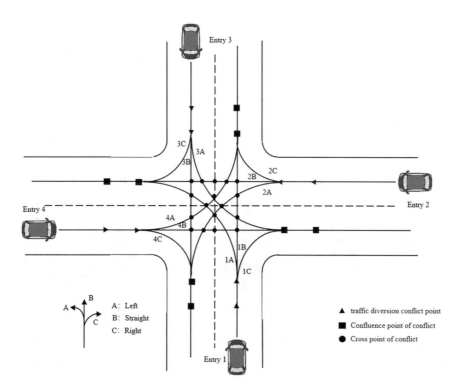

Fig. 9.33 Summary of two-way single-lane intersection conflict forms

Table 9.11 Classification of conflict points at no-signal intersections

	1A	1B	1C	2A	2B	2C	3A	3B	3C	4A	4B	4C
1A		F	F	J	H	0	0	J	H	J	J	0
1B	F		F	J	J	H	J	0	0	H	J	0
1C	F	F		0	0	0	H	0	0	0	H	0

Table 9.12 Rule-base of vehicle-to-vehicle conflict at no-signal intersections

Conflict matching	Conflict description	Conflict category	Priority	Go ahead
(1A,1B)	(Turn left, go straight in the same direction)	F	Not sure	Preceding vehicle
(1A,1C)	(Turn left, turn right in the same direction)	F	Not sure	Preceding vehicle
(1A,2A)	(Turn left, turn left on the right side)	J	2A > 1A	2A
(1A,2B)	(Turn left, go straight on the right side)	H	2B > 1A	2B
(1A,3B)	(Turn left, go straight in the opposite direction)	J	3B > 1A	3B
(1A,3C)	(Turn left, turn right in the opposite direction)	H	1A > 3C	1A
(1A,4A)	(Turn left, turn left on the left side)	J	1A>4A	1A
(1A,4B)	(Turn left, go straight on the left side)	J	4B > 1A	4B
(1B,1A)	(Go straight, turn left in the same direction)	F	Not sure	Preceding vehicle
(1B,1C)	(Go straight, turn right in the same direction)	F	Not sure	Preceding vehicle
(1B,2A)	(Go straight, turn left on the right side)	J	1B > 2A	1B
(1B,2B)	(Go straight, go straight on the right side)	J	2B > 1B	2B
(1B,2C)	(Go straight, turn right on the right side)	H	1B > 2C	1B
(1B,3A)	(Go straight, turn left in the opposite direction)	J	1B > 3A	1B
(1B,4A)	(Go straight, turn left on the left side)	H	1B > 4A	1B
(1B,4B)	(Go straight, go straight on the left side)	J	1B > 4B	1B
(1C,1A)	(Turn right, turn left in the same direction)	F	Not sure	Preceding vehicle
(1C,1B)	(Turn right, go straight in the same direction)	F	Not sure	Preceding vehicle
(1C,3A)	(Turn right, turn left in the opposite direction)	H	3A > 1C	3A
(1C,4B)	(Turn right, go straight on the left side)	H	4B > 1C	4B

Fig. 9.34 Traffic order at no signal intersection

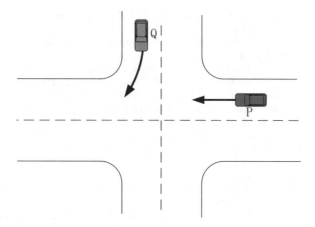

The traffic order of conflicting vehicles can be determined based on the traffic rule-base of no-signal intersections of Table 9.12.

A conflict scene of two vehicles as an example is shown in Fig. 9.34. It is assumed that the vehicles P and Q are going to pass the intersection, where the vehicle P goes straight, and the vehicle Q turns right. So there is a conflict between the two vehicles. The traffic order is determined from the point of view of the two vehicles in this case.

From the point of view of the vehicle P, it goes straight, and the other vehicle turns right on the right side. It can be seen that this situation (Go straight, turn right on the right side) is satisfied from Table 9.12. In this case, the vehicle P has the priority to use the road, that is, the vehicle P goes ahead, and the vehicle Q gives way. From the point of view of the vehicle Q, it turns right, and the other vehicle goes straight on the left side. It can be seen that this situation (Turn right, Go straight on the left side) is satisfied from Table 9.12. In this case, the vehicle Q has a low priority to use the road, so the vehicle Q gives way and the vehicle P goes ahead. In summary, it can be seen that the traffic orders of conflicting vehicles are also uniquely determined from the view of two vehicles.

Bibliography

1. Wang J, Wang X (2017) Study on the system framework and key technology of Intelligent connected vehicle. J Chang'an Univ (Soc Sci Ed) 19(6):18–25
2. Wang J, Wu C, Li X (2011) Research on architecture and key technologies of Internet of vehicles. Microcomputer Inform 27(4):156–158
3. Cheng G, Guo D (2011) Current situation and development research of Internet of Vehicle. Mobile Commun 35(17):23–26
4. Xu Q (2014) Research on traffic information collection and accident identification based on internet of vehicles .Beijing Jiaotong University
5. Van Der Voort M, Dougherty M, Watson S (1996) Combining Kohonen maps with ARIMA time series models to forecast traffic flow. Transp Res C Emerg Technol 4(5):307–318

6. Okutani I, Stephanedes YJ (1984) Dynamic prediction of traffic volume through Kalman filtering theory. Transp Res B Methodol 18(1):1–11
7. Wu Z (2018) Research on road network state prediction based on deep learning. Beihang University
8. Xu Z (2012) Research and implementation of short-term traffic flow forecasting based on support vector machine. South China University of Technology
9. Cong Y, Wang J, Li X (2016) Traffic flow forecasting by a least squares support vector machine with a fruit fly optimization algorithm. Proc Eng 137:59–68
10. Dixon MF, Polson NG, Sokolov VO (2019) Deep learning for spatio-temporal modeling: dynamic traffic flows and high frequency trading. Appl Stoch Model Bus Ind 35:788–807
11. Koesdwiady A, Soua R, Karray F (2016) Improving traffic flow prediction with weather information in connected cars: a deep learning approach. IEEE Trans Veh Technol 65(12): 9508–9517
12. Polson NG, Sokolov VO (2017) Deep learning for short-term traffic flow prediction. Transp Res C Emerg Technol 79:1–17
13. Tan X, Yin K, Li M, Guo W, Wang L, Huang Y (2017) Signal timing estimation using mobile navigation data. J Transp Syst Eng Inf Technol 17(2):60–67
14. Zhou B (2016) Urban traffic signal control based on connected vehicles simulation platform. Zhejiang University
15. Nie L (2017) Research on intelligent traffic signal control methods in VANET. Wuhan University
16. Tong Y, Zhao L, Li L et al (2015) Stochastic programming model for oversaturated intersection signal timing. Transp Rese C Emerg Technol 58:474–486
17. Li L, Su X, Wang Y et al (2015) Robust causal dependence mining in big data network and its application to traffic flow predictions. Transp Res C Emerg Technol 58:292–307
18. Li L, Su X, Zhang Y et al (2015) Trend modeling for traffic time series analysis: an integrated study. IEEE Trans Intell Transp Syst 16(6):3430–3439
19. Hao P, Ban X, Bennett KP et al (2012) Signal timing estimation using sample intersection travel times. IEEE Trans Intell Transp Syst 13(2):792–804
20. (Jeff) Ban X, Herring R, Hao P, Bayen AM (2009) Delay pattern estimation for signalized intersections using sampled travel times. Transp Res Rec J Transp Res Board 2130(2130): 109–119
21. Kerper M, Wewetzer C, Mauve M (2012) Analyzing vehicle traces to find and exploit correlated traffic lights for efficient driving. Educ Cult 28(3):310–315
22. Fayazi SA, Vahidi A, Mahler G et al (2015) Traffic signal phase and timing estimation from low-frequency transit bus data. IEEE Trans Intell Transp Syst 16(1):19–28
23. Wang Q, Tan X, Zhang S (2006) Signal timing optimization of urban single-point intersections. J Traffic Transp Eng 2:60–64
24. Lv B, Niu H (2010) Signal timing optimization of single-point intersections under random conditions. J Traffic Transp Eng 10(6):116–120
25. Han C et al (2004) A real-time short-term traffic flow adaptive forecasting method based on ARIMA model. J Syst Simul 16(7):1530–1532
26. Yao Y, Cao F (2006) Short-term traffic flow forecasting based on ARIMA. Technol Econ Areas Commun (TEAC) 3:105–107
27. Kalman RE (1960) A new approach to linear filtering and prediction problems. J Basic Eng Trans 82:35–45
28. Peng D (2009) Basic principle and application of Kalman Filter. Software Guide 8(11):32–34
29. Dong J, Hu S (1997) Research progress and development of chaotic neural networks. Inf Control 5:360–368
30. Abdi J et al (2012) Forecasting of short-term traffic-flow based on improved neuro fuzzy models via emotional temporal difference learning algorithm. Eng Appl Artif Intell 25(5):1022–1042
31. Leng Z et al (2013) Short-term traffic flow forecasting model of optimized BP neural network based on genetic algorithm. In: Proceedings of the control conference

32. Chen X et al (2008) Short -time traffic flow prediction based on BP Neural Network. Technol Highway Transp 3:115–117
33. Ma J et al (2009) Research of urban traffic flow forecasting based on neural network. Acta Electron Sin 37(5):1092–1094
34. Guo J et al (2014) Adaptive Kalman filter approach for stochastic short-term traffic flow rate prediction and uncertainty quantification. Transp Res C 43:50–64
35. Zhang J, Wang X (2010) Traffic flow prediction method based on non-linear hybrid model. Comput Eng 36(5):202–204
36. Min W, Wynter L (2011) Real-time road traffic prediction with spatio-temporal correlations. Transp Res C 19(4):606–616
37. Vlahogianni EI et al (2014) Short-term traffic forecasting: where we are and where we're going. Transp Res C Emerg Technol 43:3–19
38. Zhang LL et al (2014) Research on Short-term Traffic Flow Forecasting for Junction of Isomerism Road Network based on Dynamic Correlation. Proc Soc Behav Sci 138:446–451
39. Zou Y et al (2014) A space–time diurnal method for short-term freeway travel time prediction. Transp Res C Emerg Technol 43:33–49
40. Zeng R (2012) Prediction model of short-term traffic volume based on spatially and temporally correlated state of multi-sections of section. Chang'an University
41. Gao W et al (2011) Short-term traffic flow forecasting based on spatiotemporal characteristics of traffic flow and RBF Neural Network. J Transp Inform Saf 29(1):16–19
42. Malandraki C (1989) Time dependent vehicle routing problems [microform]: formulations, solution algorithms and computational experiments. Northwestern Univ 326(6):14–16
43. Malandraki C, Daskin MS (1992) Time dependent vehicle routing problems: formulations, properties and heuristic algorithms. Transp Sci 26(3):185–200
44. Ichoua S et al (2003) Vehicle dispatching with time-dependent travel times. Eur J Oper Res 144(2):379–396
45. Chabini I, Lan S (2002) Adaptations of the A* algorithm for the computation of fastest paths in deterministic discrete-time dynamic networks. Intell Transp Syst IEEE Trans Intell Transp Syst 3(1):60–74
46. Zhong M (2006) Discussion of designing cost function of A* algorithm. J Wuhan Eng Inst 18(2):31–33
47. Zheng Z (2016) Research on the cause, connotation and business model of sharing economy. Mod Econ Res 411(3):32–36
48. Tang T, Wu X (2015) Shared economy: a subversive economic model under the background of "Internet +". Sci Dev 12:78–84
49. Lei S, Chi C (2010) Sharing vehicle—the transformation of automobile consumption pattern. Shanghai Auto 12:1–3
50. Wang H (2016) Design and verification of robust headway control algorithm for ACC vehicles. J Shanxi Datong Univ (Nat Sci) 1:12–13
51. Ning Y, Zhao J, Zhang B et al (2017) Simulation based on intelligent control of vehicle following distance. Comput Simul 34(9):146–150
52. Gao F, Xiao L, Wang J (2010) Lag delay robusting control of platoon automated vehicles. J Beijing Univ Aeronaut Astronaut 36(10):1153–1157
53. Xu L, Huang Y (2016) Platoon stability based on different safety space policing of Adaptive Cruise Control. Sci Technol Eng 29:132–140
54. Darbha S, Rajagopal KR (1998) Intelligent cruise control systems and traffic flow stability. Transp Res C Emerg Technol 7(6):329–352
55. Wang L (2014) The characters research of lane-changing behavior in urban section. Beijing Jiaotong University
56. Wang M (2016) Analysis of the influence vehicle lane-changing behavior on traffic flow. Jilin University
57. Chen W (2015) Dynamic characteristics of lane-changing behavior and effect on the vehicular flow. Qingdao University of Technology

58. Peng J, Fu R, Shi L et al (2011) Research of driver's lane change decision-making mechanism. J Wuhan Univ Technol 33(12):46–50
59. Cao S (2009) Study on the model and impact of lane-changing vehicles for city road. Huazhong University of Science and Technology
60. Yu H, Tseng HE, Langari R (2018) A human-like game theory-based controller for automatic lane changing. Transp Res C Emerg Technol 88:140–158
61. Zhang R, You F, Chu X et al (2018) Lane change merging control method for unmanned vehicle under V2V cooperative environment. China J Highway Transp 31(4):180–191
62. Li X, Qu S, Xia Y (2014) Cooperative lane-changing rules on multilane under condition of cooperative vehicle and infrastructure system. China J Highway Transp 27(8):97–104
63. Luo Y, Xiang Y, Cao K et al (2016) A dynamic automated lane change maneuver based on vehicle-to-vehicle communication. Transp Res C 62:87–102
64. Shladover SE, Desoer CA, Hedrick JK et al (2002) Automated vehicle control developments in the PATH program. IEEE Trans Veh Technol 40(1):114–130
65. Treiber M, Kesting A (2013) Traffic flow dynamics. Springer, Berlin
66. Liu G, Hou D, Li K et al (2004) Safety alarm algorithm for vehicle active collision avoidance system. J Tsinghua Univ (Sci Technol) 44(5):697–700
67. Liu Z, Liang W (2003) Advances in urban traffic signal control. J Highway Transp Res Dev 20(6):121–125
68. Liu M, Lu G, Wang Y et al (2014) Preempt or yield? An analysis of driver's dynamic decision making at unsignalized intersections. Saf Sci 65:36–44
69. Lu G, Li L, Wang Y, et al (2014) A rule based control algorithm of connected vehicles in uncontrolled intersection. In: International conference on intelligent transportation systems. IEEE, pp 115–120
70. Levin MW (2017) Congestion-aware system optimal route choice for shared autonomous vehicles. Transp Res C Emerg Technol 82:229–247
71. Ma J, Li X, Zhou F et al (2017) Designing optimal autonomous vehicle sharing and reservation systems: a linear programming approach. Transp Res C Emerg Technol 84:124–141
72. Zhang C, Zhang Y (2017) Study on vehicle route problem under dynamic road system. J Transp Eng Inform 15(2):112–118
73. Fagant DJ, Kockelman KM (2014) The travel and environmental implications of shared autonomous vehicles, using agent-based model scenarios. Transp Res C Emerg Technol 40: 1–13
74. Chen TD, Kockelman KM, Hanna JP (2016) Operations of a shared, autonomous, electric vehicle fleet: Implications of vehicle & charging infrastructure decisions. Transp Res A Policy Pract 94:243–254
75. Sheng Q, Zheng P, Sun J (2018) Research on vehicle routing problem with time windows under dynamic road network. Logist Technol 37(10):36–47
76. Jiang R, Long C, Li M (2019) Research on safety big data collection in vehicle networking. Electron Test 15:84–86
77. Song L, Hu M, Zhu G (2017) Research on the economic model of sharing in the field of transportation. Logist Eng Manag 39(9):123–124
78. Li Y (2015) Research on time-dependent popular route recommendation. Northeastern University
79. Wang D, Wang YP, Yu GZ (2017) Research on multi-vehicle cooperative lane changing based on Internet of vehicles. Beihang University
80. Liu MM, Wang YP, Lu GQ (2015) Research on decision model of straight driving at unsignalized intersection based on risk perception. Beihang University
81. Wang PW, Huang L, Wang YP et al (2013) Cooperative adaptive cruise control algorithm considering communication delay. J Highway Transp Res Dev 30(S1):150–154

Chapter 10
Typical Intelligent Transportation Applications

CHU Duanfeng and CAO Yongxing

Intelligent transportation system (ITS) is a new development direction of modern transportation management and construction. It is a hot and frontier field of transportation engineering. The development and application of intelligent transportation technology has greatly improved the traditional traffic organization and management mode. It provides more and better solutions to alleviate traffic congestion, improve traffic safety, and reduce traffic pollution. ITS is an effective means of coordinating "people-car-road (environment)," gradually achieving the goal of safety, efficiency, energy conservation, environmental protection, and comprehensive traffic. In recent years, with the country's strong advocacy and the joint efforts of intelligent transportation practitioners, a large number of intelligent transportation springs up. The system is applied in the practice of traffic construction, organization, and management. This chapter describes four intelligent intersections with typical characteristics. General application system introduces the background, system composition, and function of each system. In each system, an instance analysis is provided.

10.1 Urban Traffic Status Monitoring and Index Evaluation System

In recent years, urban road congestion and road traffic conditions have become more complex. The total amount of traffic is getting bigger and bigger, and the probability of occurrence of accidents increases, seriously threatening people's lives and property safety. To improve urban road traffic safety level, road traffic safety

CHU Duanfeng (✉) · CAO Yongxing
Intelligent Transportation Systems Research Center, Wuhan University of Technology, Wuhan, Hubei, China
e-mail: chudf@whut.edu.cn

© Tsinghua University Press 2022
W. Yunpeng et al. (eds.), *Intelligent Road Transport Systems*,
https://doi.org/10.1007/978-981-16-5776-4_10

management departments need to master the state of urban traffic safety and monitor it and understand traffic safety risks. Relevant departments can prescribe the right approach, taking targeted measures to improve the level of urban road traffic safety.

10.1.1 System Profile

How to make urban road traffic decision-making management scientific, refined, and informationized is very urgent. How to describe the traffic operation state more precisely instead of the traditional signal light of only red, yellow, and green color. Using accurate numerical value to quantify road traffic congestion and providing scientific basis for government decision-making management becomes the inevitable trend of ITS.

It is necessary to realize scientific quantification of the road traffic operation state by measuring real-time traffic status so as to make government traffic decision management informatization, refinement, and sensitization. Beijing, Shanghai, and other cities have developed a numerical description of the state of real-time road traffic—a road traffic index, i.e., in a specified space and time norm to show the running state of urban road traffic accurately.

By establishing a complete set of basic theory, calculation method, application system, and deliver congestion index based on road traffic index, it can make a quantitative evaluation of the overall or regional operation of urban road traffic, analyze and predict short-, medium-, and long-term traffic development trend, estimate the degree and scope of traffic congestion, and formulate scientific and reasonable traffic management for traffic management departments. All these are important to make decision-making, provide quantitative basis of traffic organization and traffic law enforcement, and provide dynamic traffic travel information service to the public.

All countries around the world attach great importance to traffic safety by strengthening management, developing new technologies and methods to improve road traffic safety. Traffic safety is also a hot research point in traffic circles. New theories and research techniques have been developed. Domestic research on the related index system of urban road traffic safety is mainly focused on traffic safety evaluation index system and evaluation method.

10.1.1.1 History and Current Situation of Traffic Data Acquisition and Preprocessing

Traffic data is the basis of traffic condition monitoring and forecasting. Its timeliness, accuracy, and comprehensiveness will directly determine the effect of traffic condition monitoring and prediction. The following is a brief introduction of researching history and current situation of traffic data acquisition and its preprocessing methods.

In 1976, some scholars first mentioned traffic flow and occupation rate data obtained from expressway induction coil data, in which error data recognition through single parameter threshold method. First, a fixed upper and lower limit threshold for each traffic parameter was set, and values exceeding the corresponding threshold are regarded as error data, and the error data is processed by culling method. In 1990, in order to overcome the problem of unreasonable data omission caused by the single parameter threshold method assuming that various traffic parameters are independent of each other, a combined parameter identification method based on traffic flow theory is proposed by scholars to establish the relationship between traffic density and traffic flow. The template defines the acceptable area of traffic parameters, considers the traffic data outside the area as error data, and adopts the culling method to deal with error data. In 2002, according to traffic flows from induction coils on the Northern Virginia Highway, quantity, speed, and occupation rate data, some scholars have proposed an idea to use historical trend method and exponential smoothing method to repair fault data.

In 2012, Chinese scholars divided the fault data into missing data and error data by obtaining 5 min fixed sampling interval traffic flow for urban main road vehicle detector. If traffic is significantly smaller than a sampling interval in the historical trend value, it is regarded as missing data, and the combined parameter method based on traffic flow theory is used to identify errors data.

By analyzing the development history and present situation of traffic data acquisition and its preprocessing research field, the following five points can be obtained.

1. According to the obtained type of traffic data, the existing traffic data acquisition and its preprocessing methods can be divided into two kinds—local traffic parameter data acquisition and its preprocessing and interval traffic parameter data acquisition. Traffic parameters acquisition and its preprocessing methods can be divided into single location vehicle detector data preprocessing and multi-continuous consideration.
2. The existing methods of data acquisition and preprocessing of traffic parameters are mainly focused on highway, expressway, and ETC which are continuous flow road type. Research aiming at the urban main road isometric cut-off road types is less.
3. Among the identification and repair of the conservation law of traffic flow in existing traffic parameter data acquisition and preprocessing methods, the only two literatures not only have their own defects but also lack data preprocessing of vehicle detection at a single location device.
4. Existing data acquisition and preprocessing methods of interval traffic parameters are mainly focused on research based on GPS data vehicle study on the method of obtaining road travel time data.
5. At present, inductive traffic control system, road toll system, vehicle tracking and positioning system, and other related services systems have been widely used in various types of roads in China and will continue to accumulate a lot of data to study the new traffic data acquisition and preprocessing methods. A data source is

helpful to improve existing traffic condition monitoring under low-cost condition database for measurement and prediction.

10.1.1.2   History and Present Situation of Traffic Condition Monitoring

The early road traffic monitoring system mainly takes emergency as the monitoring object. With the rapid development of social economy and the rapid increase in traffic demand, the morning and evening peak time on roadways is longer, more roads are congested. As a result, the demand for monitoring traffic congestion is increasing, and the content of traffic condition monitoring is extended to include traffic automatic detection of traffic events and automatic detection of traffic congestion. History and the present situation of the traffic event automatic detection research field can be summarized as follows.

1. A number of AID (Automatic Incident Detection) have been developed by relevant scholars since the 1960s, which can be divided into pattern recognition-based algorithms (Goferman, Pattern Recognition, Monica algorithms, etc.), algorithms based on statistical theory (including Bayesian algorithms, HIOCC algorithms, ARIMA algorithms, etc.), algorithms based on mutation theory (represented by McMaster algorithms), and algorithms based on artificial intelligence.
2. Most AID algorithms use special vehicle detectors, such as induction coils, to obtain traffic flow, speed, or percentage. Based on the rate data, the input variables of the algorithm mainly include the space-time comparison of the measured traffic data, the measured traffic data and its comparison of predicted values, etc.
3. Different AID algorithms have their unique advantages and disadvantages, so as to further improve the detection effect of traffic events, some researchers have tried to apply data fusion technology to traffic event detection and made some progress. Hence AID calculation is the research trend in this field.
4. Relevant system data have the advantages of low cost and wide coverage; if they can be used for AID purposes, the algorithm can greatly improve the effect of traffic event state monitoring under lower cost conditions.

Overall, China's current urban road evaluation index system is relatively mature, but not much attention is paid to the measures taken after evaluation. The starting point of traffic safety evaluation index is to evaluate the present situation of traffic safety level. It is estimated that the monitoring index of road traffic safety condition is to prevent and avoid possible adverse traffic conditions in the future. The safety evaluation index starts from each subsystem of the road traffic system and carries on the urban road traffic safety from the inside. The monitoring index is designed for the road traffic management department, so that relevant departments can efficiently and quickly take some measures to eliminate the hidden dangers affecting traffic safety. The monitoring index of urban roads traffic safety state starts from the macro-management level and combines the urban economic level and the road environment condition together while the evaluation index is from the micro-influencing factors of road traffic safety. They complement each other, inseparably.

10.1.2 System Composition

10.1.2.1 Factors Influencing Traffic Safety in Cities

This section deals with urban roads in terms of driver behavior, traffic flow, weather conditions, road conditions, and management organization factors. The influencing factors of traffic safety state are analyzed. Among them, the driver behavior analyzes the driver information receiving and processing. The traffic flow mainly analyzes the driving characteristics of large vehicles and the speed of traffic flow to traffic safety. The weather factor analyzes the influence of temperature, humidity, visibility, and precipitation on the normal running. The vehicle analyzes the impact of plane alignment, longitudinal section, and linear continuity on road traffic accidents. Management factors analyze the influence of traffic safety equipment installation, driver behavior supervision, and management on traffic safety.

1. Analysis of driver behavior impact.

    The main participants in the road traffic system are motor vehicle drivers, pedestrians, and non-motor vehicle drivers (electric vehicles, bicycles). The factors that cause traffic accidents are different. Among human factors, motor vehicle drivers account for a large proportion of accidents.

    (a) Driver's information processing characteristics.

    When the road traffic system is running normally, the driver needs to constantly judge the road condition and driving conditions, and respond to this, operating the vehicle to continue driving. In essence, it can be thought of as a mental acquisition of information and process information according to feedback.

    By receiving external environmental information, the driver transmits it to the central nervous system from the sensory organ and responds to its exercise organs, then manipulate vehicles, and continue or change the operation of vehicles. If the response of the sensory organs is biased, it will lead to abnormal operation of vehicles. This information must be fed back to the central nervous system for correction and then transmitted to the motor organ to fix order. The driver's mood, health, and fatigue affect driving safety to a great extent. Whether the information is correctly received has a great impact on the following action. The driver has a certain timeliness in the processing of information. If the information cannot be processed quickly and accurately, it is likely that a traffic accident will occur.

    (b) Driver's driving ability and fatigue characteristics. Drivers' reasonable operation mainly affected by their driving ability.

Driving ability refers to the correct judgment, reaction, and manipulation ability of the driver. When the driver is tired, driving ability will go down a lot. After 12 hours of continuous driving, the probability of serious traffic accident is 1.5 times than that of 8 hours. In addition, traffic accidents caused by drivers driving for more than 7 h in a row account for about 1% of the total number of traffic accidents 1/3. Statistics also show that driver fatigue causes 40% ~ 70% of accidents.

The fatigue degree of the driver is related to the amount of information received by the driver during the driving process. Road traffic system is a complex, real-time changing system. When the traffic is complicated, the excessive tension of the driver's mental activity can cause fatigue to appear early; if the amount of information is insufficient, long monotonous operation will also make the driver feel tired earlier. When road traffic conditions suddenly change, the driver often does not have enough psychological preparation and cannot deal with the emergency correctly.

Affected by fatigue, the driver's attention, judgment ability, visual acuity, and accuracy of speed perception decline and can lead to narrow vision, faster pulse, higher blood pressure, increased reaction time, dyskinesia, and so on. So when the driver is tired, it is more likely to cause traffic accidents. The right thing to do is to stop and rest immediately or take other actions to restore normal driving capacity when you find yourself tired. Experiments show that a few short breaks during driving are more efficient than a long rest at the same period of time, and the reasonable allocation of work and rest time can effectively avoid driving fatigue.

2. Analysis of traffic flow factors.

   (a) Analysis of the influence of large-scale vehicles on road safety.
       The general permissible speed of urban main roads in China is 40 ~ 60 km/h; this has certain requirements for the performance of the vehicle. Traffic participants on urban roads are complex and diverse, the performance gap of the model is large, which has a great impact on urban road traffic safety. All types of vehicles in the road traffic system and kinetic energy are different, which will produce some transverse and longitudinal interference, which is also the cause of traffic accidents. The impacts of large vehicles on urban road traffic safety are as follows.

       • Poor reliability.
           The reasonable configuration of vehicle assembly can ensure the continuous operation of the vehicle in the road traffic system. The main assembly of large vehicle is less, the performance of the whole vehicle is unstable, and the safety and reliability are poor. Large vehicles on roads take a lot of space to keep running, it is easy to collide with other vehicles, and it increases hidden dangers.
       • Poor braking performance.

Because the mass of heavy vehicle is generally large, huge inertia can lead to longer braking distance. The deviation of large-scale vehicle, tail-flick, rear-end and collision are easy to occur. So, we are going to enhance the active and passive safety of vehicles that cannot be ignored to improve the level of road traffic safety.

- Poor manipulation stability.

The stability of vehicle operation includes the stability of longitudinal driving and the stability of transversal channel. The phenomenon of fast migration and swing occurs when heavy vehicles move, which is mainly caused by the lag of technical performance of heavy vehicle chassis.

- The speed is discrete.

According to relevant statistics, the maximum speed difference between large and small vehicles that can be reached is 60 km/h, which has a significant impact on urban road traffic safety.

(b) Analysis of the influence of speed on road operation safety.

When drivers are driving vehicles, they simply respond to the road condition information and estimate surrounding safety in time, and then operate the vehicle. But when the speed increases, time left for the driver's reaction time is greatly reduced, which greatly increases the possibility of the nervous system making false judgments and the occurrence of traffic events. This increases the probability of accidents. In addition, the time and distance required to brake turning adds as the increase of speed. The collision speed of traffic accident is also larger than the driving speed. When the vehicle is being driven at a normal or slow speed, the driver can start to brake to slow down from the long distance of obstacle. Whereas the speed difference between the highspeed vehicle and obstacle which occurs collision is much larger than the original speed difference. Speeding gives the driver the illusion that he cannot correctly estimate the distance between the front and the rear. The speed of the vehicle is estimated, the probability of the traffic accident is greatly increased, and the damage of the traffic accident is also increased.

3. Analysis of the influence of weather factors

The reliability of human physiology and psychology to various weather reactions is the most important manifestation of road traffic safety. Certain amount of air pressure, temperature, and humidity in the fixed environment can affect the driver's senses and even cause adverse reactions in the body. In addition, rain, haze, heavy snow, and other weather phenomena will affect the visibility of vehicles while driving, resulting in slippery ground, traffic environment congestion, and so on, which increases the chances of the driver making a wrong decision. A large number of driver behavior experiments have shown that meteorological factors can have an impact on human physiology and psychology.

Through the above analysis, various climatic factors and different performance levels in terms of the driver's psychology which affects the vehicle operating state, road conditions, and traffic environment have different mechanisms of action: weather changes the road friction coefficient, visibility and vehicle driving stability will be affected. In some geographical climates, haze, heavy snow, heavy rain, and so on are closely related to traffic accidents, which are destructively increased, resulting in huge property losses and casualties. Therefore, in addition to improving road conditions and driver safety awareness, it is also of great significance to investigate the probability of traffic accidents under unfavorable weather conditions.

4. Analysis of road factors.

When the driver is driving on the city road, the driver's visual judgment is most directly affected. It is the plane alignment and cross section of the road. The actual driving speed of the driver is the observation of the pavement condition and the three-dimensional line form, traffic conditions after the decision. Road alignment is directly related to traffic safety and urban roads are not only designed. It is necessary to consider the dynamic driving requirements, but also to consider the psychological and physiological state of the driver while driving continuously, and to ensure the linearity of the road continuity. From the point of view of road conditions, the data of many urban road traffic accidents show that unreasonable city road route combination and ergonomics design may cause road traffic accidents. Designing a road should take into account a comprehensive study considering the urban road function, land utilization rate, natural environment, climatic conditions, traffic safety, and so on, and fully consider safety, management convenience of transportation, which reduces road safety hidden danger as far as possible, eliminate accident black spot, and improve road traffic safety fundamentally.

5. Analysis of the impact of management factors on traffic safety.

Traffic safety management is the life of road development. Road traffic system is a complex, real-time system, so road traffic management is also a huge work, has an important influence on everyone's life, and property, and needs managers and participants in different industries of society to coordinate with each other, establish perfect laws, regulations, and management system, and strengthen urban road operation and management.

Road traffic safety management can be divided into the management of the subject and object of road traffic system. Participants including drivers, pedestrians, passengers, and their unsafe behavior will lead to traffic safety risks, and even directly lead to traffic accidents, thereby leading to urban road traffic safety state deterioration. The road traffic system consists of traffic environment and road condition. These will lead indirectly to judgment error, operation error, and information receiving error. Transport carriers, roads conditions, weather conditions, and management decision-making factors may lead to changes in traffic safety conditions, as shown by traffic events. Therefore, the probability of traffic accidents increases greatly when unsafe behavior occurs in the participants, and the design of means of transportation. The design of road and the design of driving workspace affect the decision-making of road traffic safety management, the selection of drivers, personnel management factors such as education, training mechanisms, and road traffic safety education for other traffic participants. There exist management decision-making factors in the aspect of traffic safety, total control, safety supervision and inspection, as well as traffic accident prevention.

10.1.2.2   Urban Road Traffic Safety Monitoring Index System

This section explains the principles and methods of constructing the indicator system. Then the complicated road traffic safety system is divided into traffic system. The system is divided into three parts: subject, object, and non-main participant.

1. Monitoring of road traffic safety.

The urban road traffic safety state refers to the traffic safety level of a city in a certain period of time, with the following attributes.

  (a) Predictability. The state of road traffic safety points to the future, that is, pointing out the potential hidden dangers in road traffic safety. These hidden dangers are likely to develop into traffic accidents.
  (b) Social Property. The sociability of road traffic safety state is one of the attributes of road traffic safety state, as well as the whole society. The sociality of road traffic safety state is one of its attributes, which is closely related to the whole society and involves everyone's life, work, and travel. When the road traffic safety is elevated to public safety, all participants in social activities must be taken seriously. This level of improvement has disrupted traditional transportation. Therefore, the definition of casualties and economic losses to strengthen the safety awareness of traffic activity participants, and make the more main body from the traffic management department formulate rules and regulations.
  (c) Objectivity. The objective existence of road traffic safety state refers to the uncertainty of traffic participants and the complexity of road conditions, and road traffic safety risks always exist, which determines the objectivity of traffic safety risks.
  (d) Instability. The traffic system contains many complex and integrated factors (driver's unsafe behavior, pedestrian's behavior violations, vehicle insecurity, road hazards, and environmental factors) that are the result of these uncertainties. People in social traffic activities are in an unstable state of safety. Time, place, and destruction of road traffic accidents, the degree, the scope, and whether the traffic hazards are transformed into traffic accidents make road traffic safety unstable. Due to the existence of uncertain traffic participants such as people and vehicles, the traffic state of the environment cannot be predicted. Therefore, it is very important to study the current situation of urban road traffic safety, correct remedy measures, and timely countermeasures and predict the possible situation in the future.
  (e) Dynamic. The change in terms of the influencing factors on traffic safety state, the subject of road traffic participation, the road traffic system is bound to occur.

The comprehensiveness of the social system and the intricacy of the road traffic system illustrate the various state of road traffic safety in cities. In the road traffic system, different regions, and different time periods, the influence weight of some elements or indicators is diverse due to different regions and time periods, so the geographical background and time background should be planned and limited as follow:

(a) Geographical background. In the index system of urban road safety monitoring, the regional background is country, city, and province. The division of regional level is in the middle of China's administrative division. The administrative units of urban regions are relatively complete and suitable for urban transportation. More attention is paid to improvement of safety and that facilitates collection and collation of data, which is more experimental and practical than county-city level comparison.

(b) Time background. The time background level of road traffic safety state research can be divided into weeks, months, quarters, and years.

Data availability and physical monitor ability take into account drivers, vehicles, roads, etc. In the complex road traffic system, the factors affecting the state of urban road traffic safety require careful analysis.

Monitoring and control can effectively grasp the state of urban road traffic safety. Therefore, road traffic safety needs monitoring. The object is the key factors in the man-car-road system: driver error, weather factors, vehicle performance, exterior driving environment, etc. It is the existence of these factors that lead to traffic accidents.

Road traffic system is a complex subsystem of social system, which is affected by economic development, climatic conditions, geographical location, and production. Many factors include a variety of terrain and climate, as well as a territory with a large latitude span, which requires the urban road traffic safety monitoring index system to have certain applicable conditions. Therefore, road traffic safety monitoring should follow the limitation of regional background level, that is the period and region corresponding to the monitoring index system should be set up. Taking into account our practice, the government's comprehensive, feasible implementation of the index system and road traffic safety regional monitoring keep consistent, the provincial or prefecture level of the administrative region shall be determined.

2. Principles and methods for the construction of monitoring index systems.

Construction of urban road traffic safety risk monitoring index system to track, evaluate, and study road traffic safety status has certain benefits. Many factors need to be considered in the design of the road traffic safety risk monitoring index system, and there are many monitoring factors or indicators available. Therefore, the selection of indicators should follow certain principles and be representative.

(a) Scientific and implementable

The scientific nature is based on the theory of system science, and the non-majority information is collected and analyzed to find out the occurrence interval and distribution range of abnormal data. Scientific is reflected in the understanding of indicators, index data, the acquisition should be scientific and reasonable. Implementation refers to the use of modern intelligent transportation technology from monitoring equipment or networks traffic information data obtained in the network that can be processed, and these indicators can reliably and effectively predict the safety of urban road traffic, the weak link of the whole level.

(b) Combining qualitative and quantitative principles

Road traffic safety status index includes quantitative index and qualitative index. On the basis of qualitative analysis, the characteristics of road traffic system are measured by quantitative index, and the historical data are objectively quantified and reflect urban road traffic safety state, in order to make the monitoring results more scientific and objective. Through the function of the index, the quantitative calculation, expression, and quantification of qualitative indexes make the evaluation more convenient. For hard-to-reach indicators data or some missing index data can also be supplemented by certain methods.

(c) Real-time and predictive combination

The state of road traffic safety is dynamic, so the corresponding monitoring refers to the standard should also have real-time performance to truly reflect the characteristics of the traffic system represented by the index. On the other hand, indicators may also have design indicators should be as forward-looking as possible, traffic management departments try to make good use of indicators including design indicators and realize the fact that measures are taken through indicators to deal with security risks in time.

(d) Integration and independence

The actual index system operation extension has certain limits to the index quantity. This requires the indicators to be representative and comprehensive, and to grasp the core problems in the urban road traffic system, but there must also be some independence, that is, the index design can reflect the corresponding characteristics of the influencing factors, and there cannot be too many indicators between them that will have an impact on the monitoring results.

(e) The stability and strain are combined

Stable road traffic safety monitoring index can objectively reflect the hidden dangers of urban safety. However, different geographical and temporal contexts may adjust indicators to ensure their application, and the weight of indicators also varies with the situation.

(f) Government order is combined with test-ability.

The relevant laws and regulations of the government refer to the design of road traffic safety status indicators guidelines. Test-ability means that the selected index is easy to quantify and then easy to calculate. By means of qualitative indicators, we can obtain the monitoring results within the monitoring index system. In addition, the testing results should be specific and in the legal system.

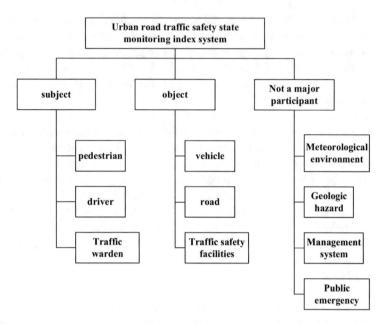

Fig. 10.1 Monitoring index system of urban road traffic safety

3. Construction of monitoring index system.

The road traffic system is mainly divided into the main body of participation, the object of participation, the organization and management of the traffic system, and generally from these three parties, to establish traffic safety monitoring indicators, as shown in Fig. 10.1.

4. Methods of index screening and evaluation.

In order to overcome the subjectivity of the initially determined index system, a better correlation index, and for some indexes with less monitoring function of traffic safety state, other high efficiency indexes are screened twice. At present, commonly used the methods of index screening are mainly analytic hierarchy process, principal component analysis, gray cluster method, rough set method information sensitivity screening method, and expert consultation method.

10.1.2.3 Domestic and Foreign Road Traffic Index Summary

The Texas Institute of Transportation defines travel time index (TTI), road congestion index (RCI), and other quantitative indicators to analyze key highways. The travel time index is the ratio of actual travel time to free flow travel time. For example, TTI = 1.35 denotes that travel time in free flow is 20 min and the actual travel time is 27 min. Road Congestion Index is the ratio of vehicle miles to lane miles, which is calculated after weighting to measure road traffic density and describes the intensity and persistence of congestion. When the index is greater than or equal to 1, it indicates that road network congestion is unacceptable. By calculating the ratio of the average travel speed to the corresponding free flow speed, the U.S. Highway Capacity Manual (Highway Capacity Manual, HCM2000) divides the highway service level into six grades to evaluate the road operation and reflect the average perception of all drivers to the traffic flow they are in, where A represents a free flow situation where the ratio is greater than or equal to 90%, and the represents a ratio of less than 33%; and the delay is serious. The U.S. Federal Highway Administration uses congestion severity as a measure to quantify congestion, which is defined as the total vehicle delay per million km. Based on American HCM2000, some research institutions integrate the relevant indicators and use limited samples to establish a congestion index system, which include travel speed, driving speed, delay rate, driving speed ratio, delay ratio, etc. In this way, they can judge road traffic congestion.

Domestic traffic field also has some theoretical research and practical accumulation of traffic index. The Beijing Traffic Congestion Index is calculated according to a time period of 15 min. This calculation is combined with road grade, travel speed, and the kilometers of driving. It reflects the traffic congestion level of the road network from a macro perspective. In this case, the value range is 0~10, which is divided into five levels. In this situation, 0~2 (including 0 and 2), 2~4 (including 2 and 4), 4~6 (including 4 and 6), 6~8 (including 6 and 8), and 8~10 (including 8 and 10) correspond to the levels of being smooth, being basically smooth, being slightly congested, being moderately congested, and being heavily congested. The higher the value, the more serious the traffic congestion is. Zhejiang Province has also introduced a traffic operation index, which comprehensively reflects road network traffic. As with the Beijing Traffic congestion Index, the range of values is 0 ~ 10, which is unblocked to 5 levels of serious congestion. Based on the standards of Zhejiang Province, the traffic operation indexes in the major cities in Zhejiang provinces, which include Hangzhou, are successively launched and applied. Shenzhen has established a traffic index calculation model based on the average diving speed, travel time, and the rating from experts. This city gradually calculates the traffic index of different spatial ranges such as road sections, gateways, districts, and the whole city from point to line and from line to surface. In this case, the value ranges from 0 to 5, which correspond to 5 levels: being smooth, being basically smooth, being slow, being slightly congested, and being congested. The larger the traffic index, the longer the travel time compared to smooth conditions (such as early morning hours). For example, in the level of being congested, the time spent on the road is at least twice as long as that in the smooth condition. In addition, companies such as Gaude Software and Beijing Siwei also introduced congestion delay index and traffic congestion index, which are used for horizontal comparison and the analysis of traffic operation among cities in China.

10.1.3   Example: Application of Shanghai Road Traffic State Index

According to the above contents, the application case of Shanghai road traffic state index is expounded. In 2016, using historical data of the state index, Shanghai made analysis and judgment of its road traffic condition of working days of 2014–2015, to provide reference for traffic decision and management.

For this reason, the peak period of the working day is defined—morning peak: 7: 00–10:00, noon peak: 14:00–16:00, evening peak: 16:00–19:00. We follow the statistics in the 2011–2014 Road Traffic Congestion Analysis Report. The three index areas, in which the congestion index is greater than 50, the accumulated congestion time exceeds 1 h, and the working time exceeds 100 days during 1 year during the peak period, are defined as heavily congested areas. Compared with the road traffic congestion in Shanghai in 2014, the analysis of this in 2015 is as follows.

10.1.3.1   Urban Express Network

Frequent congestion areas in the expressway network during the peak working day of 2015: no change in spatial location; there was a slight decrease (3), namely: outer ring (Wuzhou-Tongji), inner outer ring (Shanghai-Yu-Hujia), and outer ring. Outside (Shanghai-Yu-Jiyang); congestion days increased, with an average increase of 14 days; extreme congestion index greater than 70 reduction. On the whole, the space position of frequent congestion of expressway has not changed, the number of congestion days has increased, and the degree of extreme congestion has been decreased.

10.1.3.2   Ground Road Network

Frequent congestion areas of the ground road network during the peak working day of 2015: no change in spatial location; volume slightly increased, respectively: Tian Lin, Sichuan North Road business district, Ruihong New City; congestion days have increased, an average increase of 7 days; the extreme congestion index of more than 70 is basically the same as in 2014; the congestion range is expanding from the inner ring to the central ring. On the whole, the space position of frequent congestion on the ground road has not changed, the number of congestion days has increased slightly, and the congestion range is presented by the inner ring.

10.2 Network Control System for Key Operating Vehicles

10.2.1 System Profile

With the development of China's national economy and highway construction, the number of road passenger transport and cargo transport is increasing rapidly. The accident rate of corresponding road transportation is high, and 80% of the major road traffic accidents occur in operating enterprises. An important issue that need to be addressed urgently is improving the supervision level of key operating vehicles and reducing the loss of life and property of the public.

In order to strengthen the monitoring of key operating vehicles, reduce road traffic accidents, and realize multi-department joint supervision and law enforcement, National Network Control System for Key Operating Vehicles came into being. The network joint control system for key operating vehicles refers to a dynamic monitoring and monitoring system for operating vehicles based on satellite positioning system technology, which is established by road transportation management agencies at all levels and related enterprises.

This system includes the public service platform of vehicle dynamic information regarding the national road transport, the supervision platform of local road transport management agency, the enterprise monitoring platform of road transport and the social monitoring platform. The system is in charge of the dynamic information of key operating vehicles, which include the chartered vehicles for tourism, three or more types of buses and the vehicles transporting hazardous chemicals, fireworks and firecrackers, and civilian explosives. The dynamic information of the key operating vehicles is connected to the network, and vehicles movement and dangerous driving are monitored in real time. Besides, it monitors the movement trajectory and the dangerous driving behavior of such vehicles and collects the relevant data. The system has the functions of issuing the information such as early warning, scheduling to the vehicles, and receiving the information such as emergency calls and offline reminders from the vehicles. The network joint control system for operating vehicles uses information technology to manage the scattered and mobile operating vehicles, effectively solving the problem of "being invisible, inaudible, and uncontrollable" in road transportation supervision.

10.2.1.1 Development Background and History of Network Control System for Key Operating Vehicles

1. Road Transport Security and Movement Requirements.

   In 2009, the Ministry of Transport adopted effective technical approach for the joint supervision of operating vehicles across regions and departments to solve the problem that the provincial road transport management departments could not supervise vehicles in other provinces. Besides, in order to ensure the transport security of the Shanghai World Expo and the effective supervision of the key operating vehicles entering Shanghai, the Ministry of Transport decided to integrate the existing dynamic monitoring resources of the industry and build the network joint control system for operating vehicles. These provide the effective technical support for cross-regional and cross-departmental joint supervision of key operating vehicles.

In April 2010, the system was officially launched, with access to 30 provincial regulatory platforms involving more than 800 GPS operators and enterprise monitoring platform. Since the opening of the system in Shanghai World Expo, Guangzhou Asian Games, Shenzhen Universiade Road transport security work has played an important role in services and security. During the period of the Shanghai World Expo, a total of 1.155 million times of cross-regional trains were forwarded to Shanghai from the national platform. Additionally, during the period of the Guangzhou Asian Games, a total of 427,000 times of cross-regional vehicles were forwarded to the Guangdong platform.

2. Dynamic Regulatory Requirements for Large Operating Vehicles.

The accidents caused by large-scale vehicles are of high severity, which can easily cause mass death and injury. It has always been one of the key concerns of the work of road traffic safety. In November 2009, Zhang Dejiang, then Vice Premier of the State Council, was in "Special Information" No. 1618. "The use of modern information technology for the implementation of vehicle transport safety management, is an important measure to strengthen the construction of road traffic safety." In July 2010, the State Council issued the "Notice of the State Council on Further Strengthening the Work Safety of Enterprises" (Issued by the State Council [2010] No. 23). This notice indicates that a satellite positioning device with a driving record function should be installed and used in the certain vehicles within 2 years. These include the vehicles transporting hazardous chemicals and civilian explosives, the chartered vehicles for tourism, and the three or more types of buses.

In order to implement the principle of Document No. 23 of the State Council, the Ministry of Transport cooperates with the Ministry of Public Security, the State Administration of Work Safety, and the Ministry of Industry and Information Technology. They jointly issued the "Notice on Strengthening the Dynamic Supervision of Road Transport Vehicles" (Issued by the Ministry of Transport [2011] No. 80). This document clearly requires that the newly- produced vehicles mentioned above must be equipped with standard satellite positioning devices. The Ministry of Industry and Information Technology will not publish the product announcements of the vehicles that do not meet the regulations. Furthermore, the public security will not examine their qualifications and the transportation departments will not issue the road transport certificates for them. The road transportation management department will suspend the qualification examination of vehicles that have not installed on-board terminals or have not been connected to the national network joint control system for key operating vehicles. In terms of setting standards, the Ministry of Transport issued two key requirements, which are Vehicle Terminal Technical Requirements of Vehicle Satellite

Positioning System for Road Transport and Platform Technical Requirements of Vehicle Satellite Positioning System for Road Transport, on February 28, 2011, and March 25, 2011. The promulgation of these two standards has laid a solid technical foundation for the unified regulation of the operation of the national network joint control system for key operating vehicles.

In March 2014, the Ministry of Transport, the Ministry of Public Security, and the General Administration of Safety Supervision jointly issued the Dynamic Supervision of Road Transport Vehicles Management measures, wherein the dynamic supervision and management of road transport vehicles are clearly defined. In January 2015, the Ministry of Transport issued the National Measure for the Assessment and Management of the Network Joint Control System for Key Operating Vehicles. Besides, the revised version of it was issued in September 2016. A revised version of the method is presented. Consequently, the technical level, application degree and standardization degree of the network joint control system for key operating vehicles have been further improved due to the implementation of this measure.

10.2.1.2 System Framework of Networked Control System for Key Operating Vehicles

The network control system of key operating vehicles in China adopts the system structure of longitudinal classification and horizontal docking. Longitudinal classification means the system is divided vertically into ministerial public exchange platform, provincial monitoring platform, and regional and enterprise monitoring platform; horizontal docking refers to different levels of platform docking traffic, public security, safety supervision, environmental protection, and other government departments and related enterprises information resources to achieve information sharing and joint supervision. The system framework and architecture are shown in Fig. 10.2.

1. Ministerial Public Exchange Platform.

   The National Public Exchange platform is responsible for cross-regional information exchange of roaming vehicles between provinces and information sharing between relevant departments of the State Council. The national public exchange platform is the first data center of the national key operating vehicle dynamic information, through collecting the vehicle dynamic information of each provincial monitoring platform, the vehicle transportation information reported by the enterprise and the transportation administration database, and establishes key operating vehicle static information and dynamic information database to form a data center covering the national key operating vehicles; the national public exchange platform is the data exchange center, which adopts a unified data exchange standard. The national public exchange platform can realize data exchange with provincial monitoring platforms and realize information sharing with other ministries and commissions. The national public exchange platform

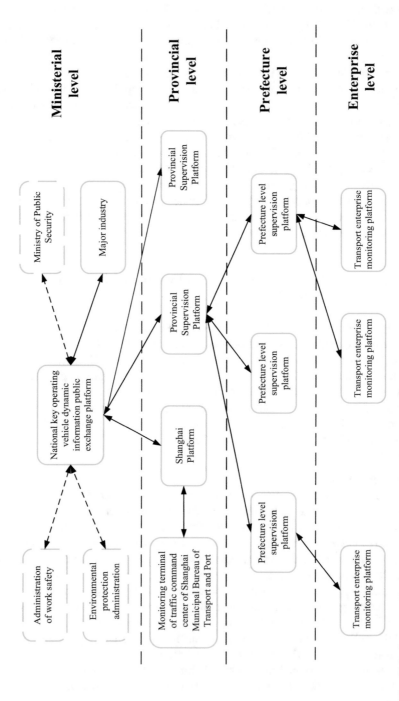

Fig. 10.2 System architecture of networked joint control system for operating vehicles

and the operation monitoring center of the system can make necessary analysis and statistics on the operation of the connected provincial monitoring platform.
2. Provincial Regulatory Platforms.

The provincial monitoring platform is responsible for cross-regional information exchange of roaming vehicles between regions and letters between relevant local departments. The provincial monitoring platform is responsible for transmitting the dynamic information of key operating vehicles to the national public exchange platform and receiving the whole country's dynamic information of cross-provincial vehicles transmitted by public exchange platform.
3. Regional Regulatory Platforms.

To realize the supervision of the key operating vehicles within the jurisdiction, supervise and examine the access of the vehicles of the transport enterprises within the jurisdiction, combined with industry management needs to conduct comprehensive supervision of local and foreign vehicles operating within their jurisdiction.
4. Enterprise-Level Monitoring Platform.

To realize the safety supervision of the vehicle, to transmit the positioning data to the superior supervision platform in time, and to ensure the data, the authenticity, accuracy, and effectiveness of the upload data, timely correction and handling of speeding, fatigue driving, and other illegal driving behavior.

10.2.1.3 Main Functions of Networked Control System for Operating Vehicles

Real-time monitoring function: the network control system of operating vehicles can communicate with the vehicle terminal in real time through the monitoring platform to the vehicle position, driving track, driver driving status, driving behavior, and other information for real-time monitoring.

Scheduling management function: in order to ensure the special capacity requirements in large-scale activities and other scenarios, the operating vehicle network control system, transportation tasks can be uniformly scheduled and distributed through real-time detection of the capacity usage of transport enterprises in different regions. And then implement cross-regional, cross-departmental joint scheduling and control.

Data playback function: the operating vehicle network control system can detect the running condition of the vehicle at the same time. The status data are collected and sorted to complete dangerous driving behaviors such as vehicle mileage, driver speed up, sharp turn, etc. The data can provide basis for driver assessment and enterprise transportation efficiency assessment.

Remote control function: when the system detects the driver's dangerous driving behavior or driving state, it can pass through the vehicle terminal. The driver is warned and the vehicle is warned and reported at the monitoring center. The operator at the monitoring center can adjust the information of vehicle camera and other sensors, and monitor the running state of the vehicle remotely.

10.2.1.4   Main Features of Networked Control System for Operating Vehicles

In the key technology, the network control system of operating vehicles has solved the problem of cross-regional and cross-departmental information exchange and sharing. What's more, it realizes the integration application of vehicle data and the mastery of vehicles motion dynamics and distribution. The network joint control system fully integrates the existing provincial road transport monitoring system resources to complete the key operating vehicles in the provinces. On the one hand, the cross-regional exchange of dynamic information of key operating vehicles is realized; on the other hand, as an all-open system, the data exchange channel is established. Now the information communication between the same area and different government management departments lays the foundation for the multi-department in the coordination office, emergency linkage, and other parties.

The networked control system realizes the effective combination of dynamic and static information of vehicles. The collection of traditional road transportation information is a summary of static data, which cannot provide real-time information to managers. Through the network joint control system, the vehicle dynamic position information, the vehicle transportation administration information, and the vehicle cargo transportation information can be forwarded to the corresponding platform in real time, so that the receiving platform can not only clearly understand the vehicle's driving track, but also know the cargo information and attribute information of the vehicle.

The network control system realizes the unified centralization of data at the ministerial level, can effectively grasp the overall operation of the national road transport industry, strengthen the supervision of the road transport industry, improve the information management level and decision analysis ability of the road transport industry, and provide data support for modern logistics industry, emergency command system, network congestion analysis, traffic economic operation analysis, and so on.

10.2.2   System Composition and Service Functions

The network control system of key operating vehicles is mainly composed of two parts: vehicle mobile terminal and monitoring center at all levels. The vehicle terminal transmits the vehicle positioning information and the driver status monitoring information to the monitoring center through the wireless communication system, and the monitoring center can also transmit the dispatching instructions to the vehicle terminal through the wireless communication system, so as to realize vehicle monitoring and scheduling.

10.2.2.1 On-Board Mobile Terminal

On-board mobile terminal consists of microprocessor, data memory, satellite positioning module, vehicle status information acquisition module, wireless communication transmission module, real-time clock, data communication interface and other host modules, as well as satellite positioning antenna, wireless communication antenna, emergency alarm button, voice message reading device, and other external equipment. The main functions of the vehicle mobile terminal are as follows.

1. Positioning function.

   The terminal can provide real-time location status information such as time, longitude, latitude, speed, elevation, and direction, which can be stored to the terminal and uploaded to the monitoring center by wireless communication. At the same time, the terminal can receive the location request of one or more monitoring centers to upload the location information, and can suspend the real-time reporting of the corresponding centers according to the requirements of the monitoring centers. When the communication is interrupted (blind area), the terminal can store not less than 10,000 positioning information in the way of first in, first out, and upload the stored location information after resuming the communication. The terminal also supports uploading location information using time, distance interval, or external event trigger. When the terminal is dormant, the location information should also be uploaded at a certain time interval, and the interval of time and distance can be set by the monitoring center. The terminal can also automatically upload location data to the location mode and interval set by alarm vehicle or key vehicle monitoring center.

2. Communication Function.

   Car mobile terminals support a variety of wireless communication networks based on 2G, 3G, and 4G. When the wireless network in the location of the vehicle supports packet data transmission, the vehicle terminal should first choose the packet data transmission mode; when the location does not support packet data transmission, it can switch to short message mode to transmit data; when the local wireless communication network is impassable, Beidou satellite communication mode can be adopted according to the need.

3. Information Acquisition Function.

   (a) Driver identity information. Terminal support through the IC card to collect driver qualification information, and upload to the monitoring center.
   (b) Electronic waybill. The terminal supports the collection and display of electronic waybill information, and upload to the monitoring center.
   (c) Vehicle CAN bus data. A terminal supports the acquisition of vehicle parameter information through a CAN bus, and upload to the monitoring center.
   (d) Vehicle loading status. The terminal determines the loading status of the vehicle by means of the interface of the vehicle load state detection device or manual input (no load, half load, full load), and upload to the monitoring center.

(e) Image information. The terminal has the function of image information acquisition and storage, and supports monitoring center control, timing, and event-triggered mode to achieve image information collection, storage, upload, and retrieval upload function.

(f) Audio information. The terminal has the function of collecting and storing audio information, and supports monitoring center control and event trigger to achieve audio information collection, compression, storage, upload, and retrieval upload function.

(g) Video messages. The terminal has video information collection and storage function, and supports monitoring center control and event trigger to achieve video information collection, compression, storage, upload, and retrieval upload function.

4. Driving Record Function.

The terminal has the function of vehicle driving record and supports the functions of real-time upload, conditional retrieval upload, and data interface export of driving record data.

5. Warning Function

When the terminal triggers the warning, it should immediately upload the warning information to the monitoring center or send the short message prompt information to the designated mobile phone as needed, and can receive the instruction of the monitoring center to cancel the warning. The warning mobile phone number can be set remotely by the monitoring center. Terminal warning function is divided into manual alarm and automatic warning. Manual alarm is the alarm triggered by the driver according to the actual situation on the spot, including when there is an emergency such as robbery, traffic accident, vehicle failure, and so on, the driver uploads the alarm information to the monitoring center by triggering the emergency alarm button, and closes the voice reading module. If the terminal has image, video, and audio acquisition function, the function should be enabled immediately. Manual alarm should have the function of preventing misoperation.

The terminal is triggered according to the conditions set by the monitoring center. The specific content includes the following 8 types.

(a) Automatic alert means that the driver does not do anything to the terminal. The terminal is triggered according to the conditions set by the monitoring center.

(b) Area reminder: triggered when a vehicle enters or leaves a restricted area; the monitoring area can be set remotely by the monitoring center.

(c) Route deviation warning: triggered when the vehicle leaves the set route; the monitoring route can be set remotely by the monitoring center.

(d) Over-speed alert: the terminal can be triggered by a preset speed threshold or by receiving information from the monitoring center, to remind the driver that he is currently speeding.

(e) Fatigue driving warning: when the vehicle or driver's continuous driving time exceeds the fatigue driving time threshold, the fatigue driving time threshold can be set remotely by the monitoring center.

(f) Battery under-voltage warning: terminal detection vehicle battery voltage below the preset value triggered. At the same time, the terminal must use the built-in backup battery power instead of taking electricity from the vehicle battery.

(g) Power off reminder: the terminal is triggered when the main power supply is cut off. Timeout warning: when the stop time exceeds the system preset time.

(h) Terminal fault alert: triggered when the external equipment connected to the terminal host is abnormal, and uploaded to the monitoring center.

6. Human-Computer Interaction.

The terminal has the function of human-computer interaction and can interact with the driver. The terminal should be able to provide information to the driver through voice reading equipment and display equipment combined with signal lights or buzzers, and the driver can operate the terminal by means of keys, touch screens, or remote control.

7. Information Services.

The terminal supports the monitoring center to send the information directly and the driver to report the information actively; it can prompt the driver to the dispatching information, logistics information, and so on through the display equipment or the voice reading equipment, and the driver can return the response information to the monitoring center by a keystroke. The terminal can store at least 50 records of all information types and support information query function.

8. Multi-Center Access.

The terminal supports connecting two or more monitoring centers at the same time and can obtain the information sent by the monitoring center. The terminal should regularly connect to the set monitoring center and obtain the information sent by it according to the set time interval.

10.2.2.2 Monitoring Platform

The monitoring center is divided into government supervision platform (Government Monitoring and Management Platform) and enterprise monitoring platform (Enterprise Monitoring and Management Platform). The government supervision platform is based on computer system and communication information technology. The system platform for the management of vehicle terminal and access platform is realized by satellite positioning technology. The enterprise monitoring platform is a satellite positioning system platform built by the enterprise or commissioned by the third-party technical unit. Based on the computer system, the vehicle terminal and user in the service range are managed by accessing the communication network. The system platform of safe operation monitoring is provided to realize real-time monitoring of vehicle safety operation in the platform.

Government platforms are connected by special line networks or Internet VPN, and enterprise platforms and government platforms can be connected by Internet or special line networks. The vehicle terminal is connected with the enterprise platform or the government platform through the wireless communication network.

1. Government Regulatory Platform

Through the platform interface and statistical analysis function, the government platform mainly realizes the data walk to the superior platform and the lower-level government. The management of government platform and the supervision and service of enterprise platform and its basic function include the following points.

(a) Access platform management. Access platform management includes access platform configuration management, information query, and assessment functions. Among them, access platform configuration management has access platform parameter configuration, access platform parameter query, and access platform parameter statistics and other basic functions. Access platform information query has the basic situation of the platform: Platform online vehicles, platform history online vehicles, platform off-line vehicles, platform running log and platform inspection log and other query functions. The assessment of access platform includes automatic check-up of platform, manual check-up of platform, dynamic data transmission of platform, online of platform vehicle, dynamic data transmission and quality, etc. It has the access platform by day, week, month, season, and year to carry on the examination function.

(b) Report export function. All query results and statistical analysis results in the platform support EXCEL export.

(c) Vehicle data timing function. Regularly sends the normal report vehicle list and abnormal vehicle list to the access platform.

(d) Alarm and alarm management. The government platform shall have the function of receiving alarm information reported by the access platform, including emergency alarm, deviation route alarm, overspeed alarm, area alarm, fatigue driving alarm, etc. When an alarm is generated, the vehicle dynamic position information and static information and related information can be prompted and displayed by sound, light, picture, and text. If the subordinate regulatory platform or the enterprise platform does not report the alarm processing information within the agreed time, the regulatory platform shall automatically send an alarm disposal request to it.

(e) Basic information management. The government platform has the functions of network docking and data exchange with local transportation and government information systems. The basic data management should have the function of querying and managing all kinds of vehicles, employees, and transportation enterprises.

(f) Dangerous goods vehicle/business management function. The dangerous goods vehicle management function has the dangerous vehicle inquiry and the dangerous goods vehicle statistics function. The dangerous goods

transportation enterprise management includes the dangerous goods transportation enterprise inquiry, the dangerous goods transportation enterprise
statistics, and the dangerous goods transportation enterprise appraisal
function.
(g) The functions of 7-shift passenger vehicle/enterprise management, tour char-
ter/enterprise management, freight vehicle/enterprise management are similar
to those of dangerous goods vehicle/enterprise management.
(h) Vehicle dynamic monitoring and management. The government platform has
the functions of vehicle real-time monitoring, single monitoring, and so on. It
provides the functions of vehicle tracking, message sending, and vehicle
photographing, supports the historical data query function of feedback message, vehicle driving record data and photo, can play back the historical track
of the designated vehicle, and supports the prompt of vehicle events at the
historical track point.
(i) Statistical analysis. The government platform should have the total number of
access platforms, the number of online platforms, the number of vehicles
entering the platform, the statistical analysis function of online vehicle
number, and platform vehicle alarm. Among them, the statistical analysis of
vehicle management includes vehicles online statistics report, vehicle cross-
regional statistics report and vehicle alarm statistics report, and vehicle
information online analysis, vehicle online/month-on-month analysis, vehicle
alarm/month-on-month analysis, vehicle across regions, year-on-year/month-
on-month analysis, and other comprehensive analysis functions, including
regional vehicle statistics analysis, key transport vehicles statistics analysis,
enterprise vehicle terminal installation rate statistics, and regional enterprise
platform online coverage statistics.
(j) Platform operation monitoring management. It includes server status monitoring, platform resource monitoring, and other functions and can monitor the
consumption of server resources.

2. Enterprise Monitoring Platform

Some functions of the enterprise monitoring platform are the same as those of
the government platform. The characteristic functions of the enterprise platform
are as follows.

(a) Alarm information processing. The enterprise platform has the function of
processing the alarm information reported by the terminal and the alarm
information generated by the analysis of the enterprise platform. The alarm
information processing process includes alarm information confirmation,
alarm disposal, alarm processing registration, and alarm information
processing status tracking. Alarm processing can be based on different
types of alarm vehicle monitoring, photo, alarm release and send information
disposal, through the sending of information to achieve the purpose of
reminding the driver. The enterprise platform should support the real-time
transmission of alarm information and alarm processing result information to
the government platform and respond to the alarm disposal request

instruction issued by the government platform. All alarm and alarm processing information should be recorded and provided with query function.

(b) Monitoring function. Vehicle monitoring and management includes the functions of vehicle up and down line warning, vehicle scheduling, vehicle monitoring, vehicle tracking, vehicle roll call, vehicle search, area inspection, and vehicle remote control. At the same time, the enterprise platform has the function of playing back the specified vehicle history track in the specified time period and supports the joint query of multi-region and multi-time period.

(c) Management functions. These include terminal parameter configuration management (such as IP address configuration, alarm parameter configuration, area setting and route setting configuration, terminal firmware upgrade, etc.), terminal account opening, closing account, vehicle deactivation, vehicle sublease and terminal transfer, SIM card management, vehicle management, employee management, fleet management, etc. The business functions of the enterprise monitoring platform are as follows.

- Off-line alarm. Alarm when the vehicle deviates from the preset route range beyond the threshold.

    Key points of route monitoring. It supports the monitoring of the critical point time of the vehicle's driving path, that is, when the vehicle does not arrive or leave the specified position according to the specified time, it prompts in real time.

- Driver identification. Identify the driver's identity information uploaded by the terminal.

- Fatigue driving warning. When the continuous driving time of the vehicle or driver exceeds the fatigue driving time threshold, the fatigue driving time threshold can be set remotely by the monitoring center. The special business functions of passenger transport are as follows.

- Line passenger route query function.

- Monitor vehicle over-staffing by photo or video, provide direct call function to vehicle terminal, remind driver over-staffing, provide warning, record, and process over-staffing.

10.2.3 Example: Driving Behavior Monitoring and Early Warning System

This section takes Wuhan University of Technology as an example to further explain its function. The driving behavior monitoring and warning system is divided into two parts: (1) vehicle driving behavior safety auxiliary warning system and (2) remote information release and monitoring early warning. Through the two-level monitoring of the vehicle end and the monitoring center, the driver of the operating vehicle can be monitored in all directions in real time, and the

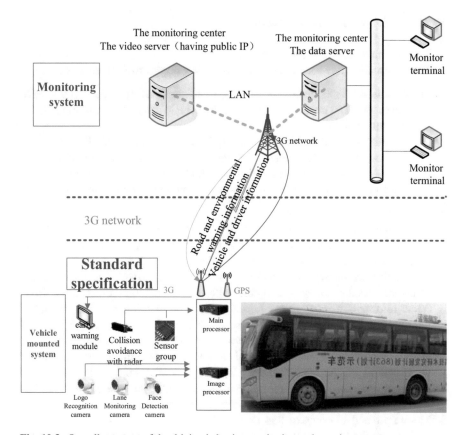

Fig. 10.3 Overall structure of the driving behavior monitoring and warning system

dangerous driving behavior can be identified, warned, and recorded. The overall structure of the system is shown in Fig. 10.3.

1. Vehicle driving behavior safety auxiliary warning system. Vehicle terminal system works by collecting driver operation, driver shape, the state of speeding, the state of not keeping safe distance, the state of lane deviation, bad driving, state identification, and early warning.
2. Remote information release and monitoring and warning system. The remote end monitoring and warning system can take regular photos of the driver, diagnose the terminal of bad driving behavior and analyze all kinds of reports by collecting the recorded data sent by the vehicle end, and issue safety warning information to all kinds of dangerous conditions. In addition, it also undertakes the function of information management of passenger transport company.

10.2.3.1 Vehicle Terminal System

1. Vehicle Status Information Collection.
   The information that the system needs to collect comes from three aspects.

   (a) Vehicle sensor or data signal sent by vehicle CAN bus after vehicle ECU processing. The position is mainly in the vehicle panel and the terminal in the instrument table.
   (b) Self-contained information processing modules, sensors (including additional sensors, cameras, millimeter-wave mines); and reach, ultrasonic probe, etc.
   (c) GPS signal acquisition module and vehicle side environment information (transmitted through wireless transmission module antenna) and so on.

2. On-board System Integration.
   The integrated hardware platform of vehicle terminal center control realizes the integration of vehicle and driver status information acquisition unit, obstacle ranging and speed measurement unit, vehicle location information, geographic information acquisition unit and road environment information acquisition unit, and completes the storage and interaction of driver operation information, driver status information, traffic sign information, road marking and lane deviation information, GPS information, and so on. The multi-channel image is mixed output to the TFT screen for voice broadcast and interface display of the early warning module; all kinds of state and early warning information are transmitted to the central master control platform in 3 ways, and all kinds of instructions of the central master control platform can be received to provide support for bad driving behavior detection and multi-way early warning.

3. Monitoring and Warning of Bad Driving Behavior.
   Through the experiment and analysis of the driver's bad driving behavior, the online detection of typical bad driving behavior and the early warning of all kinds of dangerous driving behavior are realized, including overstepping gear, glide, first step clutch and then brake, long time off-hand driving, no turn light, wrong steering wheel operation gesture, stop and stop without flame-out, turn and brake at the same time, slam on the throttle, and rapid braking of the vehicle.

10.2.3.2 Remote Information Release and Supervision Platform

The development of vehicle remote information release and supervision platform integrates information transmission and processing technology such as database technology, network communication technology, wireless communication technology, and large capacity data processing technology, specifically, a GIS spatial database for the management of geographic information, road maintenance, traffic control geographic information, road maintenance, traffic control, speed, Oracle, and other large-scale relational database platform to establish vehicle real-time status database. On the Net/J2EE platform, an enterprise-level application system based on B/S architecture is developed, including the transportation vehicle management subsystem—which realizes the management of the basic information of the vehicle, the basic information of the driver, the information of the transportation, and the data maintenance system of the transportation line GIS. The information publishing platform based on LAN and GPRS, CDMA transmission can realize real-time update of related basic data such as vehicle equipment GIS. Through the online monitoring system and the corresponding communication module to transmit a variety of forms of transport vehicle status data (text data, voice, image), real-time monitoring of transport vehicles can be achieved.

10.3 Intelligent Networked Vehicle Testing and Evaluation System

10.3.1 Background and Development

Intelligent network-connected automobile is an important part of ITS, which is promoted by new technologies such as intelligent vehicle, vehicle-road cooperation, and so on. Intelligent Networked Automobile aims to realize the full space-time real-time information interaction between vehicle and X (person, car, road, cloud, etc.) through modern sensing technology, information fusion technology, wireless communication technology, intelligent control technology, etc., so that it has high reliable functions such as environment perception, intelligent decision-making, cooperative control, and so on. Finally, it realizes the complete automatic driving under man-car-road cooperation. Different from the traditional vehicle road test field, testing and evaluation is an essential link in the development of intelligent network-connected vehicle technology. The emphasis of intelligent network-connected vehicle testing is to examine the environmental perception ability, intelligent decision-making ability, and automatic control and cooperative control ability of the vehicle. At the same time, the real time and reliability of the V2X communication system and the information security technology are also investigated. The demonstration zone for intelligent connected vehicles integrates the function of technology R&D and technology testing. Moreover, it generally contains functions such as standard customization, industrial incubation, and industrial chain cultivation. The connotation of vehicle road test field, autopilot technology test field, intelligent network vehicle, and intelligent traffic demonstration area is shown in Fig. 10.4.

Meanwhile, numerous technologies related to artificial intelligence have been applied in the research and development of the perception and decision-making of intelligent connected vehicles. Test data from real road scene is essential for the training and learning of intelligent perception and decision-making algorithms. Therefore, the intelligent connected vehicle test system can not only perform functional tests on vehicles, but its test data is also an important cornerstone to promote the development of intelligent driving technology.

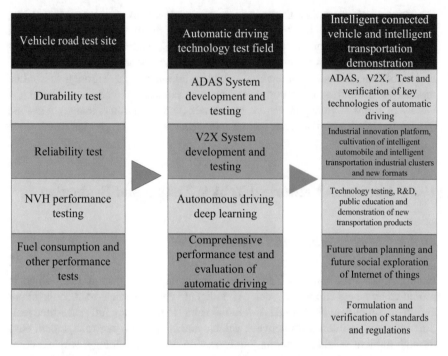

Fig. 10.4 The connotation of the three basic concepts

10.3.1.1 The Development of U.S. Intelligent Connected Vehicle (ICV) Testing

1. Establishment of Test Specifications for Intelligent Connected Vehicle.

   At the national level, the National Highway Traffic Safety Committee of the United States (NHTSA) released Preliminary Statement of Policy Concerning Automated Vehicles in May 2013. In this document, for the first time, the level of automotive automation is refined and clearly defined, and it provides guidance for the independent formulation of autonomous driving technical regulations by states. In September 2016, the U.S. Department of Transportation (DOT) officially released the Federal Automated Vehicles Policy, which proposed guidelines for the performance of autonomous vehicles and provided guidance and specifications for the safe design, development, and testing of highly automated vehicles (HAV), and it also identified 15 assessment contents, including data recording and sharing, system security, vehicle network security, post-crash behavior, object/incident detection and response, etc. In July 2017, the U.S. House of Representatives passed the Self Drive Act. Chapter 7 of the bill stipulates requirements for various participants, such as evaluation agencies, autonomous driving system providers, component suppliers, tested vehicles and on-board equipment, and participation in testing as well as information that must be submitted. In September 2017, the U.S. Department of Transportation and the

National Highway Traffic Safety Administration issued Automated Driving System 2.0: A Vision for Safety. This document replaced the federal self-driving car policy issued in 2016, reduced the content of the safety assessment from 15 to 12, and gave the safety goals and implementation methods of these 12 items. The guide encourages research and development entities to fully consider these 12 items when developing autonomous driving systems, and design a self-recording process for various evaluations, tests, and verifications. The guide encourages research and development entities to regularly publish their safety self-assessment results to the public, and demonstrate its various methods to achieve security; NHTSA will also regularly update the guide based on the development and improvement of technology to reflect lessons learned, new data, and the opinions of stakeholders.

2. Construction of a Closed Test Field for Intelligent Connected Vehicles

Currently, the U.S. Department of Transportation has designated a total of 10 autonomous driving pilot test sites, which are located in different regions such as the northeast, southwest, and south of the United States. These test sites distributed across the United States have differentiated climatic conditions and geomorphological characteristics, which make smart connected cars tested under more abundant conditions. Among them, Mcity, jointly funded and constructed by the University of Michigan and the State of Michigan, is the world's first closed test site to test V2X technology. It was officially opened in July 2015. The test site is located in Ann Arbor, Michigan, covering an area of 194 acres with the total length about 8 kilometers. A variety of roads and roadside facilities are set up in Mcity to simulate the real road environment, including high-speed experimental areas for simulating highway environment and low-speed experimental areas for simulating urban and suburban areas. Among them, the low-speed test area that simulates the urban area completely imitates the construction of ordinary towns, including two-lane, three-lane, and four-lane highways, as well as intersections, traffic lights and signs, etc., providing real road scene elements such as surfaces, signs and markings, slopes, bicycle lanes, trees, fire hydrants, and surrounding buildings.

GoMentum Station is located in Contra Costa County, California. It is the largest autonomous driving technology and connected vehicle technology test site in the United States. It is jointly constructed and operated by the local transportation authority, automobile manufacturers, communication companies, and technology companies. Now that the test site is located in the San Francisco Bay Area, close to Silicon Valley, it has attracted many technology companies, automakers, and network-connected equipment manufacturers to enter the test site.

At present, GoMentum Station has paved 32 kilometers of roads and streets, including overpasses, tunnels, railways, and other facilities and has the geological characteristics of hills, slopes, and various road surfaces to achieve various test scenarios, as shown in Fig. 10.5.

Fig. 10.5 Inner scene of GoMentum Station

10.3.1.2   The Development of Domestic Intelligent Connected Vehicle (ICV) Testing

1. Establishment of Test Specifications for Intelligent Connected Vehicle.

   In December 2017, Beijing issued the Beijing Autonomous Vehicle Road Test Management Implementation Rules (Trial) and related documents and identified 33 open test roads with a total of 105 kilometers. In March 2018, Shanghai issued the Shanghai Intelligent and Connected Vehicle Road Test Management Measures (Trial), delineating the first phase of 5.6 kilometers of open test roads; Chongqing, Shenzhen, Baoding, and other places have also issued corresponding road test rules. After initial attempts by local governments, in order to implement autonomous driving road tests in various regions, further clarify the management requirements and division of responsibilities for road tests, standardize and unify local basic test items and test procedures, on April 12, 2018, the Ministry of Industry and Information Technology, Public Security Bureau, and the Ministry of Transport jointly issued the Management Specifications for Road Testing of Intelligent Connected Vehicles (for Trial Implementation), which set strict requirements on test subjects, drivers, and vehicles respectively. The Measures proposes to the test subject the nature of the test entity, business scope, accident compensation capability, test evaluation capability, remote monitoring capability, event record analysis capability, and compliance with laws and regulations; it proposes to sign a labor contract or labor service contract, and process the test driver with autonomous driving training, no major traffic violation records, and other 8 requirements. Six basic requirements are put forward including the test vehicle registration, mandatory project inspection, man-machine control mode conversion, data recording and real-time return, specific area testing, and third-party agency testing and verification. At the same time, in order to ensure that the violating party is held accountable for traffic violations and traffic accidents during the test period, the "Management Regulations" clarifies the basis for the determination of traffic violations and accident liability, as well as the corresponding handling and punishment departments, and stipulates that the parties shall be involved after an accident. It also stipulates the reporting requirements of the parties, the test subjects, and the provincial and municipal departments after the accident.

2. Construction of a Closed Test Field for Intelligent Connected Vehicles

In June 2015, the Ministry of Industry and Information Technology of the People's Republic of China approved Shanghai to build a pilot demonstration zone for the National Intelligent Connected Vehicle (Shanghai). In June 2016, the demonstration zone was officially put into operation, becoming China's first intelligent network connection demonstration zone. Since the opening of the park, more than 200 test scenarios have been built in the Shanghai Intelligent Networking Closed Test Zone and more than 450 days and more than 5000 hours of test services have been provided to more than 40 companies.

Following Shanghai, places of Chongqing, Beijing, Hebei, Zhejiang, Changchun, Wuhan, and Wuxi have successively built smart connected vehicle test demonstration zones. Relying on local advantages and the distribution of characteristic resources, demonstration zones have actively promoted test and verification for semi-closed and open roads.

10.3.2 Test Verification Technology and Test Method

10.3.2.1 Analysis of Requirements for Test Verification of Intelligent Connected Vehicles

Intelligent networked vehicles use on-board sensors to sense the environment and integrate modern communication and network technologies to realize intelligent information sharing, and realize safe, comfortable, energy-saving, and efficient driving through processes such as intelligent decision-making and collaborative control. The research of intelligent networked vehicles and related intelligent technologies has not only become the development engine of the current combination of information technology, sensor technology, and cognitive science, but also can be applied to future smart transportation and smart cities to reduce road traffic accident rates and casualties. It can improve transportation efficiency and driving experience, and at the same time promote and drive the development and reform and innovation of the automobile industry, and provide key technical support for related research and applications in the field of national defense and security.

The technological progress and application promotion of intelligent networked vehicles need to be supported by a complete test and evaluation system. The specific requirements are mainly reflected in the following four aspects.

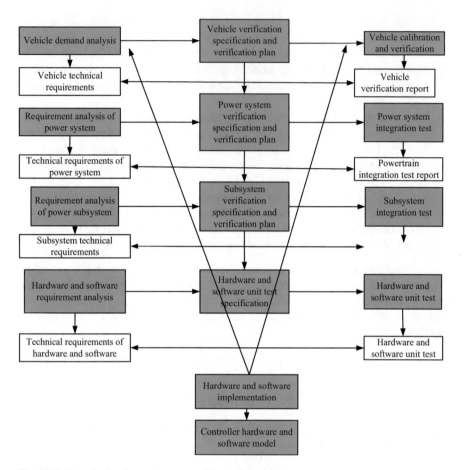

Fig. 10.6 V-mode development process of the entire vehicle

1. Requirements for Technology and Product Development.

   In the process of automobile development, the V-mode development method has been widely used. Testing and evaluation at each stage of development is the prerequisite for the realization of the V-mode development; the rapid development of intelligent networked vehicles has made it cover more and more functions. More and more user working conditions are targeted, so a lot of testing and evaluation work is needed for function and performance testing, as shown in Fig. 10.6.

2. Standard Requirements

   Different from the development of traditional cars, the performance of intelligent networked cars is highly related to the traffic environment and driving behavior. Therefore, the method of directly introducing foreign standards in the traditional automotive field is no longer applicable. It is necessary to study and formulate standards and specifications in line with China's national conditions.

3. Legal Requirements.

Intelligent networked vehicles, especially high-level autonomous and driverless vehicles, need to be supported and restricted by sound laws and regulations. On the one hand, in order to encourage technological innovation and promote technological development and industrial upgrading, it is necessary to vigorously promote autonomous driving/unmanned vehicles to enter public roads for testing and even commercial applications; but on the other hand, it is necessary to ensure that autonomous driving/unmanned vehicles entering public roads have sufficient safety, will not pose a danger to public safety, and will not interfere with normal traffic order. The test methods need to be coordinated with legal regulations to jointly promote the development of intelligent networked vehicles.

4. International Demands

Developed countries in the automotive industry such as Europe, the United States, Japan, and South Korea have all begun to build closed test grounds and pilot demonstration areas for intelligent networked vehicles, and related testing and evaluation systems and standards are being improved. We need to further accelerate the construction of intelligent networked vehicle evaluation platforms and test demonstration areas, promote the research and formulation of relevant standards as well as regulations, accelerate the docking with foreign advanced technologies, absorb advanced international experience, and promote the research and development of China's intelligent networked vehicle technology as well as industrial application.

10.3.2.2 Common Test Methods

With the continuous advancement of technology, the main content of automotive testing methods not only includes sensor-based measurement principles, methods, tools, data processing, etc., but also develops into widely used in-the-loop test methods that place the object in simulation operation. The following mainly introduces several in-the-loop test methods and real-vehicle test methods that have been common in automotive testing in recent years.

1. Model in the Loop, MIL

Model-Based System Engineering (MBSE) is a mathematical and visualization method used to solve related problems in the design of complex control, signal processing, and communication systems. It is widely used in motion control, industrial equipment, aerospace, and automotive-related applications, making it a major development and testing method.

The model provides an abstract method of the physical system, which allows engineers to ignore irrelevant details and focus on the most important parts to think about the overall design of the system. All the work in the project relies on the model to understand the complex real-world system. In MBSE, the model is an executable specification that is continuously refined throughout the development process (usually the development is guided by requirements expressed in text form, and these models are then converted into code with code generation). Compared with specifications written on paper, executable specifications can enable system engineers to have a deeper understanding of the dynamic performance of their strategies. The models are tested in the early development stage before coding, and product defects are exposed to the project, and continuous verification and testing during the development process will be applied so that engineers can focus on the research of algorithms and test cases to ensure the completeness and unambiguity of the specification. They do not have to spend time dealing with cumbersome, error-prone tasks, such as creating test fixtures. After the model is established, they will add model-based testing to ensure that the model does capture the requirements correctly.

Unlike "static" written designs, executable specifications can be evaluated during model-based testing. This can usually be done by changing a set of model parameters or input signals, or by viewing the output results or the response of the model. The sequence of simulation performed based on the model is also called model-in-the-loop testing. The test data for model-in-the-loop testing come from a test vector database or a model of the actual system. Executable specifications usually include not only functional design models and software logic, but also equipment and environment models, links to high-level requirements, and other files, and usually include verification data for automated simulation results evaluation. The results of model-in-the-loop testing can be used to verify whether the software behavior is correct and to confirm the initial requirements of the development process. The information collected through model-in-the-loop testing will become a benchmark for code verification.

2. Software in the Loop (SIL)

General software-in-the-loop testing refers to the evaluation of the code generated in the simulation or the handwritten code on the host to realize the early confirmation of the generated code. But this kind of test is only for the generated code and does not consider the correlation between the code and the model. Therefore, when a code problem is found during the test process, it is necessary to manually locate which model is the problem.

In addition, there is also a comparison test method between software-in-the-loop and model-in-the-loop, which mainly realizes the synchronous execution of model and code. This method inserts control code into the test code generated by the model to record status information and sends it to the modeling platform. After the analysis, the execution process of the model is displayed synchronously in the way of highlighting; at the same time, the test process can also obtain the currently monitored global variable information, and the tester can analyze whether the specific modeling is correct or meets the actual needs by monitoring the global variable information.

The key point of the software-in-the-loop testing method is to ensure that the generated code and model are executed synchronously, and the execution process should be intuitive enough for users to view. In addition, because additional control codes are added to the test code, there are differences between the test code and the product code, which will inevitably pose an impact on the performance of the code, so it is necessary to consider how to minimize this impact. Specifically, the following issues need to be considered when designing the test plan.

(a) The test process should accurately reflect the execution process of the model. The graphically described model and monitoring variable information must be correctly displayed during the test. When the code generated by a certain model element instance is executed, the model element instance must be highlighted. In the state diagram model, when a state transition occurs, it can automatically switch to the next state that meets the transition conditions. In addition, monitoring variable information also needs to be consistent with the code execution.

(b) The test process should be intuitive. The intermediate information of the test process should be clearly reflected in the graphical modeling. The more intuitive the test process is displayed in the modeling, the easier it is to check errors. At the same time, the display of the test process should learn from the display methods of existing tools to make it more convenient for modelers to use.

(c) The impact of the test process on the system under test should be minimized. For any kind of in-process test, the test process should minimize the impact on the system under test. At the same time, for the test code, if too much control code is inserted during the execution of the test, there will be a large inconsistency between the test code and the graphical model. Therefore, less code should be added to the automatically generated test code.

3. Hardware in the Loop (HIL)

The development of the automotive system project is a systematic project with high technology and heavy workload. The development of the entire vehicle and various parts and components is carried out at the same time. In order to ensure the progress of the project, putting the hardware in the test loop is a technical form that combines physical components and software models and is widely used in component testing or control system testing. Generally speaking, hardware-in-the-loop test systems can be divided into four types: the first type, the real controller is placed in the test loop, and the pressure or electrical signals of the remaining components are incorporated into the controller with real signals or signals simulated by the simulation environment. The control loop does not include the power loading device; the second type is to use a computer to quickly build its controller model, place the controlled object as a physical object in the simulation loop, and construct an in-loop test system. This process is also called rapid prototyping. The third type uses the dynamic characteristics of the power loading device to simulate the rest of the system to load the physical components, and the signals output by the physical components are fed back to the system model to form a system loop; the fourth type is mainly based on the second type. It incorporates the process quantity of the loop system model or the output quantity of the physical component into a larger controller to control loop.

The development of automotive hardware-in-the-loop test system is the most important measure to implement concurrent engineering to realize simultaneous development. Adopting the computer simulation test system can better solve the following problems.

(a) In the synchronous development project, it is necessary to solve the controller test in the absence of control objects and prototype vehicles in the initial development stage.

(b) Complete tests that cannot be carried out in practice or are expensive, and it is convenient to carry out accurate limit test, failure test, and the reproduction of various failures, making the test more comprehensive and complete.

(c) The dangerous situation is simulated without actual danger, and the test can be repeated and carried out automatically.

(d) The optimization of control strategy, the possible influence of each parameter, and the sensitivity of parameter changes can be verified quickly and economically, and conflicting targets can be found and coordinated early.

(e) The duplication and changes in the development process can be avoided to the greatest extent. Since the simulation has verified various operating states and functions, it avoids most of the errors in the design and greatly reduces the development risk.

(f) The cost of hardware and testing is reduced to a minimum, and the research time and development cost are greatly saved.

4. Real vehicle test method.

Model-in-the-loop, software-in-the-loop, and hardware-in-the-loop are suitable for testing controllers, components, systems, or assemblies, but when these parts or assemblies are assembled together, unexpected failures or problems often occur, so the entire vehicle must be tested and evaluated. The testing and evaluation of the whole vehicle generally requires the aid of a proving ground or large-scale testing equipment.

The automobile proving ground is the place where the whole automobile road test is carried out. In order to meet the actual testing needs of intelligent networked vehicles, the automobile test field facilities mainly include the following three aspects.

(a) Intelligent road network infrastructure. The smart road network infrastructure mainly includes participants (workers and scene dummies) and their accessory wearable devices: smart car fleets, including standard smart passenger cars, commercial vehicles, buses, taxis, etc. Vehicles need to install vehicle networking terminal equipment based on communication functions, which can realize a variety of network communications (5G/4G+, DSRC, LTE-V, etc.), and can collect real-time operating data of the monitored vehicle: road infrastructure, including bends, ramps, overpasses, tunnels, T-junctions, crossroads, crosswalks, lane markings and other traffic signs as well as street lights, traffic signs, road shoulder stones, roadside buildings, road test signal transceiver equipment, etc.

(b) Network communication environment. In the demonstration area, a comprehensive coverage of short-distance and long-distance communications such
as vehicle-vehicle, vehicle-road, vehicle-cloud computing, and road-cloud
computing will be formed, so as to realize the integration of the three
networks of in-vehicle network, inter-vehicle network, and vehicle-cloud
network. The in-car network uses mature technology to establish a complete
vehicle network; the inter-car network uses LTE-V technology, DSRC technology, and IEEE802.11 series wireless local area network protocols to form
a dynamic network; the car-cloud network uses 5G/4G+ and other communication technologies to form a dynamic network and connect wirelessly to
the Internet.

(c) Test for service support facilities. The automobile test site needs to be
equipped with a vehicle preparation room for the test unit to adjust the
vehicle, as well as a computing center for storing and processing test data, a
control center, and other equipment that provide support for testing services.

10.3.3 Case: The Closed Test Zone in Shanghai International Automobile City (F-Zone)

This section takes the closed F-Zone in the pilot demonstration zone of the National
Intelligent Connected Vehicle (Shanghai) as an example, focusing on the venues,
equipment, and testing capabilities of the closed test zone.

The F-Zone closed test zone covers an area of 2 square kilometers. It is invested
and constructed by Shanghai International Automobile City (Group) Co., Ltd., and
cooperates with the professional team of China Automotive Technology Research
Center to participate in the field service and operation. The test site has built more
than 100 test scenarios to meet various types of unmanned driving and V2X on the
basis of existing municipal roads, covering application categories such as safety,
efficiency, communications, new energy vehicles, etc.

10.3.3.1 Test Environment

1. Test area. F-Zone has built test roads with a length of 3.6 km, covering various
types of traffic roads such as T-junctions and crossroads. At the same time, there
are road facilities such as simulated tunnels, simulated tree-lined roads, and
simulated gas stations. Currently, more than 50 V2X test scenarios can be
implemented in the test area.

2. Office area. F-Zone provides 4 closed, well-equipped vehicle preparation rooms,
as well as supporting service areas such as parking, office, storage, and functional
areas such as data centers and control centers, as shown in Fig. 10.7.

Fig. 10.7 Environment of F-Zone office area

Table 10.1 Scenarios of network connection tests and quantity

Scene type	Security	Efficiency	Information service	New energy vehicle application	Communication ability test
Number	33	6	6	3	2
Ways of communication	V2V	V2I	V2P	V2C	Unlimited
Number	21	15	2	8	4

10.3.3.2 Testing Ability

F-Zone can provide two types of tests, connected type and autonomous driving type, and each type of test covers a variety of working conditions and applications.

1. Network test.

F-Zone provides more than 50 types of networking tests, covering security, efficiency, information service, new energy vehicle applications, and communication capability tests, and can be combined into a variety of custom scenarios, as shown in Table 10.1.

Among them, the safety test includes non-motor vehicle crossing warning, road slippery warning, vehicle conflict avoidance at intersections under the influence of sight distance, forward collision warning, emergency vehicle warning, emergency brake warning, red light warning, and no signal intersection traffic, accident ahead reminder, speed zone reminder, road construction reminder, blind spot warning, crossroad pass assist, overtaking assist, reverse overtaking reminder, left turn assist, pedestrian crossing warning, abnormal road warning, abnormal vehicle warning, post-accident warning, early reversing warning, following distance reminder, reverse driving reminder, restricted lane warning, vehicle size warning, night meeting car reminder, pedestrian crossing assistance, bus signal priority, and other test items.

The efficiency test items mainly include automatic parking, cooperative fleet, congestion reminder, automatic payment for arrival, green wave band traffic, dynamic lane management, etc.

Information service test items include smart parking guidance, charging/refueling reminders, in-vehicle signs, local map downloads, bus stop/outbound reminders, and information services based on ITS big data.

New energy vehicle application test items include charging map guidance, wireless charging, charging pile usage information prompts, etc.

The communication ability test items include the communication ability test under different signal blocking conditions such as tunnel traffic and tree-lined road traffic.

2. Autonomous driving test.

F-Zone provides flexible scene design, which can be combined into multi-level and multi-type custom scenes to meet the behavioral ability test of autonomous driving under normal driving conditions, and the collision avoidance ability test under dangerous conditions, as well as the exit mechanism and response. At the same time, it supports the test of the driver's ability to respond to misoperations in low-level autonomous driving.

The test content of the behavioral ability test is shown in Table 10.2.

37 dangerous driving scenarios set in collision avoidance test, as shown in Table 10.3.

The exit mechanism test refers to the test that requires the driver to take over or stop safely when the system fails or goes beyond the operational design domain (ODD). It mainly includes pre-start self-check, automatic driving mode visual prompts, detection and response to automatic driving mode (intervention and cancelation) restrictions (ODD), detecting and responding to technical failures, safe stopping, and other items.

The detection and early warning scenarios for driver misoperation include the driver leaving the steering wheel with both hands, deviating from the driving direction, dozing/sleeping, leaving the driving position, and other possible scenarios.

10.3.3.3 Test Equipment

The equipment of the closed test site in F-Zone mainly include communication network system equipment, positioning system equipment, video surveillance system equipment, road environment simulation equipment, test equipment, data acquisition equipment, etc., as shown in Table 10.4.

Table 10.2 Test items for self-driving vehicle behavior ability

Detect and respond to speed limit changes and suggested speed	Detect and respond to stopped vehicles	Go through the cross and turn	Comply with local car driving laws	Detect and respond to emergency vehicles
Leave the traffic lane and stop	Detect and respond to lane changes	Find a parking space through the parking lot	Follow the police/first responders who control traffic (as a traffic control device)	Give way to emergency vehicles at crossroads, three-way intersections, and other traffic control situations
Detect and respond to oncoming vehicles approaching	Detect and respond to static obstacles in the vehicle driving lane	Detect and respond to traffic restrictions (one-way streets, no turning, ramps, etc.)	Follow the construction workers who control the traffic mode (slow motion/stop sign bracket)	Give way to pedestrians and non-motor vehicles at intersections and crosswalks
Detect overtaking areas and no overtaking areas to overtake	Detect/respond to traffic signals and stop/yield signs	Detect and respond to the person directing traffic in the work area and unexpected/planned incidents	Citizens who respond to smart transportation after a collision	Keep a safe distance from vehicles, pedestrians, and non-motorized vehicles on the side of the road
Follow the car (including stopping and starting)	Pass a roundabout	Make appropriate first-pass decisions	Detect and respond to temporary traffic control equipment	Detect/respond to detours and/or other temporary changes in traffic mode

10.4 Traffic Organization and Management System for Large-Scale Event

In recent years, with the continuous development of urbanization and the people's living standards, more and more large-scale events have been held in major cities. Absolutely, the large number of people and vehicles attracted by the holding of large-scale events will cause huge traffic pressure on the city's daily traffic facilities and may cause congestion. Therefore, various traffic management measures are used to ensure the safety and smoothness of traffic during large-scale events. It can help minimize the impact on background traffic demand, and ensure that event participants quickly, orderly, and timely evacuate large-scale events during emergencies. Therefore, traffic organization and management system came into being.

Table 10.3 Collision avoidance test

Working conditions	Scenes
Vehicle breakdown	Off the road
The vehicle turns and loses control	
The vehicle goes straight and loses control	
Run the red light	Cross the road
Run the stop sign	
The vehicle deviates from the lane when turning	Off the road
Deviating from the lane when the vehicle is going straight	
The vehicle is off the lane when reversing	
Conflict with animals when the vehicle is turning	Animal
Encounter animals when the vehicle is going straight	
Conflict with pedestrians when the vehicle turns	Pedestrian
Conflict with pedestrians when the vehicle is going straight	
Conflict with cyclists (bicycles, motorcycles, electric bicycles, etc.) when the vehicle is turning	Non-motor vehicle
Conflict with a cyclist when the vehicle is going straight	
Conflict with other vehicles when the vehicle is reversing	Reversing
Turning vehicles conflict with vehicles traveling in the same direction	Lane change
When a stationary vehicle starts, it conflicts with a vehicle traveling in the same direction	
Vehicle changing lanes conflict with vehicles traveling in the same direction	
Vehicle drifting conflicts with vehicles traveling in the same direction	
Vehicle steering conflicts with oncoming vehicle	Driving in opposite directions
Conflict between a vehicle traveling straight and an oncoming vehicle	
When the vehicle follows the car, the steering conflicts with the preceding car	Follow the car
The vehicle approaches accelerates vehicle ahead	
The vehicle is approaching a vehicle driving at a lower and constant speed ahead	
The vehicle is approaching a vehicle that is slowing down ahead	
The vehicle is approaching a stationary vehicle ahead	
The vehicle turned left at a traffic lighted intersection and collided with the oncoming vehicle	Pass the road
The vehicle turns right at a traffic lighted intersection	
The vehicle turns left at an intersection without traffic lights and conflicts with the oncoming vehicle	
The vehicle goes straight at an intersection without traffic lights	
The vehicle turns at an intersection without traffic lights	
Avoid obstacles when turning	
Avoid obstacles when the vehicle is going straight	Off the road
Non-collision hazard	Others
Obstacles encountered when the vehicle is turning	Obstacle
Obstacles encountered when the vehicle is going straight	
Others	Others

Table 10.4 Device information

Environmental simulation equipment	Simulation equipment of road	Traffic signs board	Traffic signs
		Traffic lights	Traffic signal indication
		Road cone	Warning for keeping distance
		Floor tape	Simulated lane line
		Traffic signal control equipment	Traffic signal control
	Simulation equipment of traffic participant	Dummy	To simulate road pedestrians
		Car model	To simulate road vehicles
		Non-motor vehicle (model)	To simulate road non-motorized vehicles
		Animal model	To simulate road animals
		Others	/
	Simulation equipment of information environment	DSRC device	DSRC communication
		LTE-V equipment	LTE-V communication
		Wi-Fi communication equipment	Wi-Fi connection
		High precision map	Location targeting
		Information induction equipment	Induction display
		GPS differential base station	GPS high-precision positioning
Test equipment	Robotic driver		Driving operation
	RT-RANGE		Relative positioning
	VBOX		Data and image recording
	Four wheel alignment table		Wheel alignment
Data acquisition equipment	Vehicle data acquisition, road test data acquisition		Data collection
	Microphone		Acoustic signal collection
	Camera		Image signal acquisition
Infrastructure equipment	Spherical surveillance camera		All-round monitoring in the test area
	Microwave radar		Road monitoring
	Road feature detection camera		Junction monitoring

(continued)

Table 10.4 (continued)

Traffic violation detection equipment	Snapshot and video recording
RFID access control equipment	Access control

10.4.1 A Brief Introduction to the System

In recent years, Chinese comprehensive national strength has been continuously strengthened. With the increase of market economic activities and the improvement of people's material and cultural living standards, large-scale cultural and sports events, conventions and exhibitions, commercial promotion activities, and other large-scale events in cities have been held more frequently. Over the past several years, China has successfully hosted large-scale events such as the 2008 Beijing Olympic Games, the 2010 Shanghai World Expo, and the 2010 Guangzhou Asian Games. The holding of these large-scale events has enriched the material and cultural life of Chinese citizens, increased our international influence, and promoted economic development. However, the large number of spectators attracted by this has also brought a great impact and influence on urban traffic. As large-scale events are held in a short period of time, a large number of spectators will be concentrated in the venue and its surrounding areas, which will put great pressure on the traffic near the venue, cause road congestion and traffic accidents, and even affect the traffic and residents of the city.

In 1988, the Federal Highway Administration defined planned special events as events that occurred at a specified time and at a designated location that could cause an unconventional increase in traffic demand, such as exhibitions, entertainment activities, sports events, and holiday gatherings.

In China's first special regulation for the safety management of large-scale events, the Regulations on the Safety Management of Large-scale Social Activities in Beijing, the definition of large-scale social activities is as follows: Large-scale social activities refer to the sponsors renting, borrowing or other temporary occupation of places and venues, theatrical performances, sports competitions, exhibitions, garden parties, and other group activities held for the public.

With the holding of large-scale events, the demand for the road network has increased sharply, breaking the original balance of traffic supply and demand. To bring urban traffic back to a new balance, there are two commonly used methods: increase supply and decrease demand. Measures of increasing supply, such as improving road conditions and other hardware facilities, cannot be completed in a short period of time and will cause a waste of resources after large-scale activities are over. The demand can be reduced by reducing the scale of the event, but this method weakens the influence of the event and cannot achieve the purpose of holding large-scale events.

At present, there are relatively few domestic cases that specialize in large-scale event public transportation evacuation. Most of them are macro-expositions on large-scale event transportation organization and planning methods. Research in this area has great theoretical value.

Regarding the research on large-scale events, before the U.S. Federal Highway Administration defined planned special events in 1988, it mainly focused on the experience summary of traffic management for international super-large sports competitions. After that, progress focused on theoretical research on traffic organization management for large-scale events was made.

Foreign research on large-scale event traffic organization can be roughly divided into two categories, one is the summary of experience after the large-scale event is held; the other is the theoretical research on the problem of large-scale event traffic organization and management.

According to the traffic organization and management of the 2000 Sydney Summer Olympics, the 2002 Salt Lake City Winter Olympics, and the 2004 Athens Summer Olympics, the experience of traffic organization and management of foreign large-scale sports events is summarized as follows.

1. Organization and Management of Public Transportation

They established a dispatch center and service fleet dedicated to public transportation services for large-scale events and equipped with sufficient bus transportation capacity. For the Sydney Olympics, a private bus company composed of more than 2200 buses took on 35% of the spectator traffic for the event transportation services; the Salt Lake City Winter Olympics set up a total of 4 bus dispatch centers to ensure the normal operation of the public transportation service system during the Olympics. About 1000 buses were used exclusively for public transportation services for spectators.

A public transportation priority strategy was implemented. During the Athens Olympics, in order to ensure the priority of public transportation vehicles, special bus lanes were set up on some arterial roads, with bus priority control signals, and traffic control was implemented in the streets of the central area. Only local vehicles and public buses were allowed to pass. Dedicated bus and carpool lanes on some roads near the competition venues to give priority to ensuring smooth public transportation was set up; the 2000 Sydney Olympic Games used government subsidies to provide free ground bus or rail transportation to spectators holding tickets for the competition of the day, which increased the share rate of public transport trips.

2. Planning for Parking Lot and Transportation Hub

In order to meet the parking needs of large-scale events, a temporary parking lot is planned and constructed near the event venue, and other transportation interchange hubs are planned to alleviate traffic congestion caused by large-scale events. During the Sydney Olympics, a total of 197 temporary parking lots were planned and constructed, and shuttle buses were installed between the temporary parking lots and large-scale event venues. During the Athens Olympics, the planned transportation interchange hub was used to effectively alleviate the traffic congestion caused by the Games.

3. Transportation Needs for VIP and Other Special Personnel

   As for international large-scale events, there will be heads of state, government officials and other VIPs, event sponsors, news media reporters, and other personnel. Different from the transportation needs of ordinary audiences, these special personnel must be guaranteed to arrive at the event safely, quickly, and on time. In response to this part of the traffic requirement, the events set up individual lanes and adjust travel time to guarantee their travelling priority. And priority should be given to parking lot planning and right of way at intersections. Both the Athens and Sydney Olympics have adopted the measures and have been effectively used.

4. Traffic Demand Management

   Foreign traffic demand management measures often adopted for large-scale events include using fare strategies to adjust the distribution of traffic travel time and space; using parking fees to reduce the proportion of private car travel. The Athens Olympics uses discounts on all-day passes to reduce the traffic flow of large-scale events during peak hours; the Sydney Olympics has significantly increased the parking fees around the venues to reduce the travel of private cars and ease the traffic pressure around the venues.

   Traffic demand management measures in cities where large-scale events are held include off-peak commuting to and from work, holidays during the event, advocating remote office work at home, and prohibiting some vehicles from passing in specified areas. Both the Sydney Olympics and the Salt Lake City Winter Olympics adopted the above measures to reduce the amount of urban background traffic.

5. Traffic Information Release and ITS Application

   Utilize the Internet, TV, radio, and other methods to release traffic status information in a timely manner, and adopt the ITS to organize and manage traffic for large-scale events and monitor the operation of vehicles. The Sydney Olympic Games carried out the Olympic traffic information guide sign planning and distributed the "Traffic Information Handbook" free of charge. Both the Salt Lake City Winter Olympics and the Athens Olympics have used ITS technology extensively and played an important role in the organization and management of large-scale event traffic.

   The United States Transportation Research Council published a comprehensive report Traffic Planning and Management of Large-scale Events in 2003. The report outlines the organizational departments, managers, and technical programs related to the traffic planning and management of large-scale events.

   In September 2003, the U.S. Federal Highway Commission also published a report on the planning and management of large-scale events—Travel Management for Large-scale Events. The report is a relatively comprehensive and systematic literature on the organization and management of large-scale events. It not only analyzes and studies the definition of large-scale activities and its traffic characteristics, but also comprehensively discusses the preparatory planning of large-scale activities, the mid-term traffic operation and implementation of measures, and the later analysis and evaluation.

The domestic research on the transportation theory of large-scale events started relatively late, and only after the successful bid for the Beijing Olympic Games did relevant transportation research institutions intervene in this research field. In recent years, with the frequent holding of large-scale events, domestic research on the transportation organization of large-scale events has gradually increased. The traffic organization and management during the event played a role in guaranteeing the success of the Shanghai World Expo and the Guangzhou Asian Games.

Generally speaking, foreign research on the transportation organization and management of large-scale events started earlier and has achieved relatively rich research results. A more systematic and comprehensive large-scale event transportation organization and management method has been proposed, and it has been successfully applied to various activities. Domestic research on the transportation organization and management of large-scale events started relatively late, and there are few systematic studies on large-scale event traffic demand forecasting, transportation organization planning, traffic demand management, and transportation planning program evaluation.

10.4.2 Composition of the System

The traffic organization and management of large-scale events are different from the regular traffic organization in cities. Conventional urban traffic organization and management has a certain degree of stability and regularity. Traffic management methods are usually implemented in accordance with mature policies based on collecting historical data and investigating and analyzing current traffic. The transportation organization and management of large-scale events have the characteristics of large total volume, concentrated time and space distribution, hierarchical demand, multiple sources, and single sinks, and different types of large-scale events should often adopt different organizational management methods.

This section first introduces the principles of large-scale event traffic organization and management, and then further explains various traffic organization and management methods, including large-scale event traffic demand forecasting and management, large-scale event traffic organization planning, large-scale event traffic information release, and emergency traffic organization planning. It sets up the basic system composition of the large-scale event traffic organization and management system.

10.4.2.1   Analysis of Traffic Characteristics of Large-Scale Events

To conduct traffic demand forecasting for large-scale events and research on traffic organization and management methods, we must first analyze the traffic characteristics of large-scale events. This section starts from three aspects: traffic demand characteristics, traffic flow characteristics, and traffic organization and management characteristics. The differences between large-scale event traffic and urban normal traffic are as follows.

1. Traffic Demand Characteristics of Large-Scale Events.

The total traffic demand is large and the time is concentrated. The traffic demand induced by large-scale activities is significantly greater than the traffic demand generated by the city's land use of the same scale. During large-scale events, the amount of traffic attraction generated per unit area exceeds the traffic attraction generated by conventional land in the city by dozens or even hundreds of times.

For example, the Shenzhen Bay "Chun Jian" Stadium, where the opening ceremony of the 26th World University Games in Shenzhen was held in 2011, has an area of less than 1 km, but attracted about 50,000 participants.

Large-scale events that provide one-time services usually have a fixed start time and end time, which makes the traffic demand for large-scale events concentrated in time. For example, when the opening ceremony of the Shenzhen Universiade was dismissed, 50,000 participants had to be evacuated within one hour, and traffic demand was highly concentrated in time distribution.

The traffic demand is hierarchical. Different from the conventional urban traffic demand, the traffic demand for large-scale events has obvious priority. Especially for large-scale events with international influence, the participation of international and domestic VIP guests has a greater impact on transportation. According to the priority of traffic demand, meeting the travel needs of large-scale events at different levels while minimizing the impact on urban background traffic is an important goal of large-scale event traffic organization and management.

There are higher requirements for accessibility and punctuality.

In this case, the holding time of large-scale events is fixed, and it is necessary to ensure that the large-scale event participants arrive at the event venue on time before the event begins. Therefore, the effective traffic organization and control of the roads around the event venue are required to improve the intersections and roads that may easily prevent event participants from entering the venue on time. Besides, we are supposed to reasonably set up traffic routes so as to ensure the reliability of the connectivity of the transportation network between the venue for large-scale events and other nodes, the reliability of the traffic capacity of the road network around the venue for large-scale events, and the reliability of the travel time of event participants.

2. The Traffic Flow Characteristics of Large-Scale Events.

Compared with the conventional traffic flow in the city, the traffic flow for large-scale events has the following particularities.

(a) The temporal and spatial distribution of traffic flow is uneven, which has obvious high peaks and huge flow.

(b) The uneven distribution of traffic demand for large-scale events in time and space directly leads to the uneven distribution of traffic flow in time and space. Different from the urban background traffic, the peak traffic flow during large-scale events is huge, and the task of traffic organization and management is difficult.

(c) Unlike conventional travel, it does not have the characteristics of a balanced traffic flow distribution. When performing conventional traffic distribution, it is generally assumed that travelers are very familiar with the road network, traffic conditions, and traffic organization and management. The conventional travel route selection satisfies the users priority principle, and the traffic flow distribution model adopts a balanced distribution. Compared with regular trips, the participants of large-scale events are not familiar with the road network, traffic conditions, and traffic organization and management between the place of departure and the event venue. They cannot all choose the route with the least travel cost and the shortest travel time, so they do not meet the assumption requirements of balanced distribution.

(d) It has temporal and spatial volatility. The peak time of regular urban traffic flow is often relatively fixed, basically appearing on all road networks simultaneously, and the duration is roughly the same, such as the rush hour. The traffic peaks caused by large-scale events on the surrounding roads of the event venues have obvious temporal and spatial volatility. The traffic peaks gradually spread out and disappear from the event venues like waves along the surrounding main roads over time. Besides, different road sections or intersections will have traffic peaks of different sizes at different time periods.

(e) It has the characteristics of "multi-source single sink" traffic network flow. "Multi-source single-remittance" travel refers to the network travel with only one traffic attraction and multiple sources of traffic. Large-scale events are held at a fixed and unique location, and event participants have different starting points and unique travel destinations, thus forming a "multi-source and single-converging" transportation network.

3. Traffic Organization and Management Characteristics of Large-Scale Events.

(a) Difficulty. According to the traffic demand and traffic flow characteristics of large-scale events, the traffic flow induced by large-scale events is much higher than the normal traffic flow in cities. Conventional urban traffic organization and management can hardly meet the traffic demand of large-scale events. Therefore, the task of traffic organization and management for large-scale events is very difficult.

(b) Complexity. Participants in large-scale events have different priorities, and traffic needs are hierarchical. The transportation organization and management of large-scale events need to ensure that the transportation needs of different levels are met. Therefore, the organization's management work is complicated, and it is necessary to comprehensively adopt various traffic organization measures and management program.

(c) Amorality. The holding of large-scale events is temporary and lasts at most 10 days. The traffic planning and organization and management measures for large-scale events are also temporary. Accordingly, this feature must be considered to minimize the impact on the background traffic.

Conventional traffic management measures are formulated to alleviate traffic congestion caused by the new equilibrium. The difference is that large-scale activities are temporary. Traffic organization and management measures for large-scale activities appear after breaking the original equilibrium state and end when a new equilibrium state is not formed. The aim of the traffic organization and management of large-scale events is to alleviate the traffic pressure in an unbalanced state, which is different from the validity period of conventional traffic management measures. Therefore, the city's conventional traffic organization and management methods cannot be directly adopted in the traffic organization and management of large-scale events.

10.4.2.2 Forecast of Traffic Demand for Large-Scale Events

In order to ensure the rationality and practicability of traffic demand forecast results, the following principles should be followed when forecasting traffic demand for large-scale events.

1. Combining Theory with Practice.

   Traffic demand forecasting for large-scale sports activities does not only concern relevant prediction theory methods to make scientific and accurate predictions but also combines practice, and we try to choose simple and easy-to-understand prediction methods and models on the basis of ensuring prediction accuracy, so that they can be smoothly applied in practice.

2. Systematization and Comprehensiveness.

   The transportation demand of large-scale sports events is an integral part of the city's transportation system, and it is interconnected and influenced with other subsystems of the city's transportation system. Therefore, it is necessary to comprehensively consider various related factors when forecasting traffic demand for large-scale sports activities and conduct comprehensive and systematic research.

3. Quantitative Analysis Combined with Qualitative Analysis.

   The traffic demand forecast of large-scale sports activities lacks directly usable historical data, so it is necessary to combine quantitative analysis methods with qualitative analysis methods to make scientific, reasonable, and accurate forecasts.

10.4.2.3 Traffic Demand Management for Large-Scale Events

Traffic demand management refers to the use of urban land planning, economic means, laws and regulations, information release and other control methods by government departments to adjust the total urban traffic demand, travel structure, and traffic when resources and environment are limited. Its purpose is to achieve a balance between traffic demand and traffic supply. As an effective approach to control urban traffic travel demand, traffic demand management theories and

Table 10.5 Comparison of traffic demand management measures in Beijing, Shanghai, Guangzhou, and Jinan

	Beijing	Shanghai	Jinan	Guangzhou
Traffic control of odd and even numbers	√	√	√	√
Traffic control of license plate end number		√	√	√
"Yellow label vehicles" prohibited	√		√	√
Part of the buses closed	√	√		√
Half-day holiday for the opening ceremony	√	√	√	√
Staggered commuting	√			√
Traffic control of transit vehicles	√			
Improve public transportation security	√	√	√	√
Building construction control	√	√	√	√

measures have been widely utilized in countries around the world since the 1990s, greatly improving the operational efficiency of the transportation system.

During large-scale events, people's travel needs should be considered in the formulation of urban traffic demand management measures, meeting traffic activities services, and minimizing the impact on the original traffic. During large-scale events, urban traffic demand increases sharply, and all the cities need to take traffic demand management measures. In this section, we analyze and study the 2008 Beijing Olympic Games, the 2009 Jinan National Games, the 2010 Shanghai World Expo, the 2010 Guangzhou Asian Games, and other large-scale competitions or expositions. A summary of the comparison is provided in Table 10.5.

In order to ensure the smooth flow of traffic in large-scale competitions or expositions, all the cities have planned to set up the dedicated channels; all the cities have implemented traffic demand management and traffic control measures to ensure the normal operation of urban traffic. As seen above, the measures that can be taken in the management of traffic demand for large-scale events can be divided into three aspects: the first one is to control the total travel demand so as to ensure the traffic demand of large-scale event participants, minimizing the other needs that affect the normal progress of large-scale events; second, adjusting the structure of travel modes and guiding citizens to choose public transportation during large-scale events; the last one is to adjust the travel time and space distribution to promote the equalization of the time and space distribution of traffic volume.

10.4.2.4   Traffic Organization Planning of Large-Scale Events

The purpose of traffic organization includes improving the utilization rate of the road network, eliminating the hidden dangers of road traffic accidents, maintaining the order of traffic operation, optimizing the organization of traffic flow, and ensuring smooth and safe road traffic. The transportation organization planning for international large-scale events is a complex project that requires special planning. The transportation organization workflow is shown in Fig. 10.8.

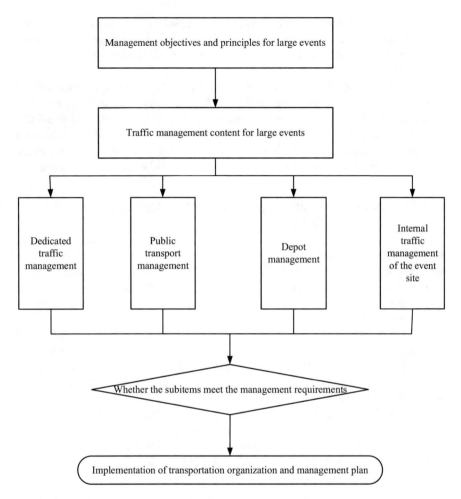

Fig. 10.8   Organization planning process of large-scale event traffic

10.4.2.5   Release of Traffic Information for Large-Scale Events during Large-Scale Events

Temporary traffic organization and management measures will have a certain impact on the daily travel of urban residents. Therefore, it is necessary to release traffic information in a timely manner so that travelers can learn about the latest traffic organization and management measures and choose the appropriate travel mode and travel route. During large-scale events, traffic information can be released before and during the trip of residents.

1. Release traffic organization and management information before residents travel. Television, radio, the Internet, brochures, newspapers, and other means are

utilized to publish traffic information so that residents can understand the various measures of urban traffic organization and management during large-scale events before they travel, and then they choose appropriate transportation modes and travel routes.

2. Release dynamic traffic information during travel. According to the basic data of the city's ITS, various methods such as VMS guidance screen, mobile phone text messages, on-board broadcasting, and on-board navigation are used to release real-time dynamic traffic guidance information to provide travelers with road condition information. Travelers can adjust their routes in time according to the dynamic information obtained during the trip to avoid entering traffic jams.

10.4.2.6 Emergency Traffic Organization and Management Plan of Large-Scale Events

In order to prevent unexpected events affecting the smooth conduct of large-scale activities, it is necessary to formulate traffic emergency organization and management plan for large-scale activities in view of possible major, general traffic events or safety accidents. The following principles should be followed in developing emergency programs.

For large-scale event personnel security, should focus on the protection of international and domestic VIP guests, taking into account other personnel traffic safety. For large-scale activity control areas, emphasis should be placed on the security areas within the activity sites, taking into account the roads and parking surrounding the activity sites.

Strict deployment and improvement of the system. Deploy a large number of police forces in important traffic distribution points, equipped with adequate intercoms and other tools to detect security risks and prevent accidents. In case of an emergency, on-site staff should process and report it in a timely manner to minimize losses as far as possible.

10.4.3 Example: Transportation Organization and Management of the Shenzhen World University Games

World University Games is abbreviated as the WUG. It is a large-scale comprehensive sports event which is only second to the Olympic Games, and it is known as the "Little Olympics." It is hosted by the International University Sports Federation. Its predecessor is the International University Games, which is limited to college students and college students who have graduated not more than 2 years (age limit is 17–28).

The 26th Summer WUG was successfully held in Shenzhen, China, from August 12 to 23, 2011, lasting 12 days. It has a total of 24 competition events and

306 sub-events, the highest in history; more than 10,000 athletes from 180 countries and regions participated in the competition. The registered staff includes four categories with a total of about 21,000 people.

Let's take the Shenzhen World University Games as an example for detailed traffic organization and management system analysis.

10.4.3.1   Traffic Demand Forecasting of Shenzhen World University Games

The demand forecast for special vehicles for registered customer groups during the WUG also includes a four-stage method, as shown in Fig. 10.9.

1. Traffic generation and distribution. Taking into account the number and importance of various service targets such as athletes, technical officials, media, general sports federation officials, and guests participating in the competition, and the number of rooms in the arranged hotel, the number of registered customer groups staying in each hotel is predicted.
2. Classification of transportation modes. All participating members will take special vehicles to the stadiums. According to the traffic service standard, the number of special vehicles produced by each hotel is predicted.
3. Traffic distribution. The traffic needs of the registered customer group are all allocated to the dedicated lane system, and the existing urban traffic model

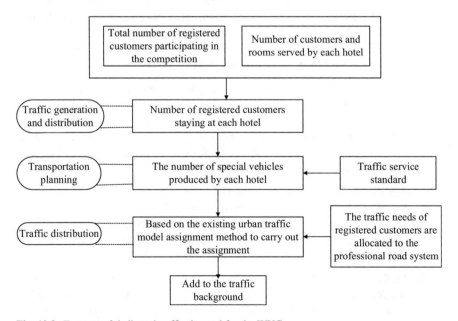

Fig. 10.9 Forecast of dedicated traffic demand for the WUG

allocation method is utilized for allocation and then superimposed on the background traffic.

During the Shenzhen WUG, free public transportation services, which include free rail transit, free regular buses, and some free taxis, will be provided for all registered personnel. They include ticket holders, certificated staff, and volunteers.

Registrants can use public transportation for free if they fail to catch the dedicated shuttle bus or other special circumstances. WUG has about 50,000 certified staff in 8 districts of the city. During this period, about 25,000 volunteers served as language assistants, dispatch assistants, and car assistants to maintain the orderly and efficient operation of the event. Volunteers for the competition live in hotels, school dormitories, etc. in Shenzhen. The number of people in the above three categories is relatively fixed. Those who visit the venues or watch the game can be treated as spectators for traffic organization and management.

10.4.3.2   Traffic Demand Management during the WUG

According to the transportation service needs of the WUG, the background traffic profile of Shenzhen, and the experience of other cities, the following traffic demand management measures have been formulated after comprehensively considering various factors.

1. Holidays for the Opening and Closing Ceremonies.

    According to the residents' travel conditions, we analyze the expected flow changes of different holiday plans based on historical data. Starting the holiday 1 day before the opening ceremony of the Universiade is more conducive to reducing the volume of social traffic on the opening ceremony. Taking all the factors into consideration, the following holiday plan is proposed.

    (a) Scope of holiday. In addition to ensuring the necessary jobs such as ensuring the operation of the city, the city' s government agencies, enterprises, institutions, and social organizations will implement the short vacations which include 4 days in opening ceremony and 3 days in closing ceremony; other social organizations in the city's administrative area can arrange their own arrangements according to actual conditions.

The holiday schedule is shown as follows.
Four days off during the opening ceremony (August 12): August 11–14;
Three days off during the closing ceremony (August 23): August 22–24.
Among them, August 11, 22, and 24 are the days for taking rest, and August 6, 20, and 21 are the days for working.

    (b) Implementation effect. During the holidays, the traffic flow of the road network dropped by an average of about 13%, reducing about 260,000 vehicles on the road. Meanwhile, it will have a certain negative impact on the production and living arrangements of enterprises and citizens.

The specific operation method is shown as follows.

(a) Policy publicity. Based on coordination with higher-level departments, neighboring provinces and cities, and news media at all levels, promotion of temporary traffic management policies for the WUG and holiday policies for the opening and closing ceremonies was carried out in a large-scale and high-density manner. The publicity work of the opening and closing ceremonies and holidays will be assigned to districts and street offices, and high-density publicity will be carried out through the Internet, radio, television, newspapers, and car owners' short messages.

(b) Notice issuance. The General Office of the Municipal Party Committee will issue a holiday notice for the opening and closing ceremonies of the city; all the district governments, neighborhood offices, and neighborhood committees will perform well in holiday notice.

2. The Traffic Control of Motor Vehicles According to Odd and Even Numbers.

(a) Limited travel time. August 4–24, 2011, 7–20 o'clock every day (opening and closing ceremonies extended to 24:00).

(b) Restricted range. The restricted area includes three areas and five roads. The three areas are the original special zone, Bao'an central area and Longgang central area; the five roads are Shuiguan Expressway (including Nanping Express), Qingping Expressway, Bao'an Avenue, Yanba Expressway, and Pingxi Road.

(c) Traffic control plan. All the motor vehicles will run on single-number single-day and double-number double-day driving according to the last Arabic numeral of the mainland-approved license plate number (including temporary license plate vehicles) (single number is 1, 3, 5, 7, 9, and double number is 2, 4, 6, 8, 0).

Identifying via appearance mark and license plate number, the vehicles that do not need to be restricted include: police cars, firefighting, ambulance, engineering rescue vehicles and vehicles of the armed police force performing missions; public transport vehicles, inter-provincial long-distance passenger vehicles, tourist passenger vehicles, large buses, and taxis; vehicles for environmental sanitation, gardening, and road maintenance; urban management, industrial and commercial, traffic, environmental law enforcement vehicles, weather monitoring vehicles, postal and rescue vehicles, and special vehicles for wrecking; vehicles of embassies, consulates and international organizations in China.

As for cars in Shenzhen and Hong Kong: On the day of the opening and closing ceremony (7:00–24:00), we can enter the three areas of the original special zone; Longgang central area, and Baoan central area, Baoan Avenue, Shuiguan Expressway (including Nanping Express), Qingping Highways, Yanba Expressway, Pingxi Road, and other roads are subject to restrictions on odd and even numbers.

As for the trucks in Shenzhen and Hong Kong: According to the principle of "east in and east out, middle in and middle out, west in and west out" that the Shenzhen government and the Hong Kong Special Administrative Region government have agreed and have been implementing vehicles on the main roads are strictly controlled, and permits should be obtained if there is a real need.

Vehicles that need certificates: Dedicated vehicles with certificates issued by the Shenzhen Universiade Organizing Committee; freight vehicles with temporary motor vehicle passing certificates issued by the Municipal Public Security Bureau; motor vehicles to ensure smooth production and operation in the city.

Management and control ideas and principles: Guidance, advice, and duty are directed at all inter-city checkpoints, second-line gateways, and important nodes (expressway toll stations and related nodes in restricted access areas).

- Propaganda Duty Point. Distributing leaflets at this type of duty point to provide explanations for drivers.
- Guiding Duty Points. Propaganda at such duty points through leaflets to guide restricted vehicles to detour.
- Implementation Effect. Reducing about 600,000 vehicles on the road, accounting for 30% of the total number of vehicles, but it will cause inconvenience to citizens' daily travel.
- The specific operation method is as follows.
- Policy announcement. Based on the coordination with higher-level departments, neighboring provinces and cities, and news media at all levels, promotion of temporary traffic management policies for the WUG and holiday policies for the opening and closing ceremonies was carried out in a large-scale and high-density manner. The publicity work of the opening and closing ceremonies and holidays is assigned to districts and neighborhood offices, and the neighborhood offices and neighborhood committees are responsible for all communities so that the publicity work is spread to every household; and high-density promotion is carried out through media such as the Internet, radio, TV, newspapers, and car owners' text messages. Communicating and coordinating with the Hong Kong Special Administrative Region government on the issue of restrictions on transit cars.
- Issuing and filing. The issuance and filing of passes for special vehicles for the WUG; the issuance and filing of passes for vehicles for clearing the port of freight vehicles and ensuring the normal production and operation of the city.
- Investigation and punishment of violations of the law. A fine of 200 yuan and 3 points will be deducted for vehicle owners who go on the road in violation of the regulations, and the automatic license plate recognition system is used to investigate and deal with illegal vehicles on the road, which play the role of electronic police "one police with multiple capabilities" so that the fixed electronic police can carry out red lights, speeding, and speeding. It can also investigate whether the relevant vehicles violate the regulations of odd and even number traffic control at the same time; the police utilize manual video capture to investigate and deal with illegal vehicles on the road.

3. Some buses are stopped.

   (a) The area of stopped driving: the administrative area in Shenzhen.
   (b) Suspension time: August 4–24, 2011, 7–20 o'clock every day (extended to 24:00 on the day of the opening and closing ceremonies).
   (c) Suspension plan: On the basis of the traffic control according to odd and even numbers, on the opening and closing ceremony of the WUG, government agencies, state-owned enterprises, and public institutions will stop 80% of their vehicles, and 50% of their vehicles in the rest of the time. Non-stop vehicles: vehicles holding a special vehicle certificate issued by the WUG Organizing Committee to guarantee the normal operation of the WUG; yellow card buses with more than 10 seats.
   (d) Implementation effect: It plays a leading demonstration role, which has good social significance, but it brings inconvenience to the government's official travel.
   (e) Specific operation methods: Policy publicity, discussions with various units, publicity of the WUG policies, and at the same time, the distribution of official car number plates of each unit is investigated and filed; illegal investigation and punishment, relevant personnel are assigned to the garage of each unit for on-site inspection.

4. Adjustment of Freight Transportation Organization.

   Drawing on the experience of Beijing, Guangzhou, and other cities, Shenzhen plans to implement measures to control transit vehicles during the WUG. Shenzhen City has proposed the goal of building an international logistics hub city. During the WUG, in order to reduce the impact on the logistics industry and at the same time ensure the transportation needs of the WUG, it is considering the implementation of freight detours on the three main passages of the WUG: Shuiguan Expressway, Qingping Expressway, and Longxiang Avenue.

   (a) Detour time: August 4–24, 2011.
   (b) Detour plan. Detour on the day of the opening and closing ceremonies (7:00–24:00): Based on the existing truck restrictions, four measures such as prohibition, restriction, regulation, and detour will be adopted to regulate the driving order, and guide freight vehicles to avoid Shuiguan Expressway, Qingping Expressway, and Longxiang Avenue. Trucks are allowed to bypass Shenhui Road along the alternative road. Except for the opening and closing ceremonies, other race days (7:00–20:00) detour Qingping Expressway and Longxiang Avenue: the detour route is the same as the opening and closing ceremonies; Shuiguan Expressway: it regulates trucks to drive in two outer lanes, depending on traffic conditions, and the rescue lane can be used as a backup lane for freight.
   (c) Implementation effect. Reducing the impact of freight traffic on WUG transportation may also have a greater impact on the freight industry.
   (d) Specific operation methods. Adopting TV, newspaper, radio, and on-site consultations for publicity; port publicity: distributing brochures in the

three port areas of Yantian, West, and Dachan Bay; as for port transit trucks, contacting and coordinating Hong Kong freight industry organizations to provide freight companies in Hong Kong promote with drivers; propagating at toll gates where trucks usually pass. Announcing detour routes in accordance with legal procedures and formulating traffic emergency plans for truck detours during the WUG. On the road, the police used the method of blocking vehicles on the spot to investigate and deal with illegal vehicles on the road.

10.4.3.3   Transport Organization Planning for the WUG

First of all, the WUG' s transportation guarantee and command system will be established. Each dispatch center and transportation service team will be staffed with clear job responsibilities. Then, dedicated transportation planning, public transportation planning, transportation site planning, internal transportation planning of competition venues, emergency transportation organization, and opening and closing ceremonies transportation organization planning are carried out respectively.

10.4.3.4   Summary After the WUG

The Shenzhen Universiade has ended. The WUG Transportation Guarantee has a total of 66,803 vehicle trips, transported 386,478 people registered for the WUG, and the total mileage of vehicles reached 2,931,365 kilometers. There were no safety liability accidents and registered group ride delays and effective complaint incident. This fully realized the overall goal of "safety, punctuality, reliability and convenience."

At this year's WUG, the Shenzhen Municipal Transportation Commission and the transportation industry have invested a total of 9538 traffic security service personnel, investing 2455 various security vehicles, and opening 367 various transportation service lines such as arrivals and departures, competitions, and interviews for 21,000 people. The registered people provide transportation services such as arrival and departure, competitions, training, watching competitions, interviews, exchanges, and opening and closing ceremonies. During the event, the regular public transportation system transported 1,497,900 spectators for free, 9494 buses were operated, and the subway transported 2.066 million spectators for free.

In terms of traffic protection for the opening ceremony, it took only 27 minutes for the athletes to evacuate on-site, and only 22 minutes for the on-site evacuation of the athletes at the closing ceremony; the time for the athletes to the competition and training venues did not exceed the 60 minutes specified by the International Sports Federation, and they all were safe and punctual. The average evacuation time for spectators by bus was 25 minutes, which was 20 minutes less than the promised 45 minutes, and both reached the international leading level. The input of traffic service personnel, drivers, vehicles, intelligent command and dispatch, delivery of vehicles, mileage, delivery service punctuality, and safety rates all set the highest level in previous WUG, reaching or exceeding the international level of large-scale competitions.

During the WUG, after the implementation of the traffic demand management policy in Shenzhen, about 800,000 vehicles were reduced on the road every day. Policies such as reducing traffic on odd and even numbers and prohibiting traffic on yellow-label vehicles, which mainly reduce traffic, have reduced road network traffic flow by about 25%. In addition, the implementation of two small holidays has reduced road network traffic by about 30%. After the establishment of a dedicated channel for the WUG and the implementation of traffic demand management, the service level of the road network has increased by 41.4% compared with the current situation, and the vehicle speed has increased by 42.3%.

References

1. Yang Z (2017) A brief analysis of traffic congestion based on Shanghai road traffic state index [J]. Traffic Transport 12:7–11. (Published in Chinese)
2. Payne HJ, Helfenbein ED, Knobel HC (1976) Development and testing of incident detection algorithms[R], vol 2. Res Methodol Detailed Results, Virginia
3. Jacobson LN, Nihan NL, Bender JD (1990) Detecting erroneous loop detector data in a freeway traffic management system[J]. In: Transportation research record 1287. TRB, National Research Council, Washington, D.C., pp 151–166
4. Smith BL, Scherer WT, Conklin JH (2003) Exploring imputation techniques for missing data in transportation management systems[C]. In: Presented at the 82nd TRB Annual Meeting (CD-ROM). Transportation Research Board, Washington D. C
5. Zheng T (2012) Prediction and application of urban road traffic flow [D]. South China University of Technology, Guangzhou. (Published in Chinese)
6. Qi L (2013) Research on traffic state monitoring and prediction method based on multi-source data [D]. Jilin University, Jilin, pp 1–223. (Published in Chinese)
7. Qiaojun X (2016) Construction of monitoring index system of urban road traffic safety status [D]. Southeast University, Jiangsu, pp 1–61. (Published in Chinese)
8. Rui Z (2014) Research on freeway traffic safety risk evaluation method [D]. Changan University, Shanxi. (Published in Chinese)
9. Shu F (2007) Experimental research on driving fatigue based on the cockpit [D]. University of Science and Technology of China, Anhui. (Published in Chinese)
10. Yang Z (2016) Introduction and application cases of Shanghai road traffic state index [J]. Traffic and Transport 3:16–18. (Published in Chinese)
11. Matthew GK, Kepaptsoglou K, Anthony Stathopoulos A (2004) Decision support system for special events public transport network planning: the case of the Athens 2004. In: Summer Olympics, TRB Annual Meeting CD-ROM
12. Sattayhatewa P, Smith RL (2003) Development of Parking Choice Models for Special Events. In: TRB Annual Meeting CD-ROM
13. Organization "Athens 2004" Transportation Authority. Athens 2004 Olympic games transportation plan, Athens, Greece, 2002
14. Karlaftis M, Kepaptsoglou. (2004) A model for deriving optimal headways and bus types for the Athens 2004 Olympic dedicated bus lines. In: 2 National Greek Conference for Research in Transportation
15. Abduh OIY (1997) Special events macroscopic simulation model with application to the hajj area in Saudi Arabia[D]. University of Pittsburgh, Pittsburgh

16. Zhagn Y (2003) Modeling traffic impact under special events[D]. University of Akron, Akron
17. Wenqun X (2001) Geographic information system GIS Digital City construction guide [M]. Hope Electronic Publishing House, Beijing, p 9. (Published in Chinese)
18. Bin X, Han W (2005) Design and development of urban digital traffic management information system based on GIS [J]. Comput Digital Engineering 33:16–18. (Published in Chinese)
19. Zhu Yin L, huapu. (2005) Design of Intelligent Traffic Management Information System Based on C/S and B/S hybrid system structure [J]. Highway Traffic Technology 22(11):147–151. (Published in Chinese)
20. Xiaoqin Z (2004) Research on network-based management information system [J]. Comput Eng Appl 5:224–226. (Published in Chinese)
21. Simei G (2004) Design and implementation of traffic management information system based on J2EE [D]. Jilin University, Jilin, pp 1–72. (Published in Chinese)
22. Hongjun C (2006) Research on key Technologies of Traffic Organization and Management for large-scale events [D]. Southeast University, Nanjing, pp 1–171. (Published in Chinese)
23. Shenggen J (2005) Research and Design of Monitoring System of intelligent traffic command center based on information platform [D]. Sichuan University, Sichuan, pp 1–93. (Published in Chinese)
24. Application and development of the networked joint control system of national key operating vehicles, Baidu Library.[EB/OL]. https://wenku.baidu.com/view/c7a23ed128ea81c758f5782f.html(Reading date July 1, 2017)
25. JT/T 794—2011, Technical requirements for on-board terminal of road transport vehicle satellite positioning system [S]. (Published in Chinese)
26. JTT 796—2016, Technical requirements for the satellite positioning system platform of road transport vehicles [S]. (Published in Chinese)
27. Wu C. Large-scale vehicle driving behavior monitoring and early warning technology and equipment [R]. (Published in Chinese)
28. Zhijun C (2012) Integrated design and implementation of monitoring system for dangerous driving behaviors of large vehicles [J]. Wuhan University of Technology, Wuhan. (Published in Chinese)
29. Renjing Z, Junyi C (2018) The development status of intelligent networked automobile test ground at home and abroad [J]. Beijing Auto 01:7–11. (Published in Chinese)
30. Liu Tianyang Y, Zhuoping XL et al (2017) The development status and construction suggestions of the intelligent networked automobile test field [J]. Automotive Technol 01:7-11,32. (Published in Chinese)
31. NHTSA. Federal Automated Vehicles Policy[S]
32. NHTSA. Automated Driving System 2.0:A Vision for Safety[S]
33. Briefs U (2015) Mcity Grand Opening. Research Review 46:3
34. Eustice R (2016) University of Michigan's work toward Autonomous Cars [S]
35. Jianhua Y, et al. Detailed explanation of U.S. self-driving car test area [EB/OL]. http://www.sohu.com/a/216658721_733088
36. On the second anniversary of "online", a "small achievement" reached by the Shanghai Intelligent Connected Vehicle Demonstration Zone [EB/OL]. http://www.sohu.com/a/233003711_114877
37. The Ministry of Industry and Information Technology interprets the four main points of the intelligent networked vehicle road test specification [EB/OL]. https://www.d1ev.com/news/zhengce/66808
38. Baidu. Apollo Pilot Safety Report[M]
39. Jinghua C (2016) Huang Xiaobin, Li Jie. Discussion on V2X communication Technology for Intelligent Networked Vehicles [J]. Telecom Technol 05:24–27. (Published in Chinese)
40. Keqiang L, Yifan D, Shengbo L et al (2017) The development status and trends of intelligent connected vehicle (ICV) technology [J]. J Automob Safety Energy Conserv 8(01):1–14. (Published in Chinese)

41. China Electronics Information Industry Development Research Institute (2017) Test and evaluation Technology for Intelligent Connected Vehicles [M]. People Post Press, Beijing. (Published in Chinese)
42. Vision Planning of National Intelligent Connected Vehicle (Shanghai) Pilot Demonstration Zone [EB/OL]. http://www.anicecity.org/smart_network/front/site/news/newsDetailByType. jhtm l?type=2
43. National Intelligent Connected Vehicle (Shanghai) Pilot Demonstration Zone Test Service [EB/OL]. http://www.anicecity.org/smart_network/front/site/news/newsDetailByType.jhtml? type=31
44. Pei G (2009) Video-based automatic detection of traffic incidents [D]. Southwest Jiaotong University, Chengdu. (Published in Chinese)
45. Qi L (2013) Research on traffic state monitoring and prediction method based on multi-source data [D]. Jilin University, Changchun. (Published in Chinese)
46. Mingtao L (2009) Research on multi-step prediction method for traffic parameters of express-way locations on short time scale [D]. Jilin University, Changchun. (Published in Chinese)
47. Guiyan J, Ande C, Shifeng N et al (2011) Dynamic predictability analysis method of traffic data sequence based on BP neural network [J]. J Beijing Univ Technol 37(7):1019–1026. (Published in Chinese)
48. Yang Z, Cheng H, Yi Z et al (2016) 上Introduction to Haishi road traffic state index and application cases [J]. Traffic and Transport 32(3):16–18. (Published in Chinese)
49. Shujian Z, Jingfeng Y (2014) Research on the calculation method of traffic congestion evaluation index at home and abroad [J]. Highways Trucks 1. (Published in Chinese)
50. Zun Z (2014) Research on optimization methods of traffic evacuation decision-making in different scenarios for urban large-scale activities [D]. Shenyang, Northeastern University. (Published in Chinese)
51. Abrams SH (2000) Moving crowds in Chicago: baseball and the 4th of July [J]. Transportation Research Board, Washington. D.C.
52. Black J (2004) Strategic transport planning. Demand Analysis of Transport Infrastructure and Transport Services for the 27th Summer Olympiad Held in Sydney. Australia. 2000. J Transport Eng Inform [J] 2(2):14–30
53. Matthew GK, Kepaptsoglou K, Stathopoulos A (2004) A Decision Support System for Special Events Public Transport Network Planning: The Case of the Athens 2004 Summer Olympics. TRB Annual Meeting[CD].
54. Zhang Y (2003) Modeling Traffic Impact Under Special Events[D]. University of Akron, Akron
55. Sattayhatewa P, Smith RL (2003) Development of parking choice models for special events. In: TRB Annual Meeting [CD]
56. Bovy P (2002) Mega Sports event transportation and Main mobility management issue. In: Proceedings of the transport and exceptional public events conference, Paris, France
57. Black J (2004) The Analysis of Demand for Transport Infrastructure and Transport Service: The 27th Olympic Held in Sydney 2000[C]. In: The first China Olympic Transportation Forum, Beijing, pp 73–88
58. Transportation Planning and Management for Special Events[EB]. http://trb.org/news/blurb_ detail.asp?ID=1327.2005-8-5
59. Latoski SP, Dunn WM (2003) Managing Travel for Planned Special Events[R]. FHWA-OP-04-010.
60. The first Olympic Transportation Forum. Beijing. 2002. (Published in Chinese)
61. Tieyong Z, Tieyun X (2003) Ten Thousand Talents Engineering Forum [C]. In: Proceedings of "Beijing Transportation and Olympics". Beijing Industry University, Beijing. (Published in Chinese)
62. Luo M (2004) Research on traffic demand management and its application in Beijing Olympic traffic [D]. Beijing Industry University, Beijing. (Published in Chinese)
63. Hongjun C (2006) Research on key Technologies of Traffic Organization and Management for large-scale events [D]. Southeast University, Nanjing. (Published in Chinese)

64. Linghong G (2009) Research on traffic demand and evacuation in urban large-scale activities [D]. Wuhan University of Technology, Wuhan. (Published in Chinese)
65. Yueping Z (2008) Research on forecasting methods of urban traffic demand under large-scale events [D]. Wuhan University of Technology, Wuhan. (Published in Chinese)
66. Jun L (2009) Research on traffic information publishing method for large-scale activities [D], Wuhan, Wuhan University of Technology. (Published in Chinese)
67. Hui Z (2007) Research on forecast of traffic demand and evacuation plan for large-scale activities [D]. Wuhan University of Technology, Wuhan. (Published in Chinese)
68. Helong Z, Hongtao D (2005) Experience and Enlightenment of Transportation Organization in the Ninth National Games [EB]. http://www.chinautc.com/organization/2001/014.asp. Accessed 13 Jan 2005. (Published in Chinese)
69. Luo D, Yonghua G (2008) Analysis of the impact of traffic control measures on urban traffic during large-scale events——based on traffic management during the Guangzhou Olympic torch relay [J]. Traffic and Transportation 12:62–64. (Published in Chinese)
70. Xinlan Z, Zhu X, Tong L (2010) Comment on the traffic Organization of the Opening Ceremony of the 11th National Games [J]. City Traffic 8(2):49–54. (Published in Chinese)
71. Li M (2006) Research on the traveling trajectory and spatial-temporal distribution of visitors to the 2010 Shanghai world expo [D]. Tongji University, Shanghai. (Published in Chinese)
72. Dian L, Yonghua G (2008) Analysis of the impact of traffic control measures on urban traffic during large-scale events——based on traffic management during the Guangzhou Olympic torch relay [J]. Traffic and Transport 12:62–64. (Published in Chinese)
73. Jianjun C (2007) Traffic management strategy and evaluation research during large-scale events [D]. Beijing Jiaotong University, Beijing. (Published in Chinese)
74. Duyingying (2012) Research on traffic organization and management methods for large-scale events [D]. Chongqing Jiaotong University, Chongqing. (Published in Chinese)
75. Xiaoming L (2010) The theory and method of traffic organization planning for large-scale events [M]. Science Press, Beijing, pp 1–216. (Published in Chinese)

Printed in the United States
by Baker & Taylor Publisher Services